IET POWER AND ENERGY SERIES 72

Control Circuits in Power Electronics

Other volumes in this series:

Volume 1	**Power Circuit Breaker Theory and Design** C.H. Flurscheim (Editor)
Volume 4	**Industrial Microwave Heating** A.C. Metaxas and R.J. Meredith
Volume 7	**Insulators for High Voltages** J.S.T. Looms
Volume 8	**Variable Frequency AC Motor Drive Systems** D. Finney
Volume 10	**SF_6Switchgear** H.M. Ryan and G.R. Jones
Volume 11	**Conduction and Induction Heating** E.J. Davies
Volume 13	**Statistical Techniques for High Voltage Engineering** W. Hauschild and W. Mosch
Volume 14	**Uninterruptible Power Supplies** J. Platts and J.D. St Aubyn (Editors)
Volume 15	**Digital Protection for Power Systems** A.T. Johns and S.K. Salman
Volume 16	**Electricity Economics and Planning** T.W. Berrie
Volume 18	**Vacuum Switchgear** A. Greenwood
Volume 19	**Electrical Safety: a guide to causes and prevention of hazards** J. Maxwell Adams
Volume 21	**Electricity Distribution Network Design, 2nd Edition** E. Lakervi and E.J. Holmes
Volume 22	**Artificial Intelligence Techniques in Power Systems** K. Warwick, A.O. Ekwue and R. Aggarwal (Editors)
Volume 24	**Power System Commissioning and Maintenance Practice** K. Harker
Volume 25	**Engineers' Handbook of Industrial Microwave Heating** R.J. Meredith
Volume 26	**Small Electric Motors** H. Moczala *et al.*
Volume 27	**AC-DC Power System Analysis** J. Arrillaga and B.C. Smith
Volume 29	**High Voltage Direct Current Transmission, 2nd Edition** J. Arrillaga
Volume 30	**Flexible AC Transmission Systems (FACTS)** Y-H. Song (Editor)
Volume 31	**Embedded generation** N. Jenkins *et al.*
Volume 32	**High Voltage Engineering and Testing, 2nd Edition** H.M. Ryan (Editor)
Volume 33	**Overvoltage Protection of Low-Voltage Systems, Revised Edition** P. Hasse
Volume 36	**Voltage Quality in Electrical Power Systems** J. Schlabbach *et al.*
Volume 37	**Electrical Steels for Rotating Machines** P. Beckley
Volume 38	**The Electric Car: Development and future of battery, hybrid and fuel-cell cars** M. Westbrook
Volume 39	**Power Systems Electromagnetic Transients Simulation** J. Arrillaga and N. Watson
Volume 40	**Advances in High Voltage Engineering** M. Haddad and D. Warne
Volume 41	**Electrical Operation of Electrostatic Precipitators** K. Parker
Volume 43	**Thermal Power Plant Simulation and Control** D. Flynn
Volume 44	**Economic Evaluation of Projects in the Electricity Supply Industry** H. Khatib
Volume 45	**Propulsion Systems for Hybrid Vehicles** J. Miller
Volume 46	**Distribution Switchgear** S. Stewart
Volume 47	**Protection of Electricity Distribution Networks, 2nd Edition** J. Gers and E. Holmes
Volume 48	**Wood Pole Overhead Lines** B. Wareing
Volume 49	**Electric Fuses, 3rd Edition** A. Wright and G. Newbery
Volume 50	**Wind Power Integration: Connection and system operational aspects** B. Fox *et al.*
Volume 51	**Short Circuit Currents** J. Schlabbach
Volume 52	**Nuclear Power** J. Wood
Volume 53	**Condition Assessment of High Voltage Insulation in Power System Equipment** R.E. James and Q. Su
Volume 55	**Local Energy: Distributed generation of heat and power** J. Wood
Volume 56	**Condition Monitoring of Rotating Electrical Machines** P. Tavner, L. Ran, J. Penman and H. Sedding
Volume 57	**The Control Techniques Drives and Controls Handbook, 2nd Edition** B. Drury
Volume 58	**Lightning Protection** V. Cooray (Editor)
Volume 59	**Ultracapacitor Applications** J.M. Miller
Volume 62	**Lightning Electromagnetics** V. Cooray
Volume 63	**Energy Storage for Power Systems, 2nd Edition** A. Ter-Gazarian
Volume 65	**Protection of Electricity Distribution Networks, 3rd Edition** J. Gers
Volume 66	**High Voltage Engineering Testing, 3rd Edition** H. Ryan (Editor)
Volume 67	**Multicore Simulation of Power System Transients** F.M. Uriate
Volume 68	**Distribution System Analysis and Automation** J. Gers
Volume 69	**The Lightening Flash, 2nd Edition** V. Cooray (Editor)
Volume 70	**Economic Evaluation of Projects in the Electricity Supply Industry, 3rd Edition** H. Khatib
Volume 74	**Power Electronic Converters and Systems: Frontiers and applications** A. M. Trzynadlowski (Editor)
Volume 75	**Power Distribution Automation** B. Das (Editor)
Volume 76	**Power System Stability: Modelling, analysis and control** B. Om P. Malik
Volume 78	**Numerical Analysis of Power System Transients and Dynamics** A. Ametani (Editor)
Volume 79	**Vehicle-to-Grid: Linking electric vehicles to the smart grid** J. Lu and J. Hossain (Editors)
Volume 905	**Power system protection, 4 volumes**

Control Circuits in Power Electronics

Practical issues in design and implementation

Edited by
Miguel Castilla

The Institution of Engineering and Technology

Published by The Institution of Engineering and Technology, London, United Kingdom

The Institution of Engineering and Technology is registered as a Charity in England & Wales (no. 211014) and Scotland (no. SC038698).

First published 2016

The Institution of Engineering and Technology
Michael Faraday House
Six Hills Way, Stevenage
Herts, SG1 2AY, United Kingdom

www.theiet.org

British Library Cataloguing in Publication Data
A catalogue record for this product is available from the British Library

ISBN 978-1-84919-822-6 (hardback)
ISBN 978-1-84919-823-3 (PDF)

Typeset in India by MPS Limited
Printed in the UK by CPI Group (UK) Ltd, Croydon

Contents

Preface

Power electronics is an energy conversion technology that enables efficient and fast processing of electrical power. In power electronics, control circuits are key elements in the operation and performance of power converters. The control circuits have a major influence on the characteristics of the power converters such as in efficiency and steady-state errors (static properties) as well as in transient response, stability, and robustness (dynamic properties). The fundamental elements of a control circuit include the sensing unit, the control algorithm, and the modulation scheme. The sensing unit captures the current state of the power converter and provides the input signals to the control algorithm in the suitable form to be processed. The control algorithm makes decisions in accordance with the defined control objectives and the current state of the converter measured by the sensing unit. The modulation scheme performs the actions decided by the control algorithm by providing the firing pulses to the converter switching devices. In addition to these elements, the fabrication technology (i.e., analogue or digital control circuits) also has a large impact on the static and dynamic characteristics of power converters. For instance, fast transient response and flexibility of operation are well-known characteristics that can be attributed directly to the control circuit technology (analogue and digital technologies, respectively).

This book presents control circuits for power electronics converters. It is divided in three parts—analogue control circuits, digital control circuits, and new trends in control circuits. The first part provides an overview of the applications of power electronics in which analogue control circuits currently show superior performance. This part includes the essential background to enable a graduate engineer to design an analogue control circuit for a power converter, current topics of research interest within the area, and detailed descriptions of practical issues related to the design and implementation of the control circuits. Design examples for DC/DC converters, single-phase rectifiers with unity power factor, and three-phase grid-connected inverters for photovoltaic applications are presented in this part.

The second part presents a wide range of control solutions based on digital circuits (hardwired custom-designed digital integrated circuits and software-based architectures including microcontrollers, field programmable gate arrays, and digital signal processors), which today represent the leading technology in a broad range of applications. Digital control solutions for DC/DC converters for renewable energy processing, electronic ballasts for high-intensity discharge lamps, three-phase rectifiers with unity power factor, three-phase voltage-sourced inverters, and multilevel converters are the focus. This part will provide essential reading for anyone involved in research in design and implementation of control algorithms and modulation schemes using digital control circuits.

The final part covers emerging topics in control circuit technology and new trends in control circuits for power electronics applications. A diverse range of studies is considered from integrated switched capacitor converters to dual-core digital signal processors which include control and communication functions for networked control applications, and from intelligent drive units that enable advanced control of power switches to control design using computational intelligence techniques such as genetic algorithms and ant colonization optimization.

The Editor would like to thank all the contributors for their excellent work and cooperation in the preparation of this book. He further acknowledges with gratitude the assistance of the editorial and production staff at the Institution of Engineering and Technology.

Miguel Castilla
Vilanova i la Geltrú
March 2016

Part I

Analogue control circuits

Chapter 1

PWM-based sliding mode control schemes for DC/DC power converters

Siew-Chong Tan[1] *and Chi Kong Michael Tse*[2]

1.1 Introduction

Pulse-width modulated (PWM) DC/DC power converters operating at a high switching frequency are a class of systems having cyclically varying structures. Such systems are inherently nonlinear as the control function involves varying the relative durations of the constituent structures [1]. Existing design practice employs predominantly voltage mode and current-mode control strategies that are based on linear small-signal techniques, and the resulting performances in terms of output regulation, transient response, and stability are therefore adequate only if the converters operate within a narrow range of parameter variation [2]. For applications involving power sources and loads that are nonlinear and more widely varied, the use of linear controllers is expected to produce sub-optimal control performances and may even fail to meet the desired specifications when the condition of operation deviates significantly from the usual small-signal condition [3].

Sliding mode (SM) control is a nonlinear control method that is particularly well suited for DC/DC power converters working with power sources and loads that vary widely and nonlinearly [4], [5]. The inherent operation of SM control is a natural strategy for controlling systems having a discontinuous switching characteristic. When applied to DC/DC power converters, highly robust and versatile SM systems that give fast and consistently stable control performance will be resulted [6], [7]. Furthermore, among all available nonlinear control techniques, the SM control is arguably by far most practical for power converters, due to its simple implementation, ease of design, and low cost.

In particular, fixed-frequency PWM-based SM controllers, which are quasi-sliding mode (QSM) controllers that work with close operational resemblance to ideal SM controllers but without the complexity of using variable switching frequency, are found to be most suited for practical applications [7]. Specifically,

[1]Department of Electrical and Electronic Engineering, The University of Hong Kong, Pokfulam, Hong Kong, China
[2]Department of Electronic and Information Engineering, The Hong Kong Polytechnic University, Kowloon, Hong Kong, China

operating power converters at a constant frequency achieves simpler filter design and keeps the power converter size small. Furthermore, fixed-frequency PWM-based SM controllers can be easily implemented in either digital form by using low-end inexpensive microcontrollers (MCU), digital signal processors (DSP) or field-programmable gate arrays (FPGA), or in analogue form using a few simple discrete components. Nevertheless, the implementation of SM control in analogue form does present a few advantages over the digital form in terms of its ability to achieve a faster transient response especially for converters switch at very high frequency and the absence of issues related to discretization. For commercial applications, the PWM-based SM controllers can be fabricated as analogue integrated circuit (IC) controller chips at a very competitive cost relative to existing PWM controllers.

We will begin this chapter with a brief account of the principle of SM control and PWM-based SM control for the DC/DC converters. The key idea of connecting the duty ratio control in conventional PWM control with an equivalent SM control will be covered. Then, the derivation of PWM-based SM control schemes for DC/DC converters will be illustrated. The mathematics behind the modelling and the inherent control features will be discussed. A detailed exposition of how such kind of controllers can be practically engineered to suit specific purposes in controlling power converters will be provided. The practical aspects of the implementation are also described.

1.2 Basic sliding mode control theory

The general idea of SM control is to design a dedicated trajectory path in the control law that will direct the controlled trajectory of the state variables towards a desired origin upon collision between the trajectory and the trajectory path [4], [5]. This trajectory path is known as the sliding surface. For a single switch DC/DC converter shown in Figure 1.1, it is appropriate to have an SM controller which implements the control function

$$u = \frac{1}{2}(1 + \text{sign}(S)); \tag{1.1a}$$

$$S = \sum_{i=1}^{n} \alpha_i x_i, \tag{1.1b}$$

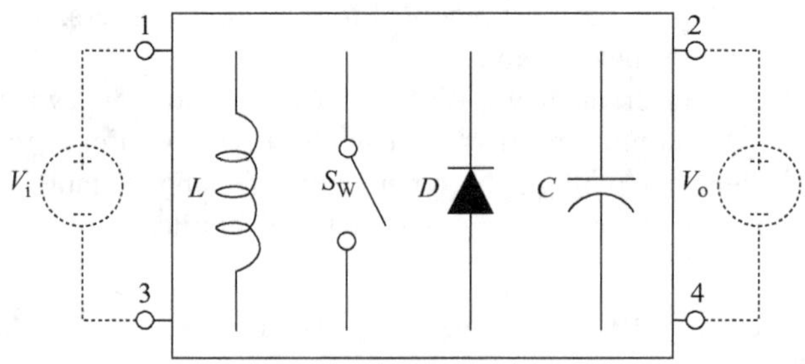

Figure 1.1 Illustration of a single switch DC/DC converter

where u is the logic state to control the converter's power switch S_W, and S is the computation of the controlled state variable's trajectory. Here, x_i denotes the desired state feedback variables (be it the output voltage or inductor current) to be controlled [7]. Under the strict governance of this control, the trajectory will be enforced to move strictly within the vicinity of a certain surface $S = 0$. The variable α_i is known as the sliding coefficient as it defines the contour of the sliding surface.

For ease of discussion, hereon, we adopt a second-order controller with the control function

$$u = \frac{1}{2}(1 + \text{sign}(S)); \tag{1.2a}$$

$$S = \alpha_1 x_1 + \alpha_2 x_2 + \alpha_3 x_3 = \boldsymbol{J}\mathbf{x}. \tag{1.2b}$$

In vector representation, $\boldsymbol{J} = [\alpha_1 \ \ \alpha_2 \ \ \alpha_3]$ and $\mathbf{x} = [x_1 \ \ x_2 \ \ x_3]^{\mathrm{T}}$. By enforcing $S = 0$, a sliding surface (plane), as shown in Figure 1.2, can be obtained.

The entire SM-control process is generally divided into two phases. In the first phase, which is known as reaching phase, the controller will drive the trajectory (regardless of the starting position) to converge the sliding surface (see Figure 1.2a). This is achieved through compliance of the hitting condition [4], [5]. The necessary and sufficient condition for satisfying the hitting condition is that the resulting control u produces the state variables x_1, x_2, and x_3 and consequently a controlled trajectory S, which satisfies the following inequality:

$$S\frac{dS}{dt} < 0. \tag{1.3}$$

The inequality (1.3) is a partial result of the Lyapunov second theorem on stability [4], [5], of which the Lyapunov function candidate is

$$V(S) = \frac{1}{2}S^2. \tag{1.4}$$

The compliance of (1.3) signifies that at a position not within the vicinity of the sliding surface, the state trajectory S is continuously being attracted and is always converging the sliding surface. Upon completion of the reaching phase when the trajectory arrives within a small vicinity of the sliding surface (see Figure 1.2a), the system enters sliding phase, which is the second phase of the control process. The controller will give a series of switching actions such that the trajectory is maintained within a small vicinity of the sliding surface and is concurrently directed towards the desired reference at origin O to achieve steady-state operation (see Figure 1.2b). This is possible by satisfying the so-called existence condition and stability condition [4], [5]. The existence condition of the sliding phase can be determined by inspecting only the local reachability condition [4], [5].

$$\lim_{S \to 0} S\frac{dS}{dt} < 0. \tag{1.5}$$

Moreover, the control action and sliding coefficients must be designed to comply with the stability condition. Failure to achieve this will lead to an SM

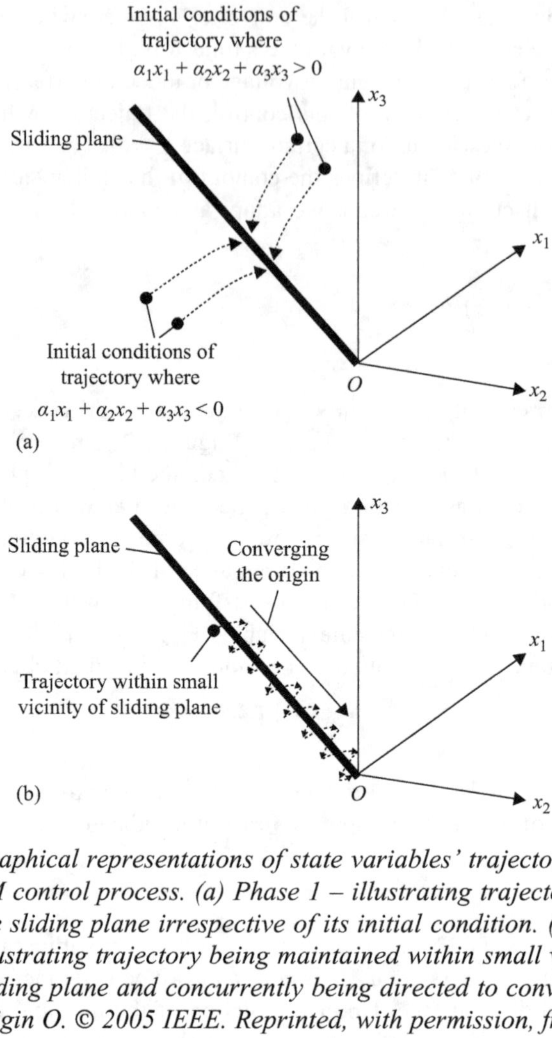

Figure 1.2 Graphical representations of state variables' trajectory behaviour in SM control process. (a) Phase 1 – illustrating trajectory converging the sliding plane irrespective of its initial condition. (b) Phase 2 – illustrating trajectory being maintained within small vicinity from the sliding plane and concurrently being directed to converge to the origin O. © 2005 IEEE. Reprinted, with permission, from Reference 8

system which is unstable. In general, the stability of a system is obtained by ensuring that the eigenvalues of the Jacobian of the system at the steady-state region have negative real parts.

1.3 PWM-based SM control

Conventionally, the SM controller is implemented based on the control law described in (1.1a) and (1.1b). Equation (1.1a) is a signum function and is easily realized using a switch relay. SM controller of this nature is known as hysteresis modulation (HM)-based SM controller. Equation (1.1b) is used for computing the instantaneous state-variable trajectory S and is realized through analogue or digital

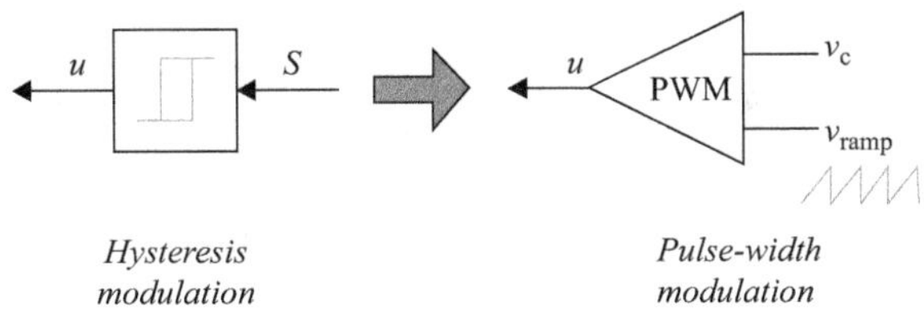

Figure 1.3 Simplified HM and PWM structure. © 2005 IEEE. Reprinted, with permission, from Reference 8

computer. The direct implementation of this control law, however, results in converters that are switched at very high and uncontrolled frequency [3], [4]. Alternatively, it is possible to achieve SM like control using PWM-based approach via equivalent SM control theory. This allows inexpensive SM controller that operates the converter at a constant frequency to be achieved.

Figure 1.3 shows the idea of the PWM-based SM controller, where PWM is used in lieu of HM without destroying the SM control properties. This requires the relationship of the two control techniques to be established. Two key results are useful here. First, in the SM control, the discrete control input (gate signal) u can be theoretically replaced by a smooth function known as the equivalent control signal u_{eq}, which can be formulated using the invariance conditions by setting the time differentiation of (1.1b) as $\dot{S} = 0$ [4]. Second, at a high switching frequency, the equivalent control is effectively a duty-cycle control [6], [7], [9], [10]. Since a duty cycle is also basically a smooth analytic function of the discrete control pulses in PWM, a PWM-based SM controller can be obtained by mapping the equivalent control function onto the duty-cycle function d of the pulse-width modulator, i.e. $d = u_{eq}$.

In the following section, we will demonstrate how PWM-based SM control schemes for voltage mode and current-mode control of the converters can be derived.

1.4 PWM-based SM voltage control

Since the PWM-based controller is a translated form of the HM-based controller, the model for the HM-based controller must first be derived. The first step of designing an HM-based SM controller is to develop the state-space description of the plant in terms of the desired control variables (i.e. voltage and/or current). In this illustration, a second-order proportional-integral-derivative (PID) SM voltage controller is applied to the three basic DC/DC converters operating in CCM.

Figure 1.4 shows the schematic diagrams of the family of sliding-mode voltage controlled (SMVC) buck, boost, and buck-boost converters with the HM-based PID SM voltage controller. Here, C, L, and r_L denote the capacitance, inductance, and instantaneous load resistance of the converters, respectively; i_C, i_L, and i_R denote the instantaneous capacitor, inductor, and load currents, respectively; V_{ref}, v_I, and v_O denote the reference, instantaneous input, and instantaneous output voltages, respectively; β denotes the feedback network ratio; and $u = 0$ or 1 is the switching state of power switch S_W.

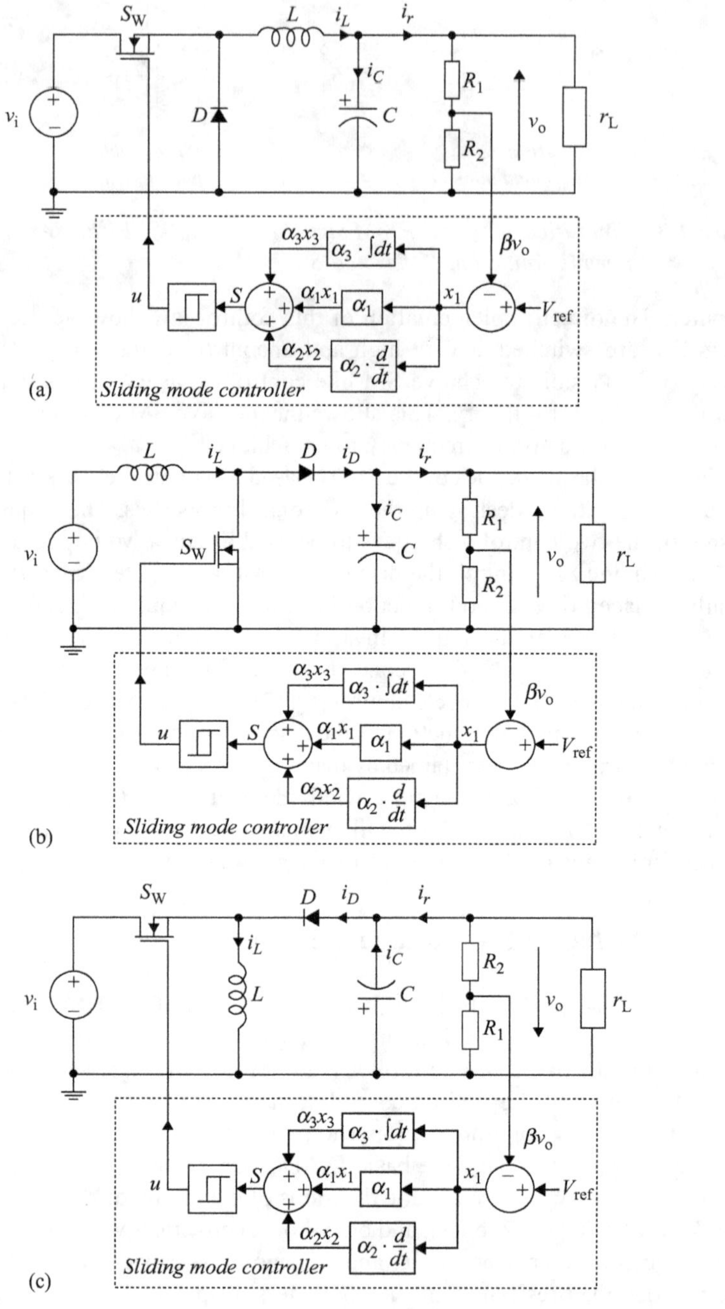

Figure 1.4 Schematic diagrams of SMVC converters with conventional HM-based PID SM controller. (a) SMVC buck converter; (b) SMVC boost converter; and (c) SMVC buck-boost converter.

Here, the control variable $\mathbf{x}$ is

$$\mathbf{x} = \begin{bmatrix} x_1 \\ x_2 \\ x_3 \end{bmatrix} = \begin{bmatrix} V_{\text{ref}} - \beta v_o \\ \dfrac{d[V_{\text{ref}} - \beta v_o]}{dt} \\ \displaystyle\int [V_{\text{ref}} - \beta v_o] dt \end{bmatrix} \tag{1.6}$$

where x_1, x_2, and x_3 are the voltage error, the voltage error dynamics (or the rate of change of voltage error), and the integral of voltage error, respectively. Substitution of the converters' behavioural models under continuous conduction mode (CCM) into (1.6) produces the following control variable descriptions: $\mathbf{x}_{\text{buck}}$, $\mathbf{x}_{\text{boost}}$, and $\mathbf{x}_{\text{buck-boost}}$ for buck, boost, and buck-boost converter, respectively.

$$\mathbf{x}_{\text{buck}} = \begin{bmatrix} x_1 \\ x_2 \\ x_3 \end{bmatrix} = \begin{bmatrix} V_{\text{ref}} - \beta v_o \\ \dfrac{\beta v_o}{r_{\text{L}} C} + \displaystyle\int \dfrac{\beta (v_o - v_i u)}{LC} dt \\ \displaystyle\int x_1 dt \end{bmatrix}; \tag{1.7}$$

$$\mathbf{x}_{\text{boost}} = \begin{bmatrix} x_1 \\ x_2 \\ x_3 \end{bmatrix} = \begin{bmatrix} V_{\text{ref}} - \beta v_o \\ \dfrac{\beta v_o}{r_{\text{L}} C} + \displaystyle\int \dfrac{\beta (v_o - v_i) \bar{u}}{LC} dt \\ \displaystyle\int x_1 dt \end{bmatrix}; \tag{1.8}$$

$$\mathbf{x}_{\text{buck-boost}} = \begin{bmatrix} x_1 \\ x_2 \\ x_3 \end{bmatrix} = \begin{bmatrix} V_{\text{ref}} - \beta v_o \\ \dfrac{\beta v_o}{r_{\text{L}} C} + \displaystyle\int \dfrac{\beta v_o \bar{u}}{LC} dt \\ \displaystyle\int x_1 dt \end{bmatrix}. \tag{1.9}$$

where $\bar{u} = 1 - u$ is the inverse logic of u, and is used particularly for modelling the boost and buck-boost topologies. The time differentiation of (1.7)–(1.9) produces the state-space descriptions required for the controller design of the respective converters.

For buck converter:

$$\begin{bmatrix} \dot{x}_1 \\ \dot{x}_2 \\ \dot{x}_3 \end{bmatrix} = \begin{bmatrix} 0 & 1 & 0 \\ 0 & -\dfrac{1}{r_{\text{L}} C} & 0 \\ 1 & 0 & 0 \end{bmatrix} \begin{bmatrix} x_1 \\ x_2 \\ x_3 \end{bmatrix} + \begin{bmatrix} 0 \\ -\dfrac{\beta v_i}{LC} \\ 0 \end{bmatrix} u + \begin{bmatrix} 0 \\ \dfrac{\beta v_o}{LC} \\ 0 \end{bmatrix}; \tag{1.10}$$

Table 1.1 Descriptions of SMVC buck, boost, and buck-boost converters operating in CCM

Type of converter	A	B	D	v
Buck	$\begin{bmatrix} 0 & 1 & 0 \\ 0 & -\frac{1}{r_L C} & 0 \\ 1 & 0 & 0 \end{bmatrix}$	$\begin{bmatrix} 0 \\ -\frac{\beta v_i}{LC} \\ 0 \end{bmatrix}$	$\begin{bmatrix} 0 \\ \frac{\beta v_o}{LC} \\ 0 \end{bmatrix}$	u
Boost	$\begin{bmatrix} 0 & 1 & 0 \\ 0 & -\frac{1}{r_L C} & 0 \\ 1 & 0 & 0 \end{bmatrix}$	$\begin{bmatrix} 0 \\ \frac{\beta v_o}{LC} - \frac{\beta v_i}{LC} \\ 0 \end{bmatrix}$	$\begin{bmatrix} 0 \\ 0 \\ 0 \end{bmatrix}$	$\overline{u}$
Buck-boost	$\begin{bmatrix} 0 & 1 & 0 \\ 0 & -\frac{1}{r_L C} & 0 \\ 1 & 0 & 0 \end{bmatrix}$	$\begin{bmatrix} 0 \\ \frac{\beta v_o}{LC} \\ 0 \end{bmatrix}$	$\begin{bmatrix} 0 \\ 0 \\ 0 \end{bmatrix}$	$\overline{u}$

For boost converter:

$$\begin{bmatrix} \dot{x}_1 \\ \dot{x}_2 \\ \dot{x}_3 \end{bmatrix} = \begin{bmatrix} 0 & 1 & 0 \\ 0 & -\frac{1}{r_L C} & 0 \\ 1 & 0 & 0 \end{bmatrix} \begin{bmatrix} x_1 \\ x_2 \\ x_3 \end{bmatrix} + \begin{bmatrix} 0 \\ \frac{\beta v_o}{LC} - \frac{\beta v_i}{LC} \\ 0 \end{bmatrix} \overline{u}; \tag{1.11}$$

For buck-boost converter:

$$\begin{bmatrix} \dot{x}_1 \\ \dot{x}_2 \\ \dot{x}_3 \end{bmatrix} = \begin{bmatrix} 0 & 1 & 0 \\ 0 & -\frac{1}{r_L C} & 0 \\ 1 & 0 & 0 \end{bmatrix} \begin{bmatrix} x_1 \\ x_2 \\ x_3 \end{bmatrix} + \begin{bmatrix} 0 \\ \frac{\beta v_o}{LC} \\ 0 \end{bmatrix} \overline{u}; \tag{1.12}$$

Rearrangement of the state-space descriptions (1.10)–(1.12) into the standard form gives

$$\dot{\mathbf{x}} = \mathbf{A}\mathbf{x} + \mathbf{B}v + \mathbf{D}; \tag{1.13}$$

where $v = u$ or $\overline{u}$ (depending on topology). Results are summarized in Table 1.1.

Next, by adopting the general SM control law given in (1.2a) and (1.2b), illustrations on how PWM-based SM controllers can be derived for the buck and the boost converters are provided.

For the buck converter, equating $\dot{S} = \boldsymbol{J}\mathbf{A}\mathbf{x} + \boldsymbol{J}\mathbf{B}u_{\text{eq}} + \boldsymbol{J}\mathbf{D} = 0$ yields the equivalent control function

$$\begin{aligned} u_{\text{eq}} &= -[\boldsymbol{J}\mathbf{B}]^{-1}\boldsymbol{J}[\mathbf{A}\mathbf{x} + \mathbf{D}] \\ &= -\frac{\beta L}{\beta v_i}\left(\frac{\alpha_1}{\alpha_2} - \frac{1}{r_L C}\right)i_C + \frac{\alpha_3 LC}{\alpha_2 \beta v_i}(V_{\text{ref}} - \beta v_o) + \frac{v_o}{v_i} \end{aligned} \tag{1.14}$$

where u_{eq} is continuous and $0 < u_{eq} < 1$. Substitution of (1.14) into the inequality gives

$$0 < u_{eq} = -\frac{\beta L}{\beta v_i}\left(\frac{\alpha_1}{\alpha_2} - \frac{1}{r_L C}\right)i_C + \frac{\alpha_3 LC}{\alpha_2 \beta v_i}(V_{ref} - \beta v_o) + \frac{v_o}{v_i} < 1. \qquad (1.15)$$

Multiplication of the inequality by βv_i gives

$$0 < u_{eq}^* = -\beta L\left(\frac{\alpha_1}{\alpha_2} - \frac{1}{r_L C}\right)i_C + LC\frac{\alpha_3}{\alpha_2}(V_{ref} - \beta v_o) + \beta v_o < \beta v_i. \qquad (1.16)$$

Finally, the translation of the equivalent control function (1.16) to the duty ratio d, where $0 < d = \frac{v_C}{\hat{v}_{ramp}} < 1$, gives the following relationships for the control signal v_C and ramp signal $\hat{v}_{ramp}$ for the practical implementation of the PWM-based SM controller:

$$v_C = u_{eq}^* = -\beta L\left(\frac{\alpha_1}{\alpha_2} - \frac{1}{r_L C}\right)i_C + LC\frac{\alpha_3}{\alpha_2}(V_{ref} - \beta v_o) + \beta v_o \qquad (1.17)$$

and

$$\hat{v}_{ramp} = \beta v_i. \qquad (1.18)$$

For the boost converter, equating $\dot{S} = \boldsymbol{J}\mathbf{A}\mathbf{x} + \boldsymbol{J}\mathbf{B}\bar{u}_{eq} = 0$ yields the equivalent control function

$$\begin{aligned}\bar{u}_{eq} &= -[\boldsymbol{J}\mathbf{B}]^{-1}\boldsymbol{J}[\mathbf{A}\mathbf{x}] \\ &= \frac{\beta L}{\beta(v_o - v_i)}\left(\frac{\alpha_1}{\alpha_2} - \frac{1}{r_L C}\right)i_C - \frac{\alpha_3 LC}{\alpha_2 \beta (v_o - v_i)}(V_{ref} - \beta v_o)\end{aligned} \qquad (1.19)$$

where $\bar{u}_{eq}$ is continuous and $0 < \bar{u}_{eq} < 1$. Substitution of (1.19) into the inequality gives

$$0 < \bar{u}_{eq} = \frac{\beta L}{\beta(v_o - v_i)}\left(\frac{\alpha_1}{\alpha_2} - \frac{1}{r_L C}\right)i_C - \frac{\alpha_3 LC}{\alpha_2 \beta (v_o - v_i)}(V_{ref} - \beta v_o) < 1. \qquad (1.20)$$

Since $u = 1 - \bar{u}$, which also implies $u_{eq} = 1 - \bar{u}_{eq}$, the inequality can be rewritten as

$$0 < u_{eq} = 1 - \frac{\beta L}{\beta(v_o - v_i)}\left(\frac{\alpha_1}{\alpha_2} - \frac{1}{r_L C}\right)i_C + \frac{\alpha_3 LC}{\alpha_2 \beta (v_o - v_i)}(V_{ref} - \beta v_o) < 1. \qquad (1.21)$$

Multiplication of the inequality by $\beta(v_o - v_i)$ gives

$$0 < u_{eq}^* = -\beta L\left(\frac{\alpha_1}{\alpha_2} - \frac{1}{r_L C}\right)i_C + LC\frac{\alpha_3}{\alpha_2}(V_{ref} - \beta v_o) + \beta(v_o - v_i) < \beta(v_o - v_i). \qquad (1.22)$$

Finally, the translation of the equivalent control function (1.22) to the duty ratio d, where $0 < d = \frac{v_C}{\hat{v}_{ramp}} < 1$, gives the following relationships for the control

Table 1.2. Control equations of PWM-based SMVC buck, boost, and buck-boost converters operating in CCM. K_{p1} *and* K_{p2} *are calculated using* $K_{p1} = \beta L(\frac{\alpha_1}{\alpha_2} - \frac{1}{r_L C})$ *and* $K_{p2} = LC\frac{\alpha_3}{\alpha_2}$

Type of converter	v_C	$\hat{v}_{ramp}$
Buck	$-K_{p1}i_C + K_{p2}(V_{ref} - \beta v_o) + \beta v_o$	βv_i
Boost	$-K_{p1}i_C + K_{p2}(V_{ref} - \beta v_o) + \beta(v_o - v_i)$	$\beta(v_o - v_i)$
Buck-boost	$-K_{p1}i_C + K_{p2}(V_{ref} - \beta v_o) + \beta v_o$	βv_o

signal v_C and ramp signal $\hat{v}_{ramp}$ for the practical implementation of the PWM-based SM controller:

$$v_C = u_{eq}^* = -\beta L\left(\frac{\alpha_1}{\alpha_2} - \frac{1}{r_L C}\right)i_C + LC\frac{\alpha_3}{\alpha_2}(V_{ref} - \beta v_o) + \beta(v_o - v_i) \tag{1.23}$$

and

$$\hat{v}_{ramp} = \beta(v_o - v_i). \tag{1.24}$$

Using a similar approach, the PWM-based SM control equation for the buck-boost converter can be derived. Table 1.2 gives the control equations of all the three PWM-based SMVC converters and Figure 1.5 shows the respective schematic diagrams.

A close inspection of the control equations reveals that the control signal is actually load dependent. Thus, for the controller to have good regulation performance against load changes, the instantaneous value of r_L should be fed back. However, this would require additional sensors and cumbersome computations, which complicate the controller. On the other hand, the dependence and sensitivity of v_C on the load can be minimized by a proper design of sliding coefficient such that $\frac{\alpha_1}{\alpha_2} \gg \frac{1}{r_L C}$.

Next, let us consider how the existence conditions of SM control operation are obtained for the converters. To ensure the existence of SM operation, the local reachability condition (1.5) must be satisfied. This can be expressed as

$$\begin{cases} \dot{S}_{S\to 0^+} = \mathbf{JAx} + \mathbf{JB}v_{S\to 0^+} + \mathbf{JD} < 0 \\ \dot{S}_{S\to 0^-} = \mathbf{JAx} + \mathbf{JB}v_{S\to 0^-} + \mathbf{JD} > 0 \end{cases}. \tag{1.25}$$

For the buck converter, substitution of $v_{S\to 0^+} = u = 1$, $v_{S\to 0^-} = u = 0$ and the matrices in Table 1.1 into (1.25) gives

$$\begin{cases} -\alpha_1\dfrac{\beta i_C}{C} + \alpha_2\dfrac{\beta i_C}{r_L C^2} + \alpha_3(V_{ref} - \beta v_o) - \alpha_2\dfrac{\beta v_i}{LC} + \alpha_2\dfrac{\beta v_o}{LC} < 0 \\ -\alpha_1\dfrac{\beta i_C}{C} + \alpha_2\dfrac{\beta i_C}{r_L C^2} + \alpha_3(V_{ref} - \beta v_o) + \alpha_2\dfrac{\beta v_o}{LC} > 0 \end{cases}, \tag{1.26}$$

which can be combined to give

$$0 < -\beta L\left(\frac{\alpha_1}{\alpha_2} - \frac{1}{r_L C}\right)i_C + LC\frac{\alpha_3}{\alpha_2}(V_{ref} - \beta v_o) + \beta v_o < \beta v_i. \tag{1.27}$$

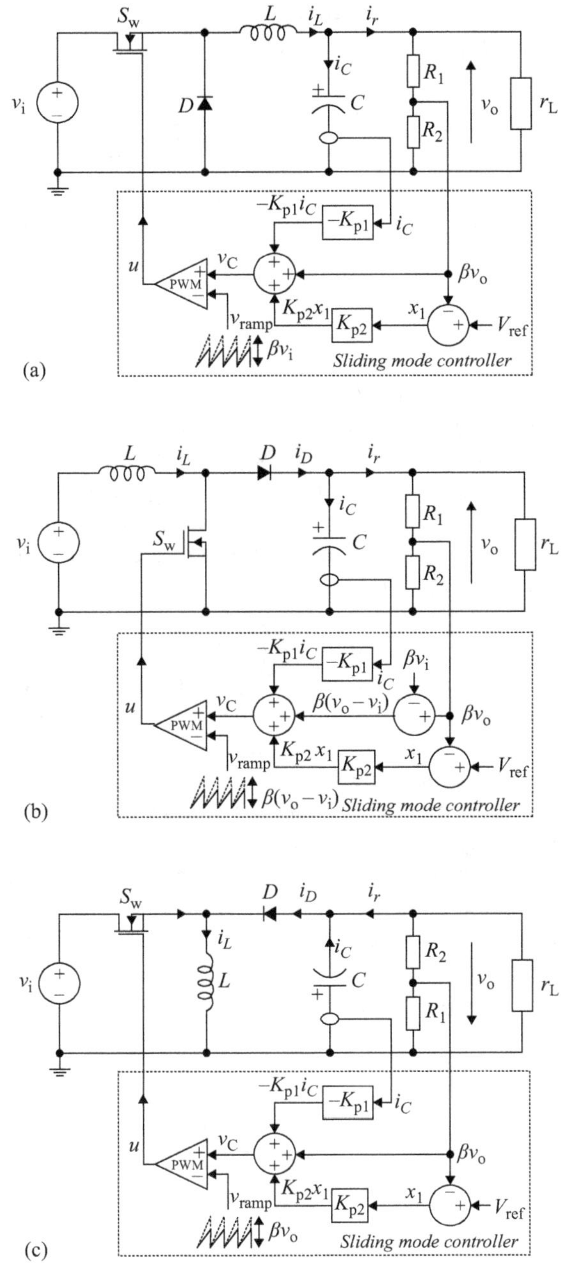

Figure 1.5 Schematic diagrams of the PWM-based PID SMVC converters. (a) SMVC buck converter; (b) SMVC boost converter; and (c) SMVC buck-boost converter. © 2005 IEEE. Reprinted, with permission, from Reference 11

For the boost converter, substitution of $v_{S\to 0^+} = \bar{u} = 0$, $v_{S\to 0^-} = \bar{u} = 1$ and the matrices in Table 1.1 into (1.25) gives

$$\begin{cases} -\alpha_1 \dfrac{\beta i_C}{C} + \alpha_2 \dfrac{\beta i_C}{r_L C^2} + \alpha_3 (V_{ref} - \beta v_o) < 0 \\ -\alpha_1 \dfrac{\beta i_C}{C} + \alpha_2 \dfrac{\beta i_C}{r_L C^2} + \alpha_3 (V_{ref} - \beta v_o) - \alpha_2 \dfrac{\beta v_i}{LC} + \alpha_2 \dfrac{\beta v_o}{LC} > 0 \end{cases}, \tag{1.28}$$

which can be combined to give

$$0 < \beta L \left(\frac{\alpha_1}{\alpha_2} - \frac{1}{r_L C} \right) i_C - LC \frac{\alpha_3}{\alpha_2} (V_{ref} - \beta v_o) < \beta (v_o - v_i). \tag{1.29}$$

The derived existence conditions for the buck, boost, and buck-boost converters are tabulated in Table 1.3. To comply with the existence condition, the sliding coefficients of the controller must be properly selected to comply with the inequalities. In the process of such a design, the hitting condition will be indirectly complied since the existence condition is a subset of the hitting condition in the region where trajectory is within a small vicinity of the sliding surface.

The inequalities in Table 1.3 provide only the general information for the existence of SM. It does not reveal how parameters can be selected. For this purpose, the Ackermann's Formula can be employed [12], which allows the selection of sliding coefficients based on the desired dynamic properties. With this method, the stability condition of the system is automatically satisfied without needing to solve for the eigenvalues of the Jacobian matrix of the system in steady state and ensuring that it does not contain negative real parts.

In our example, the equation relating the sliding coefficients to the dynamic response of the converter during SM operation can be easily found, which results in a linear second-order equation with three possible types of responses: under-damped, critically damped, and over-damped. In the case of under-damped response converters, the desired settling time $T_S = 5\tau$ s (1% criteria), where τ is the natural time constant, can be set by tuning α_1/α_2 using

$$\frac{\alpha_1}{\alpha_2} = \frac{10}{T_S} \tag{1.30}$$

Table 1.3 Existence conditions of buck, boost, and buck-boost converters operating in CCM

Buck	$0 < -\beta L \left(\frac{\alpha_1}{\alpha_2} - \frac{1}{r_L C} \right) i_C + LC \frac{\alpha_3}{\alpha_2} (V_{ref} - \beta v_o) + \beta v_o < \beta v_i$
Boost	$0 < \beta L \left(\frac{\alpha_1}{\alpha_2} - \frac{1}{r_L C} \right) i_C - LC \frac{\alpha_3}{\alpha_2} (V_{ref} - \beta v_o) < \beta (v_o - v_i)$
Buck-boost	$0 < \beta L \left(\frac{\alpha_1}{\alpha_2} - \frac{1}{r_L C} \right) i_C - LC \frac{\alpha_3}{\alpha_2} (V_{ref} - \beta v_o) < \beta v_o$

and the desired damping ratio ξ can be set using

$$\frac{\alpha_3}{\alpha_2} = \frac{25}{\xi^2 T_S^2} \tag{1.31}$$

where

$$\xi = \sqrt{\frac{\left[\ln\left(\frac{M_P}{100}\right)\right]^2}{\pi^2 + \left[\ln\left(\frac{M_P}{100}\right)\right]^2}} \tag{1.32}$$

where M_P is the percentage of the peak overshoot.

1.5 PWM-based SM current control

The SM current controller employs both the output voltage error and the inductor current error as the controlled state variables. The incorporation of the output voltage error information allows the output voltage to be accurately regulated, whereas the inductor current error information allows the inductor current to follow closely the desired reference inductor current. Similar to existing current-mode controllers, the monitoring and tracking of the inductor current reference is the key to achieving a fast dynamical response in converters with non-minimum phase characteristic.

As in typical PWM current-mode control, the instantaneous reference inductor current profile i_{ref} of this controller is generated using the amplified output voltage error, i.e.

$$i_{\text{ref}} = K(V_{\text{ref}} - \beta v_o) \tag{1.33}$$

where K is the amplified gain of the voltage error. A large value of K is chosen to improve dynamic response and to minimize the steady-state voltage error in the system. When a very tight regulation is required, an integral term can be included to give $i_{\text{ref}} = K_I \int (V_{\text{ref}} - \beta v_o)dt + K_P(V_{\text{ref}} - \beta v_o)$. For this discussion, we will derive the PWM-based SM current control (SMCC) scheme for the boost converter using the reference current based on (1.33).

Here, the SM control equations are as given in (1.2a) and (1.2b). The controlled state variables are the current error x_1, the voltage error x_2, and the integral of the current and the voltage errors x_3, which are expressed as

$$\mathbf{x} = \begin{bmatrix} x_1 \\ x_2 \\ x_3 \end{bmatrix} = \begin{bmatrix} i_{\text{ref}} - i_L \\ V_{\text{ref}} - \beta v_o \\ \int [x_1 + x_2]dt \end{bmatrix} \tag{1.34}$$

where i_L denotes the instantaneous inductor current. An integral term has been introduced into the SM current controller as an additional controlled state variable to reduce steady-state errors. Using the boost converter's behavioural model for CCM and the time differentiation of (1.34), we obtain the dynamical model of the system as

$$\dot{\mathbf{x}} = \begin{bmatrix} \dot{x}_1 \\ \dot{x}_2 \\ \dot{x}_3 \end{bmatrix} = \begin{bmatrix} -\frac{\beta K}{C} i_C - \frac{v_i - \bar{u} v_o}{L} \\ -\frac{\beta}{C} i_C \\ (K+1)[V_{\text{ref}} - \beta v_o] - i_L \end{bmatrix}. \tag{1.35}$$

The equivalent control signal of the SM current controller when applied to the boost converter is obtained by solving $\dot{S} = \boldsymbol{J}\dot{\mathbf{x}} = 0$, which gives

$$u_{\text{eq}} = 1 - \frac{K_2}{v_o} i_C - \frac{v_i}{v_o} + \frac{K_1}{v_o}[V_{\text{ref}} - \beta v_o] - \frac{K_3}{v_o} i_L \tag{1.36}$$

where

$$K_1 = \frac{\alpha_3}{\alpha_1} L(K+1), \quad K_2 = \frac{\beta L}{C}\left(K + \frac{\alpha_2}{\alpha_1}\right), \quad \text{and} \quad K_3 = \frac{\alpha_3}{\alpha_1} L \tag{1.37}$$

are the fixed gain parameters in the controller. In PWM-based form, the equations of the control law is given by

$$\begin{cases} v_c = G_s K_1[V_{\text{ref}} - \beta v_o] - G_s K_2 i_C - G_s K_3 i_L + G_s(v_o - v_i) \\ \hat{v}_{\text{ramp}} = G_s v_o \end{cases}. \tag{1.38}$$

Here, a factor of $0 < G_s < 1$ has been intentionally introduced for down scaling the equation to a practical magnitude level that conforms to the chip-level's voltage standard of implementation. Moreover, recalling that the boost-type converter cannot operate with a switching signal u that has a duty cycle $d = 1$, a small dead-time circuitry is required to ensure that the duty cycle of the controller's output is always less than one. In the proposed controller, this is satisfied by multiplying the logic state u_{PWM} of the pulse-width modulator with the logic state u_{CLK} of an impulse generator using a logic AND operator. Figure 1.6 shows an overview of the PWM-based SM current controller for the boost converter.

The existence condition can be derived as

$$\begin{cases} \alpha_1\left[-\frac{\beta K}{C} i_C - \frac{v_i}{L}\right] - \alpha_2 \frac{\beta}{C} i_C + \alpha_3([K+1][V_{\text{ref}} - \beta v_o] - i_L) < 0 \\ \alpha_1\left[-\frac{\beta K}{C} i_C - \frac{v_i - v_o}{L}\right] - \alpha_2 \frac{\beta}{C} i_C + \alpha_3([K+1][V_{\text{ref}} - \beta v_o] - i_L) > 0 \end{cases}. \tag{1.39}$$

With SMCC, the motion equation, which composes both the current and voltage state variables, is highly nonlinear. The stability analysis involves first deriving the ideal sliding dynamics of the system, then performing an analysis on the equilibrium point, and finally deriving the condition for maintaining stability [14].

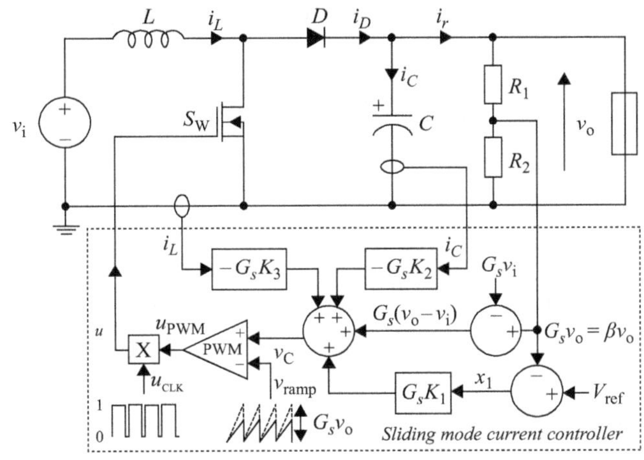

Figure 1.6 Schematic diagram of the PWM-based SMCC boost converter. © 2005 IEEE. Reprinted, with permission, from Reference 13

The replacement of $\bar{u}$ by $\bar{u}_{\text{eq}}$ in the original boost converter's description under CCM operation converts the discontinuous system into an ideal SM continuous system:

$$\begin{cases} \dfrac{di_L}{dt} = \dfrac{v_i}{L} - \dfrac{v_o}{L}\bar{u}_{\text{eq}} \\ \dfrac{dv_o}{dt} = \dfrac{i_L}{C}\bar{u}_{\text{eq}} - \dfrac{v_o}{r_{\text{L}}C} \end{cases}. \tag{1.40}$$

Then, the substitution of (1.36) into (1.40) gives

$$\begin{cases} \dfrac{di_L}{dt} = \dfrac{v_i}{L} - \dfrac{v_o}{L}\dfrac{K_2\dfrac{v_o}{r_{\text{L}}} - v_i + K_3[V_{\text{ref}} - \beta v_o] - K_3[i_{\text{ref}} - i_L]}{K_2 i_L - v_o} \\ \dfrac{dv_o}{dt} = \dfrac{i_L}{C}\dfrac{K_2\dfrac{v_o}{r_{\text{L}}} - v_i + K_3[V_{\text{ref}} - \beta v_o] - K_3[i_{\text{ref}} - i_L]}{K_2 i_L - v_o} - \dfrac{v_o}{r_{\text{L}}C} \end{cases} \tag{1.41}$$

which represents the ideal sliding dynamics of the SMCC boost converter. Assume that there exists a stable equilibrium point on the sliding surface on which the ideal sliding dynamics eventually resides. At equilibrium, we have $\frac{di_L}{dt} = \frac{dv_o}{dt} = 0$. Thus, the steady-state equation is

$$I_L = \frac{V_o^2}{V_i R_{\text{L}}} \tag{1.42}$$

where I_L, V_o, V_i, and R_{L} represent respectively the inductor current, output voltage, input voltage, and load resistance at their steady-state points.

Next, linearization of the ideal sliding dynamics described in (1.41) around the equilibrium points given in (1.42) results in

$$\begin{cases} \dfrac{d\tilde{i}_L}{dt} = a_{11}\tilde{i}_L + a_{12}\tilde{v}_o \\ \dfrac{d\tilde{v}_o}{dt} = a_{21}\tilde{i}_L + a_{22}\tilde{v}_o \end{cases} \tag{1.43}$$

where

$$\begin{cases} a_{11} = \dfrac{K_3 V_i R_L}{K_2 L V_o - L V_i R_L} \\ a_{12} = \dfrac{K_1 \beta V_i R_L - 2K_2 V_i + \dfrac{V_i^2 R_L}{V_o}}{K_2 L V_o - L V_i R_L} \\ a_{21} = \dfrac{K_2 V_i - \dfrac{V_i^2 R_L}{V_o} - K_3 V_o}{K_2 C V_o - C V_i R_L} \\ a_{22} = \dfrac{\dfrac{K_2 V_o}{R_L} - K_1 \beta V_o}{K_2 C V_o - C V_i R_L} - \dfrac{1}{R_L C} \end{cases} . \tag{1.44}$$

The system is assumed to operate at steady state, i.e. $v_i = V_i$, $r_L = R_L$, $V_{ref} - \beta V_o = 0$, and $I_{ref} - I_L = 0$, and that $I_L \gg \tilde{i}_L$ and $V_o \gg \tilde{v}_o$. The characteristic equation of this linearized system can be written as

$$s^2 - (a_{11} + a_{22})s + a_{11}a_{22} - a_{12}a_{21} = 0. \tag{1.45}$$

The system will be stable if the following conditions are satisfied

$$\begin{cases} a_{11} + a_{22} < 0 \\ a_{11}a_{22} - a_{12}a_{21} > 0 \end{cases} . \tag{1.46}$$

For $a_{11} + a_{22} < 0$, the condition for stability is

$$\frac{K_3 C V_i R_L - K_1 \beta L V_o + L V_i}{K_2 V_o - V_i R_L} < 0 \tag{1.47}$$

and for $a_{11}a_{22} - a_{12}a_{21} > 0$, it is

$$\begin{aligned} &2K_3 V_o^3 (K_2 - K_1 \beta R_L) + V_i V_o^2 K_2 (K_1 \beta R_L - 2K_2) \\ &\quad + V_i^2 V_o R_L (3K_2 - K_1 \beta R_L) - V_i^3 V_L^2 > 0. \end{aligned} \tag{1.48}$$

The existence condition (1.39) and the stability conditions (1.47) and (1.48) are the basis for the designing the control gains of the SM current controller.

1.6 Practical implementation and design issues

The approaches of deriving PWM-based SM voltage controller for the buck, boost, and buck-boost converter, and PWM-based SM current controller for the boost converter have been discussed. The same approaches can be applied to derive PWM-based SM controllers for other types of DC/DC converters. In this section, the various aspects related to the implementation of the SM controller are discussed.

First, the choice of system's state variables, i.e. voltage, current, and their derivatives and/or integrals, is important in that it affects not only the control performance but also the complexity of the implementation. These are a few issues that should be considered.

1. The higher the number of state variables, the higher the degree of system's controllability. However, more sensing and/or computations of the state variables are required.
2. The use of voltage state variables is easier because of the simplicity of implementing voltage sensors. However, the choice for their derivatives may require the use of noise sensitive differentiators. Alternatively, indirect means of sensing such state variables are available, e.g. dv_o/dt can be found by sensing the current of the output capacitor. Yet, this will lead us to the subsequent question as to whether it is convenient to perform such sensing. On the other hand, an integral term of the controlled variable x is often required to reduce the steady-state error of the system, which may otherwise exceed the regulation requirement.
3. The sensing of capacitor current to find the derivative of voltage variable is not always the best option. This is particularly true for the case where the capacitor is a filter capacitor. The addition of a current sensor in its current flow path will increase the impedance and therefore deteriorate the filtering process.
4. For the sensing of the filtering capacitors' currents, very low impedance current transformers are normally required. However, for sensing currents that have DC average components, resistors or sophisticated Hall-effect sensors may be required. These may affect the overall efficiency or cost of the DC/DC converters.

Second, similar to conventional controllers, the physical limitation of the analogue devices, e.g. bandwidths, propagation time delays, slew rates, and saturation limits, of the SM controller should be properly noted. They are the key factors affecting the healthy operation of the SM controller.

Finally, other than the implementation of the main SM controller compensation circuit, a practical controller for the power converter should include all associated peripheral circuits and protection features. Figure 1.7 gives the schematic diagram of a practical design of the PWM-based SM controller that is suitable for IC implementation for the buck converters. This controller is based on the original PWM-based SM controller given in Table 1.2 and Figure 1.5a.

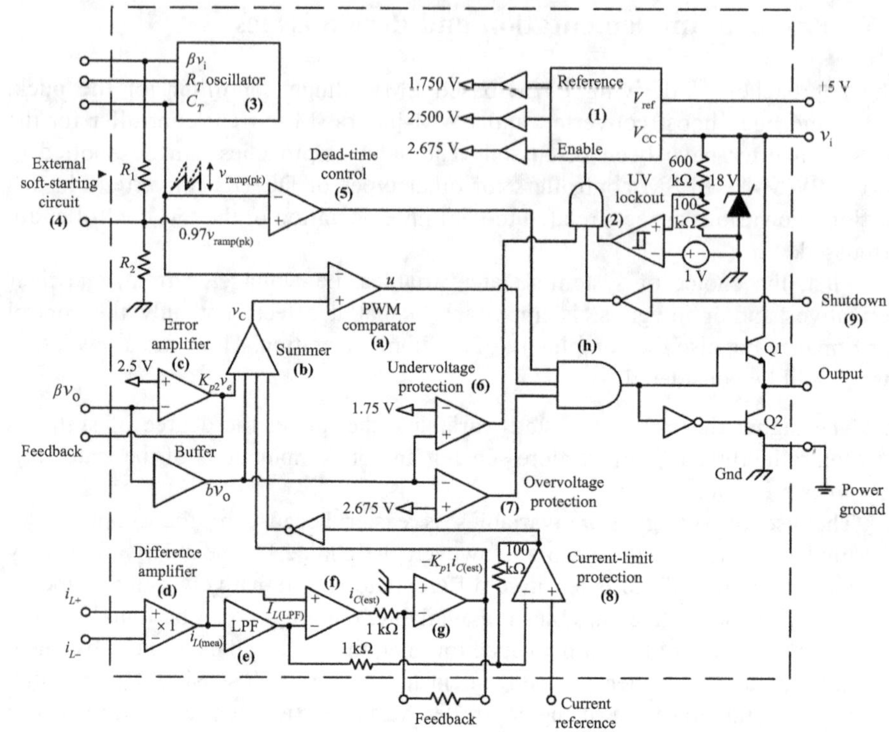

Figure 1.7 A practical design of the PWM-based SM controller for buck converters for IC implementation

The main PWM-based SM compensation circuit involves the following components.

- PWM comparator (see point (a)) – For generating the switching control signal u using the variable ramp input v_{ramp} and the control signal v_{c}.
- Summer (point (b)) – The control signal v_{c} is obtained by summing the amplified capacitor current $-k_{p1}i_{C(\text{est})}$, the amplified voltage error $k_{p2}v_e$, and the sensed output voltage βv_o.
- Error amplifier (point (c)) – The amplified voltage error $k_{p2}v_e$ is obtained from this circuit where $v_e = 2.5 - \beta v_o$. The value of k_{p2} is determined by the feedback network implemented.
- Inductor current sensing and capacitor current estimation circuit (from point (d) to (g)) – This circuit is used to first sense the inductor current before processing it to derive an estimated value of the capacitor current without the need for an additional capacitor current sensor. The process is as follows. Inductor current information $i_{L(\text{mea})}$ is obtained with the difference amplifier (point (d)), of which i_{L+} and i_{L-} are respectively connected to the incoming and outgoing nodes of

the current sensor that is placed in series with the inductor. Then, $i_{L(\mathrm{mea})}$ is passed through a low-pass filter (LPF) (point (e)) to retain only its slow moving component $I_{L(\mathrm{LPF})}$, which is used for current-limit protection of the converter. Concurrently, both $i_{L(\mathrm{mea})}$ and $I_{L(\mathrm{LPF})}$ are fed to the difference amplifier (point (f)) to derive an estimated value of the capacitor current $i_{C(\mathrm{est})}$. This signal is then fed to an inverting amplifier to generate $k_{p1} i_{C(\mathrm{est})}$.

- Logic driving circuit (point (h)) – This circuit is composed of a three-input AND gate, a NOT gate and a pair of push-pull NMOS Q1 and Q2. The AND gate serves to perform an AND logic on the input signals from the dead-time control (point (5)), the switching control signal (point (a)), and the output overvoltage protection (point (7)). The output from the AND gate is used to drive Q1, and Q2 complementary through the NOT gate. The output of the Q1 and Q2 pair is connected to switch S_{W} of the converter. A logic 1 from the output implies the turning on of the switch and logic 0 implies the turning off of the switch.

The associated peripheral circuits and protection features are as follows.

- Reference regulator – The purpose of the reference regulator is to generate the appropriate voltage sources to power up all the different circuit components in the controller and also to provide precise voltage levels for internal referencing (refer to point (1) of Figure 1.7). The power of the reference regulator is provided by the input voltage source of the converter v_{i}.
- Input under-voltage lockout – With this function, the controller is shut down in the event that the input voltage source v_{i} is below a certain required voltage (refer to point (2) of Figure 1.7). For this given controller, the under-voltage lockout is activated when the input voltage goes below 7 V. An input voltage clamping circuit is also included to clamp the maximum input voltage at 18 V, thus protecting the controller from a very high v_{i}.
- Variable slope, fixed-frequency oscillator (refer to point (3)) – This circuit is required for generating a ramp signal $v_{\mathrm{ramp(pk)}}$ with variable amplitude that changes according to the input or output voltage, in accordance to the requirement of the SM control equations, which is part of the SM control feature.
- Provision for external programmable soft-starting circuit (refer to point (4)) – A resistor-capacitor (RC) circuit can be connected to point (4) to charge the ramp up from zero to the steady-state ramp voltage amplitude. The purpose is to provide for a soft open-loop control startup to bring the converter from rest to steady-state operation before any closed-loop control is deployed. The time period of the soft-start operation is determined through the choice of RC value.
- Dead-time control (refer to point (5)) – The function of the dead-time control is to limit the maximum duty cycle of the controller to less than 100%. This ensures that the converter is switched off momentarily at every switching cycle.
- Output undervoltage protection (point (6)) – The function of this mechanism is to instantaneously shut down the controller in the event when the output voltage suddenly falls from the nominal value to an abnormally low value.

This will in turn shut down the operation of the converter and protects it from output short circuit fault. For the given controller in Figure 1.7, the under-voltage protection is set at 30% below the 2.5 V reference voltage.

- Output overvoltage protection (point (7)) – This mechanism instantaneously shut down the controller in the event when the output voltage rises above an acceptable rated value. For the given controller, the overvoltage protection is set at 7% above the 2.5 V reference voltage.
- Current-limit protection (point (8)) – The function of this mechanism is to limit the current flow of the converter to a preset current reference value. When the current flow of the converter is below the reference value, the protection mechanism has no influence on the operation. However, in the event when the current flow rises to the value of the reference value, the protection mechanism will generate a signal that will limit the duty cycle operation, consequently limiting the maximum current flow.
- External shutdown (point (9)) – The external shutdown allows the disabling of the controller and therefore the converter through external control signals. This provides a means of turning on/off of the power supply and can be used for emergency shutdown.

Figure 1.8 shows a photograph of a buck converter with a fabricated PWM-based SM controller IC.

Experimental results based on the above described circuit are shown in Figure 1.9. As shown in the results, the dynamic behaviour of output voltage ripple are basically similar (i.e. critically damped) for all three operating load conditions. This demonstrates the advantage of the SM controller in terms of robustness in the dynamic behaviour at different operating conditions and uncertainties.

Figure 1.8 Photograph of a buck converter with the fabricated PWM-based SM controller IC implementation

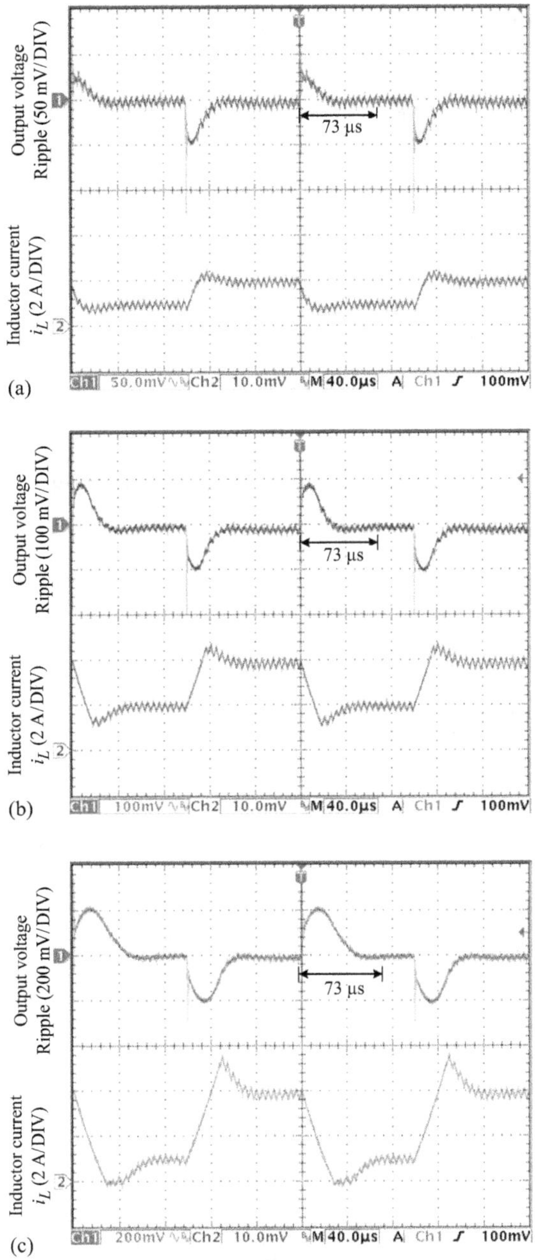

Figure 1.9 Experimental waveforms of the output voltage ripple and inductor current of a 24 V input, 12 V output, 48 W buck converter, with a 20 kHz bandwidth PWM-based SM controller operating at 5 kHz step load resistance change. (a) 6 Ω and 12 Ω; (b) 3 Ω and 6 Ω; and (c) 3 Ω and 12 Ω. © 2005 IEEE. Reprinted, with permission, from Reference 8

1.7 Conclusions

This chapter gives a brief description of how various current and voltage control PWM-based SM control schemes for common types of DC/DC power converters are derived. These control schemes are based on fixed-frequency operation and works with close operational resemblance to ideal SM controllers. They are practical and highly suitable for DC/DC applications involving power sources and loads that are nonlinear and widely varied. The practical aspects of the implementation are also described including noteworthy issues like control variable choices, design, and the necessary peripheral circuits and protection means required in transforming the control schemes into real-life practical controller ICs.

References

[1] R. W. Erickson, S. Ćuk, and R. D. Middlebrook, "Large-signals modelling and analysis of switching regulators," *IEEE Power Electronics Specialists Conference Record,* pp. 240–250, 1982.

[2] C. K. Tse and K. M. Adams, "A nonlinear large-signal feed-forward feedback control for two-state DC–DC Converters," *IEEE Power Electronics Specialists Conference Record,* pp. 722–729, 1991.

[3] S. C. Tan, Y. M. Lai, and C. K. Tse, "General design issues of sliding mode controllers for DC–DC converters," *IEEE Transactions on Industrial Electronics*, vol. 55, no. 3, pp. 1160–1174, 2008.

[4] V. Utkin, J. Guldner, and J. X. Shi, *Sliding Mode Control in Electromechanical Systems*. London, UK: Taylor & Francis, 1999.

[5] C. Edwards and S. K. Spurgeron, *Sliding Mode Control: Theory and Applications*. London, UK: Taylor & Francis, 1998.

[6] R. Venkataramanan, A. Sabanovic, and S. Ćuk, "Sliding mode control of DC-to-DC converters," *Proceedings of IEEE Conference on Industrial Electronics, Control and Instrumentations*, pp. 251–258, 1985.

[7] S. C. Tan, Y. M. Lai, and C. K. Tse, *Sliding Mode Control of Switching Power Converters: Techniques and Implementation*, Boca Raton: CRC Press, 2011.

[8] S. C. Tan, Y. M. Lai, C. K. Tse, and M. K. H. Cheung, "A fixed-frequency pulse width-modulation based quasi-sliding mode controller for buck converters," *IEEE Transactions on Power Electronics*, vol. 20, no. 6, pp. 1379–1392, 2005.

[9] H. Sira-Ramirez, "A geometric approach to pulse-width modulated control in nonlinear dynamical systems," *IEEE Transactions on Automatic Control*, vol. 34, no. 2, pp. 184–187, 1989.

[10] L. Martinez, A. Poveda, J. Majo, *et al*., "Lie algebras modeling of bidirectional switching converters," in *Proceedings of European Conference on Circuit Theory and Design* (ECCTD'93), vol. 2, pp. 1425–1429, 1993.

[11] S. C. Tan, Y. M. Lai, and C. K. Tse, "A unified approach to the design of PWM based sliding mode voltage controller for basic DC–DC converters in continuous conduction mode," *IEEE Transactions on Circuits and Systems I*, vol. 53, no. 8, pp. 1816–1827, 2006.
[12] J. Ackermann and V. Utkin, "Sliding mode control design based on Ackermann's formula," *IEEE Transactions on Automatic Control*, vol. 43, no. 2, pp. 234–237, 1998.
[13] S. C. Tan, Y. M. Lai, C. K. Tse, L. Martinez-Salamero, and C. K. Wu, "A fast response fixed-frequency sliding mode controller for boost-type converters with a wide range of operating conditions," *IEEE Transactions on Industrial Electronics*, vol. 54, no. 6, pp. 3276–3286, 2007.
[14] L. Martinez-Salamero, J. Calvente, R. Giral, A. Poveda, and E. Fossas, "Analysis of a bidirectional coupled-inductor Ćuk converter operating in sliding mode," *IEEE Transactions on Circuits and Systems I*, vol. 45, no. 4, pp. 355–363, 1998.

Chapter 2
Synthetic-ripple hysteretic controllers for DC/DC converters

Santanu Mishra[1]

A DC/DC converter generates a regulated DC output from a semi-or-un-regulated DC input. It uses switching devices to process power flow from input source to the output load. The periodic operation of these switches is controlled by a feedback control circuit. Therefore, a DC/DC converter can be thought of as a sampled system with a sampling frequency equal to the switching frequency of the switches. The transient response of a DC/DC converter is compared with its switching period. Transient response is important because the faster is its transient response, the faster will be the speed of the converter in load or line dynamic situations.

Transient response of a DC/DC converter can be improved by including the converter dynamic variables, such as the inductor current or output voltage, in the pulse width modulation (PWM) generation process. Very fast transient response can be realized by using both the peak and the valley of the ripple to determine the switching instants, leading to the class of hysteretic PWMs, so named because the modulator is a hysteresis comparator. For hysteretic PWM-controlled converters, the response to load or line transients can be within a switching cycle as opposed to multiple switching periods in case of conventional PWM controllers.

In order to generate a hysteretic PWM signal, inductor current ripple or capacitor voltage ripple needs to be sensed and compared to a reference signal. For proper hysteretic PWM generation, the sensed converter variables need to be triangular, piece-wise linear and noise-free. For modern DC/DC converters obtaining an ideal piecewise linear ripple for hysteretic PWM generation is difficult due to following reasons: (a) In order to keep efficiency high, use of a large valued sensing resistor for current sensing is avoided, (b) voltage regulation requirement is very stringent with strict regulation limits, which forces the output ripple to be negligibly small, (c) high switching frequency of the converter leads to high frequency interference during current or voltage sensing [1, 2]. In order to solve these problems, converter switching waveforms are utilized to generate a noise-free ripple for hysteretic modulator. This chapter discusses the basic theory, advantages

[1]Department of Electrical Engineering, Indian Institute of Technology Kanpur, Kanpur, Uttar Pradesh, India

and implementation of this new class of hysteretic ripple modulators. Various analogue implementations of this scheme are described in this chapter.

Section 2.1 provides a brief introduction to hysteretic control for DC/DC converter. The implementations of these building blocks are described in Section 2.2. Here all the non-idealities of the building blocks, which can lead to corrupt hysteretic operation are identified. Fundamentals of synthetic ripple generation and its usages are explained in Section 2.3. Modelling of synthetic ripple modulators are covered in Section 2.4. Section 2.5 explains methods to improve the dynamic of response of a synthetic hysteretic modulator even further.

2.1 Hysteretic controllers for DC/DC converters

Ideally, a triangular modulating signal is required for a hysteretic modulator. The switching instances of the converter are decided by the intersections of the modulating signal (V_{mod}) and a DC reference, called the command signal (V_{cmd}). The peak-to-peak amplitude of the modulating signal should be larger than the minimum realizable hysteretic amount of the comparator. For a DC/DC converter, any converter variable, with a triangular shape, can be used for hysteretic modulation.

If the capacitor voltage ripple is used as a modulating signal, it is known as a voltage hysteretic modulator. The block diagram of this control method is shown in Figure 2.1(a). For example, in case of a buck converter, the output voltage ripple can be used for hysteretic modulation, as long as it has a triangular piecewise linear profile. In this case, the switching instances of the converter are decided using output voltage ripple and a DC command signal.

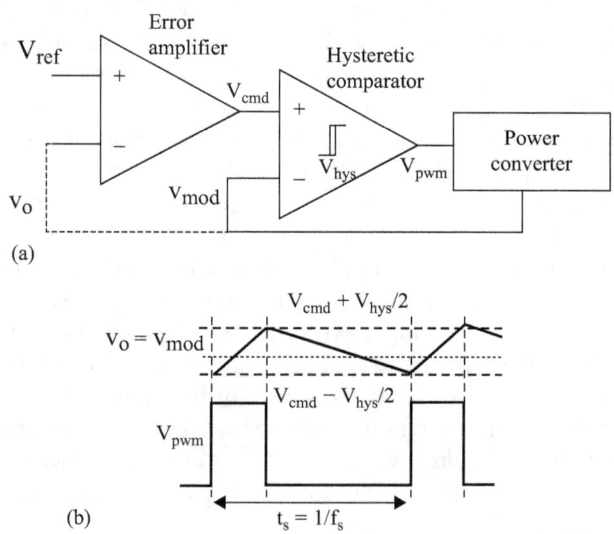

Figure 2.1 Conventional voltage hysteretic control for a power converter. (a) Schematic and (b) waveforms

Depending on regulation requirements, there are two options to provide the command signal: (a) it can be provided externally as a DC reference. For this case, there is no direct control over the output voltage of the converter. (ii) A voltage error amplifier can be used to generate the command signal. It is commonly referred to as V^2 control when implemented for a buck converter [3]. As the voltage error amplifier has direct control over the output voltage, the second strategy is used when regulation requirements are stricter. In either case, the hysteretic window limits the magnitude of the output voltage ripple. It is also important to note that there is no external clock that decides the switching frequency. The switching frequency is decided purely based on converter variables (which affect the slope of the voltage ripple) and the hysteretic window.

Similarly, inductor current ripple is also a popular choice for modulation signal, due to its natural piecewise linear wave-shape. It is commonly referred to as a current hysteretic modulator. The block diagram and operating waveforms are shown in Figure 2.2. The only difference in this case, as compared to Figure 2.1, is that instead of the voltage ripple the current ripple is used as a modulating signal. Therefore, the peak-to-peak inductor ripple and the switching frequency are limited by the hysteretic window. Similar to a voltage hysteretic modulator, the DC command signal (V_{cmd}) can be either supplied externally or using a voltage error amplifier.

One of the problems with the current hysteretic modulator is how to sense the inductor current in an inexpensive manner. Normally, current sense resistors are used to sense the inductor current. As modern DC/DC converters are required to have higher efficiency to support improved battery usage and reduce heat dissipation, they are required to use smaller current sense resistors. This makes the current sense output noisy and error prone.

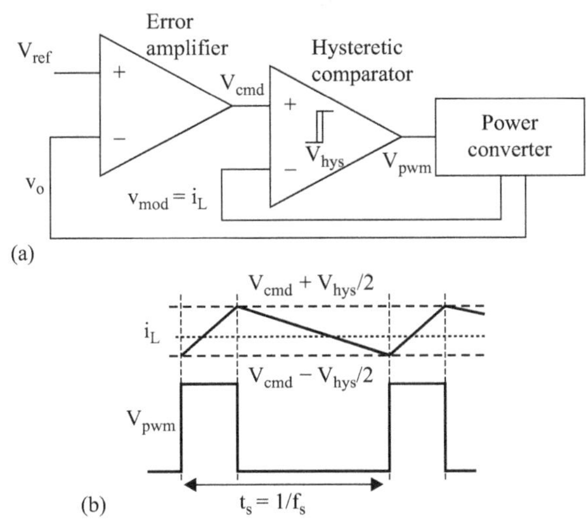

Figure 2.2 Conventional current hysteretic control for DC/DC converter. (a) Schematic and (b) waveforms

2.2 Building blocks and non-idealities

This section explains the major building blocks of a hysteretic modulator–based DC/DC converter. The basic objective is to understand the factors that affect the ideal operation of a hysteretic modulator.

2.2.1 Converter

Many factors affect ideal triangular waveform generation in a converter. This section explains some of these factors. A DC/DC converter uses an inductor as a boosting element or as a filter element. For example, in a boost converter, the inductor is a boosting element. In a synchronous buck converter the inductor is part of a second-order low-pass filter. Due to volt-second balance at the terminals of an ideal inductor, the voltage across the inductor has a zero mean value. In fact, if the average voltage across an ideal inductor is not zero, the current through it will continually increase leading to its magnetic saturation.

Typically, two types of current waveforms are encountered through an inductor in a DC/DC converter. As shown in Figure 2.3, when the average inductor current (I_L) is high compared to the peak-to-peak ripple (Δi_L), the inductor current wave shape is a triangular wave with a DC offset. In this case, inductor current ripple exhibits two distinct slopes in a switching cycle, and it is referred to as continuous conduction mode (CCM) operation. However, as average inductor current becomes smaller, the inductor current ripple dominates, and it exhibits a three-slope profile. This is due to the fact that for a part of the switching cycle the inductor current goes to zero. This mode of operation is commonly referred to as discontinuous conduction mode (DCM). DCM operation is a case where the inductor current is no longer triangular.

The aforementioned behaviour provides an understanding of the use of an inductor current ripple for hysteretic modulator. Under CCM operation, the hysteretic operation is straight-forward as the inductor current has a naturally two-slope profile. However, under DCM operation, the inductor current no longer provides two distinct slopes of proper hysteretic controller operation. However, it

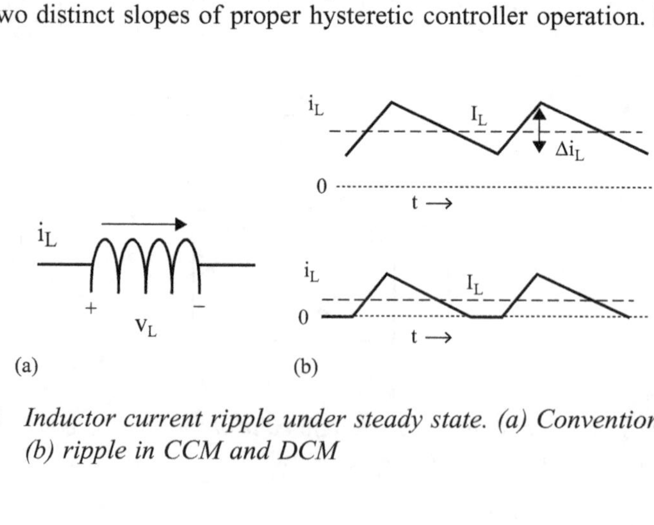

Figure 2.3 Inductor current ripple under steady state. (a) Convention and (b) ripple in CCM and DCM

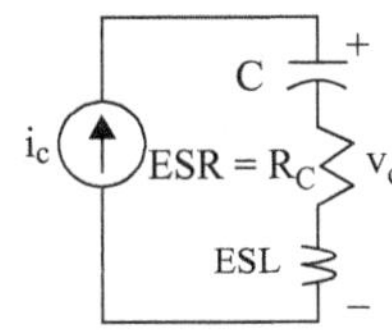

Figure 2.4 General circuit model to determine capacitor ripple in a DC/DC converter

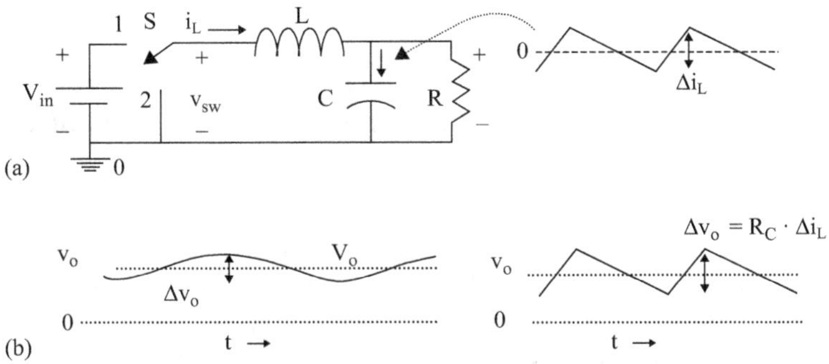

Figure 2.5 Voltage ripple across the capacitor for a buck converter with ideal capacitor. (a) Current through the capacitor, (b) output voltage ripple with ideal capacitor and with a capacitor with dominant ESR value

should be noted that in special designs, the three-slope profile of the inductor current is exploited to reduce the switching frequency of the converter leading to higher light load efficiency [4].

In order to sense the capacitor voltage ripple and extract a perfect piecewise linear waveform, the capacitor voltage can be directly sensed. However, depending on the converter topology and model of the capacitor, voltage ripple of various wave shapes are obtained. Figure 2.4 shows a general circuit model to determine voltage ripple across the capacitor. In a DC/DC converter, if the capacitor is assumed ideal, it only carries an AC current with zero average value. The current through the capacitor can be derived using steady-state waveforms of the converter. This AC current flows through the capacitor to generate ripple of various shapes.

In order to illustrate the above point, let's take the example of a buck converter. The inductor current of the buck converter has a DC part and a switching frequency ripple part. The DC part of the inductor current flows through the load. The current through the capacitor is the switching frequency ripple in the inductor current, which is triangular under CCM operation, as shown in Figure 2.5(a). This current when flows through an ideal capacitor (e.g. ceramic capacitor) leads to a parabolic ripple at the output. However, if the equivalent series resistance (ESR) of the capacitor is higher (e.g. electrolytic capacitor), the output voltage ripple is piecewise linear as shown in Figure 2.5(b). Depending on the capacitor model, the ripple shape will also vary.

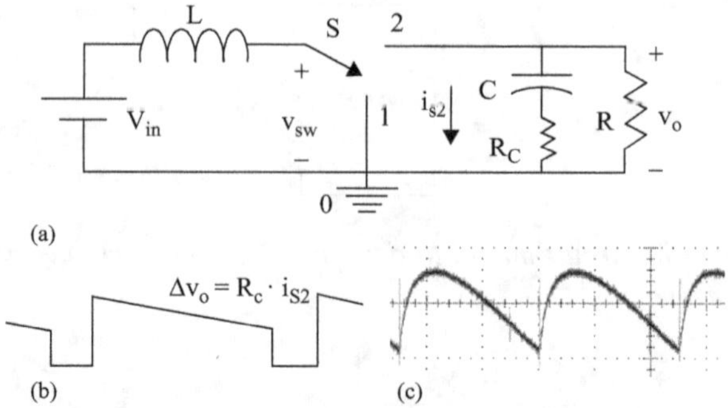

Figure 2.6 Voltage ripple for a boost converter: (a) Schematic, (b) output ripple with a capacitor large ESR (R_c) and (c) output ripple with a pure capacitor with small ESR (R_c)

A similar conclusion can be drawn for a boost converter. Figure 2.6(a) shows the schematic of a boost converter. Using this schematic the steady-state waveform of the current through the capacitor has a profile similar to that shown in Figure 2.6(b). Ideally, the current through the capacitor is a square wave, if inductor current ripple is neglected. Therefore, a capacitor with a high ESR will result in a square wave ripple. Figure 2.6(b) shows the voltage ripple when the inductor current ripple is not neglected and the output capacitor has a high ESR. Obviously, this cannot be used for a hysteretic modulator as the profile is not piecewise linear. When the capacitor is ideal, the output ripple is as shown in Figure 2.6(c).

2.2.2 Carrier generation circuit

The next important block in a hysteretic modulator is the carrier generation circuit. This circuit can either be a sensing network or an elaborate network that generates the triangular modulation signal for a hysteretic comparator. For a current hysteretic controller, the inductor current is sensed and for voltage hysteretic controller, a capacitor voltage is sensed and fed to the hysteretic comparator. Figure 2.7 shows three different methods to sense inductor current.

Figure 2.7(a) shows the simplest form of inductor sensing using sense resistor. As the sense resistor has to be small, an amplifier is cascaded to this signal to amplify the current sense signal. The sensed current is given by

$$v_{mod} = A \cdot r_L \cdot i_L(t) = A \cdot r_L \cdot (I_L + \Delta i_L) \tag{2.1}$$

Here 'A' is the gain of the current sense amplifier. It should be noted that the output of the sense circuit contains both DC (I_L) as well as a switching frequency ripple (Δi_L). It is essential that the bandwidth of the amplifier is higher than the switching frequency to avoid attenuation of the switching frequency component in the current. The major drawback of this method is that the converter has a constant conduction loss due to the sense resistor, and it may not be acceptable in many modern applications.

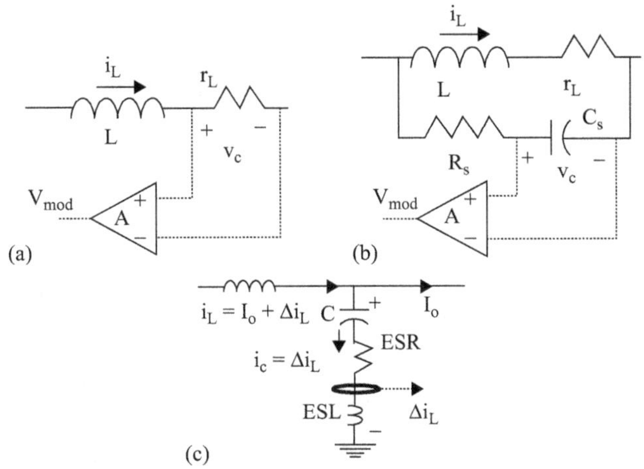

Figure 2.7 Current sensing in a DC/DC converter: (a) sense resistor–based current sensing, (b) inductor DCR–based current sensing and (c) current transformer (CT)–based current sensing

The DC resistance (DCR) of the inductor can also be used to sense the inductor current. This strategy avoids the use of an additional resistor for current sensing since r_L is the parasitic element of the inductor. The inductor DCR–based current sensing is shown in Figure 2.7(b). The expression for sensed current is given by

$$v_{mod} = A \cdot r_L \cdot \frac{(1 + sL/r_L)}{(1 + sR_sC_s)} \cdot i_L(t) \tag{2.2}$$

If the design parameters are selected such that $L/r_L = R_sC_s$, the sensed signal is same as a resistor-based sensing. Therefore, both the low- and high-frequency signals will exhibit same gain, given by $(A \cdot r_L)$. Practically, there will always be minor mismatch in component selection, which leads to a slightly different low- and high-frequency gain of the modulation signal. This is a major factor to affect linearity of the sense network. Thermistors can be used to reduce gain error due to thermal variation in component values.

Inductor current information can also be obtained using a CT. In order to use this, generally a capacitor current can be sensed. The advantage is that as the average current through the capacitor is zero, it does not saturate the CT. Figure 2.7(c) shows the use of CT to sense the inductor current for a buck converter. As the output capacitor carries the high-frequency component in inductor current, the output of the CT is proportional to the inductor current ripple.

2.2.3 Hysteretic controller

Once a triangular piecewise linear modulation signal is generated, it can be used with a hysteretic controller for PWM generation. Figure 2.8(a) shows a hysteretic controller which uses a standard comparator with positive feedback to generate the

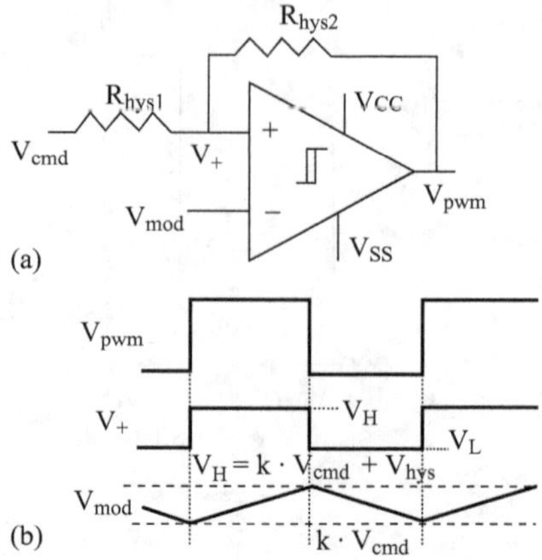

Figure 2.8 (a) Hysteretic comparator implementation using a positive feedback, and (b) its operating waveforms. The parameter k is dependent on R_{hys1} and R_{hys2} as given in equation (2.3a)

hysteretic limits. The hysteretic limits are based on a command signal. Depending on the values of R_{hys1} and R_{hys2}, the hysteretic window is decided. As can be noted, from this explanation, the switching frequency of the converter can be varied by varying the hysteretic windows.

The expression for the hysteretic limits in terms of circuit parameters are given by

$$V_H = V_{cmd} \cdot \frac{R_{hys2}}{(R_{hys1} + R_{hys2})} + V_{cc} \cdot \frac{R_{hys1}}{(R_{hys1} + R_{hys2})} \tag{2.3a}$$

$$V_L = V_{cmd} \cdot \frac{R_{hys2}}{(R_{hys1} + R_{hys2})} + V_{ss} \cdot \frac{R_{hys1}}{(R_{hys1} + R_{hys2})} \tag{2.3b}$$

It can be noted that by changing R_{hys1} or R_{hys2}, the hysteretic window can be controlled. Here V_{CC} and V_{SS} are the bias supplies for the comparator. When V_{SS} is zero, the modulator signal has a V_L proportional to only V_{cmd} as shown in Figure 2.8(b). When the command signal is derived from an error amplifier, as shown in Figures 2.1 and 2.2, it can have high-frequency noise.

A more elaborate hysteretic comparator can be used, as shown in Figure 2.9(a), to mitigate a noisy command signal. In this case, the PWM comparators are succeeded by an RS latch. The latch helps in a cleaner PWM generation even if there is noise in the command or modulation signal. This is especially true when the hysteretic window is small. Figure 2.9(b) shows the practical hysteretic comparator signals. It can be seen that all the signals contain high-frequency noise, and it may not be possible to filter the noise as it will add delay in the feedback network. Therefore, a latch will ensure a single transition and reduce chattering in the PWM output.

Commutation delay is nothing but the delay in the comparator and the feedback system. It directly affects the PWM signal. Figure 2.10 shows the commutation

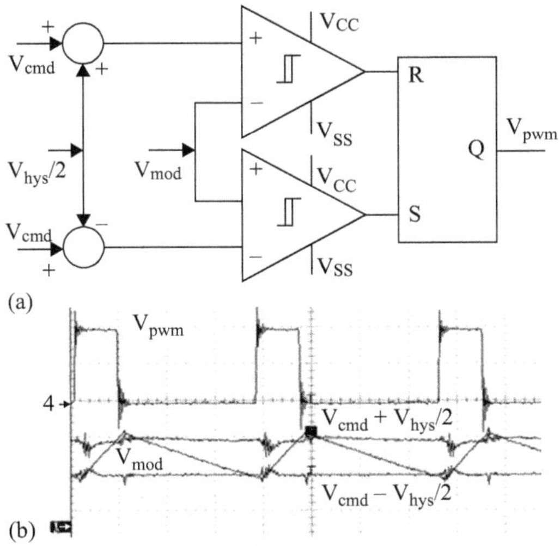

Figure 2.9 Hysteretic comparator with noise immunity feature for low hysteresis voltage levels. (a) Schematic, (b) waveforms

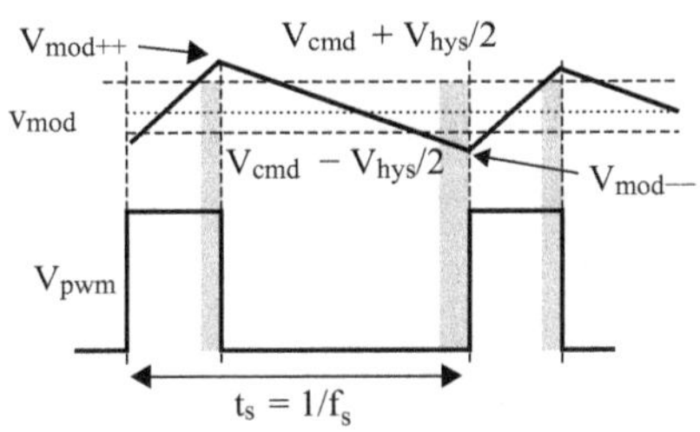

Figure 2.10 Commutation delay of a hysteretic comparator

delay as a shaded area during PWM transition. The overall impact of commutation delay is a net increase in the PWM output. In an unregulated system (without a voltage feedback), the delay can affect the regulation of output voltage. The second effect of commutation delay is a reduction in switching frequency. This can be explained by realizing that commutation delay creates a virtual hysteresis window ($=V_{mod++} - V_{mod--}$) which is larger than the actual hysteretic window. All other parameters remain unchanged, the higher the hysteretic window, the lower the switching frequency.

2.2.4 *Voltage feedback*

As identified earlier, a voltage loop results in a tighter control of the output voltage, and in a hysteretic controller it generates the command (V_{cmd}) signal. The error amplifier or voltage compensator is fed with a reference signal (V_{ref}) and the output

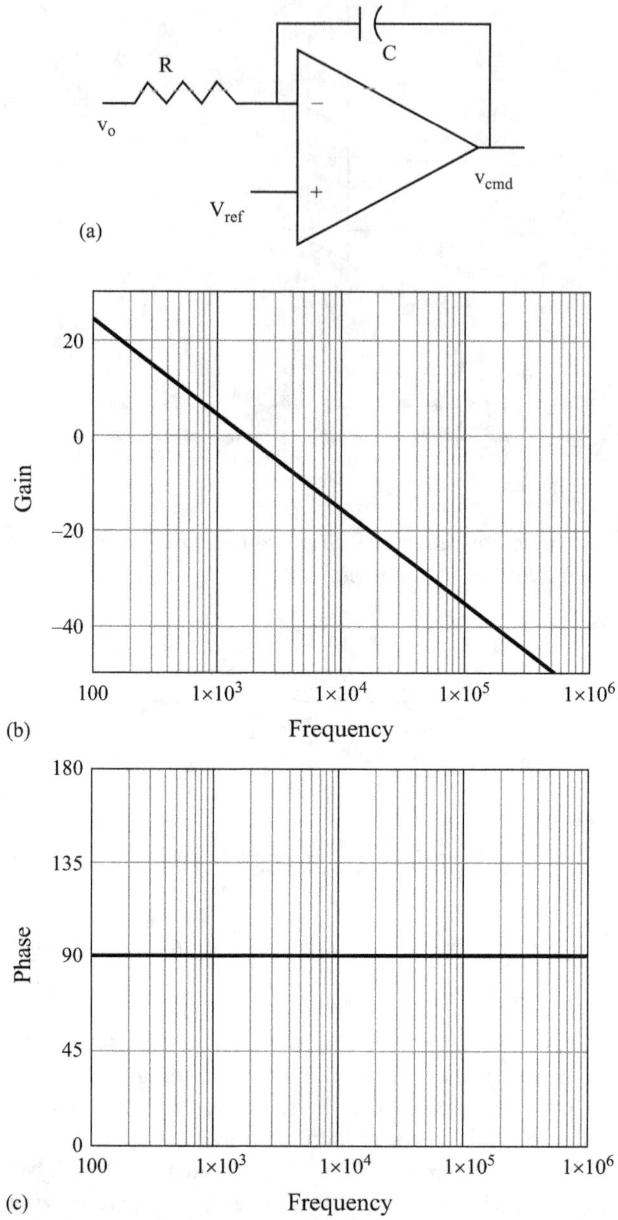

Figure 2.11 Type 1 Compensator. (a) Implementation, (b) gain plot and (c) phase plot

voltage. If the loop gain is high, the error amplifier forces the output voltage to follow a given reference in a feedback system.

There are three types of compensator commonly used in DC/DC converters. Figure 2.11(a) shows a type 1 compensator, which is an integrator. It adds a pole at low frequency in the feedback system. This is used when speed of the feedback is not of importance. As the reference voltage (V_{ref}) is fixed, the transfer function of the compensator is given by

$$\left.\frac{\hat{v}_{cmd}(s)}{\hat{v}_o(s)}\right|_{\hat{v}_{ref}=0} = -\frac{1}{sRC} = \frac{1}{\omega RC}\angle 90 \tag{2.4}$$

The gain and phase plots of the above expression are plotted in Figure 2.11(b,c). The value of R and C can be used to regulate the crossover frequency which is about 2 kHz in the given example. The compensator has a very high low-frequency gain and results in negligible steady-state error between the reference and the output voltage. When the converter has a superior dynamic requirement, either type 2 or type 3 compensators are used.

The schematic of type 2 compensator is shown in Figure 2.12(a), and it is bode plots are shown in Figure 2.12(b,c). Its transfer function is given by

$$\left.\frac{\hat{v}_{cmd}(s)}{\hat{v}_o(s)}\right|_{\hat{v}_{ref}=0} = -\frac{1}{R_i \cdot C_f} \cdot \frac{\left(1+\dfrac{s}{\omega_z}\right)}{s \cdot \left(1+\dfrac{s}{\omega_p}\right)} \text{(assuming } C_f \gg C_{hf}\text{)}$$

$$\text{where, } \omega_z = \frac{1}{R_f \cdot C_f};\ \omega_p = \frac{1}{R_f \cdot C_{hf}} \tag{2.5}$$

There are three factors that are controlled in a type 2 compensator: the flat band gain, one pole and one zero are the design variables. The low-frequency gain of this compensator is high and results in a minimal steady-state error.

Similarly, a type 3 compensator is used in many double pole plants as shown in Figure 2.13. Its transfer function is given by

$$\left.\frac{\hat{v}_{cmd}(s)}{\hat{v}_o(s)}\right|_{\hat{v}_{ref}=0} = -\frac{1}{R_i \cdot C_f} \cdot \frac{\left(1+\dfrac{s}{\omega_{z1}}\right)\cdot\left(1+\dfrac{s}{\omega_{z2}}\right)}{s \cdot \left(1+\dfrac{s}{\omega_{p1}}\right)\cdot\left(1+\dfrac{s}{\omega_{p2}}\right)} \text{(assuming } C_f \gg C_{hf}\text{)}$$

$$\text{where, } \omega_{z1} = \frac{1}{R_f \cdot C_f}\ \omega_{z2} = \frac{1}{(R_i + R_{i1}) \cdot C_{i1}}\ \omega_{p1} = \frac{1}{R_{i1} \cdot C_{i1}}\ \omega_{p2} = \frac{1}{R_f \cdot C_{hf}}$$

$$\omega_{co} = \sqrt{\omega_z \cdot \omega_p} \tag{2.6}$$

It exhibits three poles and two zeroes as shown above. Depending on the plant transfer function, one of the above three compensators is selected for regulation. The output of the error amplifier contains a DC, switching frequency component and higher order noise. Depending on the gain of the compensator at switching frequency, the output voltage ripple is amplified at its output.

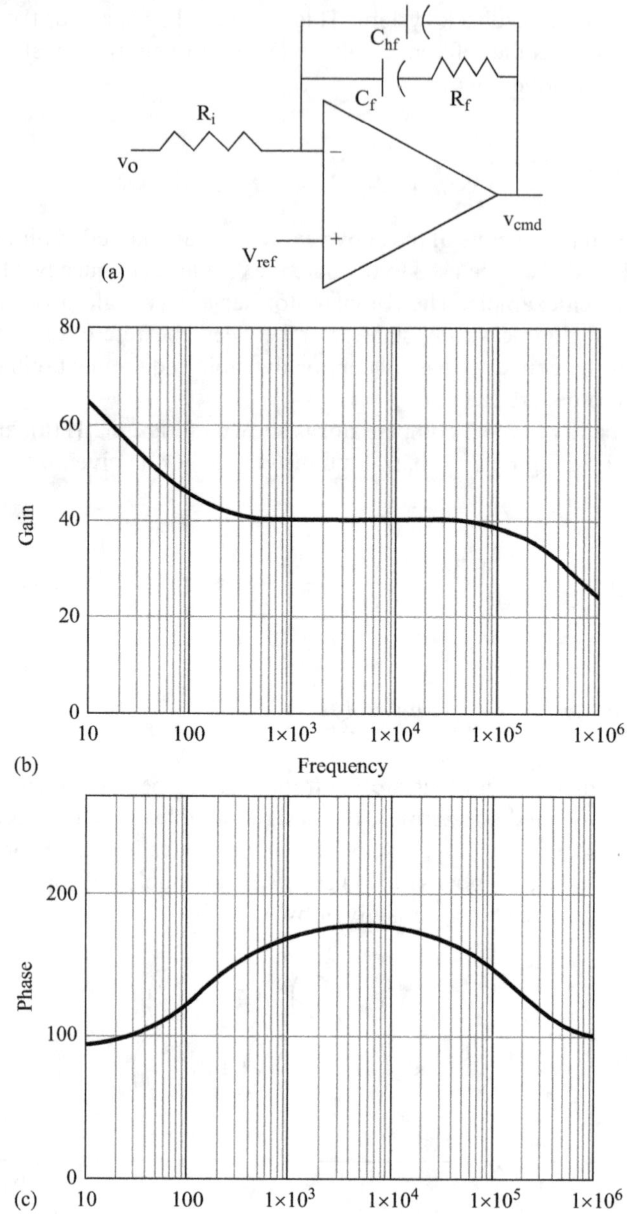

Figure 2.12 Type 2 compensator. (a) Implementation, (b) gain plot and (c) phase plot

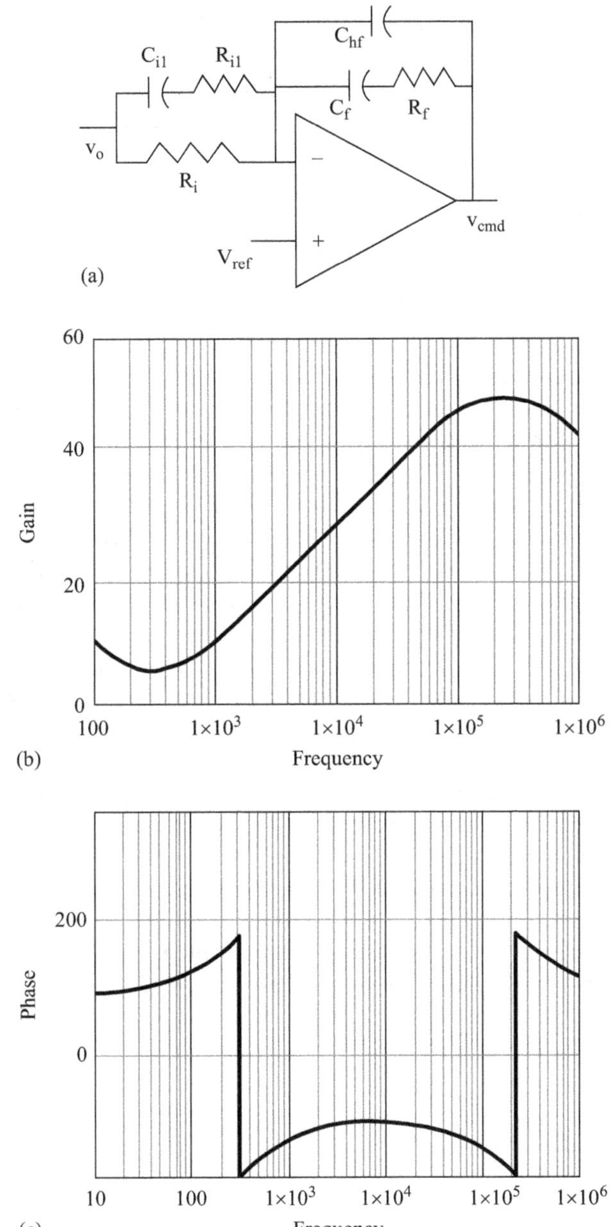

Figure 2.13 Type 3 compensator. (a) Implementation, (b) gain plot and (c) phase plot

2.3 Synthetic carrier generation circuit

The preceding sections explain the factors that affect ideal operation a hysteretic controller. Some of the factors are due to the converter or sense circuit non-idealities, while others were due to non-idealities of the control loop. While little can be done about the non-idealities in control circuits apart from making it faster or using more ideal components, it is possible to generate an ideal piecewise linear waveform from converter waveforms. Overall idea is to generate a triangular modulation signal, using converter waveforms, which contains important regulation information. In fact, a variety of carrier or modulation signals can be generated using converter waveforms. Triangular signals can be generated from virtually any AC converter waveforms and are integrated to virtually any converter variable that needs to be controlled. This strategy can bring the benefits of current-mode controllers to those controlled variables carrying insignificant or non-ideal ripple [5–7].

As shown in Figure 2.14, a typical converter-derived hysteretic PWM architecture or synthetic ripple modulator consists of a filter, an algebraic summer and a hysteretic comparator. The input to the filter can be a function of the voltages or currents of converter. Normally, switching voltages are preferred as they are easier to sense as compared to switching currents. The filter could contain low-pass or band-pass blocks, depending on whether waveforms are square or triangular. The filter can also contain algebraic blocks. The filter output, v_{tri} contains not only a triangular component but also low-frequency information useful for shaping the converter's frequency response. The adder circuit that bonds v_{tri} to a controlled variable (v_{offset}) can often be realized implicitly in the filter or the comparator, as shown by the dashed line. The inputs to the hysteretic comparator are modulation signal, which contains a converter-derived piecewise linear waveform and a command signal, which acts as a reference to the comparator. The inverting and non-inverting terminals are decided based on implementations of the circuit to ensure negative feedback.

In this section, various methods to implement a synthetic ripple generator (SRG) will be discussed and its performance will be evaluated. The filter and the summer of the synthetic ripple generation circuit can be implemented in active or passive domain. Both these methods of generation and its impact on the hysteretic

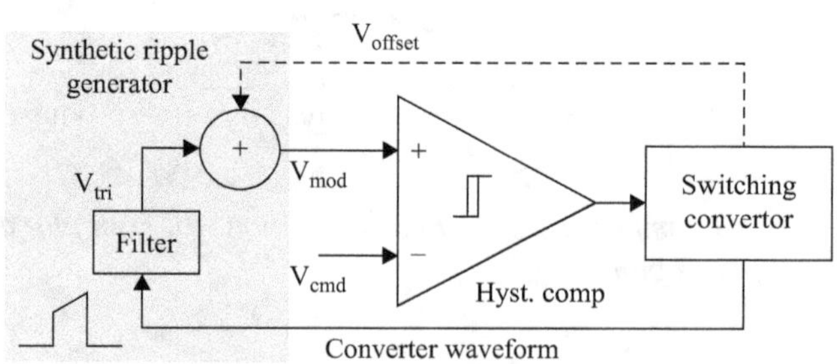

Figure 2.14 General structure of a synthetic ripple modulator

controller are explained below. For the following analysis, it is assumed that the hysteresis comparator is realized as given in Figure 2.9.

2.3.1 *Passive filtering technique*

One way to realize the filter and summer, in Figure 2.14, is to implement a conventional passive low-pass or band-pass filter. Based on the complexity of the filter, it can be broadly classified as follows.

The simplest form of implementation of a filter is shown in Figure 2.15(a), and its application to a buck converter is shown in Figure 2.15(b). The filter is supplied with the switch node and the output voltage of the converter. Note that the summer is implicitly a part of the modulation signal. The modulation signal carries a DC part and a switching frequency part. The DC part can be obtained by the following equation. The buck converter output voltage is regulated based on this expression.

$$\begin{aligned} V_{cmd} &= \langle V_{sw} \rangle_{avg} = V_o + r_L \cdot I_o \\ &\Rightarrow V_o = V_{cmd} - r_L \cdot I_o \end{aligned} \tag{2.7}$$

It is important to note that under ideal case when converter losses are neglected, the output voltage of the converter (V_o) can be kept constant by maintaining a fixed command signal (V_{cmd}). The output voltage of the converter is dependent on the inductor resistor (r_L) as well as the command signal. The AC part of the modulation signal is an ideal piecewise linear curve, which is used for hysteretic modulation, and

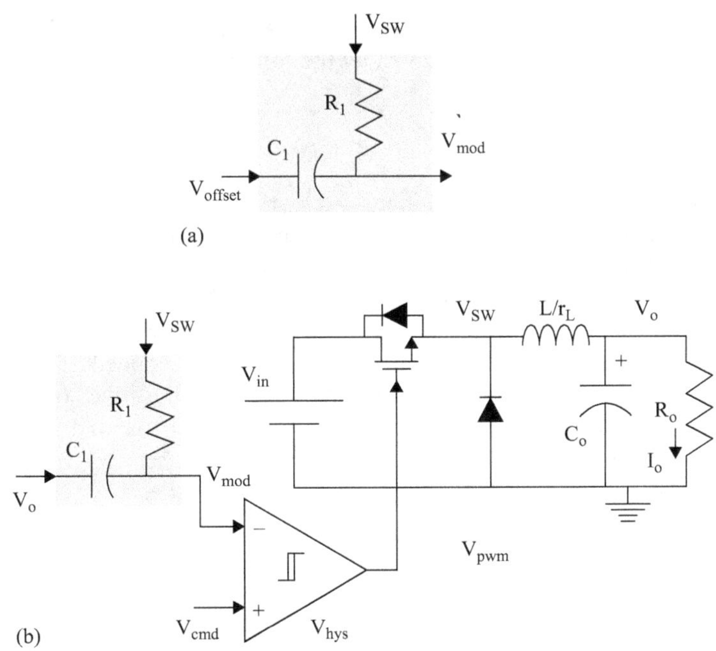

Figure 2.15 (a) Schematic of type-1 SRG and (b) application to a buck converter

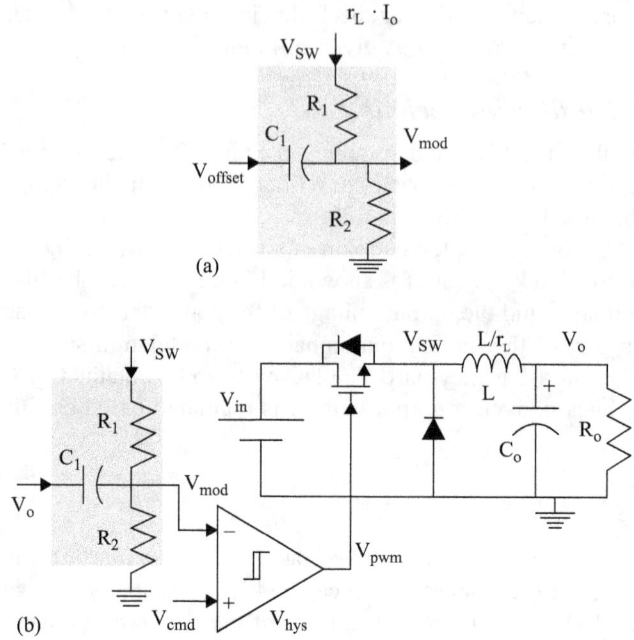

Figure 2.16 (a) Schematic of a type-2 SRG and (b) application to a buck converter

it decides the switching function. An approximate expression for the switching frequency of the buck converter is given by the following expression.

$$f_s = \frac{(V_{in} - V_o) \cdot V_o}{V_{in} \cdot V_{hys} \cdot R_1 \cdot C_1} \tag{2.8}$$

The switching frequency can be increased by reducing the hysteretic window or reducing the values of the filter parameters (R_1 and C_1). Note that the time constant (R_1C_1) is selected much larger than the desired switching period. The switching frequency has a strong dependence on inputs and outputs of the converter. One of the drawback of this strategy is that the V_o has a direct dependence on converter parasitics.

Better regulation over the output voltage can be obtained if the filter is implemented as shown in Figure 2.16(a). Application of the filter to a buck converter is shown in Figure 2.16(b). In this case, a smaller command signal can be used to realize a larger output voltage by controlling the feedback parameters R_1 and R_2. The steady-state output voltage is given by

$$V_{cmd} = \langle V_{sw} \rangle_{avg} \cdot \left(\frac{R_2}{R_1 + R_2} \right) = (V_o + r_L \cdot I_o) \cdot \left(\frac{R_2}{R_1 + R_2} \right)$$

$$\Rightarrow V_o = \left(\frac{R_1 + R_2}{R_2} \right) . V_{cmd} - r_L \cdot I_o \tag{2.9}$$

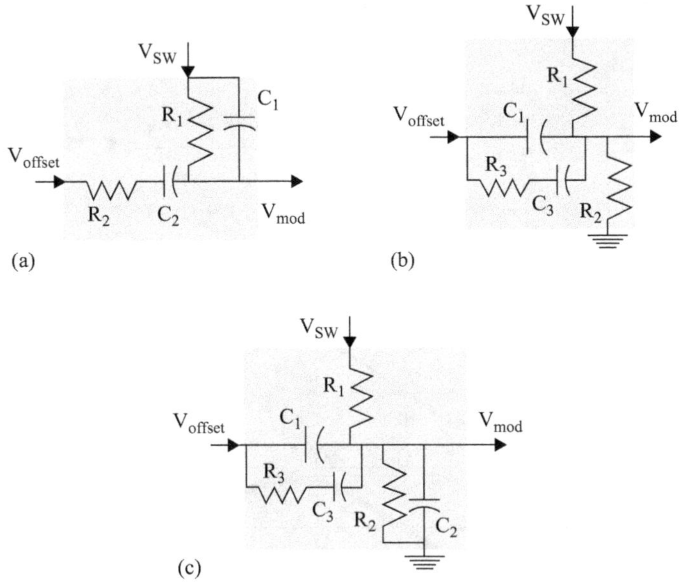

Figure 2.17 Example schematics of higher order SRG

As the term in the bracket in RHS of the above equation is always larger than one the dependence of the output voltage on parasitic resistance (r_L) reduces.

Higher order filters, as shown in Figure 2.17, can also be used to realize a passive SRG. The selection of an SRG is based on requirements of better control over output voltage. It is also worth mentioning that the switching frequency dependence on input and output voltage can be reduced in higher order filters.

Application of a higher order SRG to a boost converter is shown in Figure 2.18. In this case, the output is always higher than the input voltage. Therefore, a command signal is applied with an offset (V_{offset}), and it is always higher than the input voltage as shown in the schematic and its response. The low-frequency behaviour of the output voltage is described as

$$V_o = V_{cmd} \tag{2.10}$$

This implementation is also an example where the command signal is added to the feedback signal of a converter and is compared with another feedback variable (output voltage). In other words, V_{mod} and V_{cmd} both contain feedback terms. This is unlike the cases presented in Figures 2.15 and 2.16 where command signal is a DC reference and is applied to the non-inverting terminal of the hysteretic controller, and all the feedbacks are applied to the filter.

2.3.2 Active filtering technique

Active filters can also be used to realize the filter block in Figure 2.14. Analogue components have higher tolerance and lead to more variation in design variables in

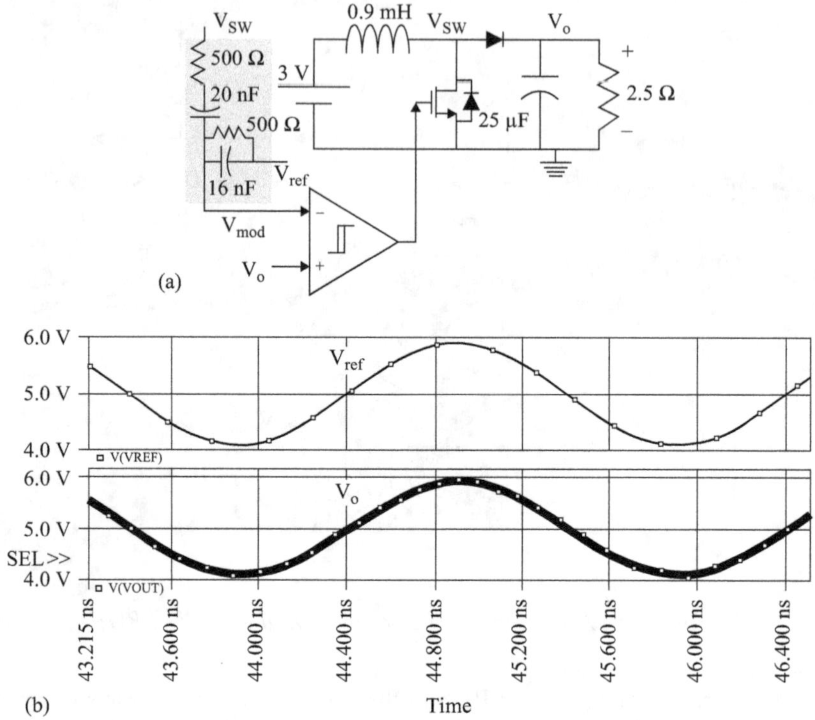

Figure 2.18 Application of a higher order SRG to a boost converter. (a) Schematic and (b) waveforms

mass production. Active filter has the advantage of having better control over design variables while generating the synthetic ripple. This leads to smaller variations in switching frequency. This is important when the synthetic ripple is small. Figure 2.19 shows one such ripple synthesizer. It consists of a transconductance amplifier (g_m-amplifier) feeding an RC filter. The inputs to the transconductance amplifier are the converter waveforms as well as an offset voltage. The voltage across the output filter is the synthetic ripple being sought, which is an ideal piecewise linear signal. This ripple can be stacked on to another DC offset to generate the modulating signal for the hysteretic comparator. Using an offset at the output of the transconductance amplifier has the advantage of creating a unipolar modulation signal. As the ripple generated at V_{mod} is dependent on circuit parameters, it can be controlled to be higher than the natural ripple of the converter. The resistance R_{cmod} absorbs the DC output of the transconductance amplifier. It is related to the average input to the g_m-amplifier, and therefore R_{cmod} can be used to control the low-frequency performance of the converter.

Figure 2.20(a) shows an implementation of an SRG for a buck converter. The $V_{offset1}$ and $V_{offset2}$ are both selected to be equal to the output voltage. Therefore, the input to the g_m-amplifier is essentially the voltage across the filter inductor (L).

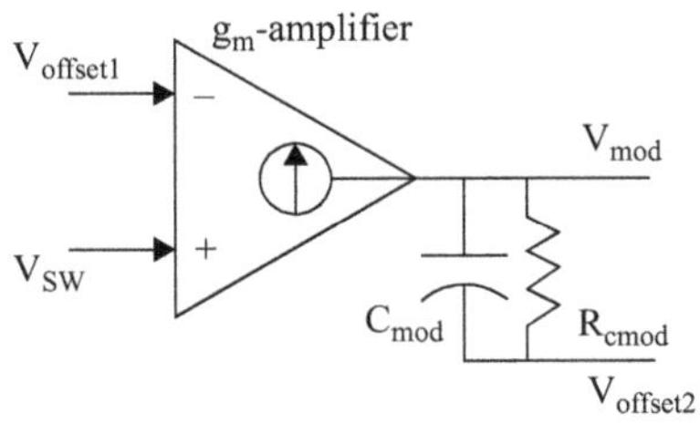

Figure 2.19 Active synthetic ripple generation technique

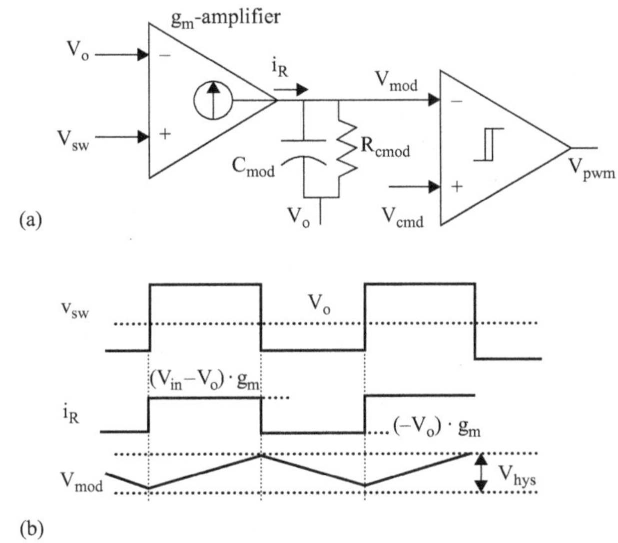

Figure 2.20 Implementation of an SRG for a buck converter (as shown in Figure 2.16(b)). (a) Schematic and (b) operational waveforms

The operational waveforms of the g_m-amplifier for a buck converter are shown in Figure 2.20(b). When the switch node voltage is high, the output current of the amplifier is $(V_{in}-V_o) \cdot g_m$ and in the rest of the interval, it is $(-V_o \cdot g_m)$. This assumes a CCM operation for the buck converter. The DC part of the output current i_R flows through the resistor (R_{cmod}) and is noted as I_R. The AC part of i_R flows through the capacitor (C_{mod}) to generate the ripple. Therefore, the following equation is obtained.

$$V_{cmd} = V_{mod} = V_o + I_R \cdot R_{cmod} \tag{2.11}$$

As the average current of the g_m-amplifier is dependent on the drop across the inductor, the following relationship is obtained.

$$V_o = V_{cmd} - I_o \cdot (r_L \cdot g_{mMod} \cdot R_{cmod}) \tag{2.12}$$

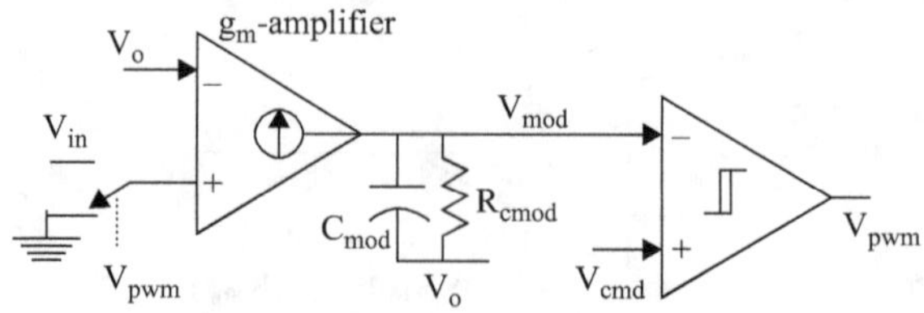

Figure 2.21 Alternate way to realize an active ripple synthesizer

In the above equation, r_L is the DCR of the inductor and g_{mMod} is the transconductance of the transconductance amplifier. If the converter non-idealities are neglected, the output voltage (V_o) equals the command signal (V_{cmd}). However, when r_L is not neglected, the output voltage droops (from a no load value of V_{cmd}) as the output current (I_o) is increased. The term in the bracket is called droop resistance, which decided the slope of this dependence between load current and output voltage. This drooping behaviour is quite useful reducing output capacitors when converters have high transient loads and need to meet tighter regulation [8]. In this case, even with a smaller r_L, a larger droop resistance can be realized.

The switching frequency can be derived by calculating the slopes of the modulation signal (V_{mod}) and is given by

$$f_s = \left(1 - \frac{V_o}{V_{in}}\right)\frac{V_o}{V_{hys}} \cdot \frac{g_{mMod} \cdot L}{C_{mod}} \tag{2.13}$$

An alternate way to realize the circuit in Figure 2.20(a) is given in Figure 2.21. In this case, a switching circuit is implemented to replicate the switching behaviour of a buck converter as shown in Figure 2.20. This circuit has the advantage that only low-frequency signals (V_{in} and V_o) are sensed by the ripple synthesizer. With the absence of any high-frequency switching signal fed directly from the converter, this implementation is relatively noise-free and easy to implement in an IC-based design.

2.4 Load current feedforward

The inclusion of switching frequency variables to derive the synthetic ripple has the advantage of including inductor current information in ripple derivation. For example, the average switch node voltage of a buck converter carries information about the load current because of the DCR of the inductor. When the load current increases, the average voltage of switch node increases to account for the DC voltage drop across the inductor as shown in the expression below.

$$V_o = V_{sw} - I_L \cdot r_L = V_{sw} - I_o \cdot r_L \tag{2.14}$$

Here, V_{SW} is the average of the switch node voltage and I_L is the average inductor current. The equation shows a natural dependence of switch node voltage on the output current. However, as the inductor current cannot change instantaneously, it limits the speed of the controller when switch node voltage is used as a control variable.

However, there is a way to provide instantaneous transient information to the controller. Let us explain this phenomenon for a buck converter. If a high ESR ($=R_c$) output capacitor is used for a buck converter, the output voltage exhibits a sudden change in magnitude, when a load step is applied. This can be fed to the hysteretic controller for improved dynamic response.

In order to see this response for a buck converter with an active synthetic ripple modulator, as given in Figure 2.20(a), is applied with an instantaneous change in load. In this controller implementation, the output voltage is added to the synthetic ripple. Therefore, any sudden change in output voltage leads to an instantaneous impact on the modulation signal (v_{mod}). This sudden change is given by $\Delta I_o \cdot R_c$, where R_c is the ESR of the output capacitor and ΔI_o is the applied load step.

Figure 2.22 shows the response of the converter to a step-down transient for a buck converter, which is asserted asynchronously during the beginning of the buck switch off time. The load transient forces the output voltage to respond almost instantaneously and, therefore, affects the modulation signal, which in turn leads to an increased off time (Δt_{off}) of the bottom switch. Note that a load transient instigates a response by the modulator within a switching cycle. This large signal response is known as load current feedforward. The higher the ESR of the output capacitor, higher will be the load feedforward. However, under steady state, it will also lead to higher output voltage ripple.

Load current feedforward is possible for converters where the capacitor voltage (which is used to derive a synthetic ripple) and load voltage are related. This leads to a direct control of the output voltage during a load transient event.

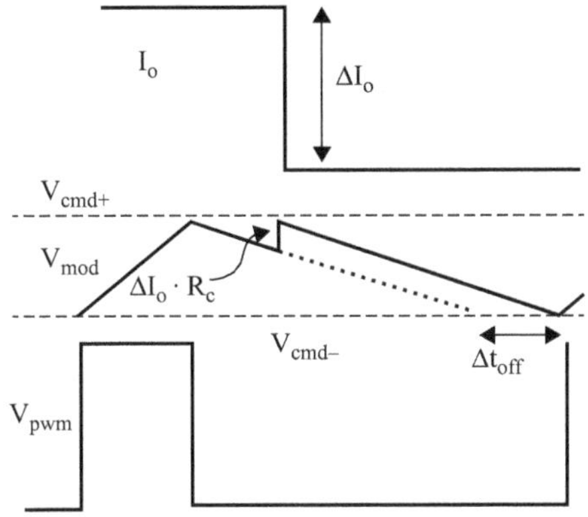

Figure 2.22 Feedforward of load transient and its impact on controller response for a buck converter

2.5 Linear model development

In order to use the synthetic ripple generation block and the hysteretic comparator in a voltage-regulated converter system with a voltage compensator, linear models of the blocks are essential. Derivations of these models are explained in this section.

2.5.1 Modelling an SRG

2.5.1.1 Method 1

A linear frequency domain model can be derived for the ripple regulator by using the block diagram in Figure 2.23. The model describes the basic frequency domain characteristic of the ripple regulator.

As the model is linear, separate transfer function can be derived to describe the output of the ripple regulator (v_{mod}) in terms of its inputs (v_{sw} and v_{offset}). Therefore, the generalized overall transfer function is given by

$$v_{mod} = G_{ao}(s) \cdot v_{offset} + G_{ad}(s) \cdot v_{sw} \tag{2.15}$$

The above equation can be used to develop models of a passive SRG for a buck converter as given in Figures 2.15 to Figure 2.17. General expressions of the gain blocks with $v_{offset} = v_o$ are given by the following equations. Table 2.1 shows the various parameters in terms of passive component values.

$$G_{ao}(s) = \left.\frac{v_{mod}}{v_{offset}}\right|_{v_{sw}=0} = \frac{s \cdot (k_o k_t \cdot s + k_o + k_p)}{(k_o + k_a) \cdot k_t \cdot s^2 + (k_o + k_a + k_p + (1+\alpha) \cdot k_t) \cdot s + (1+\alpha)} \tag{2.16a}$$

$$G_{ad}(s) = \left.\frac{v_{mod}}{v_{sw}}\right|_{v_{offset}=0} = \frac{k_q \cdot s^2 + (k_r + k_t) \cdot s + 1}{(k_o + k_a) \cdot k_t \cdot s^2 + (k_o + k_a + k_p + (1+\alpha) \cdot k_t) \cdot s + (1+\alpha)} \tag{2.16b}$$

This method is helpful in deciding the filter transfer function to obtain a simple low cost filter realization. Once the above equations are derived, v_o and v_{sw} can be

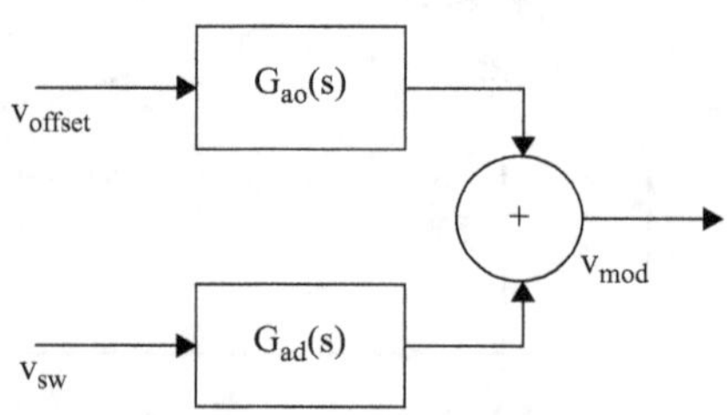

Figure 2.23 Small-signal model of a ripple regulator

Table 2.1 Parameters for various passive SRGs

Filter	k_o	α	k_P	k_t	k_a	k_q	k_r
Type 1 (Figure 2.15(a))	R_1C_1	0	0	0	0	0	0
Type 2 (Figure 2.16(a))	R_1C_1	R_1/R_2	0	0	0	0	0
Higher order (Figure 2.17(a))	0	0	R_1C_2	R_2C_2	R_1C_1	$R_1R_2C_1C_2$	R_1C_1
Higher order (Figure 2.17(b))	R_1C_1	R_1/R_2	R_1C_3	R_3C_3	0	0	0
Higher order (Figure 2.17(c))	R_1C_1	R_1/R_2	R_1C_3	R_3C_3	R_1C_2	0	0

related to drive a final expression relating v_o and v_{mod}. For a buck converter, it is given by

$$v_o(s) = v_{sw}(s) \cdot \left[\frac{Z_o}{sL + Z_o}\right], \text{ where } Z_o = \frac{R_o}{1 + sR_oC_o} \tag{2.17}$$

2.5.1.2 Method 2

If v_o and v_{sw} are related, as in case of a buck converter, a second method can be used to derive a linear model that relates v_o and v_{mod}. This uses the voltage across the inductor to establish the relationship. For the active filter–based SRG, as shown in Figure 20(a), this model can be derived as follows. The first step is to derive a relation between inductor current (i_L), transconductor current (i_R), and $v_{mod}(s)$.

$$i_R(s) = (sL + r_L) \cdot g_m \cdot i_L(s) \tag{2.18}$$

It can be used to derive an expression for $v_{mod}(s)$ as shown below

$$v_{mod}(s) = Z_m \cdot i_L(s) + v_o(s)$$
$$\text{where } Z_m = (r_L + sL) \cdot g_m \left(\frac{R_{cmod}}{1 + sR_{cmod}C_{mod}}\right) \tag{2.19}$$

Therefore, the transfer function relating $v_o(s)$ and $v_{mod}(s)$ are given by

$$G_c(s) = \frac{v_o(s)}{v_{mod}(s)} = \frac{v_o(s)}{v_{cmd}(s)} = \frac{Z_o}{Z_o + Z_m} \tag{2.20}$$

The above equation relates v_o and v_{mod}, and it can now be used to design a voltage mode control once the hysteretic controller is modelled.

2.5.2 Modelling a hysteretic controller

The hysteretic comparator produces a switching pulse defined by the edges of the ripple in the v_{mod} signal intersecting the hysteretic limits defined by the command signal (V_{cmd}) signal. The control law of the hysteretic controller can be expressed as a block with input–output characteristic with an infinitely large slope (as shown by the grey dotted curve in Figure 2.24(a)), which in turn represents an ideal op-amp,

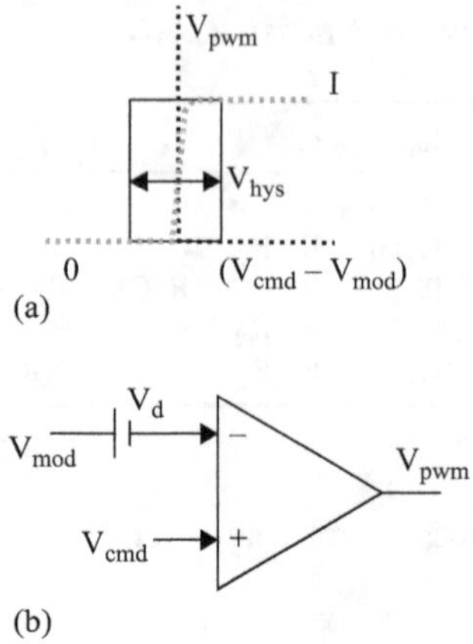

Figure 2.24 Average model of a synthetic ripple hysteretic modulator. (a) Model of an ideal op-amp (b) ideal op-amp with an offset models a hysteretic comparator

as shown in Figure 2.24(b). Under ideal operation, the low-frequency component of V_{mod} equals the low-frequency component of V_{cmd}.

In order to model the commutation delay of the hysteretic controller, as shown in Figure 2.10, a DC offset is added to either the positive or negative terminal depending on the rising slope delay or falling slope delay. Now the model can be combined with small-signal model of the converter to develop a complete small-signal linear model of the system and can be used to develop feedback controllers.

2.6 Conclusions

Modern converters have non-ideal ripple in inductor current and output voltage for ideal hysteretic controller implementation. This chapter explains various methods to generate synthetic ripple from converter operating waveforms using analogue blocks. As this ripple is near ideal, they lead to proper hysteretic controller operation. Either active or passive networks can be used to implement a ripple generator. Various methods to generate synthetic ripple are explained in this chapter. Their analogue implementations are explained. Models of the synthetic ripple modulator are derived, which can be used in feedback design.

References

[1] S.-H. Jung, N.-I. Kim, and G.-H. Cho, "Class D audio power amplifier with fine hysteresis control," *Electron. Lett.*, vol. 38, pp. 1302–1303, 2002.

[2] S. Mishra and K. Ngo, "Dynamic characterization of the synthetic ripple modulator in a tightly regulated distributed power application," *IEEE Trans. Ind. Electron.*, vol. 56, no. 6, pp. 1164–1173, 2009.

[3] W. Huang, "A new control for multiphase buck converter with fast transient response," *Proc. IEEE Appl. Power Electron. Conf.*, pp. 273–279, 2001.

[4] SNVA170B, Application Note AN-1487: Current Mode Hysteretic Buck Regulators, 2006.

[5] M. Castilla, L.G. de Vicuña, J.M. Guerrero, J. Matas, and J. Miret, "Designing VRM hysteretic controllers for optimal transient response," *IEEE Trans. Ind. Electron.*, vol. 54, no. 3, pp. 1726–1738, 2007.

[6] K.D.T. Ngo, S.K. Mishra, and M. Walters, "Synthetic-ripple modulator for synchronous buck converter," *IEEE Power Electron. Lett.*, vol. 3, no. 4, pp. 148–151, 2005.

[7] K.D.T. Ngo, S.K. Mishra, and S. Thekkevalappil, "Pulse width modulator with carriers derived from converter waveforms," *Electron. Lett.*, vol. 41, no. 3, pp. 152–154, 2005.

[8] S.K. Mishra, "Design oriented analysis of modern active droop controlled power supplies," *IEEE Trans. Ind. Electron.*, vol. 56, no. 9, pp. 3704–3708, 2009.

Chapter 3

One-cycle controlled power inverters

Giovanni Petrone[1] *and Giovanni Spagnuolo*[1]

One Cycle control is a technique matching many features that are typical of the linear control methods with some others that characterise high performances non-linear control approaches. Indeed, it ensures a constant switching frequency with a high promptness and noise rejection. The One Cycle control technique has recently found applications in many fields related to the renewable energies and to the power factor control, so that the customers' interest is increasing. In this chapter the basic elements of the One Cycle control technique are reviewed and some issues related to its wide applications are discussed. A special emphasis is given to the DC/AC single-phase conversion.

3.1 Introduction

In the last decade the digital control has gained the largest part of the power electronics market. The control of switching converters has been traditionally performed through analogue solutions implemented by means of dedicated integrated circuits (ICs). Nevertheless, the increase of the complexity of power architectures, made of switching converters interacting each other, and the request for more advanced functions have led the producers to the adoption of digital devices. These ones do not only accomplish the basic control functions, but run algorithms for the power management of the complex power system, implement strategies for diagnostic and prognostic purposes, and allow to communicate with the external world in an age of the internet-of-things. Apart from the aforementioned features, the main key of the success, especially for consumer applications, is its inherent flexibility and programmability. In this way, any mass producer of electronic systems can buy a very large quantity of the same digital device in order to use the same component in different applications and products. This gives to it a high contractual power against the components' producer, so that a lower component cost is gotten.

[1]Department of Information and Electrical Engineering and Applied Mathematics, University of Salerno, Fisciano, Salerno, Italy

In spite of these undeniable advantages, analogue control still has rooms for being used in many applications, especially when a control action with high performances and robustness is needed. In addition to these features, the adoption of analogue controllers also ensures a better protection of the intellectual property against competitors and lower product costs. The price to pay is in the high expertise the designer needs to have before approaching any analogue control design problem.

The analogue linear control techniques are well known and implementable by means of a large variety of ICs available on the market. Non-linear control ensures the best performances, especially when fast response and simple circuitry are needed. This is the case, for instance, of photovoltaic (PV) applications, which require a considerably high dynamic response from the controller in order to track the fast irradiance variations, with a strong constraint represented by the costs. To this regard, in Reference 1, it has been demonstrated that the best performances in PV applications Maximum Power Point Tracking (MPPT) are obtained through the joint adoption of digital and analogue resources embedded into a unique device. Indeed it was shown that, in the MPPT application, the Sliding Mode Control (SMC) exhibits superior performances with respect to linear control, but it suffers from some problems related to the analogue-to-digital (A/D) signal conversion. The well-known discretisation chattering problem is due to the finite sampling leading to an undesirable phenomenon of oscillations having finite frequency and amplitude. It may appear in the application of many other control techniques requiring the comparison of signals acquired on the switching converter and converted through an A/D system. In addition, especially whenever cheap digital devices have to be used for increasing the cost competitiveness of the product, the control performances of the technique implemented through analogue devices are significantly better than those ones ensured by the digital controller, due to the time needed for processing the information. The computation speed is evidently important in applications, e.g. PV harvesting for sustainable mobility, where the reaction to the changes of exogenous variables is a key factor.

The superiority of non-linear control techniques in many applications has been widely demonstrated in literature. A significant example is just the SMC, which unfortunately shows the drawback of a non-constant switching frequency when it is applied to a power converter. This effect has detrimental consequences in the control design itself. One Cycle Control (OCC), instead, cumulates the advantages of SMC, especially in terms of promptness, with a constant switching frequency of the converter it is applied to. It was firstly introduced in 1991 [2] and, since then, many applications of this control technique have appeared both in literature and on the market, after its patenting [3] and commercialisation [4]. It has been applied to the control of DC/DC converters but especially for the grid connection of renewable energy sources and in power factor correction (PFC).

In this chapter an overview of the working principle and of the main applications of OCC is firstly given. The case of single-phase inverters is specially treated, with an emphasis to the applications to renewable energy systems.

3.2 OCC: operating principle and applications overview

In a traditional PWM-based linear control, the driving signals for the switches are obtained by comparing a fixed sawtooth waveform with a modulating signal coming from a linear compensation network. The main advantage of the PWM-based linear control is that the system stability and its dynamic performances can be well predicted by deriving the small-signal model of each block and thus the whole system can be analysed by representing it as a cascade of functional blocks in the frequency domain [5]. The main limitation of such an approach is in the fact that any control action is firstly processed by the compensation network and subsequently applied to the converter. The dynamic performances can be improved by acting directly on the PWM sawtooth waveform. In this way part of the control action does not involve the compensation network and it is applied to the switches modulation more promptly. In these cases, the switches modulator and the controller are seen as a whole and they cannot be analysed separately. The absence of the PWM block is a feature of many non-linear controls techniques, such as SMC, hysteretic control, constant on-time and constant off-time control, as well as OCC. The OCC integrates a variable, a voltage, a current or a combination of them, in order to force its average value to be equal to some control reference. It was introduced in Reference 2 by referring to the output voltage control of a basic buck DC/DC converter operating in continuous conduction mode.

In Figure 3.1, the elements in the grey box represent the basic OCC block, including one resettable integrator, one SR flip-flop (FF), one clock generator and a comparator.

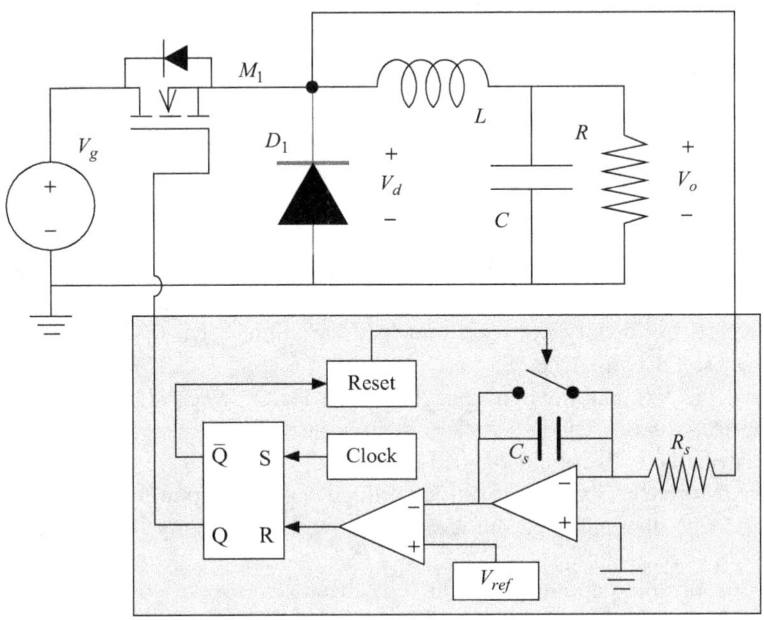

Figure 3.1 OCC controlled buck converter

The output voltage of the buck converter is the average value of the diode voltage, namely, the area under each pulse of the diode voltage divided by the switching period. In formulae

$$<v_d>_{T_s} = f_s \int_0^{T_s} v_d(t)dt = f_s \int_0^{DT_s} v_g(t)dt \tag{3.1}$$

where $<v_d>_{T_s}$ is the average value of the diode voltage $v_d(t)$, $v_g(t)$ is the converter input voltage, D is the duty cycle, T_s and f_s are the switching period and frequency, respectively. Pulses at the constant frequency f_s start the integration of the diode voltage periodically and synchronously with the MOSFET turning ON. The integrated diode voltage is compared with a threshold v_{ref}, so that the MOSFET is turned OFF and the integrator is reset as soon as these two signals assume the same value. The following control equation, fulfilled cycle-by-cycle, describes the OCC modulation of the switching ON time

$$v_{ref} = \frac{1}{R_s C_s} \int_0^{DT_s} v_d(t)dt \tag{3.2}$$

The condition $R_s C_s = T_s$ ensures that the OCC guarantees the condition $<v_d>_{T_s} = v_{ref}$ in every switching period. If the supply voltage $v_g(t)$ varies slowly, with respect to the switching frequency, the integration can be simplified by the product between the instantaneous value of the duty cycle and the input voltage. Thus, the OCC equation holding in the low-frequency range is obtained

$$v_{ref} = d(t) \cdot v_g(t) \quad \rightarrow \quad d(t) = \frac{v_{ref}}{v_g(t)} \tag{3.3}$$

Equation (3.3) explains the non-linearity of the OCC operation: it balances the intrinsic converter non-linearity introduced by the buck converter: $v_o(t) = d(t) \cdot v_g(t)$. Equation (3.1) reveals that the larger the input voltage, the shorter will be the time interval needed by the integrated diode voltage for reaching the threshold, so that the duty cycle assumes a small value. Instead, a small input voltage corresponds to a large duty cycle value. One of the advantages offered by the OCC is in the fact that the duty cycle value can change in just one switching period if the reference voltage is changed stepwise. It is also able to reject disturbances affecting the input buck voltage, but it does not react to a load variation, because of the output filter dynamics the OCC neglects.

Since the first publication concerning the OCC principles [2], described by referring to the basic DC/DC buck converter shown in Figure 3.1, many applications, especially in the field of grid-connected systems, have been presented in literature. Moreover, many innovative products have been put on the market by the inventors, who also patented the idea [6], through a company [4] having the core business in OCC technique.

PFC is a leading application, which has been also supported by a dedicated IC, produced by International Rectifier [7] and available on the market at a price lower

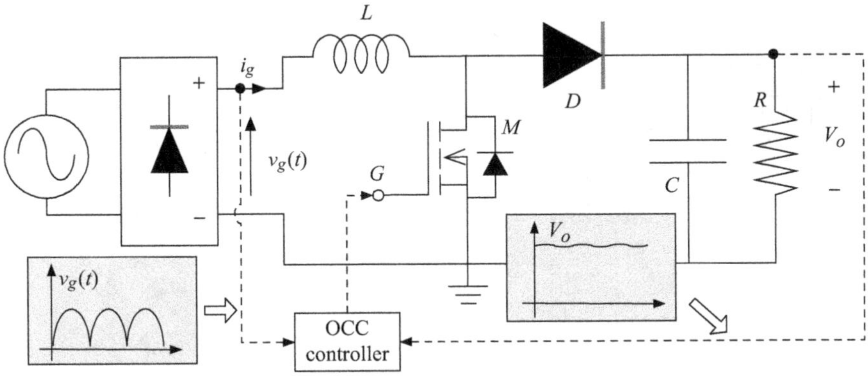

Figure 3.2 PFC boost single phase

than 1$. Such an IC allows to have a continuous conduction mode boost type PFC with an average current mode control at a programmable switching frequency up to 200 kHz. The product data sheet puts into clear evidence that an high power factor and a low total harmonic distortion are achieved with a drastic reduction in component count. The OCC method for PFC applications has been applied to single phase [8] as well as to three-phase systems. In single phase, the idea for achieving a resistive behaviour at the DC/DC converter input terminals is usually explained by referring to a boost converter, shown in Figure 3.2. By assuming that the switching converter operates in discontinuous conduction mode (DCM), the average value of the inductor current is expressed as

$$< i_L >_{T_s} = \frac{d^2 T_s V_o v_g}{2L(V_o - v_g)} \tag{3.4}$$

where T_s is the switching period, d is the duty cycle, L and V_o are the boost converter input inductance and the almost constant output voltage, respectively. As shown in Figure 3.3, the rectified grid voltage v_g, which appears at the input of the boost converter, is a time varying quantity.

In order to have an average inductor current in phase with the input voltage, the switching converter must behave at its input as it were a pure resistor, namely

$$< i_L >_{T_s} = \frac{v_g}{R_e} \tag{3.5}$$

Condition (3.5) holds provided that the control action is able to keep the quantity $d^2/(V_o - v_g)$ constant.

The OCC circuit shown in Figure 3.3 allows to implement such a control action. The clock generator fixes the switching frequency. Indeed, at the beginning of every switching period T_s, the clock signal enables the FF whose output turns on the controlled switch of the boost converter. During this phase, the switches (S_1, S_2)

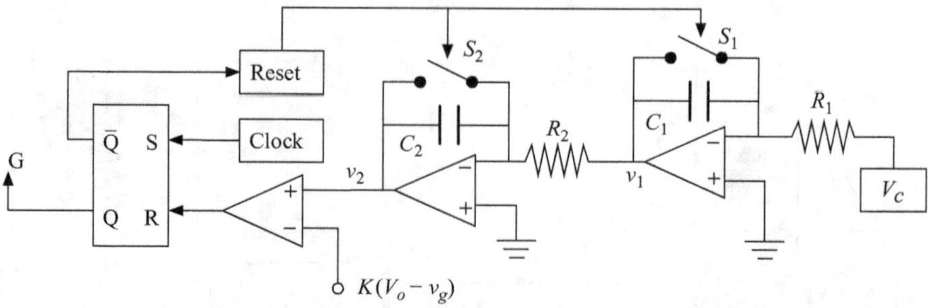

Figure 3.3 PFC analogue control circuitry

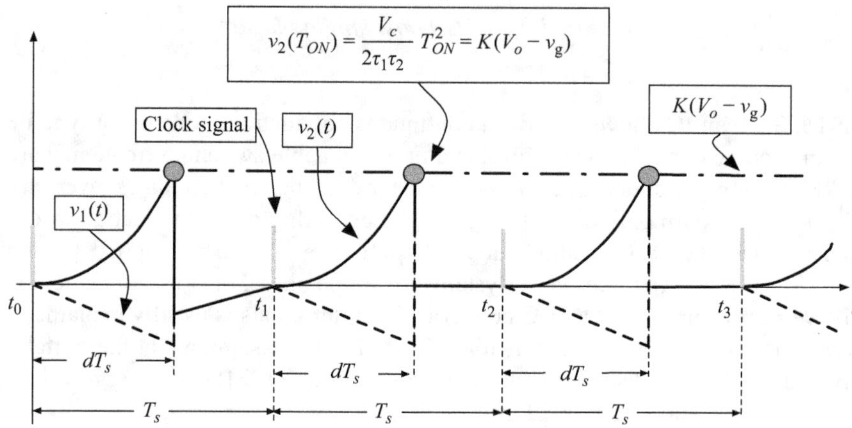

Figure 3.4 Typical waveforms of the OCC signal for PFC applications

of the two resettable integrators are turned off and the constant control signal V_c is integrated twice. Thus

$$v_1(t) = -\frac{1}{\tau_1}\int_0^t V_c d\tau \tag{3.6}$$

$$v_2(t) = -\frac{1}{\tau_2}\int_0^t v_1(\tau)d\tau = \frac{V_c}{2\tau_1\tau_2}\cdot t^2 \tag{3.7}$$

where τ_1 and τ_2 are the time constants of the two integrators.

By referring to the scheme shown in Figure 3.3 and to the waveforms shown in Figure 3.4, the control action is described as follows: when the OCC signal v_2 reaches the value $K(V_o - v_g)$, the FF Reset signal is activated and the integration process is stopped. At the same time, the boost converter controlled switch is turned off. As a consequence, the following condition is assured for T_{ON}

$$v_2(T_{ON}) = \frac{V_c}{2\tau_1\tau_2}T_{ON}^2 = K(V_o - v_g) \tag{3.8}$$

where K is the gain of the voltage sensors used for acquiring both V_g and V_o. Being T_{ON} related to the duty cycle and the switching period, thus $d = \frac{T_{ON}}{T_s}$, the OCC equation (3.8) is expressed as

$$\frac{d^2}{\left(V_o - v_g\right)} = \frac{2\tau_1\tau_2 K}{V_c T_s^2} \tag{3.9}$$

The fact that all the terms on the right side of (3.9) are constant ensures that the OCC objective (3.5) through (3.4) is achieved. Indeed, by equating (3.4) and (3.5) and by exploiting (3.9), the equivalent resistance seen at the input of the boost converter is

$$R_e = \frac{V_c L T_s}{K V_o \tau_1 \tau_2} \tag{3.10}$$

Thus, R_e is proportional to the control voltage V_c, whose value is varied for changing the input power, and the input power is in turn changed by varying V_c, so that the output DC voltage V_o is regulated [8].

OCC shows superior performances also in interfacing power systems to the AC mains. Some papers in literature put into evidence the ability of the control technique to have a sinusoidal current, with a low harmonic distortion, injected into the grid in phase with the grid voltage. Such a function can be performed without using a Phase Locked Loop (PLL) device, thus with significant advantages when a weak grid must be approached. Furthermore, a proper design of the OCC parameters allows also to achieve other main control functions. For instance, in PV applications, the MPPT function can be obtained at the same time. This feature is very interesting especially to the aim of developing simple and robust single-stage PV inverters, which minimise the length of the part list and the number of stages of the power processing system. Indeed both the functions of source control, thus MPPT, and grid connection by fulfilling power quality standards can be embedded into an OCC controlled unique power converter.

The following sections give more details about these issues, thus it is focused on the features offered by the OCC technique in the field of grid-connected inverters, especially for applications to renewable energy systems.

3.3 OCC inverters for PV applications

In both single-phase and in three-phase PV applications the main control objective is to impose a sinusoidal profile to the currents injected into the grid. The amplitude and phase of such currents depend on the system operating mode, which can be in connection with the grid or feeding isolated loads and with connection to a microgrid. Due to the need to control both the harmonic content and the power factor of the current injected into the grid, the use of current controls is very popular in distributed generation systems. Usually the current loop is designed by using resonant integrators or, alternatively, standard integrators in multiple rotating frames [9], [10]. In literature, the digital implementation of these techniques, profiting from the flexibility and robustness of such solution, is preferred to the analogue one. The current reference is obtained by using the phase information of the utility voltage or current. Generally, this is obtained by employing either a PLL device or a current phase observer digital technique. Unfortunately, especially in poor grid, the PLL routine requires a high implementation effort. Indeed, the compensation of the grid non-idealities and of the harmonic distortions requires the adoption of complex algorithms, i.e. Discrete Fourier Transform. As a consequence, a considerable amount of the digital computational resources dedicated to the PLL implementation are exploited by such algorithms [11]. As a further drawback, PLL-based approaches can exhibit instability if the controller frequency range approaches the bandwidth of the system, which in turn depends on the actual value of the grid impedance [12].

For these reasons, in applications where the reliability and the cost constraints do not allow to adopt powerful digital systems, analogue controllers are much more appreciated. Thus, the OCC is a very good candidate for controlling DC/AC converters for PV applications. The advantage of avoiding PLL circuits sums to OCC response promptness, simplicity, intrinsic stability and universal adaptability. In the following sections, the issues related to both single-phase and three-phase OCC implementations are analysed in detail.

3.3.1 OCC for single-phase PV inverters

The typical structure of a single-phase OCC buck-derived single-stage PV inverter is shown in Figure 3.5. Its typical power rating considered in the literature is in the

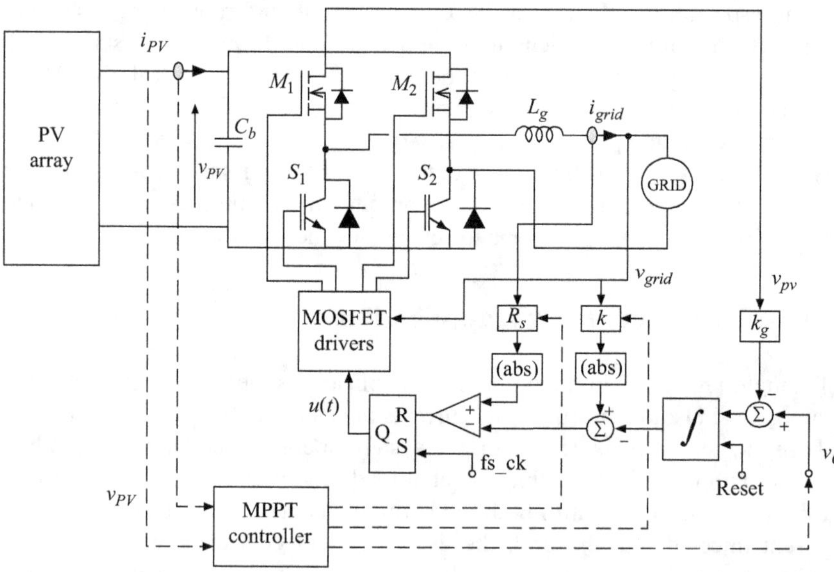

Figure 3.5 Typical OCC circuit for a unipolar single-stage PV inverter

range of 3–6 kW, so that it is dedicated to PV residential applications. The power stage input terminals are directly connected to the PV source. This solution has the advantage of avoiding a preliminary DC/DC conversion stage: the latter being aimed at performing the MPPT function and at adapting the voltage/current levels of the PV field to the DC/AC converter input rating. As a result, the efficiency of the power processing system chain reaches a high value. The drawback of the single-stage solution is in the need of a suitable design of the PV field in order to meet the voltage constraint imposed by the use of a buck-derived DC/AC converter. Depending on the required power rating, the PV array must be composed of one or more strings of PV modules. In turn, the modules are connected in series to reach an operating voltage that is higher than the maximum value of the grid voltage. In this way, the DC/AC step-down operating mode is assured. More flexibility should be achieved by using a switching DC/AC converter embedding a transformer, working at the switching frequency or at the line frequency, but at the price of increasing the complexity of the power stage or decreasing its conversion efficiency.

The inverter is driven by using the unipolar pulse width modulation (UPWM) in which M_1 and M_2 commutate at the switching frequency, while S_1 and S_2 commutate at the grid frequency. The main advantages of the UPWM full-bridge inverters, with respect to the bipolar pulse width modulation (BPWM), are in a higher DC voltage utilisation, in a smaller current ripple in the filter inductor, and in a higher conversion efficiency. As a drawback, in the transformer-less grid-connected PV systems, the UPWM inverters exhibit a time-varying common-mode voltage, thus a leakage current flows through the parasitic capacitance between the PV array and the ground. This parasitic current might be the cause of undesired activation of the PV protection systems and it accelerates the PV modules degradation [13]. For these reasons the BPWM is usually preferred in PV inverters. Recently, some topologies that guarantee no variable common-mode voltage even with UPWM full-bridge inverter [14] have been developed. This makes them suitable to be used in combination with the OCC also for transformer-less PV inverters.

Figure 3.5 reveals that the UPWM requires that the scaled values of voltage and current of the utility grid are firstly rectified and subsequently combined in the OCC circuit. In order to have a unitary power factor at the AC terminals, the OCC in a UPWM single-phase grid-connected inverter has to fulfil the following analogue control equation [15]

$$R_s \cdot i_{grid}(t^*) = k \cdot v_{grid}(t^*) - \frac{1}{\tau}\int_0^{t^*} v_m(t)dt \tag{3.11}$$

where $t^* = d \cdot T_s$ is the instant in which the inverter switches M_1 and M_2 are turned off, $\tau = R_s C_s$ is the time constant of the resettable integrator and T_s is the inverter switching period. Figure 3.6 shows the points marked with circles corresponding to the time instants where (3.11) is fulfilled. The signal v_m is a linear combination of an internal reference signal V_{occ} and of the PV voltage

$$v_m(t) = V_{occ} - k_g \cdot v_{pv}(t) \tag{3.12}$$

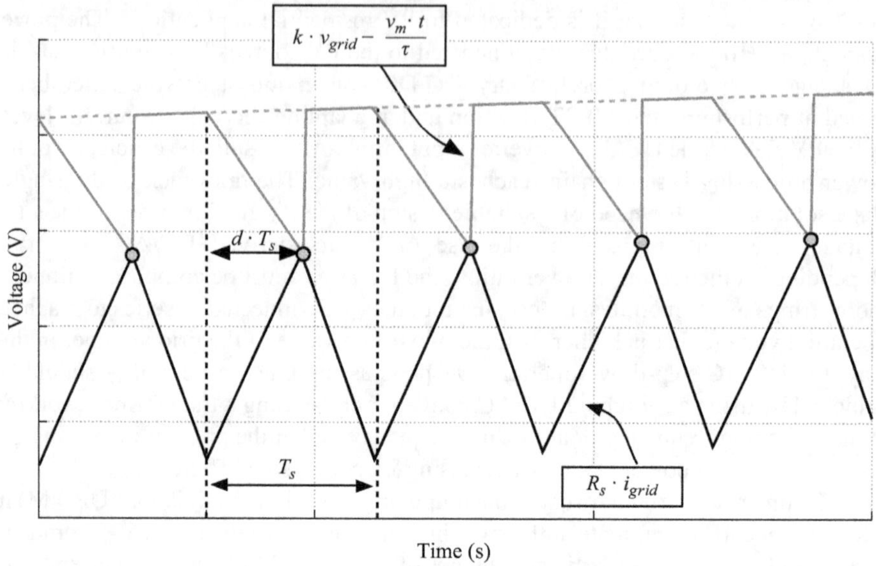

Figure 3.6 OCC signals for the unipolar single-stage PV inverter

The parameters k and k_g are the gains of the two voltage sensors, dedicated to the grid and to the PV voltages, respectively. R_s is the value of the sensing resistance used for acquiring the grid current.

Typical variations of the irradiance and temperature PV array working conditions allow to assume that the PV voltage $v_{pv}(t)$ in (3.12) is almost constant in a switching period, thus the integral in (3.11) is easily calculated and the low frequency expression of the OCC equation is

$$R_s \cdot i_{grid} = k \cdot v_{grid} - \frac{T_s \cdot v_m}{\tau} \cdot d \tag{3.13}$$

In every half cycle of the grid voltage, the input to output relationship of the buck derived inverter, which is $v_{grid} = d \cdot V_{pv}$, allows to eliminate the duty-cycle dependency in (3.13), thus the following expression of the grid current results

$$i_{grid} = \frac{v_{grid}}{R_s}\left(k - \frac{T_s \cdot v_m}{V_{pv}\tau}\right) = \frac{v_{grid}}{R_s}\left(k + \frac{k_g T_s}{\tau} - \frac{T_s \cdot V_{occ}}{V_{pv}\tau}\right) \tag{3.14}$$

Equation (3.14) reveals that the OCC law (3.11) allows to force the grid current to be proportional to the grid voltage, so that the unity power factor is achieved. A stable working condition of the non-linear OCC technique is ensured if the following conditions hold

$$V_{occ} \geq k_g \cdot V_{pv} + \frac{R_s \cdot \tau \cdot (2 \cdot V_{o,max} - V_{pv})}{2L} \qquad \tau < T_s \tag{3.15}$$

where $V_{o,max}$ is the peak value of the line voltage. The meaning of (3.15) is analysed in detail in Section 3.4, where the OCC stability issues are treated more in depth. In Reference 15 it is also explained how to select the parameters k_g, k, V_{occ} and R_s in order to assure the desired output current on the basis of the power coming from the PV field. The OCC parameters and the active power injected in to the grid are correlated through the following expression

$$P_o = \frac{V_{o,rms}^2}{R_s}\left(k + \frac{k_g T_s}{\tau} - \frac{V_{occ} T_s}{V_{pv}\tau}\right) \tag{3.16}$$

$V_{o,rms}$ being the rms value of the line voltage. The balancing between the power delivered by the PV source and the power P_o injected in to the grid provides a further equation that allows to obtain the value of the PV field voltage V_{pv}

$$P_{pv} = I_{pv} \cdot V_{pv} = P_o \tag{3.17}$$

It is well known that the PV source shows a non-linear relationship between the current I_{pv} and the voltage V_{pv} at its terminals. It is accurately expressed by using the single diode equivalent circuit of the PV generator, including a current source, representing the photo-induced current, in parallel with a diode, which models the P–N junction. Furthermore, a series and a parallel resistances model the ohmic and the recombination losses, respectively [16]. Although the PV $I_{pv} - V_{pv}$ curve obtained by using the single diode model assumes an implicit form, the following explicit expression of I_{pv} as non-linear function of the voltage V_{pv} can be obtained by means of the Lambert W-function $\mathbf{W}(\cdot)$ [16].

$$I_{pv} = F\left(V_{pv}, \overline{P}\right) = \frac{R_{hpv} \cdot \left(I_{ph} + I_s\right) - V_{pv}}{R_{spv} + R_{hpv}} - \frac{\eta \cdot V_t}{R_{spv}} \cdot W(\theta_I)$$
$$\theta_I = \frac{\left(R_{hpv}//R_{spv}\right) \cdot I_s \cdot e^{\frac{R_{hpv} \cdot R_{spv} \cdot \left(I_{ph}+I_s\right) + R_{hpv} \cdot V_{pv}}{\eta \cdot V_t \cdot \left(R_{hpv}+R_{spv}\right)}}}{\eta \cdot V_t} \tag{3.18}$$

The five single diode model parameters $\overline{\mathbf{P}} = \left[I_{ph}, I_s, \eta, R_{spv}, R_{hpv}\right]$, which are the photo-induced current, the diode saturation current and ideality factor, the series and the parallel resistances, greatly depend on the environmental condition, e.g. the irradiance and temperature. Their values are calculated on the basis of the experimental data reported in the PV module data-sheet and referring to the Standard Test Conditions (STC) [16]. The PV array non-linear model (3.18) holds when all the PV cells in the generator are of the same type, thus their parameters assume the same value, and work at the same irradiance and temperature conditions. Whenever the PV array operates in inhomogeneous conditions, the PV generator modelling is much more complicated and it involves a greater computing effort (e.g. see Reference 17).

Finally, the P_{pv} power in (3.17) is evaluated by expressing the PV current I_{pv} through the non-linear function (3.18). Figure 3.7 shows the power delivered by the

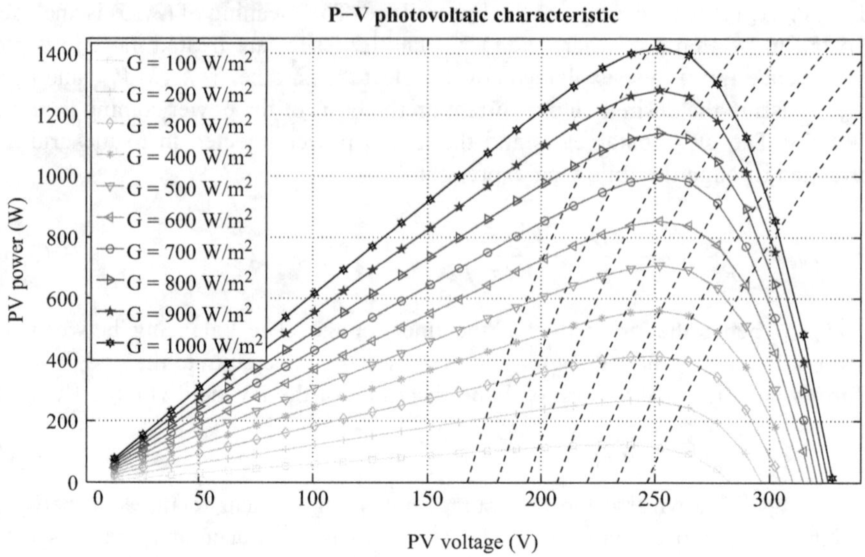

Figure 3.7 Identification of the OCC-inverter operating conditions. Continuous line: delivered PV-power for different irradiance (G) values. Dashed line: power injected into the grid for different OCC parameters.

Table 3.1 OCC parameters

V_{occ}	k	k_g	R_s	τ	T_s	L
[3.6–5.6] V	0.031	0.013	0.2 Ω	9.5 μs	30.3 μs	500 μH

PV source at different environmental conditions and the power P_o injected into the grid, which has been evaluated by using (3.16). The P_o plots refer to the US utility grid, thus with $V_{o,rms} = 110$ V and $f_{grid} = 60$ Hz, and they have been obtained by using the OCC parameters listed in Table 3.1.

The optimal performances of the OCC in terms of the grid-current regulation is confirmed by the simulation results shown in Figure 3.8, where the current waveform is compared with a scaled version of the grid voltage waveform. The two waveforms are almost perfectly in phase, so that a quasi-unitary power factor is assured. The triangular ripple affecting the grid current can be reduced by a proper choice of the output filtering inductance value.

In the OCC configuration proposed in Reference 15 the parameters are assumed to be fixed at a value specified during the design procedure. Unfortunately, such an architecture does not allow to deliver the maximum power in every environmental condition. Indeed the possible operating points of the circuit belong

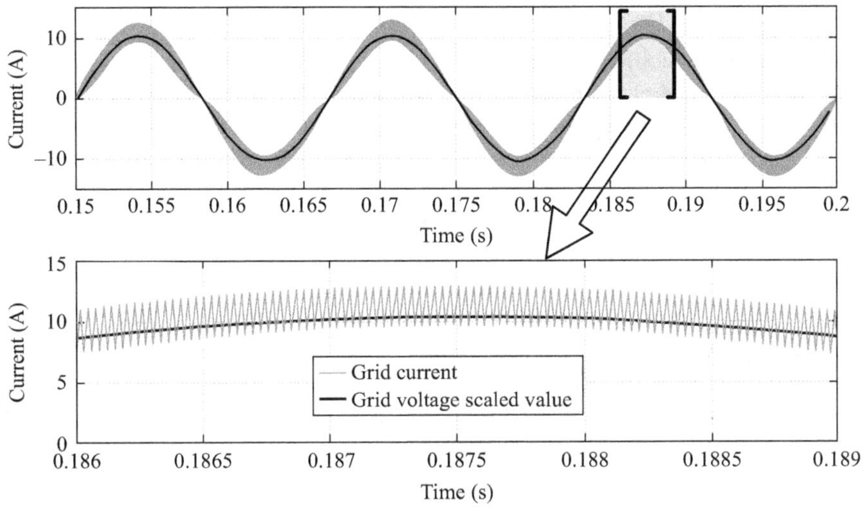

Figure 3.8 Simulated waveform of the One-Cycle controlled single-phase PV inverter

to only one black dashed curve of Figure 3.7 that does not intersect all the PV field curves, at different irradiance and temperature values, in its own Maximum Power Point (MPP). By changing the V_{occ} value, the P_o curve is moved in a wide voltage range, so that the maximum PV power for any environmental condition is achieved. By referring to Figure 3.5, the dashed line shows the signal connection that must be implemented in order to have the additional feature of adapting some of the OCC parameters in real-time. For instance, this allows to apply a Perturb and Observe (P&O) strategy for performing the MPPT function of the PV source. In this way, some of the OCC parameters can be trimmed so that the P_o curve, thus the dashed one in Figure 3.7, is moved in a way that it intersects the actual P_{pv} curve, which is one of the continuous ones in Figure 3.7 depending on the operating conditions, just in its MPP. In References 18 and 19 a multi-objective optimisation procedure has been carried out in order to select the optimal sets of the OCC parameters for different system operating conditions and a P&O MPPT strategy has been combined with the OCC in order to modify in real-time the V_{occ} parameter. A further improvement has been proposed in Reference 20 where a multivariable MPPT is used for adapting more than one OCC parameter at the same time in order to extend the operating range of the single-stage inverter. It is worth noting that the approaches mentioned earlier, which are able to change the OCC parameters, employ digital controllers. In the specific case, this is mandatory for the implementation of the MPPT function, even if some examples of analogue MPPT methods are presented in literature [16]. This would allow to have a fully analogue control, dedicated to both the PV control side and to the current injection into the AC grid.

3.3.2 OCC for three-phase PV inverters

The universal OCC configuration described in Section 3.2 can be extended to control a wide class of three-phase power converters applications: grid-connected inverters (GCI), active power filters (APF), PFC rectifiers and static VAR compensators (STATCOM), as shown in Reference 21. The mode of operation of the universal OCC configuration for a three-phase application is well described in Reference 22. Such architecture exploits the fact that, by dividing the line period into six equal regions identified by the zero-crossing points of the phase voltages, the three-phase system can be decomposed by a sequence of two boost DC/DC converter connected in parallel at the output ports corresponding to the DC-bulk voltage of the three-phase system. A single OCC core is used for controlling them but an additional multiplexing circuit must be adopted to select the switches involved in the switching process for each region of the line period. Unfortunately such architecture exhibits instability at light load or when a bidirectional mode is required [23], [24]: in such cases the modified OCC controller described in Reference 25 can be adopted. The modified control scheme not only overcomes the instability problems and allows to operate in bidirectional mode, but it also simplifies the whole OCC architecture because it allows to use BPWM inverters. In this way the rectification of the sensed signals is no more required. Such an architecture can be used for both single-phase and three-phase inverters; in the following the three-phase application is proposed.

Figure 3.9 shows a PV systems with a storage element, which includes the storage unit and the related power electronics. The three-phase converter operates as an inverter, if the PV power is higher than the power required by the battery charger, or as a rectifier if the PV power flow is too low for recharging the storage unit. By assuming that the low side switches (M_2, M_4 and M_6 in Figure 3.9) are

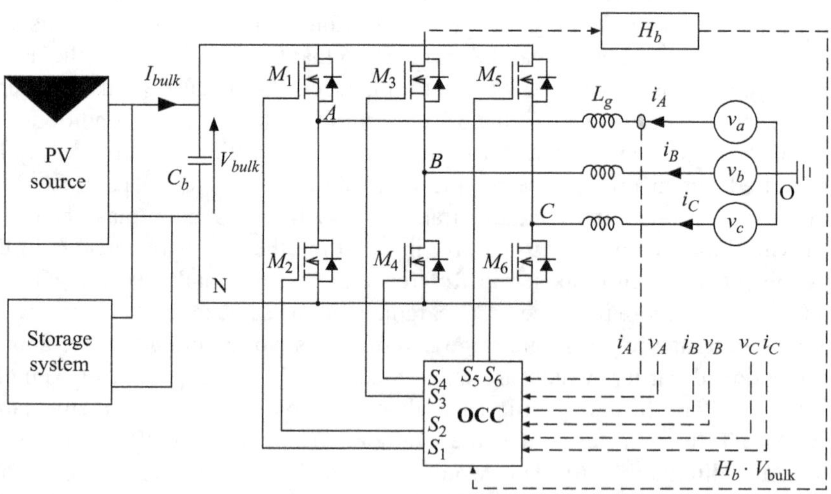

Figure 3.9 Three-phase OCC inverter operating in bidirectional mode

driven by the three duty cycle (d_A, d_B, d_C), then the average voltages at nodes A, B and C, referred to the node N, are

$$V_{nN} = (1 - d_n) \cdot V_{bulk} \quad (n = A, B, C) \tag{3.19}$$

The schematic of the modified OCC block is shown in Figure 3.10.

The three-phase modulation is obtained by comparing the output signal of the resettable integrator with the three scaled signals coming from each phase of the grid. Figure 3.10 puts into evidence that the single-phase system is obtained by using only one comparator. With respect to the basic OCC configuration, a fictitious current has been introduced in each phase by multiplying the corresponding sensed source voltages by a gain R_s/R_F. The corresponding OCC equation for each phase is

$$R_s \cdot i_n(t) + v_n(t) \cdot \frac{R_s}{R_F} = V_M(t) - \frac{1}{T_i}\int_0^{d_n Ts} V_M(t)dt \quad (n = A, B, C) \tag{3.20}$$

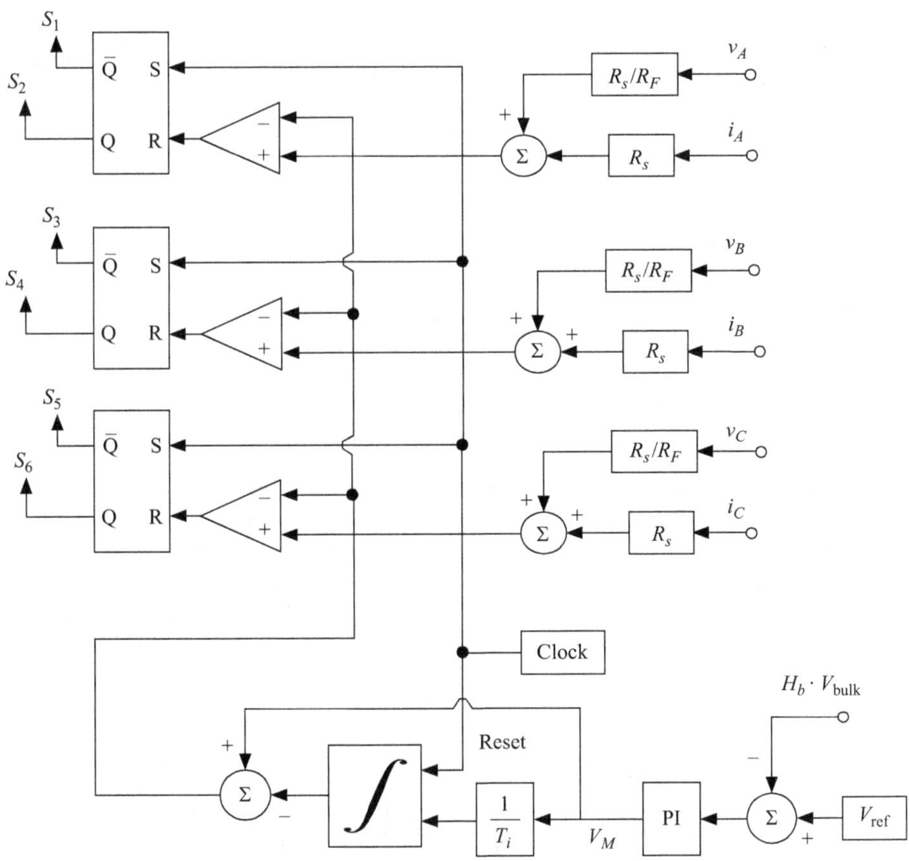

Figure 3.10 Three phases of the modified OCC block

where R_s is the current sensor gain. Instead, R_F is a fictitious resistance whose value is used to enlarge the system stability region by allowing to change the value and the sign of the power injected in to the grid. $V_M(t)$ is the control signal provided by a voltage feedback loop used to regulate the bulk voltage V_{bulk}

$$V_M(s) = \left(k_p + \frac{k_I}{s}\right) \cdot (V_{ref}(s) - H_b \cdot V_{bulk}(s)) \tag{3.21}$$

By assuming that the proportional-integral (PI) controller has a bandwidth lower than the switching frequency, the OCC equation can be approximated by means of the following low-frequency expression

$$R_s \cdot i_n(t) + v_n(t) \cdot \frac{R_s}{RF} \simeq V_M(t) \cdot \left(1 - \frac{T_s}{T_i} \cdot d_n\right) \quad (n = A, B, C) \tag{3.22}$$

The integrator time constant T_i must be settled on the basis of the following conditions [24]

$$T_i = K_1 \cdot T_s \quad \text{where } K_1 \in \left[\left|\frac{v_{o,peak}}{V_{bulk}}\right|, 1 - \left|\frac{v_{o,peak}}{V_{bulk}}\right|\right] \tag{3.23}$$

If the voltage-current displacement at the grid side has a so small amplitude to be neglected, then the total power injected into the grid is [25]

$$P_o \simeq 3\frac{v_{o,rms}^2}{R_s} \cdot \left(\frac{V_M}{K_1 \cdot V_{bulk}} - \frac{R_s}{R_F}\right) \tag{3.24}$$

The Poincaré map is useful for determining a condition for the stable operation of the OCC converter [23]; it can be expressed as

$$V_M > -\frac{T_i \cdot R_s}{L} \cdot v_{o,peak} \tag{3.25}$$

Figure 3.11 shows the simulation results concerning the OCC three-phase converter operating with a bidirectional power flux. Table 3.2 shows the OCC parameters used in the simulation. The OCC controlled converter has at its input the grid voltage at $220V_{rms}$ and at its output the bulk voltage $V_{bulk} = 800$ V. The current and voltage references are those ones shown in Figure 3.9. Initially the PV source is disconnected and the battery is charged by using the energy taken from the grid through the three-phase boost rectifier.

At the time instant 0.5 s the PV source is connected to the system and it starts to inject energy into it. A part of such an energy flux is used to charge the battery and the remaining part is injected into the grid. In this condition the OCC controller automatically adapts its parameters for driving the three-phase converter as a voltage source inverter. In this mode, both the bulk current and the phase current invert their sign, while the bulk voltage remains regulated at the desired reference value. At the time instant 0.7 s the solar irradiation reduces abruptly and the power

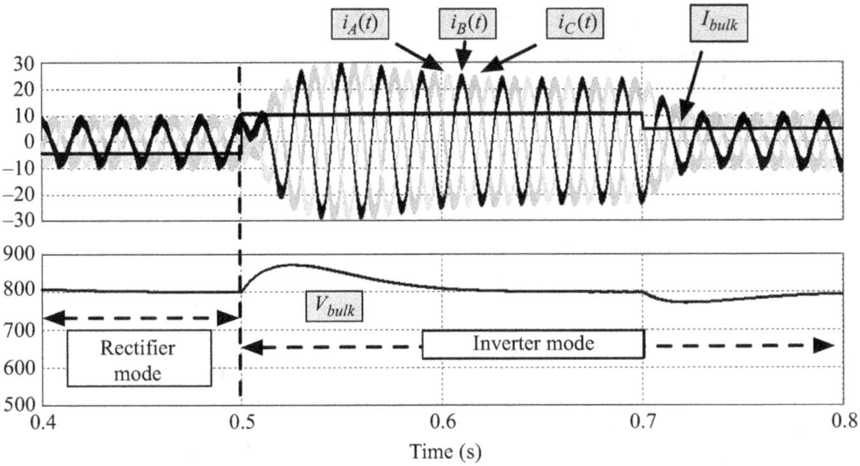

Figure 3.11 Transition from rectifier mode to inverter mode. For the current and voltage references refer to Figure 3.9

Table 3.2 OCC parameters for the three-phase application

K_1	R_s	R_F	H_b	T_s	K_p	K_I
1	0.01 Ω	1 Ω	0.01	100 μs	0.01	10

delivered by the PV source is reduced as well. In the meanwhile, the OCC controller continues to regulate the bulk voltage and the currents injected into the grid. This confirms the self-adaptivity and flexibility of the OCC controller also in this case.

The OCC has been also applied in wind turbine systems, in particular for controlling the back-to-back converters used in both doubly fed induction generators (DFIGs) and permanent magnet synchronous generators (PMSGs). The main benefits and drawbacks of the two solutions are described in References 26 and 27; in both cases the OCC achieved precise and fast speed control with respect to traditional control approaches usually based on the $d - q$ transformation. Recently some digital version of the OCC has been also developed, but these solution are out of the scope of this chapter. The reader can find further details in References 28 and 29.

3.4 OCC stability analysis by means of *Poincaré* maps

An aspect of the OCC deserving attention concerns the stability of the converter that is controlled by using this technique. In a one-cycle controlled converter the system stability depends on how the resettable integrator processes the variables involved in the control circuit.

The first possibility is that the output of the resettable integrator makes the value of the controlled variable, e.g. the duty cycle, in the actual switching period independent of the value of the same variable in previous period. In this case the OCC controller does not introduce any feedback action and the OCC behaviour is equivalent to a feedforward controller. As a consequence, the system stability is analysed by using the large and the small-signal model of the converter and by including the linearised OCC equation, which is valid for the frequency range below the switching frequency. This is the case of the step-down DC/DC converter shown in Section 3.2, where the duty-cycle value in the k-th period depends on the value assumed by the v_d voltage in the T_{ON} interval of the same switching period.

The second possible case concerns those applications in which the actual value of the controlled variable depends on the value it assumed in the previous switching period. In those cases the OCC introduces a feedback at the switching frequency that cannot be analysed by using the well-known frequency-domain analysis. Indeed, at the switching frequency the system cannot be described with its linearised model, the latter being valid only below one tenth of the switching frequency. In such cases the conditions for large signal stability analysis can be derived by using the *Poincaré* method [30].

The *Poincaré* map is a powerful tool for analysing the dynamic behaviour of non-linear systems characterised by a switching operating mode. It studies the transient of the controlled signal by taking into account the system behaviour inside the switching period. The system stability is verified by evaluating the time evolution of the controlled variable in presence of an external perturbation. The example shown in Figure 3.12 is helpful to explain this approach.

The grey waveform in Figure 3.12 shows the behaviour of the controlled variable in the two subintervals of the switching period by considering its steady-state condition. The system convergence is assured if the $\hat{x}_0$ perturbation, which has been superimposed to the controlled signal X_0 in $t = t_0$, vanishes after some cycles,

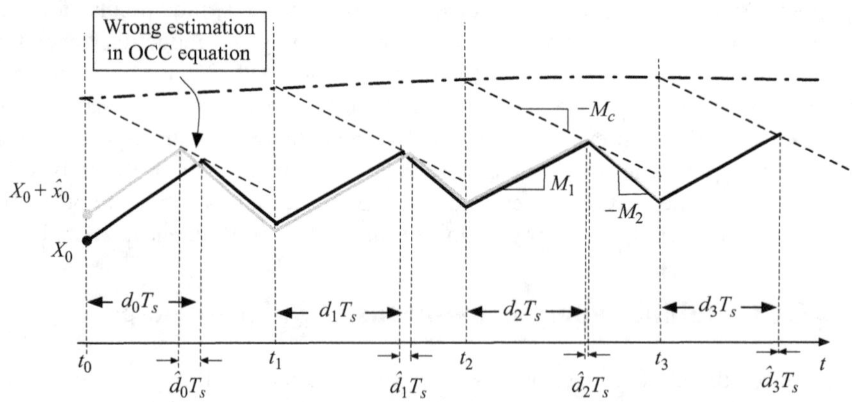

Figure 3.12 Example of a Poincaré map derived by the OCC operating conditions in presence of disturbances

as it is the case shown in Figure 3.12. The black waveform shown in Figure 3.12 is the unperturbed signal.

This method is recognised in literature as *first order discrete time analysis* because the linearity of the controlled signal within the switching period has been assumed. This hypothesis holds when the switching frequency is higher than the system bandwidth. In this case, the amplitude of the disturbance signal at the end of the switching cycle is evaluated by having the values at the starting of the switching cycle and by using geometrical considerations about the signal slopes within the switching period. The faster the attenuation of the disturbance amplitude the better the dynamic performances of the system. In particular, the transient up to the k-th cycle is estimated as follows

$$\begin{aligned} \hat{x}_{Ts} = \hat{x}_0 \cdot \mu \quad \rightarrow \quad \hat{x}_{kTs} &= \hat{x}_{(k-1)Ts} \cdot \mu \\ \hat{x}_{kTs} &= \hat{x}_0 \cdot \mu^k \end{aligned} \tag{3.26}$$

where the hats refer to small variations on the average values of the variables. The value of the μ coefficient depends on the system parameters and it can be evaluated by calculating the slopes (M_1, M_2, M_c) of the OCC signals. The convergence is assured if $\mu < 1$, thus the stability condition is used as a design equation for a suitable parametric design.

The stability analysis of the OCC single-phase PV inverter becomes clear by referring to Figure 3.13, showing the signal waveforms appearing at the OCC comparator. The triangular signal is the scaled value of the grid current thus: $x = R_s \cdot i_{grid}(t)$. The dashed signal is the combination of the scaled value of the grid voltage, thus $k \cdot v_{grid}(t)$, and of the output of the resettable integrator depending on v_m.

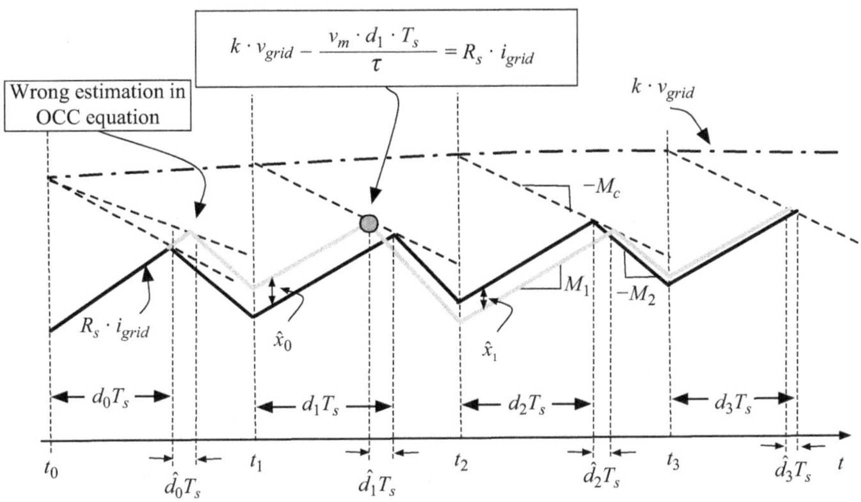

Figure 3.13 Poincaré map for a single-phase OCC inverter

A disturbance affecting the PV voltage vanishes in few switching cycles provided that the OCC parameters are suitably designed. In Figure 3.13 a $\hat{d}_0$ error appears in the duty-cycle and it propagates through the system by modifying the grid current waveform. Each grey circle represents the instant where (3.11) is fulfilled for the switching period that is just after the one where the perturbation occurs.

The application of some simple geometric principles allows to obtain the following conclusions

$$\begin{cases} \hat{x}_0 = R_s \cdot \hat{i}_0 = -\hat{d}_1 \cdot T_s \cdot (M_1 + M_c) \\ \hat{x}_1 = R_s \cdot \hat{i}_1 = -\hat{d}_1 \cdot T_s \cdot (M_c - M_2) \end{cases} \tag{3.27}$$

where the terms $\hat{i}_0$, $\hat{i}_1$ are the variations affecting the grid current in two consecutive switching periods, thus $\hat{i}_1$ can be expressed as a function of $\hat{i}_0$ as follows

$$\hat{i}_1 = \hat{i}_0 \frac{M_c - M_2}{M_1 + M_c} \tag{3.28}$$

By iterating the process over n cycles it results that

$$\hat{i}_n = \hat{i}_0 \left(\frac{M_c - M_2}{M_1 + M_c} \right)^n = \hat{i}_0 \cdot \mu^n \tag{3.29}$$

The system is stable if (3.29) converges to zero. This means that

$$|\mu| < 1 \rightarrow \text{OCC is STABLE} \tag{3.30}$$

In the case under study the slopes of the signals at the OCC input assume the following expressions

$$\begin{cases} M_1 = R_s \cdot \dfrac{V_{pv} - V_{grid}}{L} \\ M_2 = R_s \cdot \dfrac{v_o}{L} \\ M_c = \dfrac{V_m}{\tau} = \dfrac{V_{occ} - k_g \cdot V_{pv}}{\tau} \end{cases} \tag{3.31}$$

By keeping into account the condition $\tau < T_s$ and by taking the worst case values for the parameter appearing in (3.29), the stability conditions already introduced in the previous section is obtained

$$V_{occ} > k \cdot V_{pv} + \frac{(2 \cdot V_{grid,max} - V_{pv}) \cdot R_s \cdot \tau}{2L} \tag{3.32}$$

It is worth to note that the PV voltage is strongly dependent on the environmental conditions thus condition (3.32) should be assured for any operating condition. From Figure 3.7 the operating PV voltage is included in the range: $V_{pv} \in [160, 310]$ V. By considering the OCC parameters shown in Table 3.1 and the maximum value of the US grid voltage, the stability criterion (3.32) is always satisfied.

3.5 Conclusions

In this chapter some aspects concerning the OCC technique are discussed. The basic principles are reviewed and the main field of application are mentioned, also by referring to commercial products that use this control approach and that are available on the market. The second half of the chapter is focused on the application of the OCC to the DC/AC switching converters, with a special emphasis on PV systems. Issues concerning the parametric design of the control technique aimed at ensuring optimal performances both in terms of power quality and maximisation of the power extracted from the renewable energy source are discussed. Finally, those aspects related to the management of bidirectional fluxes of energy and stability issues are analysed for single-phase as well as three-phase inverters.

References

[1] E. Mamarelis, G. Petrone and G. Spagnuolo, "An hybrid digital-analog sliding mode controller for photovoltaic applications," *Industrial Informatics, IEEE Transactions on*, vol. 9, no. 2, pp. 1094–1103, May 2013.

[2] K. Smedley and S. Cuk, "One-cycle control of switching converters," *Power Electronics Specialists Conference, 1991. PESC '91 Record., 22nd Annual IEEE,* pp. 888–896, Jun. 1991.

[3] K. Smedley, "One-cycle controlled switching circuit," Jan 11, 1994, US Patent 5,278,490. [Online]. Available: http://www.google.com/patents/US5278490

[4] [Online]. Available: http://www.onecyclecontrol.com

[5] R. Erikson and D. Maksimovic, *Fundamentals of Power Electronics*, Second Edition, Kluwer academic Publishers, ISBN 0-7923-7270-0, 2000.

[6] [Online]. Available: http://www.onecyclecontrol.com/products.html

[7] [Online]. Available: http://www.irf.com/part/AIR1155SPBF

[8] Z. Lai, K. Smedley and Y. Ma, "Time quantity one-cycle control for power-factor correctors," *Power Electronics, IEEE Transactions on*, vol. 12, no. 2, pp. 369–375, Mar. 1997.

[9] D. Zmood and D. Holmes, "Stationary frame current regulation of PWM inverters with zero steady-state error," *Power Electronics, IEEE Transactions on*, vol. 18, no. 3, pp. 814–822, May 2003.

[10] V. Blasko and V. Kaura, "A novel control to actively damp resonance in input lc filter of a three phase voltage source converter," *Applied Power Electronics Conference and Exposition, 1996. APEC '96. Conference Proceedings 1996., Eleventh Annual*, vol. 2, pp. 545–551, Mar. 1996.

[11] B. C. Babu, K. Sridharan, E. Rosolowski and Z. Leonowicz, "Analysis of {SDFT} based phase detection system for grid synchronization of distributed generation systems," *Engineering Science and Technology, an International Journal,* vol. 17, no. 4, pp. 270–278, 2014. [Online]. Available: http://www.sciencedirect.com/science/article/pii/S2215098614000573

[12] M. Liserre, R. Teodorescu and F. Blaabjerg, "Stability of photovoltaic and wind turbine grid-connected inverters for a large set of grid impedance values," *Power Electronics, IEEE Transactions on*, vol. 21, no. 1, pp. 263–272, Jan. 2006.

[13] J. Bastidas-Rodriguez, G. Petrone, C. Ramos-Paja and G. Spagnuolo, "Photovoltaic modules diagnostic: an overview," *Industrial Electronics Society, IECON 2013 – 39th Annual Conference of the IEEE,* pp. 96–101, Nov. 2013.

[14] A. Datta, G. Bhattacharya, D. Mukherjee and H. Saha, "Ground leakage current elimination in a transformerless unipolar modulation based single phase grid-connected photovoltaic system," in *Power and Energy Engineering Conference (APPEEC), 2013 IEEE PES Asia-Pacific,* pp. 1–5, Dec. 2013.

[15] Y. Chen and K. Smedley, "A cost-effective single-stage inverter with maximum power point tracking," *Power Electronics, IEEE Transactions on*, vol. 19, no. 5, pp. 1289–1294, Sept. 2004.

[16] N. Femia, G. Petrone, G. Spagnuolo and M. Vitelli, *Power Electronics and Control Techniques for Maximum Energy Harvesting in Photovoltaic Systems*, First Edition, Taylor & Francis – CRC Press, p. 368. ISBN:978-1466506909, http://www.amazon.ca/Electronics-Control-Techniques-Harvesting-Photovoltaic/dp/1466506903, Dec. 2012.

[17] M. Orozco-Gutierrez, J. Ramirez-Scarpetta, G. Spagnuolo and C. Ramos-Paja, "A method for simulating large {PV} arrays that include reverse biased cells," *Applied Energy*, vol. 123, no. 0, pp. 157–167, 2014. [Online]. Available: http://www.sciencedirect.com/science/article/pii/S0306261914001962

[18] N. Femia, D. Granozio, G. Petrone, G. Spagnuolo and M. Vitelli, "Optimized one-cycle control in photovoltaic grid connected applications," *Aerospace and Electronic Systems, IEEE Transactions on*, vol. 42, no. 3, pp. 954–972, Jul. 2006.

[19] M. Fortunato, A. Giustiniani, G. Petrone, G. Spagnuolo and M. Vitelli, "Maximum power point tracking in a one-cycle-controlled single-stage photovoltaic inverter," *Industrial Electronics, IEEE Transactions on*, vol. 55, no. 7, pp. 2684–2693, Jul. 2008.

[20] G. Petrone, G. Spagnuolo and M. Vitelli, "A multivariable perturb-and-observe maximum power point tracking technique applied to a single-stage photovoltaic inverter," *Industrial Electronics, IEEE Transactions on*, vol. 58, no. 1, pp. 76–84, Jan. 2011.

[21] K. Smedley, "One-cycle control and its applications in power quality control and renewable power generation," *Power Electronics Systems and Applications*, 2004. Proceedings. 2004 First International Conference on, pp. 167–181, Nov. 2004.

[22] T. Jin, L. Li and K. Smedley, "A universal vector controller for four-quadrant three-phase power converters," *Circuits and Systems I: Regular Papers, IEEE Transactions on*, vol. 54, no. 2, pp. 377–390, Feb. 2007.

[23] C. Qiao and K. Smedley, "Unified constant-frequency integration control of three-phase standard bridge boost rectifiers with power-factor correction," *Industrial Electronics, IEEE Transactions on*, vol. 50, no. 1, pp. 100–107, Feb. 2003.

[24] K. Smedley, L. Zhou and C. Qiao, "Unified constant-frequency integration control of active power filters-steady-state and dynamics," *Power Electronics, IEEE Transactions on*, vol. 16, no. 3, pp. 428–436, May 2001.

[25] D. Ghodke, K. Chatterjee and B. Fernandes, "Modified one-cycle controlled bidirectional high-power-factor ac-to-dc converter," *Industrial Electronics, IEEE Transactions on*, vol. 55, no. 6, pp. 2459–2472, Jun. 2008.

[26] A. Yazdekhasti, A. Ketabi and I. Sadeghkhani, "One-cycle control application to wind turbine power control," *International Transactions on Electrical Energy Systems*, vol. 25, no. 10, pp. 2427–2442, Oct. 2015.

[27] M. Li and K. Smedley, "One-cycle control of PMSG for wind power generation," *Power Electronics and Machines in Wind Applications, 2009. PEMWA 2009. IEEE*, pp. 1–6, Jun. 2009.

[28] G. Petrone, M. Vitelli and G. Spagnuolo, "Digital implementation of one cycle control in back to back converters," *Power Electronics for Distributed Generation Systems (PEDG), 2012 3rd IEEE International Symposium on*, pp. 344–348, Jun. 2012.

[29] M. Barbati, C. Caluisi and C. Cecati, "One-cycle controlled active rectifier for full digital implementation," *IECON 2010 – 36th Annual Conference on IEEE Industrial Electronics Society*, pp. 625–630, Nov. 2010.

[30] S. Banerjee and G. C. Verghese, *Nonlinear Phenomena in Power Electronics: Bifurcations, Chaos, Control, and Applications, Wiley-IEEE Press*, pp. 1–472, ISBN: 978-0-7803-5383-1, Jul. 2001.

[illegible] Schiano
[30] [illegible]

Part II

Digital control circuits

Chapter 4

Digital PWM control of high-frequency DC-DC switched-mode power converters

Luca Corradini[1]

High-frequency DC-DC converters are largely employed in electronic equipment where compactness, efficiency, and dynamic performance are the important requirements to be fulfilled by the onboard power electronics. Target commercial sectors for these converters include the consumer market of mobile/handheld devices such as cell phones or tablets, computer applications with desktop and laptop power supplies and their internal DC-DC regulators, telecom/datacom and server applications with increasingly complex and distributed power supply systems, and automotive, where many DC-DC converters are used in the vehicle for lighting control and onboard electronics supply.

Digital control of such high-frequency switched-mode power converters has been slowly but steadily establishing as a valid alternative to the more traditional analog approach [1–3]. In the last decade, academic research has shown several areas in which a clever use of the digital resource leads not only to outperforming analog solutions but also to introducing entirely new features as well:

- *Digital power management* functions are facilitated by the definition of protocols based on a simple serial communication bus.
- *Superior dynamic performances* can be obtained by implementing nonlinear control functions for fast transient response.
- *Autotuning* functions, embedded into the digital controller, allow the control IC to self-tune the compensation parameters and to adapt to parametric variations of the power converter characteristics over time.
- *Efficiency improvement* is possible by automatic adjustment of the switching frequency or of the dead-time intervals, smart current sharing/redistribution in multiphase topologies, and true online efficiency optimization algorithms implemented digitally.

To be competitive in high switching frequency applications, digital controllers demand optimized implementations where processing delays are minimized and essentially limited by the technological constraints rather than by the platform

[1]Department of Information Engineering, University of Padova, Padova, Italy

architecture. Hardwired custom-designed control implementations are therefore largely preferred to software-based controllers relying on less flexible micro-controller architectures. Early examples of such optimized fully-integrated digital controllers are found in References 4 and 5. At the time of writing, digital control ICs for high-frequency switching converters are part of many vendors portfolios and hold a definite ever increasing market share.

This chapter covers a basic overview of both system-level design and implementation issues of digital controllers for high-frequency switched-mode power converters. The discussion revolves around a specific case study of great practical relevance, namely a voltage-mode-controlled point-of-load (POL) regulator based on an interleaved multiphase Buck topology. Design and simulation results are presented alongside the exposition. Section 4.1 introduces the main blocks of a digitally controlled multiphase converter. In Section 4.2 the problem of small-signal modeling of a digitally controlled converter is first briefly discussed. Next, system-level synthesis of the digital PID transfer function is treated. Section 4.3 provides an overview on the most important quantization effects, along with basic design guidelines, known as no-limit-cycling conditions, to prevent them from degrading the system's closed-loop performance. Section 4.4 is devoted to discussing the main trade-offs and design alternatives in the implementation of the main controller building blocks.

4.1 The digital control loop

In the POL application considered here as a case study, the converter operates off an input voltage source $V_g = 12\,\text{V}$ and produces a 1.5 V, 10 A regulated output. The POL converter is the four-phase, 500 kHz synchronous Buck converter topology illustrated in Figure 4.1 whose parameters are listed in Table 4.1. The multiphase approach, widely adopted in POL applications, combines the advantages of improved thermal management due to power sharing among the phases, high output current handling capability, fast dynamic performances and small, low-footprint per-phase inductors [4, 6, 7]. Furthermore, operating the converter phases in an interleaved fashion, as sketched in Figure 4.2, leads to reduced output capacitance requirements. Interleaving consists of uniformly phase shifting the gate driving signals $c_1 \dots c_4$ within the switching period T_s, in order to correspondingly shift the phase currents $i_1 \dots i_4$. This leads to a partial harmonic cancellation in the spectrum of the total current $i_\Sigma(t)$, and to a peak-to-peak ripple on $i_\Sigma(t)$ smaller than the sum of the individual phase current ripples. The harmonic cancellation mechanism leads to more relaxed filtering requirements on C with respect to the noninterleaved case. Quantitatively, ripple attenuation due to interleaving is a function of the number of phases and of the steady-state duty cycle D. In an N-phase converter and in the absence of any mismatch among the phase inductances, the equivalent switching ripple on $i_\Sigma(t)$—and, consequently, on the output voltage $v_o(t)$—has a fundamental frequency equal to Nf_s. In the considered four-phase example the effective switching rate of $v_o(t)$ is $4f_s = 2$ MHz.

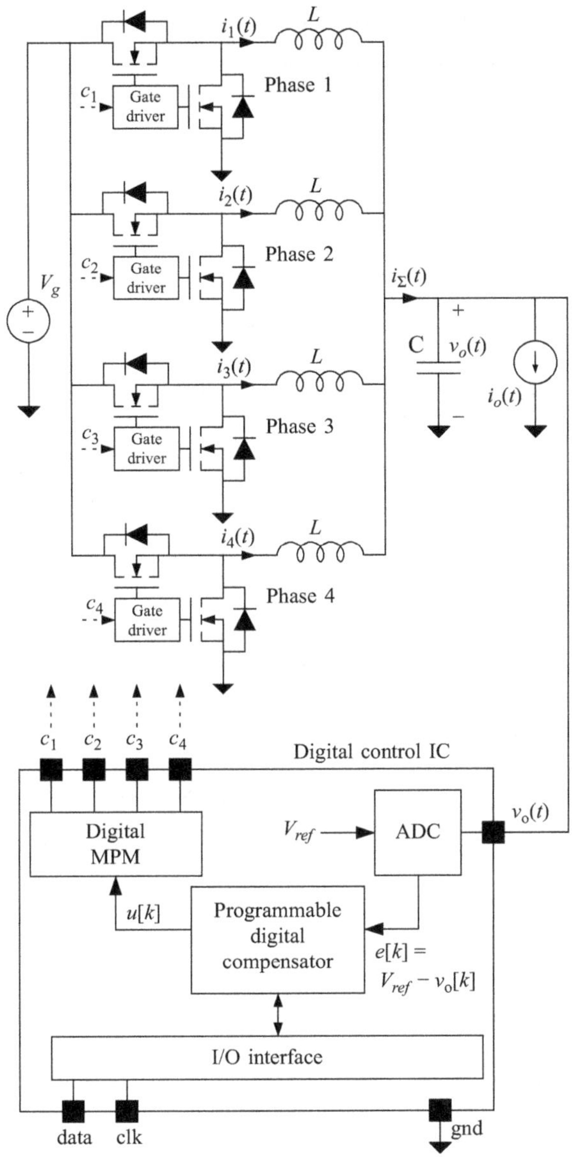

Figure 4.1 Digitally controlled four-phase POL regulator

Also illustrated in Figure 4.1 is a functional block diagram of the digital controller. Main building blocks, on which the rest of the chapter will focus, are represented by:

- a fast analog-to-digital converter (ADC), which converts the sensed voltage $v_o(t)$ into a digital sequence $v_o[k]$ or, as in Figure 4.1, directly into the digitized error $e[k]$ between the control setpoint V_{ref} and $v_o[k]$;

Table 4.1 Converter parameters

Input voltage V_g	12 V
Output voltage V_o	1.5 V
Maximum output current I_o	10 A
Switching frequency f_s	500 kHz
Per-phase inductance L	470 nH
Per-phase parasitic resistance r_L	10 mΩ
Output capacitance C	5 mF
Capacitor equivalent series resistance r_C	2 mΩ
MOSFET on-resistance $r_{ds,on}$	5 mΩ

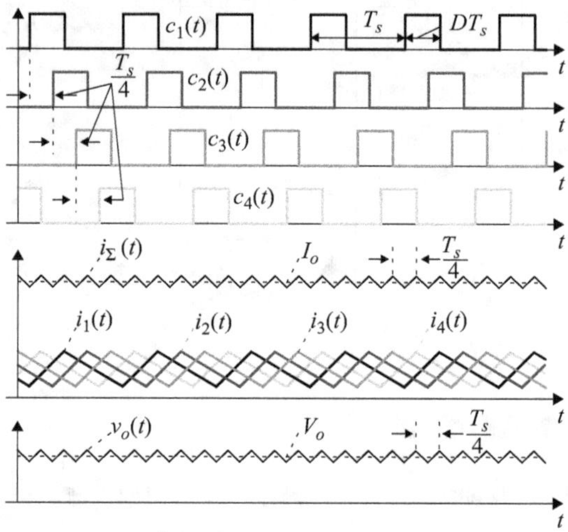

Figure 4.2 Main POL steady-state waveforms corresponding to interleaved operation

- a hardwired digital compensator, which produces a digital control command $u[k]$ based on $e[k]$ on a sampling cycle basis;
- a digital multiphase modulator (MPM), which generates the interleaved driving signals $c_1 \ldots c_4$ for the converter phases.

In addition to the above main control building blocks, the digital IC may have an I/O interface logic. This feature enables the implementation of various power management functions, including, as suggested in Figure 4.1, the possibility to program or transmit the compensation parameters via an external serial communication bus.

4.1.1 Timing diagram and controller operation

Examine now the operation of the multiphase controller with the aid of the timing diagram illustrated in Figure 4.3. The diagram illustrates the main waveforms of the

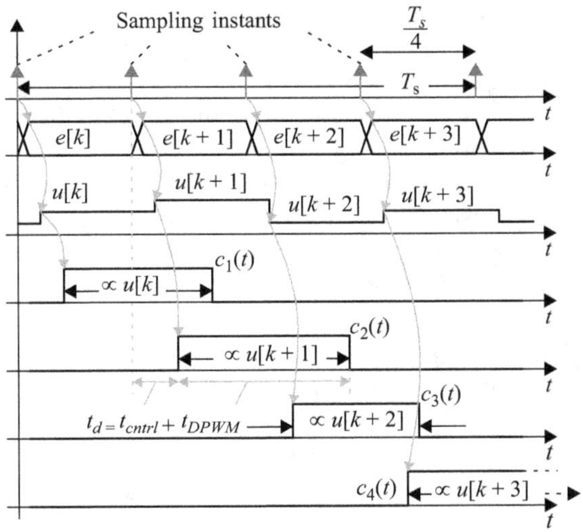

Figure 4.3 Timing diagram of the multiphase controller operation

controller over one switching interval T_s. The internal A/D converter samples the output voltage at a sampling rate equal to $f_{sampling} = 4f_s$, producing a digital sequence $e[k]$ having a sampling period $T_{sampling} = T_s/4 = 500$ ns. Neglecting, for the time being, amplitude quantization, one has

$$
\begin{aligned}
e[k] &\triangleq V_{ref} - v_o[k], \\
v_o[k] &\triangleq v_o\left(t_k = k\frac{T_s}{4}\right).
\end{aligned}
\tag{4.1}
$$

The regulation error $e[k]$ is processed by the digital compensator on a sampling cycle basis. The compensator therefore produces a digital control sequence $u[k]$ updated four times within each switching interval. Samples of $u[k]$ are then acquired by the digital MPM and used to drive the converter phases. How samples of $u[k]$ are employed to generate the phase-shifted PWM signals $c_1 \ldots c_4$ depends on the specific MPM implementation. In Figure 4.3, $u[k]$ is first time-demultiplexed and then delivered to four distinct trailing-edge digital pulse width modulators (DPWMs). With this choice, if phase 1 is modulated with a duty cycle proportional to $u[k]$, then phase 2 duty cycle is proportional to $u[k+1]$, phase 3 is to $u[k+2]$ and phase 4 is to $u[k+3]$.

4.1.2 Loop delays

Figure 4.3 also highlights the *total loop delay* t_d between each sampling event and the falling edge of the corresponding pulse width modulated signal. Delay t_d plays an important role determining the loop dynamics, as will be clarified in the next section, and it is necessary to correctly identify and calculate t_d before designing the compensation.

As shown in the figure, the total loop delay can be decomposed into a *control delay* t_{cntrl} followed by a *modulation delay* t_{DPWM}. The former accounts for both A/D conversion delay and for the computational delay of the digital compensator. Overall, t_{cntrl} represents the time delay between the sampling instant and the instant the MPM latches the corresponding sample of $u[k]$. Value of t_{cntrl} ultimately depends on how fast A/D conversion and control calculations can be carried out by the digital hardware in use. For commercial low-cost digital technological processes and for switching rates in the order of hundreds of kilohertz, t_{cntrl} can be easily reduced to a small fraction of the switching interval. In the case study example under examination, the value $t_{cntrl} = 50$ ns shall be assumed.

Modulation delay t_{DPWM}, on the other hand, can be thought of as the time interval separating the latching of the modulating signal $u[k]$ by the modulator, and the time instant in which the duty cycle is actually applied to the corresponding phase. For a trailing-edge modulation such as the one depicted in Figure 4.3, in which the modulated edge of the PWM signal is the falling one, it can be shown that [3, 8]

$$t_{DPWM} = DT_s = \frac{V_o}{V_g} \times T_s = 0.125 \times 2\ \mu\text{s} = 250\ \text{ns}. \tag{4.2}$$

In the POL example, the total loop delay is then $t_d = 50\ \text{ns} + 250\ \text{ns} = 300\ \text{ns}$.

It is worth noting explicitly that, in an analog control loop, one has $t_d = 0$ because no computational or PWM delays are present. In this case, dynamic phase response of the uncompensated plant is solely due to the power converter itself and, to a lesser extent, to the analog sensing/conditioning circuitry. The additional delay t_d therefore represents an issue specific to digital control design.

4.2 Dynamic modeling and system-level compensator design

Due to the switching ripple of $v_o(t)$ having a fundamental frequency equal to $4f_s$, sampling strategy (4.1) does not lead to aliasing-induced artifacts other than a corruption of the DC value. Even such residual aliasing effect, on the other hand, remains negligible because of the very small peak-to-peak ripple on $v_o(t)$, ensured by the output filter capacitance C. As a result, the sampled output voltage $v_o[k]$ follows well, both statically and dynamically, the *average* voltage, denoted here as $\overline{v}_o(t)$, over one switching period,

$$v_o[k] \approx \overline{v}_o(t_k). \tag{4.3}$$

Such *small-aliasing approximation* [3] leads to a strong simplification of the modeling step: conflation between averaged and sampled dynamics allows, in effect, to resort to standard continuous-time small-signal averaged models [9] rather than the exact, but less mathematically manageable discrete-time models [10]. Following this simplified approach, the POL converter is first modeled in the continuous-time domain. Next, a continuous-time compensator transfer function is designed, and finally, discretized into the digital domain. As compensation specifications, a target crossover frequency $f_c = 50\,\text{kHz}$ and a phase margin $\varphi_m = 60°$ shall be assumed.

4.2.1 Loop small-signal modeling

Dynamically, the four phases of the POL converter operate in parallel, as all exhibit the same dynamic response to their respective duty cycle command. As a consequence, the small-signal duty cycle-to-output voltage transfer function $G_{vd}(s)$ of the four-phase POL converter is simply the control-to-output transfer function of a single-phase Buck converter having an inductance equal to $L_{eq} \triangleq L/4$, and equivalent inductor resistance equal to $r_{eq} \triangleq (r_{ds,on} + r_L)/4$,

$$G_{vd}(s) \triangleq \frac{\hat{v}_o(s)}{\hat{d}(s)} = \frac{(1 + sr_C C)V_g}{1 + s(r_{eq} + r_C)C + s^2 L_{eq} C}, \tag{4.4}$$

where $\hat{x}(s)$ indicates the Laplace-transform of the small-signal component $\hat{x}(t)$ of signal $x(t)$.

To complete the small-signal modeling step, one needs to include the effect of the total loop delay t_d calculated in Section 4.1.2. This can be easily accomplished by attributing the total delay t_d to the converter dynamics,

$$G^{\dagger}_{vd}(s) \triangleq G_{vd}(s)e^{-st_d}, \tag{4.5}$$

where $G^{\dagger}_{vd}(s)$ describes the *effective dynamics* of the converter [3], which extends (4.4) by including control and modulation delays. With a unity feedback sensing gain and assuming the modulator static gain is normalized to unity, (4.5) also represents the uncompensated loop gain to be considered in the digital controller design.

4.2.2 Compensator design and discretization

Figure 4.4 illustrates the Bode diagrams of the relevant transfer functions of the compensated loop. As seen from the plots of $G_{vd}(s)$ and $G^{\dagger}_{vd}(s)$, the total loop delay introduces an additional phase rotation with respect to the plain small-signal averaged model (4.4). Nonetheless, given the crossover frequency and phase margin required and considered that the phase response of $G^{\dagger}_{vd}(s)$ at $f_c = 50$ kHz is only slightly less than –90°, the system can be compensated by a simple proportional–integral (PI) compensation of the type

$$G_{PI}(s) = K_p + \frac{K_i}{s}. \tag{4.6}$$

By imposing the desired crossover frequency and phase margin and solving for K_p and K_i, one has $K_p \approx 1.402$ and $K_i \approx 125.9 \times 10^3$. Bode diagrams of $G_{PI}(s)$ and of the final compensated loop gain $T(s) = G_{PI}(s)G^{\dagger}_{vd}(s)$ are also reported in Figure 4.4. In addition, the simulated loop gain magnitude and phase is reported at specific frequencies for comparison and as a support of the accuracy of the foregoing modeling methodology. The simulated loop gain has been determined with

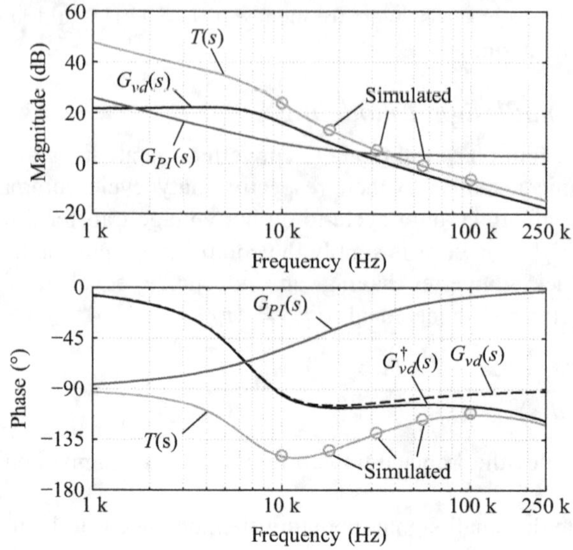

Figure 4.4 Magnitude and phase Bode diagrams of the control loop transfer functions

Matlab® employing the Middlebrook's injection approach [11], combined with a fast Fourier transform–based numerical analysis.

The last step of the system-level compensator design is the discretization of the Laplace-domain controller transfer function $G_{PI}(s)$ into a corresponding z-domain $G_{PI}(z)$. Among the many existing discretization techniques [12], standard Euler discretization will be employed here, based upon the substitution

$$s \longrightarrow \frac{1 - z^{-1}}{T_{sampling}}, \tag{4.7}$$

in which z^{-1} represents the one-sampling step delay operator. It follows, from (4.6) and from $T_{sampling} = T_s/4$, that the z-domain compensator transfer function is

$$G_{PI}(s) \rightarrow G_{PI}(z) = K_p + \frac{K_i T_s}{4} \frac{1}{1 - z^{-1}}. \tag{4.8}$$

Figure 4.5 illustrates a Matlab® simulation of the designed POL control under a 0.1 to 10 A step load variation. In the figure, the instantaneous output voltage $v_o(t)$ and the sampled output voltage $v_o[k]$ are superimposed. Observe that, as anticipated, the two signals have essentially the same dynamics. Furthermore, notice that the regulated signal is $v_o[k]$ and not the true average value of $v_o(t)$ as would happen with an analog controller. The resulting error is, in practice, negligible due to the small peak-to-peak ripple of $v_o(t)$. A second more important issue is amplitude quantization—not accounted for in the simulation of Figure 4.5—which represents

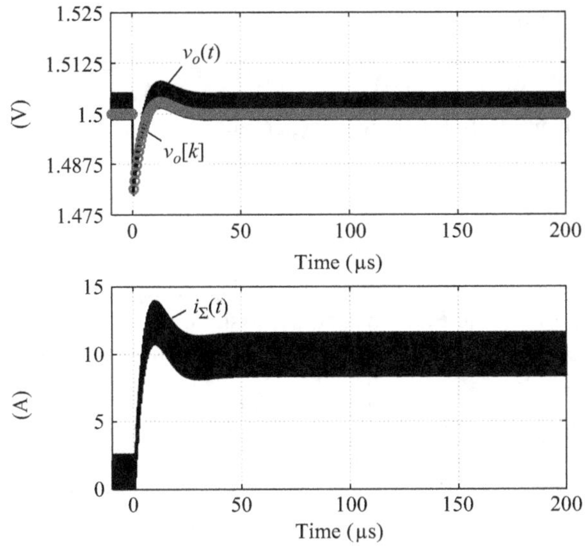

Figure 4.5 Simulated POL transient response to a 0.1 to 10 A step load variation

the true limiting factor for DC regulation accuracy and which will be addressed in the next section.

These results conclude the system-level design of the digital compensation. Specific aspects pertaining to the controller implementation are now treated, starting from quantization issues affecting the choice of the A/D and MPM resolutions.

4.3 Quantization effects and limit cycling

One usually distinguishes between three types of quantizations in a digitally controlled converter:

- *A/D quantization*, due to the analog-to-digital conversion process, and which acts on the signal $v_o[k]$ and on the digitized regulation error $e[k]$;
- *modulator quantization*, due to the modulation finite resolution and which can always be thought of as an equivalent amplitude discretization on the control command $u[k]$;
- *compensator quantization*, due to the finite arithmetic precision in which the compensator internal signals are represented.

In this section, focus is given to A/D and modulator quantizations, and on how their mutual interaction can potentially trigger the undesired phenomenon of limit cycling. The issue of compensator quantizations are briefly addressed in the next section in conjunction with implementation-related considerations.

4.3.1 A/D quantization

A/D quantization subdivides the output voltage range into several *quantization intervals* or *bins*, the amplitude of which characterizes the A/D resolution. For an $n_{A/D}$-bit ADC operating over a full-scale range V_{FS}, amplitude $q_{A/D}$ of the quantization bin is

$$q_{A/D} = \frac{V_{FS}}{2^{n_{A/D}}}. \tag{4.9}$$

Within each quantization interval, all voltage values are *indistinguishable* by the digital compensator. Therefore, choice of $n_{A/D}$ is directly related to the required DC regulation accuracy. In the POL case study, assume v_o must be regulated to within $\pm 1\%$ of its setpoint $V_{ref} = 1.5$ V and that $V_{FS} = 2$ V. Then, by inverting (4.9),

$$n_{A/D} > \log_2\left(\frac{2\text{ V}}{20\text{ mV}}\right) \to \begin{cases} n_{A/D} = 7\text{ bits}, \\ q_{A/D} = 15.625\text{ mV} \end{cases}. \tag{4.10}$$

With this choice, the digital compensator ideally regulates $v_o[k]$ to within the interval $V_{ref} \pm q_{A/D}/2$. This specific quantization bin, located across the regulation set point, is usually referred to as *zero-error bin*. Goal of the digital controller is to bring the sampled output voltage back inside the zero-error bin in response to various system disturbances and guarantee such condition during steady-state operation.

4.3.2 Modulation quantization

Modulation quantization refers to the capability of a digital modulator to only reproduce a *discrete* set of duty cycles rather than a continuous set as with analog modulators. Such effect can be equivalently described as an amplitude discretization acting on $u[k]$. An n_{DPWM}-bit modulator subdivides the switching interval T_s into $2^{n_{DPWM}}$ time slots, for an equivalent duty cycle resolution equal to

$$q_D = \frac{1}{2^{n_{DPWM}}}. \tag{4.11}$$

For instance, a 9-bit modulator would exhibit a duty cycle resolution of $q_D = 2^{-9} \approx 1.95 \times 10^{-3}$. If the modulator static gain is normalized to unity, so that $0 < u[k] < 1$, then q_D also represents the equivalent granularity on the control command $u[k]$.

Choice of the minimum acceptable value for n_{DPWM} is determined by the need to accurately position the converter output voltage inside the zero-error bin of the A/D converter. This problem is strictly connected with the issue of limit cycling.

4.3.3 No-limit-cycling design criteria

In the POL converter case study, given the finite set of duty ratios which can be reproduced by the MPM, the steady-state output voltage levels which can be regulated by the converter are

$$V_o = iq_D V_g - r_{eq} I_o, \quad i = 0 \dots 2^{n_{DPWM}} - 1. \tag{4.12}$$

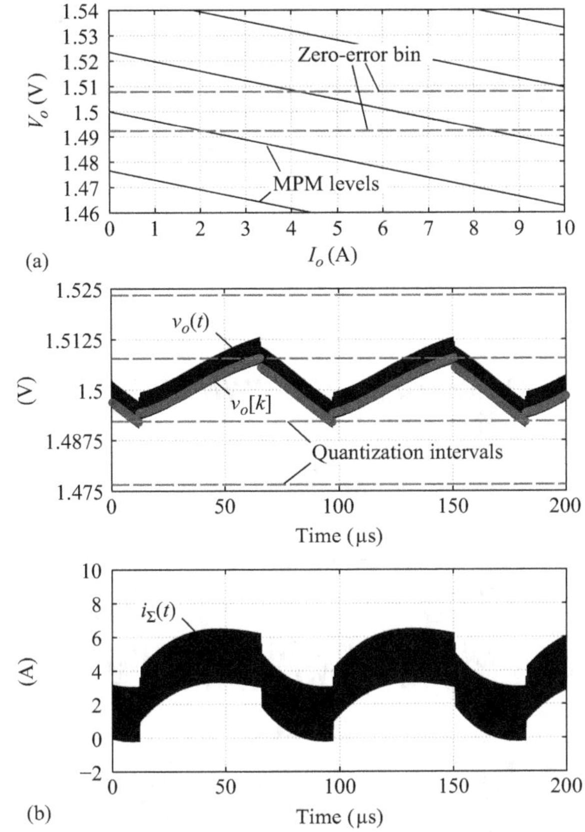

Figure 4.6 (a) A/D and modulator quantization levels for $q_D V_g > q_{A/D}$ ($n_{DPWM} = 9$), and (b) steady-state operation at $I_o = 3.5$ A exhibiting a limit cycle oscillation

This set of curves is illustrated in Figure 4.6(a), superimposed to the A/D quantization bins, for $n_{DPWM} = 9$. According to (4.12), the various curves are separated by $q_D V_g \approx 23$ mV, a quantity strictly *larger* than the A/D bin $q_{A/D} \approx 15.6$ mV.

Notice how, in the situation depicted, there are intervals of load currents for which none of the MPM levels (4.12) falls inside the zero-error bin. From Figure 4.6(a), these levels roughly comprise the $2\text{ A} < I_o < 4\text{ A}$ and $8.5\text{ A} < I_o < 10\text{ A}$ intervals. At such load levels the digital controller would be unable to null the regulation error, and the system would enter a periodic nonlinear oscillation of indefinite duration called *limit cycle*. An example of steady-state limit cycling is illustrated in Figure 4.6(b), for a load level $I_o = 3.5$ A: the system bounces back and forth across the zero-error bin without being able to regulate a constant voltage. Observe that the limit cycle amplitude is usually small, in the order of $q_{A/D}$. Nonetheless, limit cycling

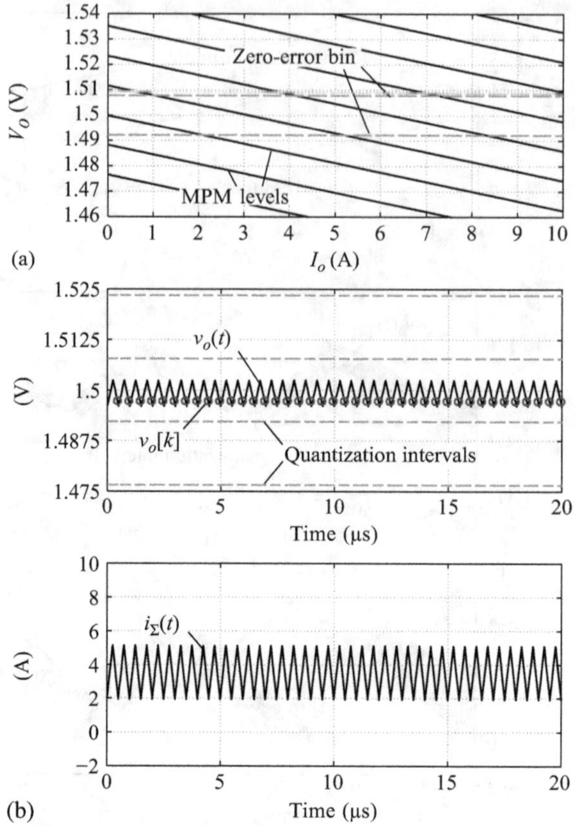

Figure 4.7 *(a) A/D and modulator quantization levels for $q_D V_g < q_{A/D}$ ($n_{DPWM} = 10$), and (b) steady-state operation at $I_o = 3.5$ A without limit cycle oscillations*

is usually considered as an undesired condition incompatible with proper converter operation.

Clearly, if the MPM bin is chosen to be smaller than the A/D bin, i.e., if

$$q_D V_g < q_{A/D}, \tag{4.13}$$

then at least one MPM quantization level falls inside the zero-error bin for any value of the load current. This situation is exemplified in Figure 4.7(a) for $n_{DPWM} = 10$, for which $q_D V_g \approx 11.7$ mV $< q_{A/D} \approx 15.6$ mV. The corresponding steady-state operation at $I_o = 3.5$ A is shown in Figure 4.7(b): the output voltage is now limit-cycle-free and correctly regulated inside the zero-error bin.

Criterion (4.13) is an example of *no-limit-cycling condition* and represents a first basic guideline for choosing the modulator resolution in a digitally controlled DC-DC converter. In this case the limit cycling issue is correlated with the presence

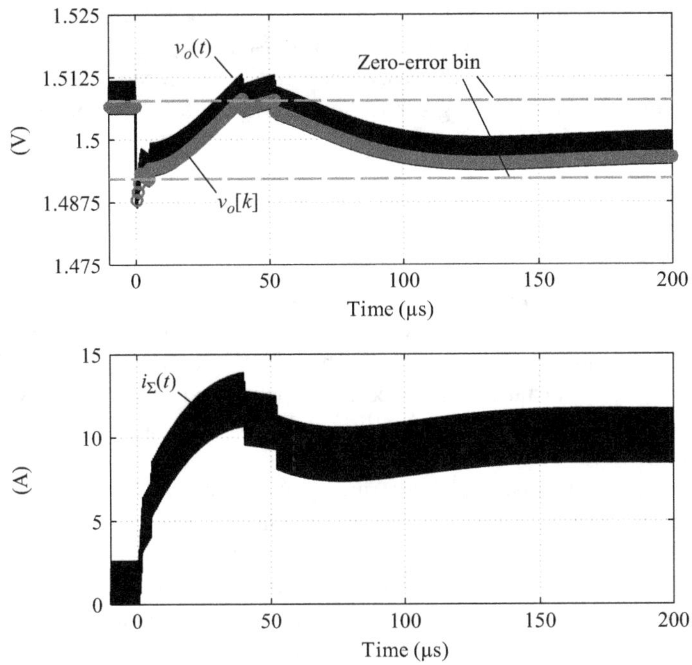

Figure 4.8 Simulated POL transient response to a 0.1- to 10-A step load variation, including ADC and modulator quantization effects

or absence of a DC solution for the feedback system. In the absence of such solution, a limit cycle would *certainly* occur. On the other hand, examples can be brought showing how (4.13) is a necessary, yet not sufficient condition to ensure limit-cycling-free operation. A broader view on no-limit-cycling design criteria, including additional static conditions as well as dynamic constraints, can be found in References 3, 13, and 14.

Assuming here $n_{DPWM} = 10$ bits as a design choice for the MPM implementation and the previously determined ADC resolution (4.10), the converter closed-loop response to the same 0.1 to 10 A step load variation previously considered is reported in Figure 4.8. Once quantizations are accounted for, the shape of the closed-loop response may qualitatively differ from what one is usually accustomed to expect from a continuous-time controller. Nonetheless, the output voltage remains tightly regulated inside the ADC zero-error bin according to the closed-loop bandwidth and phase margin specifications. After the transient, the system settles inside the zero-error bin with no additional limit cycling.

4.4 Controller implementation

With switching rates of several hundreds of kilohertz and quickly approaching the megahertz range, all main building blocks of the digital controller of Figure 4.1

demand implementations strongly optimized for speed and reduced silicon area. Although these two requirements are often self-conflicting, a number of design solutions have been devised by the researchers in the last decade. A few of these are briefly described in the following subsections as possible approaches for the digital multiphase POL controller under study.

4.4.1 Analog-to-digital converter

Among the several ADC architectures known in microelectronic design, the parallel or *flash* type is certainly the one which best suits the speed requirements of high-frequency switched-mode power supply applications, easily reducing conversion times down to a small fraction of the switching period. Other ADC structures, based, for instance, on successive approximations, pipeline conversion, or oversampling techniques, are usually much less attractive due to their larger conversion latencies [15].

In a conventional flash ADC, a string of comparators is first used to produce a thermometer-coded version of the sensed signal $v_o(t)$. An encoder successively produces the digital error signal $e[k] = V_{ref} - v_o[k]$ in, for instance, binary 2's complement format. The flash architecture, although fast, has the major drawback of requiring $2^{n_{A/D}} - 1$ comparators, with consequent disadvantages in terms of silicon area and power consumption as the required resolution increases. The *windowed flash* ADC structure illustrated in Figure 4.9 has been devised in order to address these issues [7]. In the windowed ADC, only a small voltage range across the desired setpoint V_{ref} is A/D-converted using a limited number of

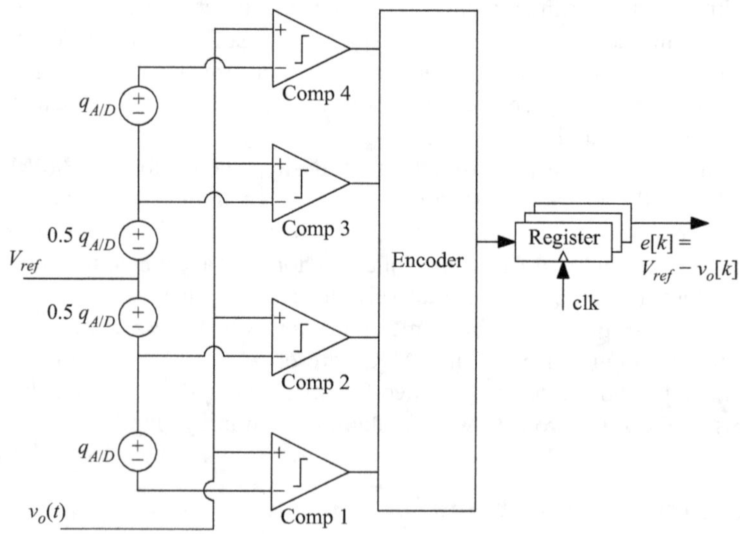

Figure 4.9 A two-bit windowed flash ADC using four comparators.
© Reproduced by permission of John Wiley & Sons, Inc. [3]

comparators. In Figure 4.9, for instance, four comparators are used to create four quantization levels centered around V_{ref}, realizing a 2-bit ADC. The resulting ADC preserves the speed of the flash concept but strongly optimizes the silicon footprint and power consumption, especially at high sampling rates. Furthermore, since the POL output voltage must be tightly regulated around the control set-point during converter operation, the loss in voltage conversion range of the windowed flash ADC with respect to the conventional flash architecture is a disadvantage of minor importance. For instance, in the 10 A transient of Figure 4.8 the output voltage only invades the two ADC bins immediately adjacent to the zero-error bin.

One situation in which $v_o(t)$ largely differs from the target V_{ref} is at start-up. Nonetheless, if V_{ref} can be ramped up smoothly—either using an analog circuitry or with the use of a dedicated digital to analog converter—then a controlled soft-start of the converter can be realized. Figure 4.10 illustrates the simulated soft-start of the digital POL designed in the previous section. In the simulation, a three-level windowed ADC is modeled, and the controller set point is ramped up from 0 V to V_{ref} in 500 μs under no-load conditions ($i_o = 0$).

Other types of ADC realizations tailored for high-frequency control of power converters have been investigated. These include the use of delay-line–based converters with windowed characteristics [1, 5, 16], ring-oscillator–based structures [17], and threshold-inverter quantization techniques [18].

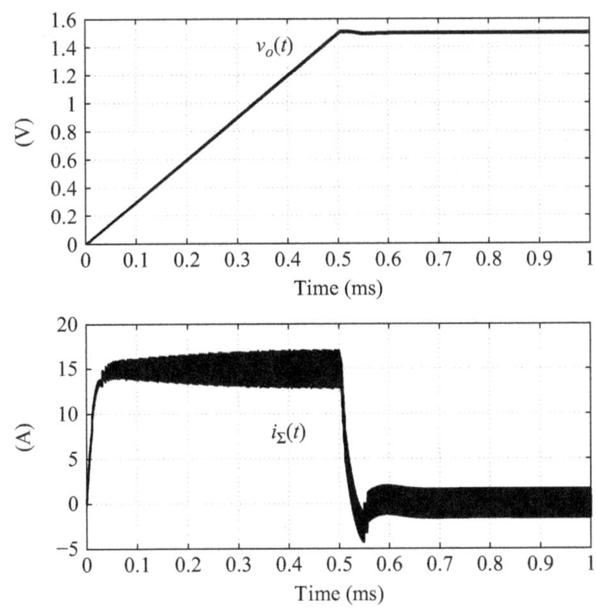

Figure 4.10 Simulated POL soft-start with a three-level windowed ADC under no-load conditions ($i_o = 0$)

4.4.2 Digital compensator

In order to obtain a fast and compact implementation of the control law, the digital compensator structure is usually realized as a hardwired solution composed by standard arithmetic and logic blocks. The compensator is typically coded in a Hardware Description Language (HDL) such as VHDL or Verilog. Taking advantage of the fully digital nature of the structure, the steps of gate-level and transistor-level realization of the compensator are then typically accomplished using a software tool for automatic synthesis.

Figure 4.11 illustrates the parallel realization of a programmable digital PI compensator. The core control law is implemented as the superposition of the proportional and integral actions by the equations

$$\begin{aligned} u_{PI}[k] &= u_p[k] + u_i[k], \\ u_p[k] &= K_p e[k], \\ u_i[k] &= \frac{K_i T_s}{4} e[k] + u_i[k-1], \end{aligned} \tag{4.14}$$

which are the time-domain counterpart of (4.8).

In addition to the basic control law represented by (4.14), a number of other provisions commonly embedded in the compensator structure are depicted in Figure 4.11 [3]:

- Compensator gains K_p and K_i can be programmed either by an external digital interface, as already suggested in Figure 4.1, or adjusted by a digital autotuner for automatic calibration of the compensator frequency response.
- Saturation elements are included both inside the local integrator loop, and in cascade with the overall PI structure. The former saturation prevents the integrator state variable to wind up to excessively large or small values during controller operation, and therefore implements a form of *integral anti-windup*

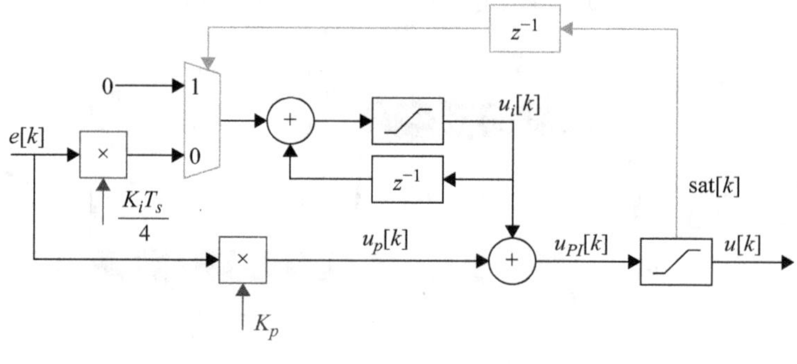

Figure 4.11 Parallel realization of a digital programmable PI compensator, including saturations and anti-windup provisions

provision [19]. The second saturation ensures that the overall control command $u[k]$ is correctly constrained within the admissible limits of the digital MPM cascaded to the compensator.

- *Conditional integration* is implemented as an additional anti-windup mechanism, by freezing the integration process upon saturation of the overall control command $u[k]$. This is accomplished by a local feedback loop activated by a saturation flag "sat". Assertion of such flag acts on a 2:1 multiplexer placed in front of the digital integrator and causes the integrator input signal to be set to zero. The integral term then remains constant until the saturation condition ceases to exist.

One strong simplification of the above implementation is illustrated in Figure 4.12 [5]. The control law is here in the form

$$u[k] = u[k-1] + \underbrace{\left(K_p + \frac{K_i T_s}{4}\right)e[k]}_{\text{LUT1}} - \underbrace{K_p e[k-1]}_{\text{LUT2}}. \tag{4.15}$$

Two look-up tables (LUT) are here employed, addressed by $e[k]$ and $e[k-1]$, respectively, which directly store the results of the multiplications involving $e[k]$ and $e[k-1]$. This realization makes use of a single adder, a single saturation block, and eliminates the use of digital multipliers entirely. Furthermore, computational latency is reduced to a few propagation delays through basic logic blocks, making the solution suitable for high-frequency applications. Programmability is not sacrificed, provided that the contents of the LUT can be loaded from an external programmable memory. Such structure is especially convenient in combination with the windowed flash ADC concept discussed in Section 4.4.1, in which the limited range of $e[k]$ allows to strongly reduce the size of the LUTs.

Word length determination for the compensator internal signals and coefficients represents the final step of the compensator logic-level design. A number of techniques, nowadays well established in the field of digital signal processing and basic digital control, have been devised to address such issue and will not be

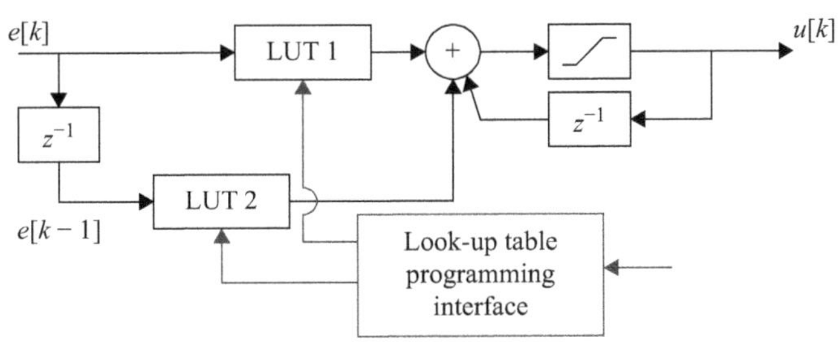

Figure 4.12 Look-up table based on digital PI compensator

repeated here. A set of guidelines and approaches for compensator implementation and word length determination for hardwired structures can be found in Reference 3.

4.4.3 Digital MPM

One possible realization of the digital MPM used in the POL converter example is depicted in Figure 4.13. The MPM is a collection of four distinct DPWMs, each processing a specific time-demultiplexed version of the control command $u[k]$,

$$\begin{aligned} u_1[nT_s] &\triangleq u[nT_s], \\ u_2[nT_s] &\triangleq u\left[nT_s + \frac{T_s}{4}\right], \\ u_3[nT_s] &\triangleq u\left[nT_s + \frac{T_s}{2}\right], \\ u_4[nT_s] &\triangleq u\left[nT_s + \frac{3T_s}{4}\right]. \end{aligned} \tag{4.16}$$

The four DPWMs operate at the same sampling rate f_s, but with time-shifted carriers in order to drive the power converter in the desired phase-interleaved fashion.

Various digital techniques for implementing the fundamental DPWM function have been disclosed in the literature, the main challenge being the inherent trade-off between silicon area, complexity, and achievable resolution [20]. The simple *counter-based* DPWM sketched in Figure 4.14(a) imitates the operation of an analog PWM, with a digital ramp substituting the analog sawtooth carrier, and a digital comparator in place of its analog counterpart. Main operating waveforms of the counter-based DPWM module are sketched in Figure 4.14(b). Clock rate of the digital ramp determines the duty cycle of such modulator as

$$q_D = \frac{f_s}{f_{clk}}. \tag{4.17}$$

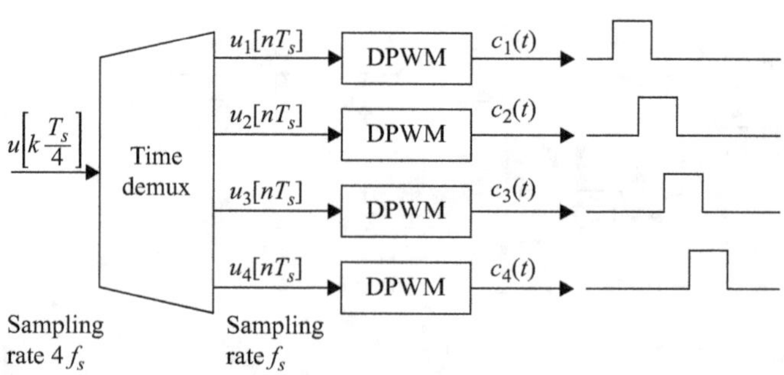

Figure 4.13 Digital MPM as a string of phase-shifted DPWMs

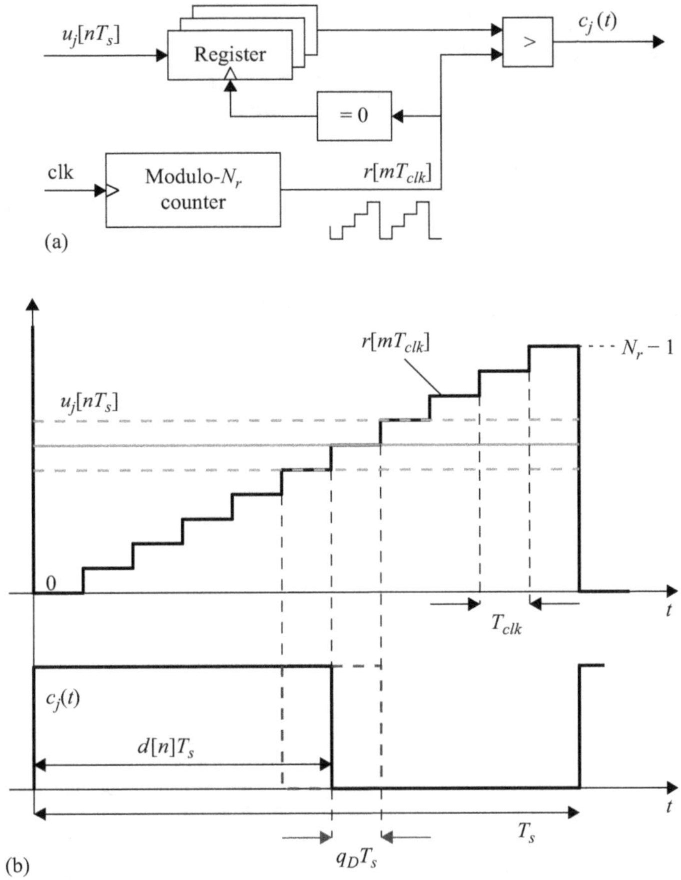

Figure 4.14 (a) Counter-based DPWM and (b) associated waveforms.

Notice that any additional bit of resolution involves doubling the DPWM clock rate. For such reason, pure counter-based DPWMs are rarely employed in high switching rate applications. For instance, a counter-based implementation of the 10-bit MPM required for the POL digital controller example would involve four distinct DPWM modules each operating at a clock rate $f_{clk} = 2^{10} \times f_s =$ 512 MHz.

An alternative DPWM architecture which achieves the required resolution without the need for a high-frequency clock is described in Figure 4.15(a) and (b). At the beginning of each modulation cycle, the DPWM output is set to logic 1. Simultaneously, a pulse is delivered to a string of $2^{n_{DPWM}}$ delay cells, each having a propagation delay $t_c = T_s/2^{n_{DPWM}}$. The rising edge of such initial pulse therefore propagates through the delay line. A $2^{n_{DPWM}}$: 1 multiplexer, controlled by the *j*th

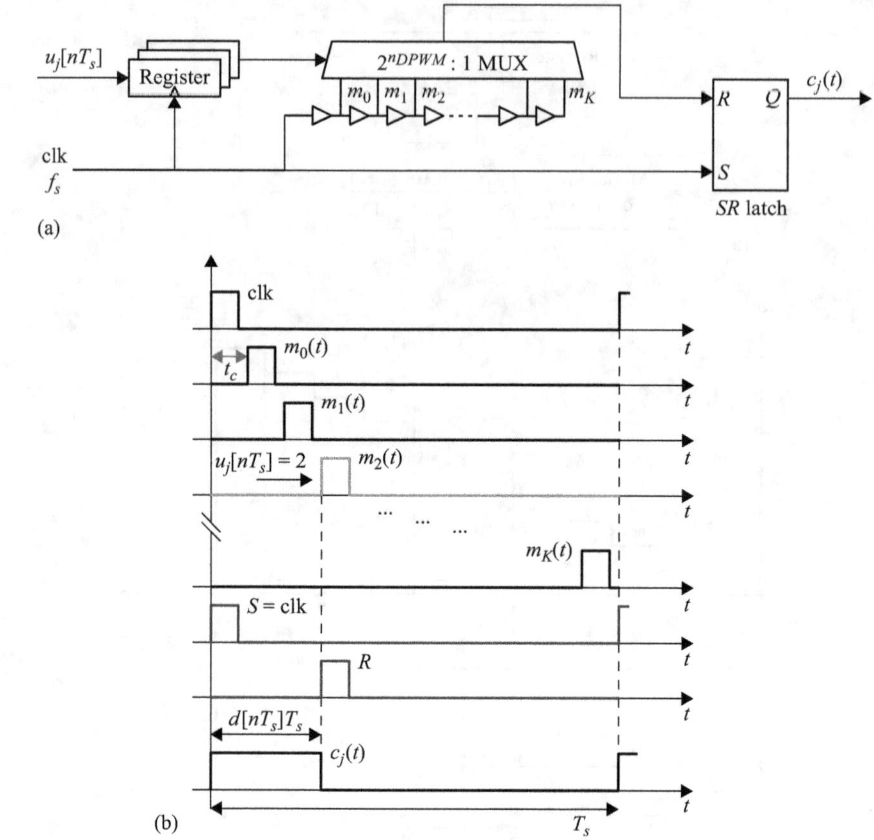

Figure 4.15 (a) Delay line DPWM and (b) associated waveforms.
© Reproduced by permission of John Wiley & Sons, Inc. [3]

time-demultiplexed signal $u_j[nT_s]$, selects which tap of the delay line resets the DPWM output node back to logic 0. In this structure, resolution is directly determined by the cell propagation delay t_c, whereas the clock frequency coincides with the switching rate.

Silicon area is one major drawback of the delay line DPWM, since the number of delay cells and the size of the multiplexer grow exponentially with n_{DPWM}. For such reason, the delay line concept is often employed in combination with the counter-based architecture. In such *hybrid* DPWM structures, one can find a design trade-off between the clock rate and overall modulator size [22].

Architectures for digital multiphase modulation alternative to Figure 4.13 have also been disclosed in the literature. For instance, a scalable, hardware-efficient MPM architecture based on the power D/A concept is disclosed in Reference 22.

4.5 Summary of Key Points

- Digital PWM control of high-frequency DC-DC converters is being increasingly adopted because of its versatility in providing advanced control, interface, and power management functions. Understanding both the potentials and the critical aspects associated with these types of controllers has become a mandatory background knowledge for the power electronics expert.
- Main building blocks of a digital controller are the A/D converter, the digital compensator, and the digital modulator. Additional features typically include a digital interface for programming and monitoring purposes.
- Dynamic modeling of the digital loop entails not only describing the dynamics of the power converter itself but also correctly quantifying the various computational, control, and modulation delays which impact the closed-loop dynamics. Once the uncompensated dynamics of the digital loop is modeled, design of the digital compensator transfer function can proceed according to standard techniques known in control theory.
- Quantization effects are related to the finite A/D and modulator hardware resolutions. Guidelines for selecting the appropriate number of A/D and modulator bits can be formulated based on the required DC regulation accuracy on one hand and on the need to prevent potentially disruptive effects due to limit cycling on the other.
- Fast A/D architectures are required to minimize the ADC conversion latency in high-frequency applications. Among these, the flash-type ADC structure is the architecture of choice. One interesting hardware-efficient realization is the windowed-flash ADC, in which only a few quantization levels around the control setpoint are realized, thus saving silicon area without compromising performance.
- Hardwired custom-designed structures are employed for the digital compensator realization. These can be simple structures composed of standard arithmetic and logic digital blocks. Strong simplifications aimed at further reducing the overall complexity can be devised.
- Implementation of DPWM cores poses design trade-offs in terms of resolution, silicon area, and required clock frequency. The basic counter-based DPWM architecture, often inadequate in high-frequency application due to an excessively large required clock frequency, can be successfully combined with a delay line structure into a hybrid architecture in which clock rate, hardware resolution, and overall footprint can be optimized.

References

[1] D. Maksimović, R. Zane, and R. W. Erickson, "Impact of digital control in power electronics," In: *Proc. 16th IEEE International Symposium on Power Semiconductor Devices*, pp. 13–22, May 2004.

[2] D. Maksimovic, R. Zane, and L. Corradini, "Advances in digital control for high-frequency switched-mode power converters," *Power Electronics Monthly*, vol. 44, no. 12, pp. 2–19, 2010, Serial no. 217, Sponsored by Xi'an Power Electronics Research Institute, China.
[3] L. Corradini, D. Maksimović, P. Mattavelli, and R. Zane, *Digital Control of High-Frequency Switched-Mode Power Converters*. Wiley-IEEE Press, 2015, ISBN: 978-1-118-93510-1.
[4] A. M. Wu, X. Jinwen, D. Marković, and S. R. Sanders, "Digital PWM control: application in voltage regulation modules," In: *Proc. 30th IEEE Power Electronics Specialists Conference* (PESC), vol. 1, pp. 77–83, July 1999.
[5] B. J. Patella, A. Prodić, A. Zirger, and D. Maksimović, "High-frequency digital PWM controller IC for DC-DC converters," *IEEE Trans. Power Electron.*, vol. 18, no. 1, pp. 438–446, 2003.
[6] Y. Panov and M. M. Jovanović, "Design considerations for 12-V/1.5-V, 50-A voltage regulator modules," *IEEE Trans. Power Electron.*, vol. 16, no. 6, pp. 776–783, 2001.
[7] A. V. Peterchev, J. Xiao, and S. R. Sanders, "Architecture and IC implementation of a digital VRM controller," *IEEE Trans. Power Electron.*, vol. 18, no. 1, pp. 356–364, 2003.
[8] R. D. Middlebrook, "Predicting modulator phase lag in PWM converter feedback loops," In: Advances in Switched-mode Power Conversion, Teslaco, 1981, pp. 245–250.
[9] R. W. Erickson and D. Maksimović, *Fundamentals of Power Electronics*, 2nd ed. Springer, US, 2001.
[10] D. Maksimović and R. Zane, "Small-signal discrete-time modeling of digitally controlled PWM converters," *IEEE Trans. Power Electron.*, vol. 22, no. 6, pp. 2552–2556, 2007.
[11] R. D. Middlebrook, "Measurement of loop gain in feedback systems," *Int. J. Electronics*, vol. 38, no. 4, pp. 485–512, 1975.
[12] G. F. Franklin, J. D. Powell, and M. L. Workman, *Digital Control of Dynamic Systems*, 3rd ed. Addison-Wesley, Menlo Park, CA, 1998.
[13] A. V. Peterchev and S. R. Sanders, "Quantization resolution and limit cycling in digitally controlled PWM converters," *IEEE Trans. Power Electron.*, vol. 18, no. 1, pp. 301–308, 2003.
[14] H. Peng, A. Prodić, E. Alarcon, and D. Maksimović, "Modeling of quantization effects in digitally controlled dc-dc converters," *IEEE Trans. Power Electron.*, vol. 22, no. 1, pp. 208–215, 2007.
[15] D. A. Johns and K. Martin, *Analog Integrated Circuit Design*, 2nd ed. John Wiley & Sons, Inc., Hoboken, NJ, 2011.
[16] A. Parayandeh and A. Prodić, "Programmable analog-to-digital converter for low-power dc-dc SMPS," *IEEE Trans. Power Electron.*, vol. 23, no. 1, pp. 500–505, 2008.
[17] J. Xiao, A. V. Peterchev, J. Zhang, and S. Sanders, "A 4μA quiescent current dual-mode digitally controlled buck converter IC for cellular phone applications," *IEEE J. Solid-State Circuits*, vol. 39, no. 12, pp. 2342–2348, 2004.

[18] L. Corradini, E. Orietti, P. Mattavelli, and S. Saggini, "Digital hysteretic voltage-mode control for dc-dc converters based on asynchronous sampling," *IEEE Trans. Power Electron.*, vol. 24, no. 1, pp. 201–211, 2009.

[19] K. Åström and T. Hägglund, *PID Controllers: Theory, Design, and Tuning*, 2nd ed. Instrument Society of America, Research Triangle Park, NC, 1995.

[20] A. Syed, E. Ahmed, D. Maksimović, and E. Alarcon, "Digital pulse width modulator architectures," In: *Proc. 35th IEEE Power Electronics Specialists Conference* (PESC), vol. 6, pp. 4689–4695, 2004.

[21] V. Yousefzadeh, T. Takayama, and D. Maksimović, "Hybrid DPWM with digital delay-locked loop," In: *Proc. 10th IEEE Workshop on Computers in Power Electronics* (COMPEL), pp. 142–148, 2006.

[22] T. Carosa, R. Zane, and D. Maksimović, "Scalable digital multiphase modulator," *IEEE Trans. Power Electron.*, vol. 23, no. 4, pp. 2201–2205, 2008.

Chapter 5

Microcontroller-based electronic ballasts for high-intensity discharge lamps

J. Marcos Alonso[1] *and Marco A. Dalla Costa*[2]

This chapter deals with the use of digital control applied to discharge lamp electronic ballasts with a special focus on high-intensity discharge (HID) lamp ballasts. HID lamp ballasts is one of the power electronics fields in which digital control is well justified. HID lamps require a complicated process of ignition, warming-up, and lamp current and power regulation. Each of these operating stages requires a precise timing during which the measurement of lamp parameters results critical to take the correct decision for a proper operation of the ballast. In addition, HID lamp ballasts deal with very high voltages, which can reach up to tens of kilovolts for hot re-ignition. These high voltages, along with the fact that HID lamps operate at high temperature and pressure, makes safety one of the important issues in HID lamp supply. In this context, microcontroller-based electronic ballasts are an excellent option to achieve all these stringent requirements.

5.1 HID lamp operation principles and modelling

5.1.1 HID lamps

High-intensity discharge (HID) lamps are still extensively used in many lighting applications owing to their good features in terms of efficacy, correlated colour temperature (CCT), colour rendering index (CRI) and life [1, 2]. Figure 5.1 illustrates the basic structure of an HID lamp.

In an HID lamp, an arc discharge is created between two electrodes located inside an arc tube. An outer bulb is used to assure an adequate operating temperature of the discharge gas, so that the expected photometric and electric characteristics can be attained. Typical gas discharges employed in HID lamps are based on the use of mercury and/or sodium in vapour phase, while xenon is employed in some special applications, such as projectors and flash lamps. Thus,

[1]Electrical Engineering Department, University of Oviedo, Gijón, Spain
[2]Electrical Center of Technology – CT, Energy Processing Department – DPEE, Group of Intelligence in Lighting – GEDRE, Federal University of Santa Maria, Santa Maria, RS, Brazil

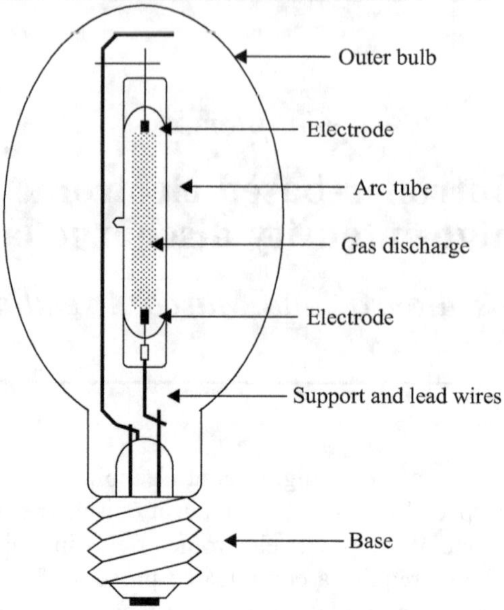

Figure 5.1 HID lamp basic structure

Table 5.1 Typical features of HID lamps

Characteristic	HPMV	HPS	Quartz MH	Ceramic MH
Wattage range (W)	50–1000	35–1000	35–3000	20–400
Efficacy (lm/W)	50–60	90–140	80–120	80–100
CCT (K)	3200–5700	1900–2200	3000–7000	3000–4500
IRC	30–50	20–60	65–90	65–90
Light colour	White	Yellow	White	White
Start-up time (minutes)	5–7	3–4	2–5	2–5
Restrike time (minutes)	3–6	1–3	5–15	3–15
Typical life (khours)	10–20	10–40	5–20	5–20

high pressure sodium (HPS) lamps are based on the use of sodium discharge, while high pressure mercury vapour (HPMV) lamps are based on mercury vapours.

An evolution of mercury-based vapour lamps is the metal halide (MH) lamp, in which additional metals (Sc, Tl, In, etc.) in the form of iodides or bromides are added to improve some of the lamp features, mainly CCT and CRI. MH lamps can be found in two types, depending on the material used for the arc tube. Originally, MH lamps were developed using a quartz tube; later, ceramic materials were employed to provide a more stable light output and colour over lamp life. Table 5.1 summarizes the typical features of HID lamps [1–4].

For many years, HID lamps have been supplied by means of electromagnetic ballasts, which operate at line frequency. However, because of the advancements of

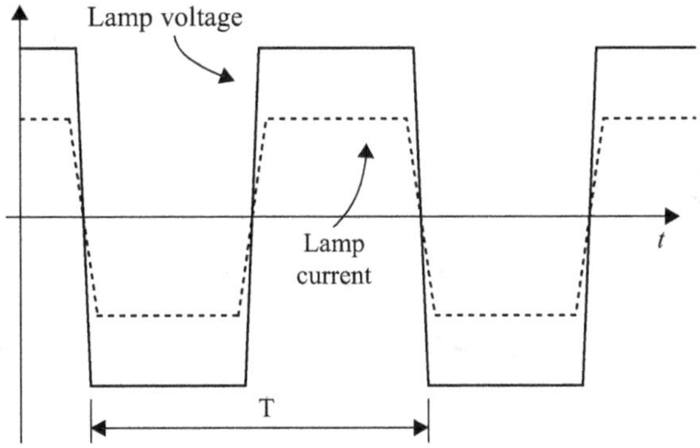

Figure 5.2 Lamp voltage and current under low-frequency square wave operation

power electronics technology, electronic ballasts are being increasingly used to supply HID lamps, thus following the trend in fluorescent lamp technology, where the normal situation is to employ electronic ballasts to supply this type of lamps.

One of the drawbacks that prevent the use of electronic ballasts for HID lamp applications is the generation of acoustic resonances inside the discharge volume when the lamp is supplied at frequencies higher than several hundreds of Hertz. Acoustic resonances appear when power is injected into the discharge at a certain characteristic frequencies [2]. When acoustic resonances take place in the discharge, the electric arc becomes unstable, fluctuates and generates annoying light flicker; effects that can even result in discharge tube fracture due to excessive wall overheating at some spots. For many years, researchers have worked on different methods to avoid acoustic resonances in HID lamps. Today, it has been accepted by most of ballast designers and manufacturers that the best method to avoid acoustic resonances in HID lamps is to supply the lamp with low-frequency square waveforms (LFSW). Figure 5.2 illustrates the typical voltage and current waveforms under this operation technique. The operating frequency must be high enough to avoid light flickering and low enough to avoid high-frequency lamp power harmonics that could generate acoustic resonances. Typical operating frequency is 200 Hz.

5.1.2 HID lamps operating requirements

Figure 5.3 illustrates the typical operation stages of an HID lamp during the start-up process [4–6]. Initially, the lamp behaves as a very high impedance, and it is necessary to apply a high voltage spike of around 3–4 kV to initiate the discharge, as illustrated as region (1) in Figure 5.3. Once the electric discharge has been initiated, the lamp voltage drops to a value much lower than the normal operation lamp voltage. This situation represents nearly a short circuit for the electronic ballast. Therefore, the ballast must operate in current source mode, region (2), limiting the current through the discharge to avoid any damage. During this interval, the lamp warm-up phase starts and the lamp voltage increases because of the evaporation of the different materials present in the discharge tube. This also means

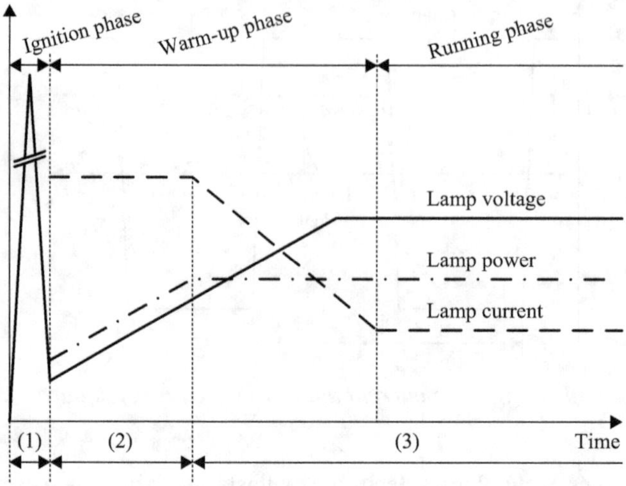

Figure 5.3 HID lamp voltage, current and power evolutions during the start-up process

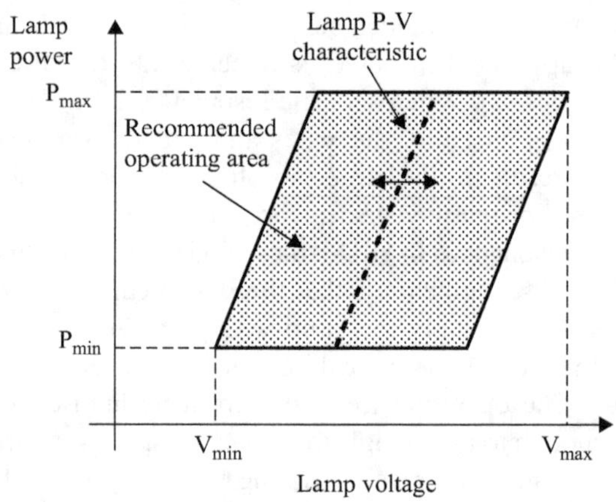

Figure 5.4 Typical recommended operating area of an HID lamp

that the lamp power will be increasing during the warm-up phase. Since it is not advisable to surpass the maximum lamp-rated power, once the lamp reaches a maximum power, the ballast will enter into the power source mode, decreasing lamp current until the nominal operation point is attained. At this point, the normal running phase is achieved. During normal operation, the lamp power is regulated to a constant value in order to assure maximum lamp life and optimal electrical and photometrical performance.

Figure 5.4 illustrates the typical lamp power versus root mean square (RMS) voltage characteristic and the recommended operating area of HID lamps [6].

In normal running operation, when steady state is reached and for slow perturbations, the lamp exhibits a power–voltage characteristic with a positive slope, as shown by the thick dashed line in Figure 5.4. This characteristic shifts from low voltage values to high voltage values depending on lamp operating conditions, especially ambient temperature and lamp aging. A safe operation area is defined by the quadrilateral region illustrated in Figure 5.4, which assures the lamp photometrical features and rated life. Therefore, the electronic ballast must guarantee the lamp operation inside this area, which must be defined for each particular HID lamp.

5.1.3 HID lamps modelling

The availability of a lamp model is essential for the correct design of an HID lamp electronic ballast. In modern power electronics design, a model is usually represented as an equivalent circuit that can be employed to perform computer simulations of the complete electronic ballast, which includes both power converter and lamp. In this way, computer simulations allow the designer to study and optimize the lamp-ballast operation under different circumstances, as ignition, steady state, transients and closed-loop operation.

The starting point to build the HID lamp model is given by its static voltage–current characteristic. An example of this characteristic is illustrated in Figure 5.5. The lamp voltage–current characteristic is usually hyperbolic-shaped, which means that as the lamp current increases, its voltage decreases. The slope of the curve represents the dynamic resistance of the lamp, which is negative. Usually, rather than employing complicated expressions to represent the lamp behaviour, a linear approximation can be made, as illustrated in Figure 5.5. This linear approximation can mathematically be expressed as follows:

$$V_{LA} = v_S - r_S I_{LA}, \tag{5.1}$$

where V_{LA} and I_{LA} are the RMS lamp voltage and current, respectively, and v_S and r_S are the HID lamp parameters that define the linear approximation of the actual

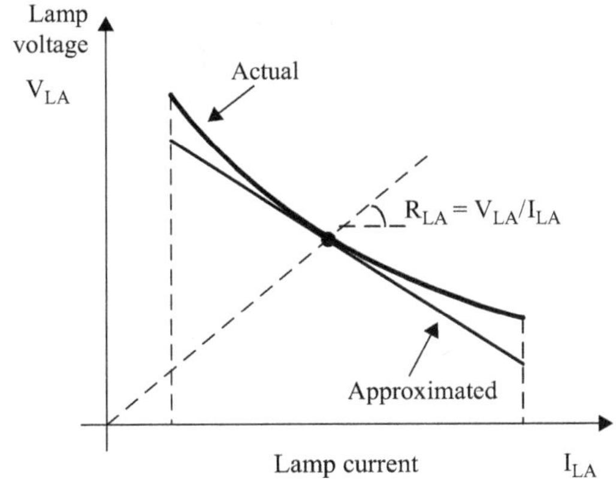

Figure 5.5 Typical HID lamp voltage–current characteristic and its linear approximation

lamp characteristic. The lamp equivalent resistance can then be calculated from (5.1) by dividing lamp voltage by lamp current:

$$R_{LA} = \frac{V_{LA}}{I_{LA}} = \frac{v_S}{I_{LA}} - r_S \tag{5.2}$$

Another important aspect of the lamp modelling relates to its dynamic behaviour. The lamp dynamic behaviour can be represented by means of its incremental impedance, which is defined by the ratio between lamp voltage and current perturbations. Deng and Cuk discovered that the lamp incremental impedance can be described by means of a first-order single-pole single-zero transfer function [7], as shown in the following:

$$Z_{LA}(s) = k\frac{s+\omega_z}{s+\omega_p} \tag{5.3}$$

where k is a lamp parameter that represents the high-frequency dynamic lamp resistance, and ω_z and ω_p are the angular frequencies of the lamp model zero and pole, respectively.

Figure 5.6 illustrates the implementation of a Simulation Program with Integrated Circuit Emphasis (SPICE)-based model that includes both the static and dynamic behaviour of the HID lamp. Table 5.2 shows the values of the sources and parameters used in the model. The circuit is based on (5.1) and (5.2). The lamp equivalent resistance is implemented by the voltage-controlled current source G1, by dividing lamp voltage by lamp resistance as given by (5.2), where the lamp resistance depends on the lamp RMS current. Thus, in order to determine the lamp RMS current, the instantaneous lamp current is measured by resistance *R_m*, then it is squared by the controlled voltage source E1, and filtered by the first-order filter *R_t* and *C_t*. Finally, the square root of the filtered voltage is calculated by the controlled voltage source E2, providing the RMS value of the lamp current, which is used as an input for calculating the lamp resistance in G1 expression.

Ben-Yaakov *et al.* demonstrated in Reference 8 that the presented lamp model exhibits a dynamic behaviour equivalent to that defined by the incremental impedance in (5.3). The parameters of the incremental impedance for this model are shown in Table 5.3, where R_{LA0} represents the static value of the lamp resistance at

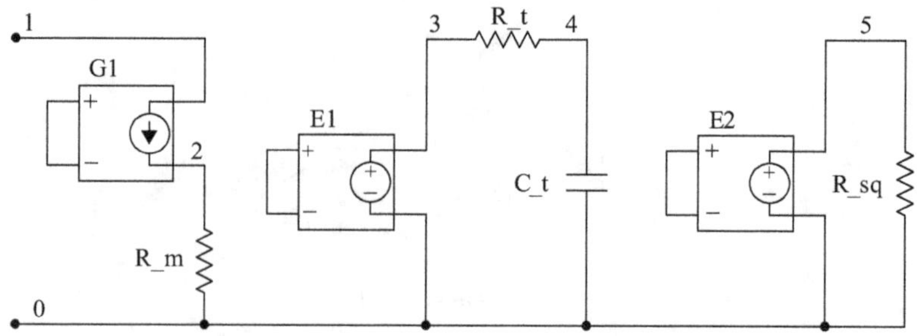

Figure 5.6 SPICE-based lamp model

Table 5.2 Values of the SPICE-based lamp model sources and parameters

Parameter	Value
G1	$\dfrac{V(1)}{\dfrac{v_S}{V(2)} - r_S}$
E1	$V(2) \times V(2)$
E2	$\text{sqrt}[V(4)]$
R_m	1
R_t	1
C_t	$\dfrac{1}{\omega_p}$
R_sq	100

Table 5.3 Incremental impedance parameters of the SPICE-based lamp model

Parameter	Value
k	R_{LA0}
ω_z	$-\dfrac{r_S}{R_{LA0}} \dfrac{1}{R_t C_t}$
ω_p	$\dfrac{1}{R_t C_t}$

each operating point. A methodology to obtain the lamp model parameters based on a step transient can be found in Reference 9.

5.2 Electronic ballasts for HID lamps

5.2.1 AC-operated electronic ballasts

Figure 5.7 illustrates the general block diagram of an AC-operated LFSW HID lamp ballast. As can be seen, HID lamp ballasts are complex converters that comprise several stages: EMI filter, rectifier, power factor correction (PFC) stage, DC-DC converter, low-frequency inverter, igniter and control circuitry. In the following, some comments on each stage are provided:

- EMI filter and rectifier: The EMI filter is essential to avoid electromagnetic interference to be transmitted through the power line and satisfy standard requirements such as EN55015 (CISPR15) [10]. The EMI filter must deal with both common mode and differential mode noise, and reduce noise levels below

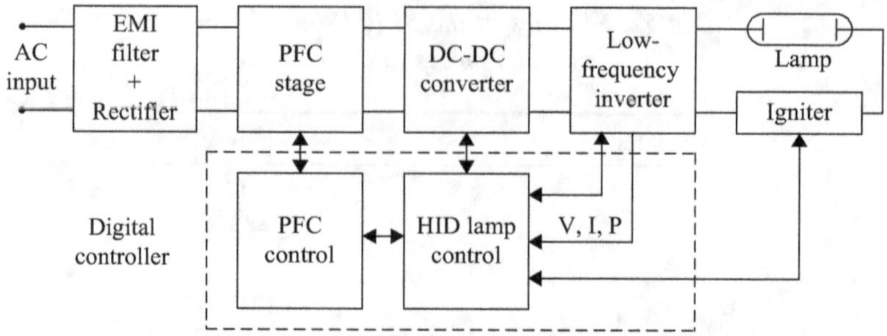

Figure 5.7 Electric diagram of an LFSW HID lamp ballast

the limits required by the standard. Usually, a second-order filter is enough, but in some cases, a fourth-order filter may be necessary [11]. A bridge rectifier is usually employed at the output of the EMI filter to rectify the AC line voltage and provide the input voltage for the PFC stage.

- PFC stage: This stage is used to assure a sinusoidal current being drawn from the line. Standards like EN61000-3-2 and Energy Star require that the current absorbed by the ballast must assure high input power factor and low total harmonic distortion. One of the most commonly used topologies for this stage is the boost converter operating either in critical conduction mode (CRM) for low power ballasts or in continuous conduction mode (CCM) for higher power levels. The typical output voltage when using a boost converter is around 380 V DC for 230 V_{rms} line voltage, which is fed to the next stage. Nevertheless, other topologies such as buck-boost or flyback converters with different output voltage levels can also be used.
- DC-DC converter: The function of this converter is to regulate the current injected to the lamp so that stable operation in all operating modes is assured. In order to assure a proper warming-up phase, this stage must maintain a constant current through the discharge after ignition. Once the maximum lamp power is reached, both voltage and current are monitored and lamp current adjusted so that constant lamp power is assured. A typical topology employed for this stage is a buck converter [5].
- Low-frequency inverter: This stage is used to invert the polarity of the DC-DC converter output so that the lamp is fed with an LFSW. The typical topology is a full bridge, implemented with four transistors. In some particular DC-DC converter topologies, such as the two-output flyback converter, a half-bridge inverter implemented with only two switches can be used, which decreases complexity and increases efficiency [12].
- Igniter: The igniter is an essential component that provides the initial high voltage necessary to start the discharge. Usually, it is placed in series with the lamp and employs a small high-voltage transformer, which is supplied on the primary side by a capacitor controlled by a SIDAC or spark-gap (SG) switch. It continuously generates high-voltage spikes until the lamp is ignited. At this moment, the igniter is deactivated by short-circuiting the capacitor, which prevents it from charging again.

Table 5.4 Summary of protections that an HID lamp ballast should incorporate

Fault condition	Recommended ballast action
Line blackout/brownout	Switch off ballast output. Re-ignite the lamp when the line is recovered
Open circuit at ballast output/no lamp	Switch off ballast output if the lamp fails to ignite within a maximum time period
Short circuit at ballast output	Switch off ballast output. Re-ignite the lamp when fault disappear
Lamp end of life/unstable operation	Switch off ballast output. Remain off until next ballast energizing

- Control circuitry: The control circuitry must generate the required signals to command all the power stages, including PFC stage, DC-DC converter, low-frequency inverter and igniter. It must generate the duty cycle and frequency patterns for the DC-DC converter depending on lamp voltage, current and power levels. The control circuitry must also implement all the required protections for the ballast, a summary of which have been gathered in Table 5.4 [5]. As can be understood, a rather complex control is necessary, which justifies the use of digital controllers, such as microcontrollers and digital signal processors.

5.2.2 DC-operated electronic ballasts

DC-operated electronic ballasts for HID lamps are usually employed in automotive applications, outdoor/indoor emergency lighting systems and renewable-energy-based lighting applications. Basically, a DC-operated electronic ballast employs the same stages as an AC-operated one, except for the rectifier and PFC stage, which are not necessary in a DC application. In some cases, the PFC stage is substituted by an upstream DC-DC converter, which performs an adaptation between the DC voltage source and the input of the DC-DC converter that supplies the current to the lamp. The remainder stages have the same features and functionality as for an AC-operated ballast.

5.3 Digital control applied to electronic ballasts

5.3.1 General control strategy applied to HID lamps

In order to attempt the complex characteristics of HID lamp operating stages, a digital control solution is very desirable, as it may incorporate the control strategies of all power stages in a single digital component. Otherwise, it would be difficult to implement all functionalities with analogue circuits or even using dedicated integrated circuits.

In addition, digitally controlled circuits have the advantages of reliability, noise immunity and flexibility [13]. They may also be simultaneously used for dimming [14, 15] and communication [16].

The typical HID electronic ballast is presented in Figure 5.8, in order to describe the general control strategy applied to HID lamps. This circuit is

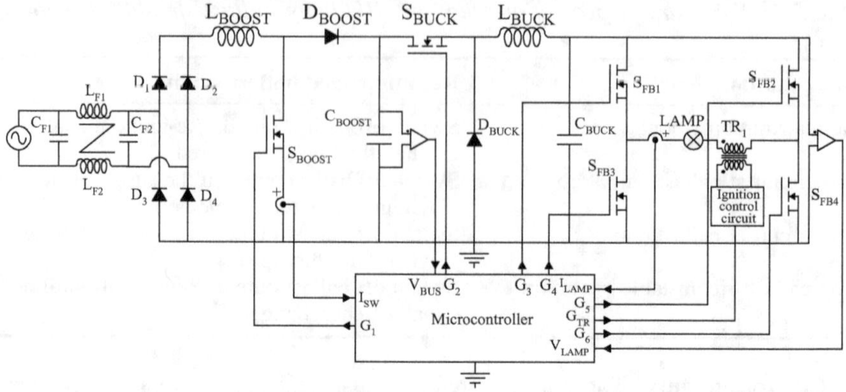

Figure 5.8 Typical HID lamp electronic ballast

composed of EMI filter, rectifier, a boost converter working in CRM as PFC stage, a buck converter as DC-DC converter, a full-bridge converter as low-frequency inverter, igniter and control circuitry. The digital control strategy is applied to the PFC stage, DC-DC converter, low-frequency inverter, igniter and protections. The details of this operation are presented in next subsections.

5.3.2 PFC converter

The most usual control strategy for the boost converter working as PFC stage is based on CRM operation, which means that the converter will be operating at the boundary between CCM and discontinuous conduction modes (DCM). In these operating conditions, the switch is turned on when the inductor current reaches zero, thus avoiding diode reverse recovery losses. Moreover, CRM operation retains the DCM advantages while maintaining a line current without dead time intervals, which reduces harmonic content and simplifies filtering [17, 18].

The CRM control is conventionally based on dedicated circuits. However, it can also be performed by the microcontroller by controlling the switch S_{BOOST} by the gate control signal G_1. The microcontroller reads the S_{BOOST} current by I_{SW} signal and the bus voltage by V_{BUS} signal, in order to perform the PFC function. Sensing signals I_{SW} and V_{BUS} can easily be generated by using a shunt resistor and a resistive voltage divider, respectively.

5.3.3 DC-DC converter

The control and design of the buck converter must take into account the lamp model $Z_{LA}(s)$ as shown in (5.3). Figure 5.9 represents the dynamical equivalent circuit of the buck converter that must be taken into account in order to design the control algorithm. The buck converter is modelled as a current source, i, in parallel with the output capacitance, C_{BUCK}, and the lamp impedance, Z_{LA}. The PFC converter and the full-bridge inverter can be neglected in this analysis. The output capacitance, C_{BUCK}, must be high enough to limit the lamp current high-frequency ripple below 5%,

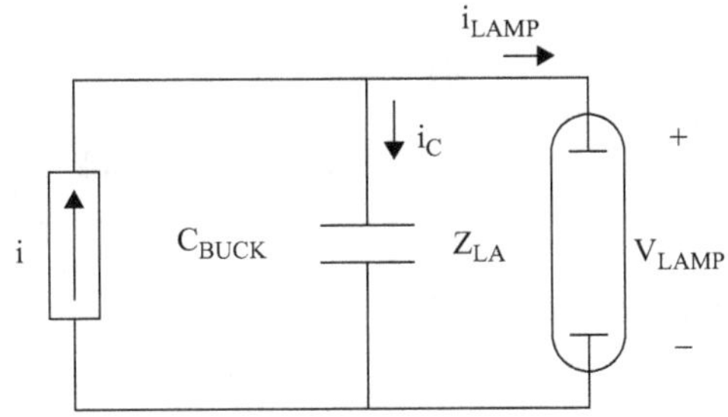

Figure 5.9 Equivalent circuit of the buck converter

in order to prevent the generation of acoustic resonances [19], and low enough so that an unstable behaviour due to the lamp incremental impedance can be avoided [20]. Therefore, the digital control loop must be designed considering this characteristic, which will regulate the lamp power and current by means of G_2 gate control signal. The lamp current control during the warm-up phase and the lamp power control during the warm-up and running phases, according to Section 5.1.2, are performed by sensing the lamp current, I_{LAMP}, and lamp voltage, V_{LAMP}, as shown in Figure 5.8. The lamp current can be sensed by a hall effect sensor or simply by a ground-referred resistor in series with switches S_{FB3} and/or S_{FB4}.

5.3.4 Low-frequency inverter

The low-frequency inverter is responsible of changing the lamp current polarity, in order to avoid the cataphoresis phenomenon. In some cases, the ignition and warm-up phases are performed in DC, which means that the digital control circuit must keep switches S_{FB1} and S_{FB4}, or S_{FB2} and S_{FB3}, closed during this interval. The control of the low-frequency full-bridge inverter in running phase is performed in open loop by the microcontroller through G_3–G_6 gate control signals.

5.3.5 Igniter

There are two types of igniters for HID lamps: cold and hot lamp igniters. Igniters for cold lamps apply a 3–4 kV in order to start the discharge, while igniters for hot lamps must apply near 30 kV to ignite the lamp. General lighting HID electronic ballasts usually are the cold lamp type, while headlamp automotive HID ballasts must be the hot lamp type. Regarding the cold lamp type, the ignition process can be performed in several ways, which will determine the action of the digital control, as explained below:

- Uncontrolled SG or SIDAC: the buck converter output voltage can be used to charge an RC circuit that switches an SG or a SIDAC element, which applies the starting voltage to the lamp through a step-up transformer (TR_1) [21]. In this case, the microcontroller does not control the starting process, but it has to detect when the lamp turns on in order to start the warm-up phase. The turn-on detection can be implemented by measuring I_{LAMP} and/or V_{LAMP}.

- Controlled SG or SIDAC: This method is very similar to the previous one; the only difference is that the microcontroller disables the starting circuit after the lamp has been ignited [5].
- Resonant ignition: This method uses a resonant tank in order to apply the starting voltage to the lamp. It is performed by changing the switching frequency of the full-bridge inverter to the resonant frequency of the resonant tank during the ignition phase [22].

5.3.6 Protections

The digital control applied to HID lamp electronic ballasts provides the flexibility of implementing protections for the correct behaviour of the circuit. Some of these protections are described below:

- Overcurrent protections: All the switches of the circuit shown in Figure 5.8 have an inherent protection against overcurrent. Switch S_{BOOST} current is sensed by I_{SW} measuring, while switches S_{BUCK} and S_{FB1}–S_{FB4} currents are sensed by I_{LAMP} measuring.
- No-load protection: If the lamp turns off during operation the control circuit must disable G_1 and G_2 gate control signals, in order to protect the circuit.
- Hot-lamp protection: For general lighting HID electronic ballasts that use cold lamp igniters, a hot-lamp protection must be implemented. When the hot-lamp is detected, the microcontroller must wait a pre-defined time before starting again the ignition phase. This time interval depends on the lamp model and can take from a few minutes to tens of minutes.
- Lamp end-of-life protection: When the lamp reaches the end of its life, its operation can become unstable. The protection circuit must detect this malfunction by sensing the lamp voltage and current, and turning off the circuit to prevent any damage.

5.4 Practical example

This section is devoted to present a real example of an HID lamp electronic ballast with digital control.

5.4.1 HID lamp ballast

The typical HID lamp electronic ballast, presented in Figure 5.8, is a very complex and expensive solution, because of the high number of controlled switches. One solution studied more in the literature is the use of integrated converters, in order to obtain a less expensive and less complex circuit [12, 21, 23–26]. Figure 5.10 shows the ballast topology presented as an example in this chapter, which was firstly introduced in Reference 21. This circuit is obtained by integrating a buck converter that provides PFC and a flyback converter that controls the lamp power. The flyback inductor includes two secondary windings that switch complementarily at low frequency in order to supply the lamp with an LFSW current. Both converters operate

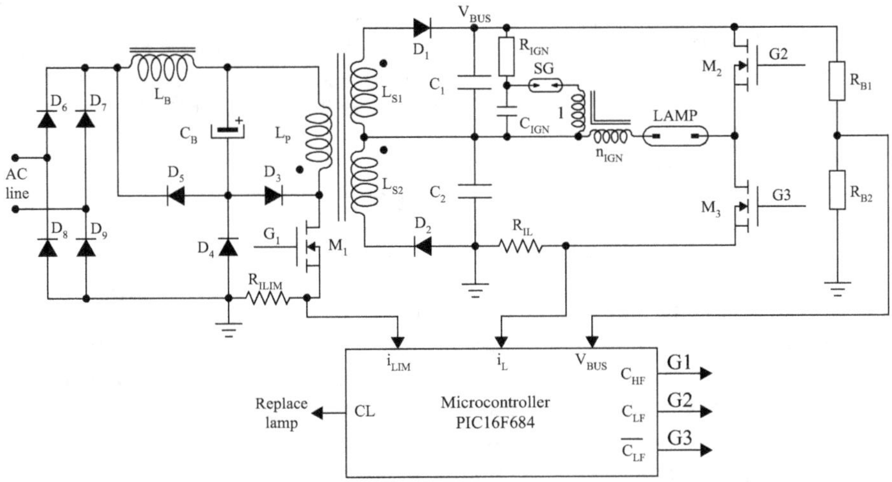

Figure 5.10 HID electronic ballast

Table 5.5 HID electronic ballast parameters

Parameter	**Value**
Mains	230 V_{rms}, 50 Hz
D_1–D_9	MUR160
M_1–M_3	BUK456
L_B	434 μH on core EF25
C_B	220 μF/200 V
L_P–L_{S1}–L_{S2}	180–315–315 μH on core EF25
C_1, C_2	220 nF/500 V
Lamp	Philips CDM-T 35 W
V_L, I_L, R_L	90 V, 0.42 A, 215 Ω
f_S, T_S	85 kHz, 11.75 μs
f_{LF}	100

in DCM and share the same switch (M_1). Capacitors C_1 and C_2 are employed to filter the AC component of the lamp current, which results in the DCM operation of the flyback converter. Additional characteristics of this circuit are detailed in Reference 21. The parameters of the HID electronic ballast are presented in Table 5.5.

5.4.2 Microcontroller PIC16F684

In this application, the PIC16F684 microcontroller from microchip has been selected owing to its good features and low cost. The PIC16F684 is an 8-bit microcontroller with an RISC microprocessor and Harvard architecture [28]. In Harvard architecture, program memory and data memory are physically separated and have their own buses to communicate with the microprocessor. This allows the microprocessor to access program and data memory simultaneously. Thus, while the current instruction

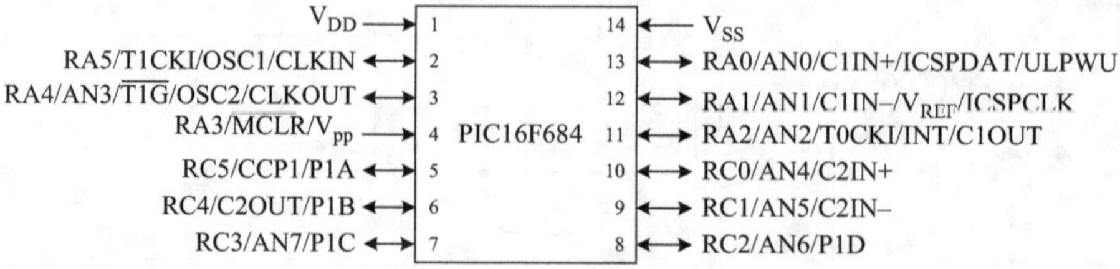

Figure 5.11 Pin diagram of the PIC16F684

is being executed, the next instruction is being fetched simultaneously from the program memory, thus making Harvard microcontrollers very efficient [28]. Additionally, one of the advantages of this microcontroller is its low cost; as per 2015 budget, the cost of this microcontroller for 100 units was 1.10€.

More specifically, the PIC16F684 is a very compact microcontroller, which features a 14-pin package, as illustrated in Figure 5.11. The main features of this microcontroller are the following:

- Operation up to 20 MHz
- Wide voltage range: 2.0–5.0 V
- 2 kB program memory, 128 B RAM, 256 B EEPROM
- 12 general application digital I/O
- 10-bit A/D with 8 channels
- 8/16 bit and watchdog timers
- 10-bit PWM module
- 2 analogue comparators

5.4.3 Control strategy

This section describes the control strategy performed by the microcontroller-based circuit, which is presented in Reference 29. Due to the requirements imposed by the lamp operation, the PIC16F684 microcontroller from microchip was chosen. The starting circuit is based on an SG that applies the starting pulses when the lamp is off. Capacitor C_{ign} is charged through R_{ign} until the SG breakdown voltage (250–300 V); then, this voltage is applied across the primary winding of the transformer and is raised according to its turn ratio, $n_{ign} = 15$. The required measurements were implemented by means of resistive sensors: the bus voltage, V_{BUS}, through a voltage divider formed by R_{B1} and R_{B2}, which corresponds to twice the lamp voltage; the lamp current, i_L, through a series resistance R_{il}; and the main switch current by the R_{ilim} series resistance. Therefore, since all measurements are ground referenced, no isolation circuit is needed in this application. Switches M_2 and M_3 are driven through a conventional half-bridge driver, IR2111, controlled by a microcontroller output port. Next subsections describe systematically the control strategy based on the flow chart presented in Figure 5.12.

5.4.4 Lamp starting

Before the lamp starting, the flyback converter operates with no load, and the output voltage, V_{BUS} ($V_{BUS} = V_{C1} + V_{C2}$) starts to rise. Here, the microcontroller

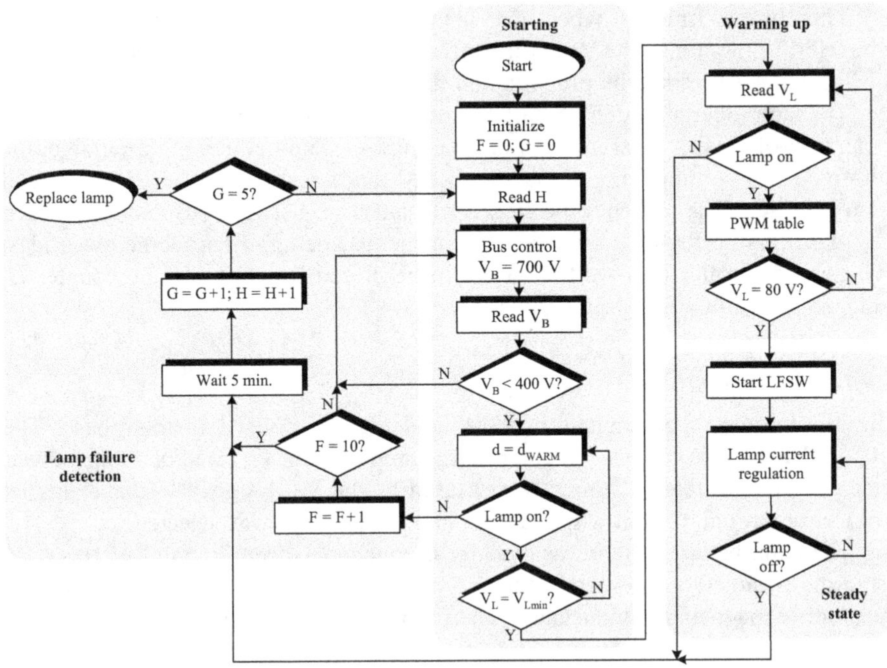

Figure 5.12 Flow chart of the control algorithm

controls the duty cycle of M_1 (*d*) in order to limit V_{BUS} below the breakdown voltage of switches M_2 and M_3 (800 V). In this stage, M_2 is closed and M_3 is open. Eventually, the SG circuit applies the starting pulse to the lamp and, if the lamp starts, V_{BUS} will fall quickly. When the microcontroller detects this voltage drop, it increases the duty cycle in order to maintain the discharge, and to warm up the lamp as fast as possible. If the starting process would fail and V_{BUS} increase again, the microcontroller would return *d* to the minimum value, and a fail counter (*F*) would be increased. It has been defined that when this fail counter reaches a value of 10, the driver must wait 5 minutes before restarting the process in order to wait for lamp cooling. Then, another counter (*G*) is increased to record this event. Should *G* reach a value of 5, the starting process would stop and an LED would indicate the necessity of a lamp replacement. These values were obtained experimentally from the typical operation of HID lamp ballasts, since there is no standard applicable nowadays. Nevertheless, since all the control is performed by a microcontroller, it would be very easy to adjust these values to any other requirement.

5.4.5 *Warm-up process*

Once the lamp has passed through the glow-to-arc transition, the lamp voltage reaches its minimum value (around 20 V). In this state, the microcontroller must control the duty cycle *d* in order to keep the DCM operation of the flyback converter.

This phase finishes when the lamp voltage reaches its nominal value (above 80 V). Again, if the lamp turns off during this phase, the driver would wait for 5 minutes to restart the process, and the counter *G* should be increased.

The lamp operation during starting and warm-up process is very unstable, and it can turn-off easily. Therefore, in order to turn these processes more reliable, the low-frequency commutation of the switches M_2 and M_3 is disabled, and only one of them is closed. The switch to be closed will alternate between M_2 and M_3 at each new starting, in order to guarantee a homogeneous ageing of both lamp electrodes. The microcontroller chooses the switch to be closed by verifying the counter *H*, which is stored in a non-volatile memory.

5.4.6 Steady state

The low-frequency commutation of M_2 and M_3 is activated in this phase. The proposed hardware allows for controlling lamp current (i_L) and/or lamp power, once the lamp voltage is indirectly measured by the V_{BUS} signal. In this example, a power control algorithm with the reference set at 35 W was selected.

The lamp-ballast system was dynamically modelled according to Reference 20, using the parameters shown in Table 5.5 and considering the effect of the igniter transformer magnetizing inductance. The following transfer function was obtained:

$$G(s) = \frac{I(s)}{D(s)} = \frac{2.454s + 26240}{9.9 \cdot 10^{-9}s^3 + 1.387 \cdot 10^{-4}s^2 + 1.612s + 9654} \quad (5.4)$$

which relates the lamp current with the switch duty cycle. The system block diagram is shown in Figure 5.13, where $G_R(z)$ is the transfer function of the digital controller, $G(s)$ represents the transfer function of the ballast-lamp system, and $H(s)$ corresponds to the feedback gain of the lamp current sensor.

The lamp current measure sampling time was chosen to be 10 ms, which corresponds to the LFSW period. Therefore, the lamp current can be measured by only one sensor resistance in series with M_3. Besides, a response time of tens of milliseconds is acceptable for this application. The proposed PI controller has only

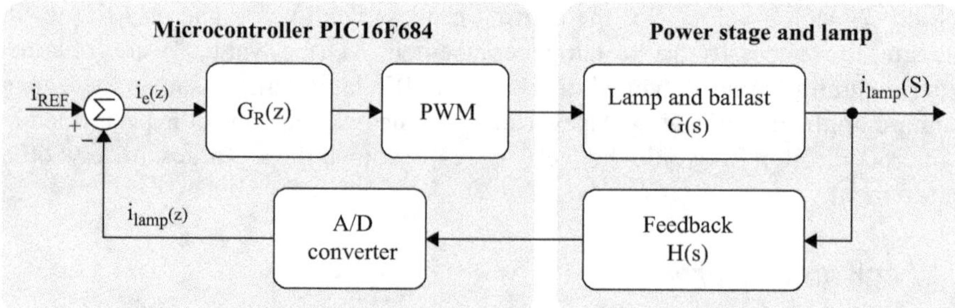

Figure 5.13 System block diagram of the circuit

a pole in the origin, and the discrete differential equation is derived using the zero-order hold discretization method, resulting in:

$$D_k = D_{k-1} + \frac{1}{8}\varepsilon_{k-1} \tag{5.5}$$

5.4.7 Protections

Besides the basic control functions described previously, the microcontroller must offer some crucial protections to ensure a reliable behaviour of the ballast, which are described below.

- Protection against no load: If the lamp turns off during any of the operating phases of the ballast, the output voltage will start to rise, reaching hazardous values. Therefore, two different protections were implemented, during the measuring routines of V_{BUS} and i_L, if the microcontroller detects the BUS voltage surpassing 750 V or if the lamp current is lower than a minimum value, the PWM signal is switched off, and the control will wait for 5 minutes before trying a new starting process.
- Over-current protections: MOSFETs M_1, M_2 and M_3 must be protected against overcurrent. The low-frequency switches, M_2 and M_3, have already an inherent protection against overcurrent, because of the lamp current measurement, i_L, once these switches are in series with the lamp. The overcurrent protection of M_1 is implemented by means of an interrupt request. If some of these protections are activated, the PWM signal is turned off, and the control waits for 5 minutes before trying a new starting process.

5.4.8 Experimental results

This section presents the experimental results of the example circuit in order to illustrate the control strategy implemented by the microcontroller-based digital control. Figure 5.14 shows the lamp voltage and current within an LFSW period. As can be seen, the lamp is operating close to its nominal operating point.

Figure 5.15 shows the line voltage and current. It can be seen that there appears a dead time in the line current, which is produced by the operation of the buck converter as PFC stage. Nevertheless, it can be proved that the attained current waveform can satisfy the harmonic standard requirements [29].

Figure 5.16 shows the lamp voltage and current during the start-up process. During the warm-up phase the duty cycle is modified according to the look-up table, so that the lamp current is adjusted accordingly to the lamp voltage level. During the warm-up phase, the lamp current is maintained at a safe value, not higher than the lamp nominal value, thus avoiding premature lamp ageing. In this way, as can be seen in Figure 5.16, the lamp reaches steady state in 2 minutes and 40 seconds approximately. This is detected by the lamp voltage reaching the predefined value of 80 V. At this moment, the microcontroller starts the LFSW operation.

Figure 5.17 shows the closed loop response of the lamp current against an input voltage perturbation. It can be seen how the lamp current is regulated in about 80 ms, which is the expected time response of the designed digital regulator.

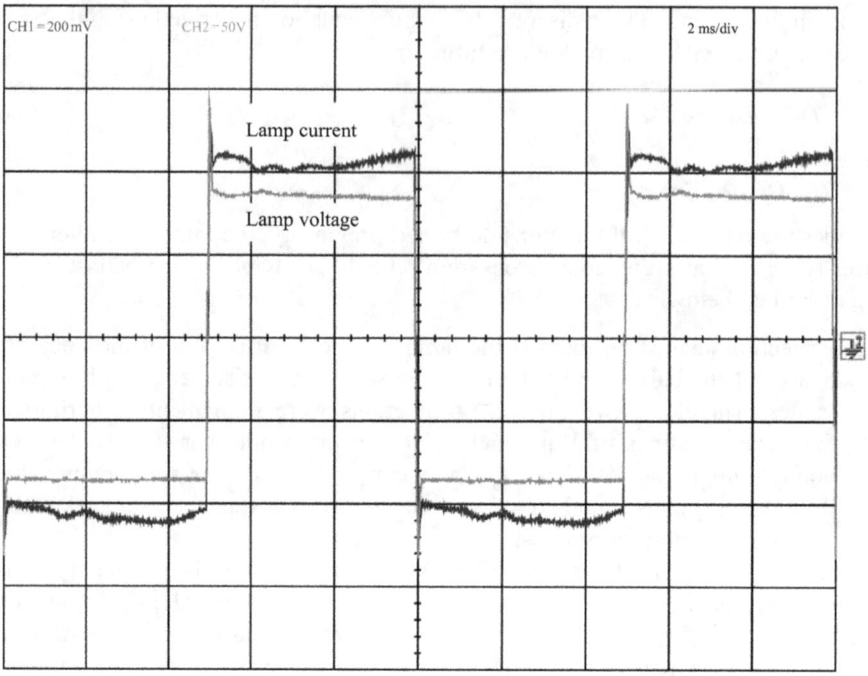

Figure 5.14 Lamp voltage and current waveforms (50 V/div, 200 mA/div; 2 ms/div)

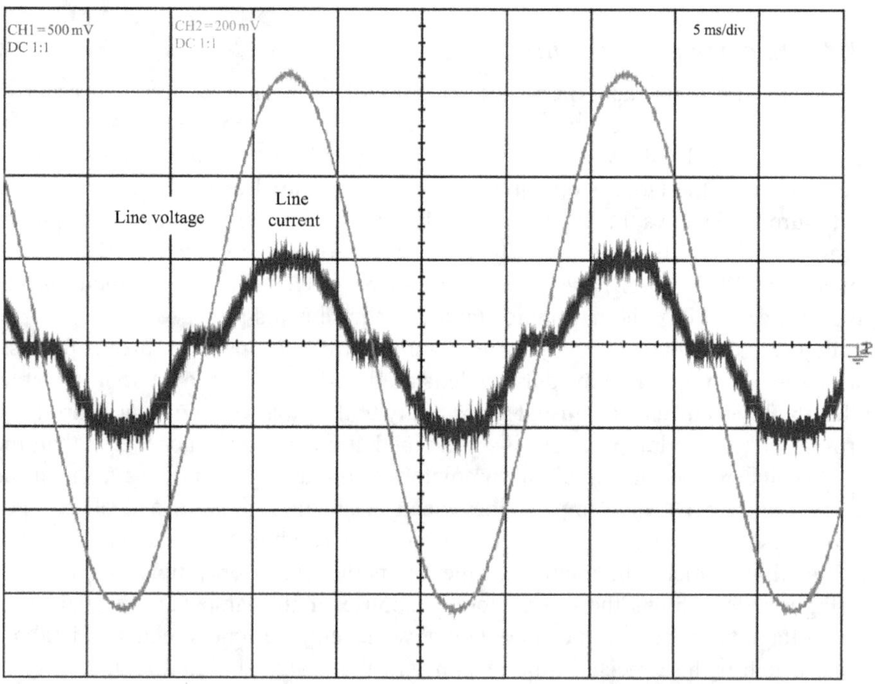

Figure 5.15 Line voltage and current (100 V/div, 0.5 A/div, 5 ms/div)

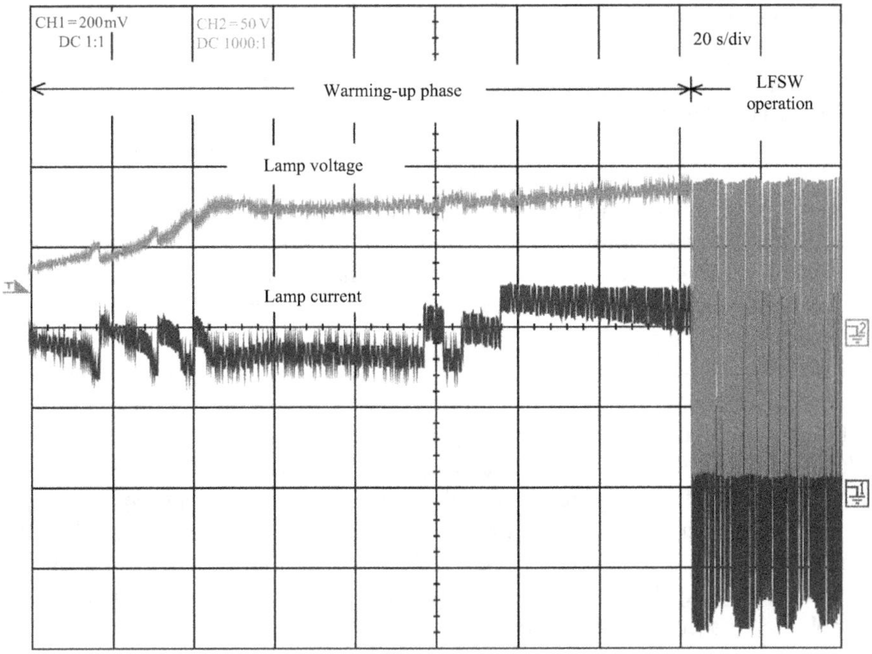

Figure 5.16 Lamp voltage and current during the start-up process (50 V/div, 0.2 A/div, 20 s/div)

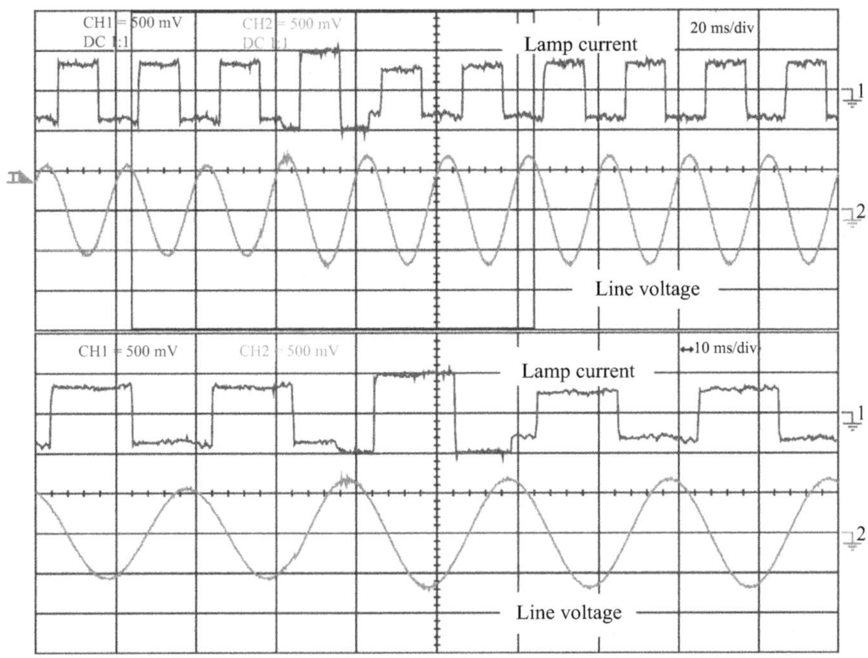

Figure 5.17 Lamp current and line voltage during a transient response (0.5 A/div, 250 V/div). Top: 20 ms/div, Bottom: zoom at 10 ms/div

5.5 Summary

In this chapter, the application of digital control to HID electronic ballasts has been presented. HID lamps are complex systems that require a good knowledge on its functionality and operating requirements before designing and implementing a power converter for its supplying. Therefore, the first part of the chapter was devoted to general issues related to HID lamp supplying. The lamp-operating principles and modelling were presented at the beginning of the chapter. The different operation stages of the lamp, ignition, warm-up phase and normal running mode, which represent a fundamental concept in the design of the lamp ballast, were exposed and explained. A lamp model that can be used for computer simulation, which includes both static and dynamic behaviour, was also presented. In addition, the main stages of general AC- and DC-operated electronic ballasts were introduced. The second part of the chapter focused on particular topics related to digital control of HID lamp ballasts. The typical structure based on boost PFC converter, buck DC-DC converter and full bridge inverter was presented and explained from the point of view of digital control. Issues related to ignition and protection were described from a practical point of view. Finally, in order to provide a practical purpose to this chapter, an example HID lamp ballast implementation was introduced and expounded. The example was based on the use of an integrated power converter that allows for the minimization of the number of controlled switches, and also for simplifying the number of tasks and algorithms to be performed by the digital processor. Details on lamp ignition, warm-up process, steady state and protections related to this ballast example were provided. Finally, some experimental results that help to understand the operation of the complete ballast were included.

References

[1] IES Lighting Handbook, 1984 Reference Volume. Illuminating Engineering Society of North America. IESNA, New York, 1984.

[2] De Groot J., Van Vliet J. The High-Pressure Sodium Lamp. *Philips Technical Library*, Macmillan Education, London, UK, 1986.

[3] HID Ballast Application Guide. General Electric. 2012.

[4] Metal Halide Lamps. Instructions for the Use and Applications. Osram, Munich, Germany, 2007.

[5] Ribarich T. 'Shedding Light on HID Ballast Control'. *Power Electronics Technology*. Penton: www.powerlectronics.com, 2006, 34–37.

[6] Alonso J. M. 'Electronic Ballasts', Chapter 22. In: *Power Electronics Handbook* (Third Edition), Edited by M. H. Rashid, Elsevier, Massachusetts, USA, 2011.

[7] Den E., Cuk S. 'Negative incremental impedance and stability of fluorescent lamps', Proceedings of Applied Power Electronics Conference, Vancouver, BC, Canada, 1997, 1050–1056.

[8] Ben-Yaakov S., Shvartsas M., Glozman S. 'Statics and dynamics of fluorescent lamps operating at high frequency: modeling and simulation', *IEEE Transactions on Industry Applications*. 2002, 38(6), 1486–1492.

[9] Alonso J. M, Dalla-Costa M. A., Cardesin J., Martin-Ramos, J. A. Garcia-Garcia, J. 'Small-signal modeling of discharge lamps through step response and its application to low-frequency square-waveform electronic ballasts', *IEEE Transactions on Power Electronics*. 2007, 22(3), 744–752.

[10] EN 55015. Limits and methods of measurement of radio disturbance characteristics of electrical lighting and similar equipment. European Standard, 2007.

[11] Input EMI Filter Design for Offline Phase-Dimmable LED Power Supplies. Texas Instruments Power Supply Design Seminar, 2012.

[12] Dalla Costa M. A., Alonso, J. M., Marchesan, T. B., Cervi, M., Prado, R. N., 'Electronic Ballasts for HID Lamps', *IEEE Industry Applications Magazine*. 2011, 17(2), 54–59.

[13] Buccella C., Cecati C., Latafat H., 'Digital control of power converters—a survey', *IEEE Transactions on Industrial Informatics*, 2012, 8(3).

[14] Zhao Y., Sun Z., 'Development micro-controller-based electronic ballast for HPS lamps with dimming control function', IPEMC Conference, 2006, 1747–1750.

[15] Stankovic A. V., Nerone L., Kulkarni P., 'Modified synchronous-buck converter for a dimmable HID electronic ballast', *IEEE Transactions on Industrial Electronics*, 2012, 59(4), 1815–1824.

[16] Barriquelo C. H., Denardin G. W., Pinto R. A., Silva M. F., Campos, A., Prado R. N. 'An intelligent system for street lighting control and measurement', Industry Applications Society Annual Meeting, 2009.

[17] Alonso J. M., Ribas J., Calleja A. J. López E., Rico M. 'An intelligent neuron-chip-based fluorescent lamp ballast for indoor applications,' *Journal of the European Power Electronics Association*, 2000, 10(3–4), 21–27.

[18] Alonso J. M., Ribas J., Coz J. J. D., Calleja A. J., Corominas E. L., Rico-Secades M., 'Development of a distributive control scheme for fluorescent lighting based on LonWorks technology,' *IEEE Transactions on Industrial Electronics*, 2000, 47(6), 1253–1262.

[19] Dalla Costa M. A., Alonso J. M., García J., Cardesín J., Rico-Secades M. 'Acoustic resonance characterization of low-wattage metal halide lamps under low-frequency square-waveform operation', *IEEE Transactions on Power Electronics*, 2007, 22(3), 735–743.

[20] Dalla Costa M. A., Kirsten A., Alonso J., Garcia J., Gacio D. 'Analysis, design and experimentation of a closed-loop metal halide lamp electronic ballast', *IEEE Transactions on Industry Applications*, 2012, 48(1), 28–36.

[21] Dalla Costa M. A., Alonso J. M., García J., Cardesin J., Lamar D. G. 'A single-stage high-power-factor electronic ballast based on integrated buck flyback converter to supply metal halide lamps', *IEEE Transactions on Industrial Electronics*, 2008, 55(3), 1112–1122.

[22] Diaz F. J., Azcondo F., Casanueva R., Brañas C., Zane R. 'Digital control of a low-frequency square-wave electronic ballast with resonant ignition', *IEEE Transactions on Industrial Electronics*, 2008, 55(9), 3180–3191.
[23] Shen M., Qian Z., Peng F. Z. 'Design of a two-stage low-frequency square-wave electronic ballast for HID lamps', *IEEE Transactions on Industry Applications*, 2003, 39(2), 424–430.
[24] Li H., Shen M., Qian Z. 'A novel low frequency electronic ballast for HID lamps', *Industry Applications Conference (IAS) Proceedings*, 2004, 1, 321–324.
[25] Có M. A., Brumatti M., Simonetti D. S. L., Vieira J. L. F. 'Single stage electronic ballast for HID Lamps', *Industry Applications Conference (IAS) Proceedings*, 2003, 1, 339–344.
[26] Yang Y., Qian Z., Wu X. 'A novel single-stage low-frequency square-wave electronic ballast for low-wattage HID lamps', *IEEE Applied Power Electronics Conference (APEC) Proceedings*, 2005, 1, 1048–1052.
[27] PIC16F684 Datasheet. Microchip. 2007.
[28] PICmicro Mid-Range MCU Family. Reference Manual. Microchip. 2004.
[29] Dalla Costa M. A., Alonso J. M., Garcia J., Kirsten A. L., Gacio D. 'Microcontroller-based high-power-factor electronic ballast to supply metal halide lamps', *IEEE Transactions on Industrial Electronics*, 2012, 59(4), 1779–1788.

Chapter 6

FPGA-based controllers for direct sliding mode control of PWM boost rectifiers

Mohamed Wissem Naouar[1] and Eric Monmasson[2]

6.1 Introduction

For industrial applications like AC adjustable speed drives, a DC voltage has to be provided. In most of the cases, a diode rectifier associated with a storage capacitor is used for this purpose. This is mainly due to the fact that it is a simple, robust and low-cost solution. However, the use of diode rectifiers causes the following disadvantages:

- High-level harmonic contents in the line currents because of their pulsating nature.
- Poor power factor.
- Unidirectional power flow.

To cope with these disadvantages, various power factor correction techniques based on active wave shaping of the line currents can be used. Among these techniques, the best solution consists in using a Pulse Width Modulation (PWM) rectifier. Its main advantages are:

- Low harmonic contents in the line currents.
- Bidirectional power flow.
- Control of power factor and possibility of working with unitary power factor.
- Control of the level of the dc-link voltage at a constant value, even if voltage sag or line voltage fluctuations occur.
- Possibility of adding active filtering functions [1], [2].
- Possibility of realizing dynamic reactive power compensation [1], [2].
- Interesting solution for high-speed motor drives applications since the dc-link voltage should always be higher than the peak line voltages.
- Interesting solution for extending the constant torque region of motor drive applications.

[1]Laboratoire de Systèmes Electriques, Ecole Nationale d'Ingénieurs de Tunis, Université de Tunis El Manar, Tunis, Tunisia
[2]Lab. Systems and Applications of Information Technology and Energy, University of Cergy-Pontoise, Cergy-Pontoise, France

Although PWM rectifiers have a more complex structure than diode rectifiers, it is expected that, as the cost of power electronics decreases and reduction of current harmonic contents are strictly enforced [3], their application will increase. It should be noted also that the use of PWM rectifiers requires the use of additional sensors in order to provide measurements for the used control algorithm.

First control algorithm implementations were realized based on analogue solutions, which were characterized by good accuracy and high bandwidth performances. However, they were sensitive to disturbances and thermal drift of analogue components. Nowadays, most of the control algorithm implementations are realized through digital solutions, which are more reliable and immunized from disturbances. Microcontrollers are the most used digital solutions for the digital control of power converters [4], [5]. In fact, they incorporate dedicated peripherals for this kind of control such as Analogue-to-Digital Converters (ADCs), Arithmetic Logic Units (ALUs), PWM modules, Timers [6]. Nevertheless, these digital solutions are characterized by a feedback loop delay and they are still limited for complex control algorithms. The hardware digital solutions such as Field Programmable Gate Arrays (FPGAs) can also be considered as an appropriate solution for the digital implementation of power converters control algorithms [7]. Indeed, they allow designing of dedicated hardware architectures, which run control algorithm tasks in only a few microseconds despite their complexity.

Having in mind the need to show the reported FPGA advantages, the authors present in this chapter the FPGA implementation of a *direct sliding mode control* for both single-phase and three-phase PWM rectifiers. The choice of the sliding mode theory was motivated by the fact that it offers a powerful analysis tool, which is well adapted for direct control of power converters. So, by means of this theory, the authors attempt to synthesize the current control loop for PWM rectifiers control algorithms. Also, the FPGA implementation of the synthesized algorithms is discussed.

6.2 Sliding mode control: theory and application for power converters control

The sliding mode theory is recognized as an efficient mathematical tool to design robust controllers for power converters. This paragraph presents the basic concepts, mathematical and design aspects of direct sliding mode control of power converters. To this purpose, let us consider the system given by (6.1), where x is the state vector (n dimensional), V is the control input vector (m dimensional), f is the system function and B is a $(n \times m)$ matrix.

$$\frac{dx}{dt} = f(x) + B(x)V \tag{6.1}$$

For the synthesis of the direct sliding mode control, a switching function $S(x)$ (m dimensional) given by (6.2) is first defined.

$$S(x) = [S_1(x) \dots S_m(x)]^t \tag{6.2}$$

where $S_k(x)_{(k=1\ \dots\ m)}$ is the kth switching function of $S(x)$.

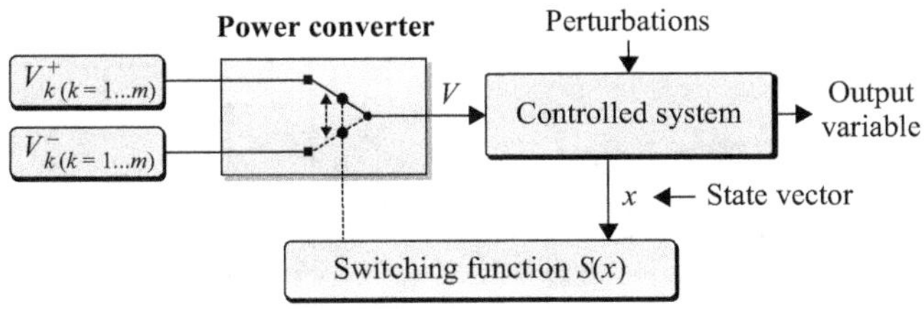

Figure 6.1 Direct sliding mode control of power converters

The set of points where the switching function is equal to zero ($S_k(x) = 0$, $k = 1 \ldots m$) is known as the *switching line* or *sliding surface*. The sliding mode control can then be achieved based on the defined switching function $S(x)$ so that the switching functions $S_k(x)_{(k=1\ldots m)}$ are directed toward their sliding surfaces. For this reason, the switching functions $S_k(x)_{(k=1\ldots m)}$ must verify conditions given by (6.3) [8], [9].

$$\begin{cases} \dot{S}_k(x) < 0 & if \quad S_k(x) > 0 \\ \dot{S}_k(x) > 0 & if \quad S_k(x) < 0 \end{cases} \quad (k = 1 \ldots m) \tag{6.3}$$

The sign of the time derivatives of switching functions depends on the applied control input vector $V = [V_1 \ \ldots \ V_{\mathrm{m}}]^t$. Consequently, in order to ensure conditions of (6.3), the control input vector is selected so that (6.4) is verified. Figure 6.1 summarizes the principle of the direct sliding mode control of power converters.

$$V_k = \begin{cases} V_k^+ & if \quad S_k(x) > 0 \\ V_k^- & if \quad S_k(x) < 0 \end{cases} \quad (k = 1 \ldots m) \tag{6.4}$$

6.3 Direct sliding mode control for single-phase PWM rectifier

6.3.1 Single-phase PWM rectifier model

Figure 6.2 shows a controlled single-phase PWM rectifier connected to the grid through an L filter. In this figure, u is the switching signal, L (respectively R) is the filter inductor (respectively the filter resistor), V_L is the drop voltage on the L filter, V_{dc} is the dc-link voltage, V_{dc}^* is the reference dc-link voltage, V_{conv} is the AC side converter voltage, i_{conv} is the AC side converter current, i_{dc} is the DC side converter current, V_g is the grid voltage, i_g is the grid current, C is the dc-link capacitor, i_c is the dc-link capacitor current and i_{Load} is the load current.

Based on Figure 6.2, the following equation can be deduced

$$V_{conv} = V_g - L\frac{di_g}{dt} - Ri_g \tag{6.5}$$

In the complex plane, (6.5) becomes

$$\underline{V}_{conv} = \underline{V}_g - jL\omega\underline{i}_g - R\underline{i}_g \tag{6.6}$$

where ω is the electrical angular frequency of the grid voltage.

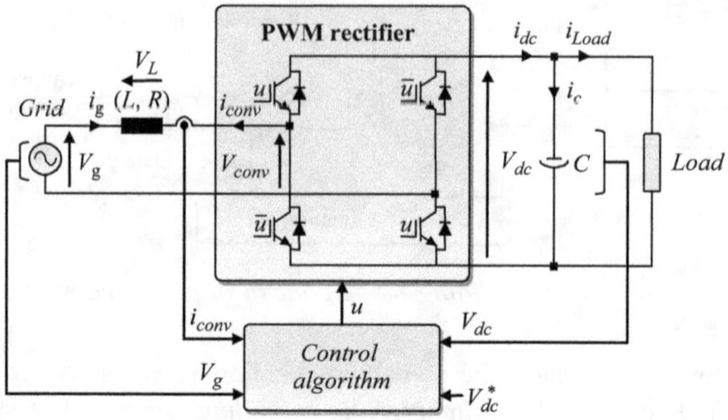

Figure 6.2 Controlled single-phase PWM rectifier

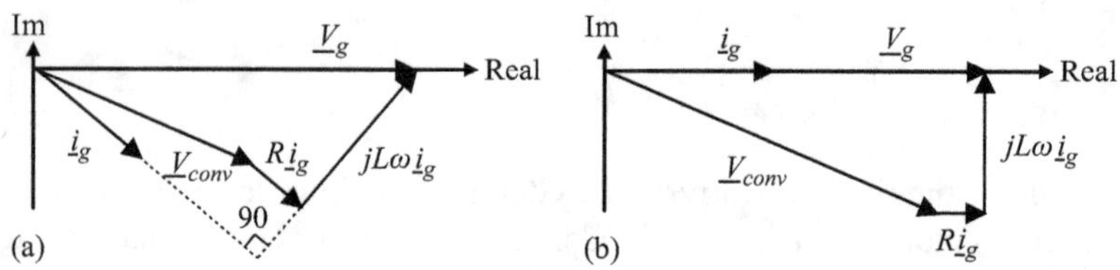

Figure 6.3 Phasor diagrams (a) general case (b) unitary power factor operation

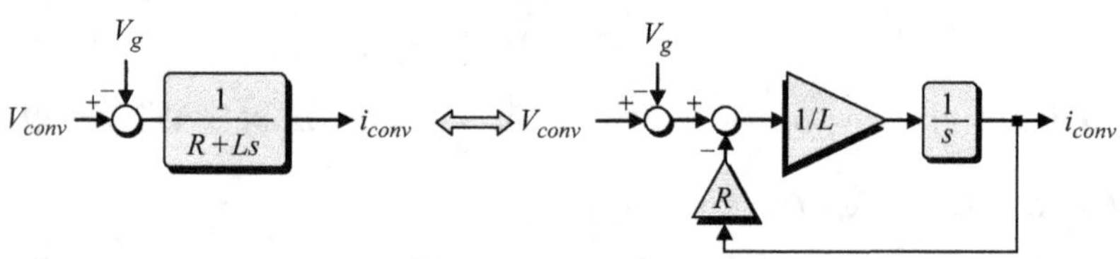

Figure 6.4 L filter model

According to (6.6) and based on the space vector representation of the grid current $\underline{i}_g$, grid voltage $\underline{V}_g$ and power converter voltage $\underline{V}_{conv}$ (in the complex plane), the phasor diagrams of Figure 6.3 are deduced. The one presented on Figure 6.3(b) is obtained when the single-phase PWM rectifier ensures unitary power factor operation.

Based on (6.5), the L filter model used to interface between the grid and the power converter of a single-phase PWM rectifier is given by (6.7) and Figure 6.4.

$$i_{conv} = \frac{1}{R + Ls}(V_{conv} - V_g) = \frac{1}{Ls}(V_{conv} - V_g - Ri_{conv}) \tag{6.7}$$

6.3.2 Steady-state operation limits

For unitary power factor operation and based on the phasor diagram of Figure 6.3(b), the magnitude V_{convm} of the converter voltage V_{conv} is expressed as in (6.8), where V_{gm} and I_{gm} are the magnitudes of the grid voltage and current, respectively.

$$V_{convm} = \sqrt{(V_{gm} - RI_{gm})^2 + (L\omega I_{gm})^2} \approx \sqrt{V_{gm}^2 + (L\omega I_{gm})^2} \tag{6.8}$$

According to Figure 6.1, the converter voltage V_{conv} can be expressed according to the logical level of the switching signal $u(t)$ and to the dc-link voltage V_{dc} as in (6.9).

$$V_{conv}(t) = V_{dc}(2u(t) - 1) \tag{6.9}$$

Since the instantaneous values of the voltage V_{conv} can be either $+V_{dc}$ or $-V_{dc}$ (according to the state of the switching control signal u that can take values 1 or 0), and according to (6.9), the V_{dc} voltage must verify the following equation:

$$V_{dc} \geq \sqrt{V_{gm}^2 + (L\omega I_{gm\max})^2} \tag{6.10}$$

where $I_{gm\max}$ is the maximum value of grid current magnitude I_{gm}.

By neglecting the effect of the resistor R in (6.5), the time derivative of the grid current can be expressed as follows:

$$\frac{di_g}{dt} = \frac{1}{L}(V_g - V_{conv}) \tag{6.11}$$

Figure 6.5 shows an example of the grid current waveforms during a kth switching period and with regard to the sign of the grid voltage V_g. In this figure, α refers to the duty cycle of the switching signal u, $I_{\max}^k$ and $I_{\min}^k$ refer respectively to the maximum and minimum current magnitudes during the kth switching period and T is the switching period. So, relying on Figure 6.5, the grid current i_g can be expressed as in (6.12) during the interval $[kT, kT + \alpha T]$.

$$i_g(t) = \frac{1}{L}(V_g - V_{dc})(t - kT) + I_{\max}^k \tag{6.12}$$

Based on (6.12) and Figure 6.5, the current ripple Δi_g^- during the interval $[kT, kT + \alpha T]$ can be expressed as follows:

$$\Delta i_g^- = I_{\max}^k - I_{\min}^k = -\frac{1}{L}(V_g - V_{dc})\alpha T \tag{6.13}$$

During the interval $[kT + \alpha T, (k+1)T]$, the grid current i_g can be expressed as follows:

$$i_g(t) = \frac{1}{L}(V_g + V_{dc})(t - kT - \alpha T) + I_{\min}^k \tag{6.14}$$

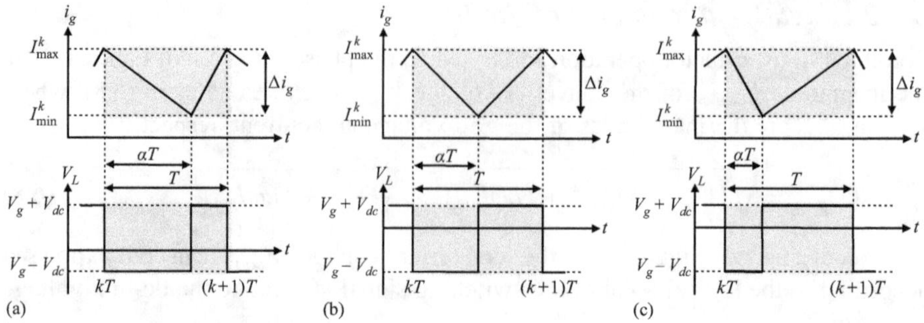

Figure 6.5 Waveforms of i_g and V_L when V_g is (a) positive (b) null (c) negative

Based on (6.14) and Figure 6.5, the current ripple Δi_g^+ during the interval $[kT + \alpha T, (k+1)T]$ can be expressed as follows:

$$\Delta i_g^+ = I^k_{\max} - I^k_{\min} = \frac{1}{L}(V_g + V_{dc})T(1-\alpha) \tag{6.15}$$

Supposing that the switching period is very small compared to the grid voltage period, the current ripples Δi_g^+ and Δi_g^- can be considered equals ($\Delta i_g^+ = \Delta i_g^- = \Delta i_g$). In this case, the switching period T can be expressed as in (6.16).

$$T = \alpha T + (1-\alpha)T = \frac{2L\Delta i_g V_{dc}}{V_{dc}^2 - V_g^2} \tag{6.16}$$

From (6.16), the maximum switching frequency F_{max} is obtained when the grid voltage is equal to zero and is expressed as follows:

$$F_{\max} = \frac{V_{dc}}{2L\Delta i_g} \tag{6.17}$$

6.3.3 Synthesis of the direct sliding mode control

For the design of the direct sliding mode control for a single-phase PWM rectifier, the switching function $S(i_{conv})$ expressed in (6.18) is defined.

$$S(i_{conv}) = i^*_{conv} - i_{conv} \tag{6.18}$$

where i^*_{conv} is the reference converter current.

The switching function $S(i_{conv})$ defines a sliding surface ($S(i_{conv}) = 0$), where the trajectory of the current i_{conv} reaches its reference i^*_{conv}. Its time derivative can be used to analyze the direction of the converter current i_{conv}. Since the reference current i^*_{conv} is considered constant each commutation period, the time derivative of the switching function $S(i_{conv})$ can be expressed as follows:

$$\dot{S}(i_{conv}) = -\frac{di_{conv}}{dt} = \frac{di_g}{dt} = -\frac{1}{L}(V_{conv} - V_g + Ri_g) \tag{6.19}$$

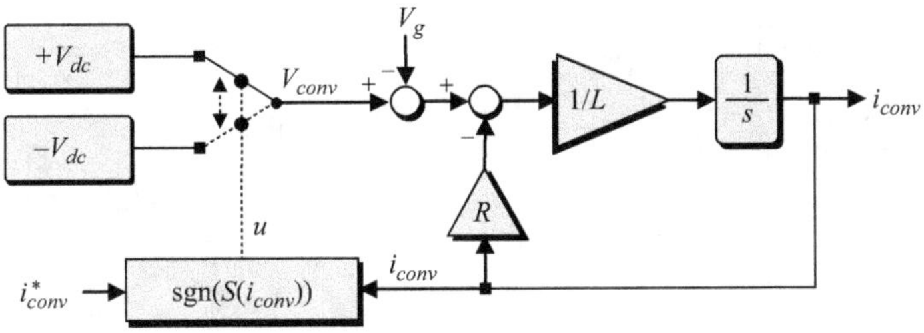

Figure 6.6 Block diagram of the direct sliding mode control

In order to verify the conditions given by (6.3), the applied converter voltage V_{conv} is equal to $+V_{dc}$ or $-V_{dc}$ depending on the sign of the switching function $S(i_{conv})$. Such control structure is described according to (6.20) and Figure 6.6.

$$\begin{cases} (u = 1 \Rightarrow V_{conv} = +V_{dc}) & if \quad S(i_{conv}) < 0 \\ (u = 0 \Rightarrow V_{conv} = -V_{dc}) & if \quad S(i_{conv}) > 0 \end{cases} \tag{6.20}$$

Based on (6.20), it can be noted that:

- When the switching function $S(i_{conv})$ is negative, the applied converter voltage V_{conv} is equal to $+V_{dc}$. For that case and based on (6.9) and (6.19), the time derivative of the switching function $S(i_{conv})$ is positive and conditions of (6.3) are verified.
- When the switching function $S(i_{conv})$ is positive, the applied converter voltage V_{conv} is equal to $-V_{dc}$. For that case and based on (6.9) and (6.19), the time derivative of the switching function $S(i_{conv})$ is negative and conditions of (6.3) are verified.

As shown in Figure 6.6, a sign comparator is used to determine the sign of the switching function $S(i_{conv})$ and to compute the logical level of the switching signal u according to (6.20). In practice, a hysteresis controller is used instead of the sign comparator in order to limit the switching frequency of the power converter. In this case, for a hysteresis band of the hysteresis controller equal to Δh and relying on (6.17), the switching frequency is variable without exceeding its maximum value $F_{\max}$ given by (6.21).

$$F_{\max} = \frac{V_{dc}}{2L\Delta h} \tag{6.21}$$

Figure 6.7(a) shows the hysteresis-based direct sliding mode control of a single-phase PWM rectifier. In this control, an additional external control loop based on a PI controller is used to control the dc-link voltage V_{dc} so that it becomes equal to its reference V^*_{dc} during steady-state operation. The PI controller computes the magnitude i_{convm} of the reference current i^*_{conv}. In order to ensure grid synchronization and unitary power factor operation, the magnitude of the reference current is multiplied by the inverted per unit feedback of the measured grid voltage V_g.

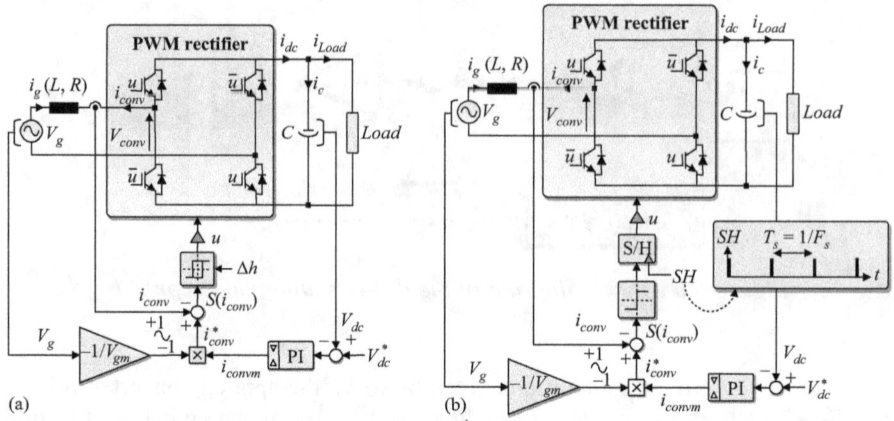

Figure 6.7 Direct sliding mode control for a single-phase PWM rectifier based on (a) hysteresis controller (b) delta modulation technique

Another solution to limit the switching frequency of the power converter can be achieved by using the delta modulation technique [10]. This technique replaces the hysteresis controller used for the current control loop by a sign comparator associated to a sample and holder as depicted in Figure 6.7(b). The obtained control is quite similar to the one presented in Figure 6.7(a), but the operating principle is different. In fact, only the error sign of the converter current error is detected by the sign comparator and the computed switching signal $u(t)$ is sampled at a fixed rate, so that the converter status is kept constant during each sampling period T_S. As a result, the switching frequency is variable and limited to the half of the sampling frequency F_S. It should be noted that in this case, the magnitude of current ripples is variable and depends on the L filter parameters, the dc-link voltage and the used sampling frequency.

6.3.4 FPGA-based controller

- Generic IP (*Intellectual Property*) module architecture

The generic IP module architecture is presented in Figure 6.8. It is typically divided into two parts: a *data-path* and a *control unit*.

The data-path contains all the required basic operators (registers, multiplexers, adders, multipliers, tables, etc.) connected through signals and aimed to perform necessary operations. The control unit is based on a *Finite State Machine* (FSM) that generates control signals and specifies how the data flows through the data-path. Note that the data transfers are synchronized to a clock signal (*Clk*) and the data initial values can be set through a reset signal (*Reset*). In order to ensure ease of use of the designed module architecture, the control unit is activated through a *Start* pulse signal. When the computation tasks of the IP module are achieved, a *Ready* pulse signal is activated in order to indicate that the data outputs are available. The designed architecture can be coded using a hardware description

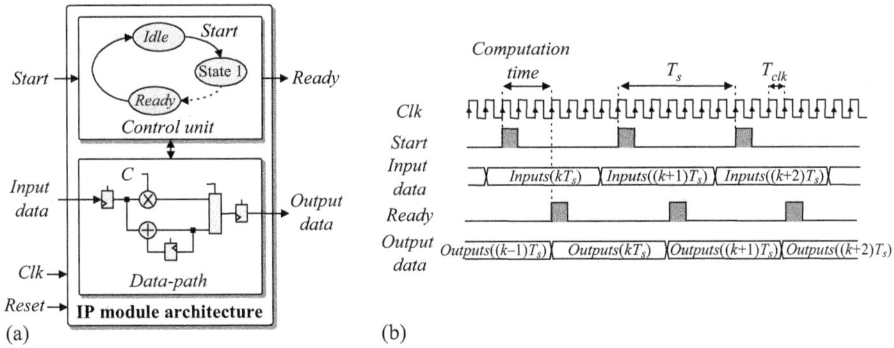

Figure 6.8 (a) General IP module architecture (b) timing diagram

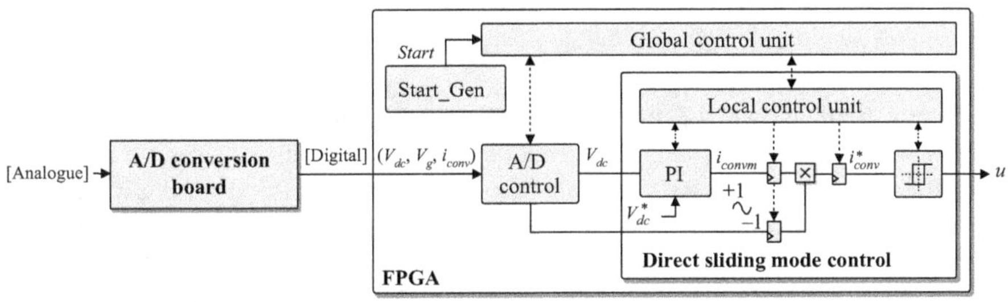

Figure 6.9 FPGA-based hardware architecture

language such as VHDL (*VHSIC* (*Very-High Speed Integrated Circuits*) *Hardware Description Language*) or Verilog HDL (*Verilog Hardware Description Language*) [11], [5].

- *Architecture design of the direct sliding mode control*

Figure 6.9 shows the hardware architecture for the FPGA implementation of the direct sliding mode control for a single-phase PWM rectifier. It is divided into two main IP modules, which are managed by a global control unit. The first one is the *A/D control* module that controls the A/D conversion process. The second one is the *Direct sliding mode control* module, which is managed by its own control unit and includes two other IP modules:

- A *PI* module that computes the current magnitude i_{convm}.
- A *Hysteresis* module that generates the switching signal u.

Note that between these two modules a multiplier and controlled registers are used in order to compute the reference current i^*_{conv}.

The *Start* signal that activates the global control unit is generated by a *Start_Gen* module and two operation modes can be achieved:

- First operation mode

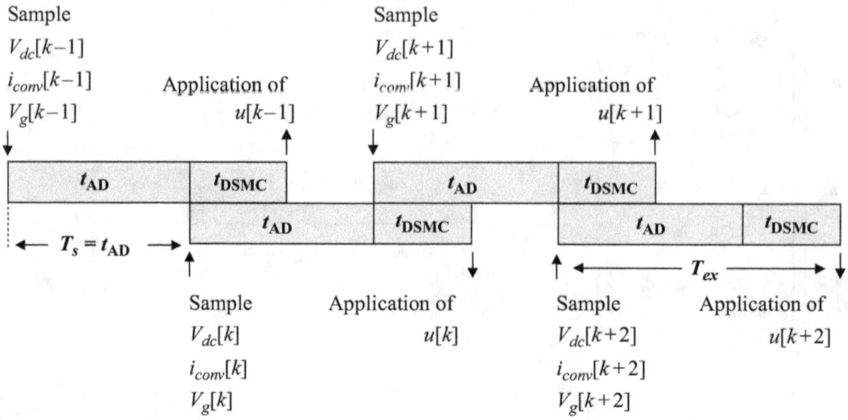

Figure 6.10 Timing diagram for the realization of a hysteresis-based control

Figure 6.10 shows the timing diagram corresponding to the first operation mode, which is aimed to realize the direct sliding mode control based on hysteresis controller. Such timing sequences can be easily done by forcing the *Start* signal of the global control unit at an active level in order to activate each computation cycle immediately after the achievement of the previous one.

For each computation cycle, the global control unit activates in parallel the A/D control module and the direct sliding mode control module. The computation time of the direct sliding mode control module (t_{DSMC}) is generally lower than the A/D conversion time t_{AD}. For this reason, the global control unit restarts a new computation cycle, immediately after the achievement of the A/D conversion process. Consequently, the sampling period T_s is equal to the A/D conversion time t_{AD} and the execution time T_{ex} is equal to the sum of t_{AD} and t_{DSMC}. When associated to fast A/D converters, the high computation capabilities of FPGA solutions allow obtaining very low execution time of only a few micro seconds. Thus, the obtained digital feedback loop can be approximated quite close to an analogue one since the effects of sampling and computing delays are negligible.

- Second operation mode

Figure 6.11 shows the timing diagram corresponding to the second operation mode, which is based on the delta modulation technique. It can be easily done based on the same architecture presented in Figure 6.9 and only a few modifications are required:

- A pulse *Start* signal with a sampling period equal to T_s is used.
- The hysteresis band is set to zero. In this case the hysteresis controller works as a sign comparator.
- The global control unit activates sequentially the A/D control module and the direct sliding mode control module.

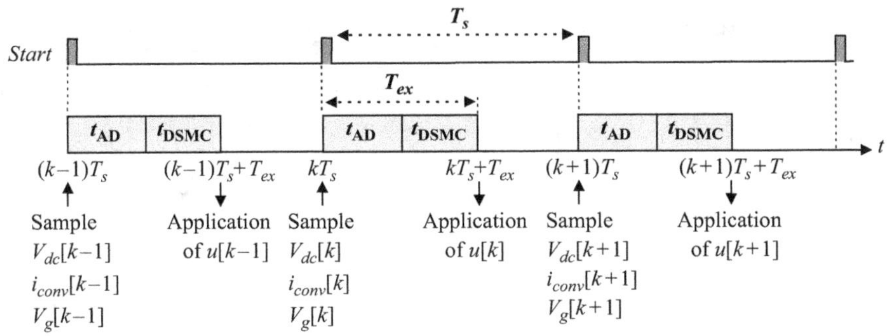

Figure 6.11 Timing diagram of the second operation mode

Table 6.1 Main data and parameters of the experimental tests

Line voltage	70 V_{RMS}
Voltage frequency	50 Hz
Filter inductor L	20 mH
Filter resistor R	0.3 Ω
DC-link capacitor C	1100 μF
Load	100 Ω
Initial dc-link voltage V_{dc}^{init}	100 V
DC-link reference voltage V_{dc}^*	200 V
PI gains (proportional, integral)	(0.07, 1.09)
(t_{AD}, t_{DSMC}, T_{ex}, T_s)	(1.34 μs, 0.24 μs, 1.58 μs, 100 μs)

As it can be noted in Figure 6.11, the computed logical level of the switching signal u is updated every sampling period T_s. Consequently, the switching frequency is variable without exceeding the half of the sampling frequency F_s.

The test of the implemented hardware architecture can be done in three steps. First, the switching states of the PWM rectifier are all set to zero. In this case, the single-phase PWM rectifier works as a diode rectifier and the dc-link capacitor is charged to its initial voltage value V_{dc}^{init}. During this step, the load is disconnected from the dc-link. Then, the hardware architecture is activated and the switching signal u is applied to the power converter. The main data and parameters of the used experimental prototype are summarized in Table 6.1.

Figure 6.12(a) shows the response of the dc-link voltage V_{dc} and the converter current i_{conv}. It can be noted that the dc-link voltage V_{dc} is well controlled and becomes equal to its reference in steady state. The transient state is characterized by an increase of the magnitude of converter current, which becomes again equal to zero in steady state. Finally, in the third step, the load is connected to the dc-link. As shown in Figure 6.12(b), the dc-link voltage V_{dc} decreases slightly when the load is connected, then the control compensates the load disturbance. It can be

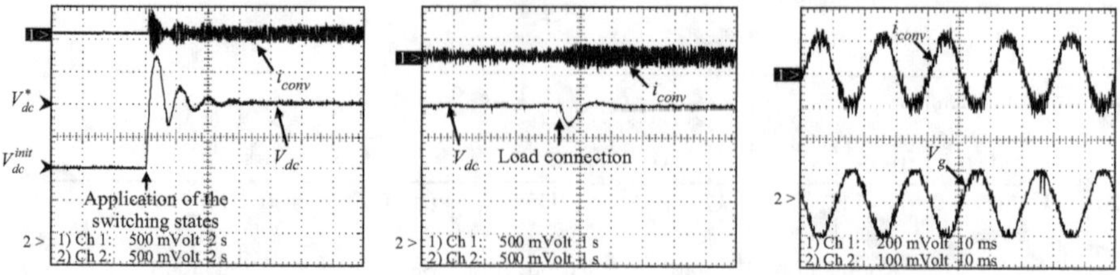

(a) (b) (c)

Figure 6.12 Response to step reference of V_{dc}^* and load connection (with direct sliding mode control based on delta modulation technique) (a–b) converter current and dc-link voltage (c) grid voltage and converter current during steady-state operation

noted also that the magnitude of the current i_{conv} increases after load connexion. Figure 6.12(c) presents converter current i_{conv} and grid voltage V_g waveforms during steady-state operation. It can be noted that the single-phase PMW rectifier is characterized by a sinusoidal current absorption and it allows the achievement of unitary power factor operation.

6.4 Direct sliding mode control for three-phase PWM rectifier

6.4.1 Three-phase PWM rectifier model

The power circuit of a three-phase PWM rectifier connected to the grid through L filters is presented in Figure 6.13. In this figure, L (respectively R) is the filter inductor (respectively the filter resistor). $V_{g(a,b,c)}$ (respectively $i_{g(a,b,c)}$) refers to the components of the grid voltage vector $\underline{V}_g$ (respectively the components of the grid current vector $\underline{i}_g$) in the natural reference frame. The currents $i_{conv(a,b,c)}$ refer to the components of the power converter current vector $\underline{i}_{conv}$ in the natural reference frame.

The power converter can apply up to eight voltage vectors $V^i_{conv(i=0\ldots7)}$ (six active vectors $V^i_{conv(i=1\ldots6)}$ and two null vectors (V^0_{conv}, V^7_{conv})). In the natural reference frame, the components $V_{conv(a,b,c)}$ of the $V^i_{conv(i=0\ldots7)}$ voltage vectors are expressed as in Table 6.2.

The power converter voltage vector $\underline{V}_{conv}$ can be adjusted in magnitude, frequency and phase via the control algorithm of the power converter. So, the converter current $\underline{i}_{conv}$ can be controlled indirectly by modifying the voltage drop on the used filter according to the applied power converter voltage vector. Moreover, the use of the filters creates a current source that provides boost features for the grid connected power converter. The phasor diagrams that characterize the operation of a three-phase PWM rectifier in general case and with unitary power factor are quite similar to those depicted in Figure 6.3. Based on the power circuit of Figure 6.13, the mathematical model of a three-phase PWM rectifier in the (d, q)

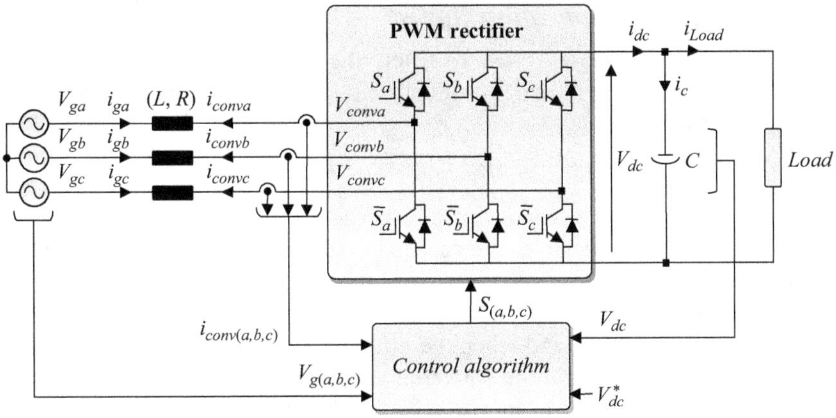

Figure 6.13 Controlled three-phase PWM rectifier

Table 6.2 Switching table of the direct sliding mode control

$\mathbf{V}_{\mathbf{conv}}$	S_a	S_b	S_c	V_{conva}	V_{convb}	V_{convc}
$\mathbf{V}^0_{\mathbf{conv}}$	0	0	0	0	0	0
$\mathbf{V}^1_{\mathbf{conv}}$	1	0	0	$2V_{dc}/3$	$-V_{dc}/3$	$-V_{dc}/3$
$\mathbf{V}^2_{\mathbf{conv}}$	1	1	0	$V_{dc}/3$	$V_{dc}/3$	$-2V_{dc}/3$
$\mathbf{V}^3_{\mathbf{conv}}$	0	1	0	$-V_{dc}/3$	$2V_{dc}/3$	$-V_{dc}/3$
$\mathbf{V}^4_{\mathbf{conv}}$	0	1	1	$-2V_{dc}/3$	$V_{dc}/3$	$V_{dc}/3$
$\mathbf{V}^5_{\mathbf{conv}}$	0	0	1	$-V_{dc}/3$	$-V_{dc}/3$	$2V_{dc}/3$
$\mathbf{V}^6_{\mathbf{conv}}$	1	0	1	$V_{dc}/3$	$-2V_{dc}/3$	$V_{dc}/3$
$\mathbf{V}^7_{\mathbf{conv}}$	1	1	1	0	0	0

synchronous reference frame, where the d axis is linked to the grid voltage vector, is expressed as in (6.22), (6.23), (6.24) and (6.25).

$$V_{convd} = Ri_{convd} + L\frac{di_{convd}}{dt} - \omega Li_{convq} + V_{gd} \tag{6.22}$$

$$V_{convq} = Ri_{convq} + L\frac{di_{convq}}{dt} + \omega Li_{convd} \tag{6.23}$$

$$P = -V_{gd}i_{convd} \tag{6.24}$$

$$Q = V_{gd}i_{convq} \tag{6.25}$$

V_{gd} and V_{gq} are the d and q components of the grid voltage vector, and V_{convd} and V_{convq} (respectively i_{convd} and i_{convq}) are the d and q components of the power converter voltage vector (respectively the power converter current vector). P and Q are respectively the active and reactive powers.

6.4.2 Steady-state operation limits

For the case of three-phase PWM rectifier, the V_{dc} voltage must verify (6.26) obtained for a maximum modulation depth and in order to conduct current with maximum magnitude $I_{gm\max}$.

$$V_{dc} \geq \sqrt{3}\sqrt{V_{gm}^2 + (L\omega I_{gm\max})^2} \tag{6.26}$$

The converter current ripples are caused by the pulsed voltage generated by the power converter. During the switching operation, the worst case that leads to a maximum converter current ripple is obtained when the applied converter voltage varies from $-V_{dc}/3$ to $V_{dc}/3$ and when the grid voltage is equal to zero. Figure 6.14 shows a waveform example of the converter current $i_{convi(i=a,b,c)}$ with regard to the applied converter voltage $V_{convi(i=a,b,c)}$. In this figure, T is the switching period, α is the duty cycle, $\Delta i_{convi(i=a,b,c)}$ is the converter current ripple and $I_{\max}^k$ and $I_{\min}^k$ are respectively the maximum and minimum converter current value during a kth switching period.

During the interval $[kT, kT+\alpha T]$, the converter current $i_{convi(i=a,b,c)}$ can be expressed as follows:

$$i_{convi}(t) = \frac{V_{dc}}{3L}(t - kT) + I_{\min}^k \quad (i = a, b, c) \tag{6.27}$$

Based on (6.27), the current ripple Δi_{conv}^+ during the interval $[kT, kT+\alpha T]$ can be expressed as follows:

$$\Delta i_{conv}^+ = I_{\max}^k - I_{\min}^k = \frac{V_{dc}}{3L}\alpha T \tag{6.28}$$

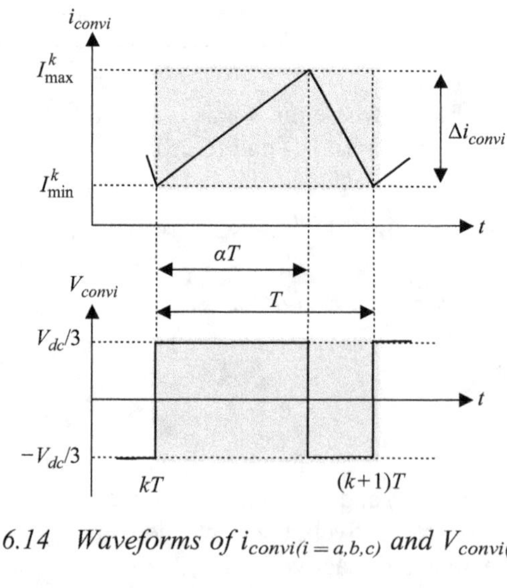

Figure 6.14 Waveforms of $i_{convi(i=a,b,c)}$ and $V_{convi(i=a,b,c)}$

During the interval $[kT + \alpha T, (k+1)T]$, the converter current $i_{convi(i=a,b,c)}$ can be expressed as follows:

$$i_{convi}(t) = -\frac{V_{dc}}{3L}(t - kT - \alpha T) + I_{\max}^{k} \quad (i = a, b, c) \tag{6.29}$$

Based on (6.29), the current ripple Δi_{conv}^{-} during the interval $[kT + \alpha T, (k+1)T]$ can be expressed as follows:

$$\Delta i_{conv}^{-} = I_{\max}^{k} - I_{\min}^{k} = \frac{V_{dc}}{3L} T(1 - \alpha) \tag{6.30}$$

Supposing that the switching period is very small compared to the grid voltage period, the current ripples Δi_{conv}^{+} and Δi_{conv}^{-} can be considered equals ($\Delta i_{conv}^{+} = \Delta i_{conv}^{-} = \Delta i_{convi(i=a,b,c)}$). In this case, the switching period T can be expressed as in (6.31).

$$T = \alpha T + (1 - \alpha)T = \frac{6L\Delta i_{convi}}{V_{dc}} \tag{6.31}$$

From (6.31), the current ripple $\Delta i_{convi(i=a,b,c)}$ can be expressed as in (6.32), where F is the switching frequency.

$$\Delta i_{convi} = \frac{V_{dc}}{6LF} \quad (i = a, b, c) \tag{6.32}$$

6.4.3 Synthesis of the direct sliding mode control

The direct sliding mode control algorithm for the three-phase PWM rectifier is synthesized in the (d,q) synchronous reference frame (the d axis is aligned with the grid voltage vector $\underline{V}_g$) and is aimed to control the active and reactive power through the control of the d and q converter current components, respectively. According to (6.24) and (6.25), and taking into account that V_{gd} voltage is constant, the active and reactive powers (P and Q) are proportional to i_{convd} and i_{convq} currents, respectively. So, to synthesize the direct sliding mode control, two switching functions $S(i_{convd})$ and $S(i_{convq})$ are defined as the difference between the reference current i_{convq}^{*} (respectively i_{convq}^{*}), and measured current i_{convd} (respectively i_{convq}) as follows

$$S(i_{convd}) = i_{convd}^{*} - i_{convd} \tag{6.33}$$

$$S(i_{convq}) = i_{convq}^{*} - i_{convq} \tag{6.34}$$

These switching functions define two sliding surfaces ($S(i_{convd}) = 0$) and ($S(i_{convq}) = 0$). The direct sliding mode control is synthesized so that the switching functions are attracted toward their respective sliding surfaces during transient state. This means that the d and q converter current components will be also directed to their references as shown in Figure 6.15. To ensure this objective, the

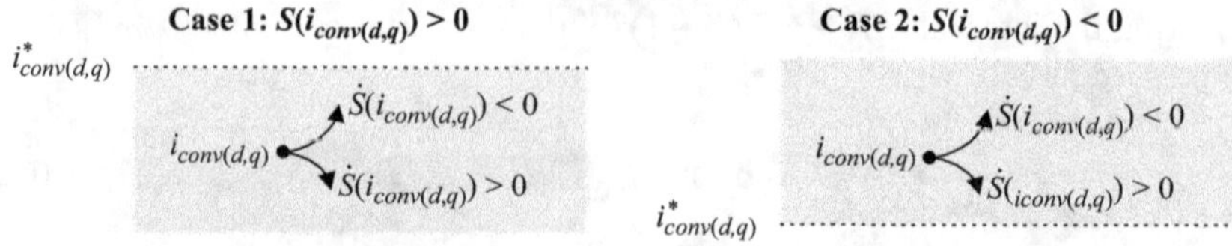

Figure 6.15 Current trajectories according to the sign of switching functions and their time derivatives

switching functions ($S(i_{convd})$ and $S(i_{convq})$) and their time derivatives must verify conditions given by (6.35) and (6.36).

$$\dot{S}(i_{convd})S(i_{convd}) \prec 0 \tag{6.35}$$

$$\dot{S}(i_{convq})S(i_{convq}) \prec 0 \tag{6.36}$$

From (6.35) and (6.36), the applied voltage vectors depend on the sign of the switching functions $S(i_{convd})$ and $S(i_{convq})$ and their time derivatives. The sign of the switching functions can be determined via sign comparators and Boolean variables $C_{(d,q)}$ as shown in (6.37) and (6.38).

$$if \quad S(i_{convd}) \succ 0 \quad then \quad C_d = 1 \quad else \quad C_d = 0 \tag{6.37}$$

$$if \quad S(i_{convq}) \succ 0 \quad then \quad C_q = 1 \quad else \quad C_q = 0 \tag{6.38}$$

Based on (6.37) and (6.38), and taking into account that the references i^*_{convq} and i^*_{convq} are kept constants during each sampling period, the time derivatives of the switching functions are expressed as in (6.39) and (6.40). Note that the drop voltage on R resistor is neglected in these equations.

$$\dot{S}(i_{convd}) = -\frac{di_{convd}}{dt} = \frac{1}{L}(V_{gd} - \omega L i_{convq} - V_{convd}) \tag{6.39}$$

$$\dot{S}(i_{convq}) = -\frac{di_{convq}}{dt} = \frac{1}{L}(\omega L i_{convd} - V_{convq}) \tag{6.40}$$

Based on (6.39) and (6.40), it can be noted that the time derivatives of the switching functions depend on

- The applied d and q components of the converter voltage vector.
- The magnitudes of the d and q components of the converter current vector.
- The grid voltage magnitude and frequency.
- The power circuit parameters.

The power converter can apply up to eight voltage vectors $V^i_{conv(i=0\ldots7)}$ (six active vectors $V^i_{conv(i=1\ldots6)}$ and two null vectors (V^0_{conv}, V^7_{conv})). The V_{convd} and V_{convq} voltages expressed in the (d, q) synchronous reference frame can be determined by applying a rotation (with an angle equal to θ_{dq}) to the α and β components of the power converter voltage vectors expressed in the stationary reference frame (α, β). As a result, the sign of the time derivatives of the switching functions $S(i_{convd})$ and

$S(i_{convq})$ depend on the θ_{dq} position. For this reason, the (α, β) plane was divided into 12 sectors (S = 0 … 11) as shown in Figure 6.16. The locus of the time derivatives of the switching functions according to θ_{dq} positions is also depicted in this figure. Note that these figures are directly deduced based on (6.39) and (6.40) through the computation of V_{convd} and V_{convq} voltages for different values of the θ_{dq} position.

In order to ensure the conditions of (6.35) and (6.36), it is necessary to simultaneously analyze the switching functions and their respective time derivatives according to Figure 6.16. This must be done for each sector in which is located the θ_{dq} position and it leads to the optimized switching table presented in Table 6.3.

For example, if the θ_{dq} is located inside the sector $(S = 0)$, four cases can be considered:

- Case 1 ($C_d = 0$ and $C_q = 0$): Based on (6.37) and (6.38), the switching functions $S(i_{convd})$ and $S(i_{convq})$ are negatives. So, the time derivatives of the switching functions must be positives and, according to Figure 6.16(b), two voltage vectors ($\underline{V}^5_{conv}$ and $\underline{V}^6_{conv}$) can be selected. The module of switching functions time derivatives imposes the dynamic response of switching functions, which indicates how rapidly the controlled currents ($i_{conv(d,q)}$) reach their references ($i^*_{conv(d,q)}$). In order to ensure more balanced dynamic for the control of ($i_{conv(d,q)}$) currents, the voltage vector $\underline{V}^5_{conv}$ is selected in Table 6.3.
- Case 2 ($C_d = 0$ and $C_q = 1$): Based on (6.37) and (6.38), the switching function $S(i_{convd})$ is negative while the switching function $S(i_{convq})$ is positive. So, the time derivatives of $S(i_{convd})$ and $S(i_{convq})$ must be positive and negative, respectively. According to Figure 6.16(b), five voltage vectors ($\underline{V}^2_{conv}$, $\underline{V}^3_{conv}$, $\underline{V}^4_{conv}$, $\underline{V}^0_{conv}$ and $\underline{V}^7_{conv}$) can be selected. As for the first case and in order to ensure balanced dynamic between d and q axis, the voltage vector $\underline{V}^3_{conv}$ is selected in Table 6.3.
- Case 3 ($C_d = 1$ and $C_q = 0$): Based on (6.37) and (6.38), the switching function $S(i_{convd})$ is positive while the switching function $S(i_{convq})$ is negative. So, the time derivatives of $S(i_{convd})$ and $S(i_{convq})$ must be negative and positive, respectively. According to Figure 6.16(b), no possible vectors can be selected. In Table 6.3, the voltage vector $\underline{V}^1_{conv}$ is selected since it imposes the required sign for the time derivative of $S(i_{convd})$ and small module for the time derivative of $S(i_{convq})$. Also, according to Figure 6.16(n), with θ_{dq} variation inside sector $(S = 0)$, the voltage vector will lead to the correct sign for the time derivatives of $S(i_{convd})$ and $S(i_{convq})$.
- Case 4 ($C_d = 1$ and $C_q = 1$): According to (6.37) and (6.38), the switching functions $S(i_{convd})$ and $S(i_{convq})$ are positives. So, the time derivatives of $S(i_{convd})$ and $S(i_{convq})$ must be negatives. The only voltage vector that leads to a negative time derivative for the switching functions is the vector $\underline{V}^1_{conv}$, that's why it was selected in Table 6.3 for this case.

The same reasoning can be applied for all the other cases. Figure 6.16(n) and 6.16(o) illustrate the behavior of switching functions time derivatives locus in response to a small variation $\Delta\theta_{dq}$ of the θ_{dq} position. It can be noted that, each voltage vector leads to the same sign of switching function time derivatives in this case.

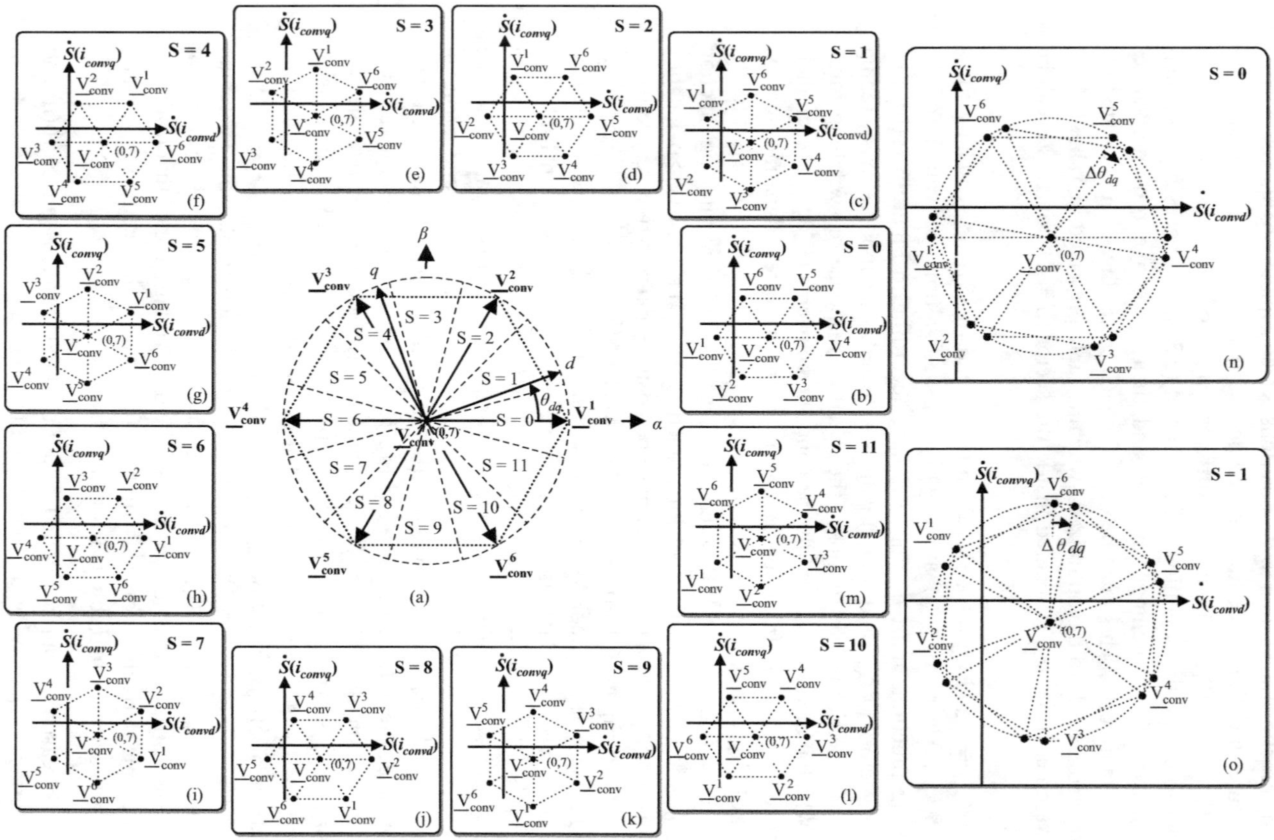

Figure 6.16 (a) Division of the (α,β) plane into 12 sectors ($S=0 \ldots 11$) (b–m). Locus of the switching functions time derivatives for (b) $S=0$ ($\theta_{dq}=0$) (c) $S=1$ ($\theta_{dq}=\pi/6$) (d) $S=2$ ($\theta_{dq}=\pi/3$) (e) $S=3$ ($\theta_{dq}=\pi/2$) (f) $S=4$ ($\theta_{dq}=2\pi/3$) (g) $S=5$ ($\theta_{dq}=5\pi/6$) (h) $S=6$ ($\theta_{dq}=\pi$) (i) $S=7$ ($\theta_{dq}=7\pi/6$) (j) $S=8$ ($\theta_{dq}=4\pi/3$) (k) $S=9$ ($\theta_{dq}=3\pi/2$) (l) $S=10$ ($\theta_{dq}=5\pi/3$) (m) $S=11$ ($\theta_{dq}=11\pi/6$) (n–o). Evolution of the locus of the switching functions time derivatives in response to a small variation $\Delta\theta_{dq}$ of the θ_{dq} position (n) $S=0$ ($\theta_{dq}=0$) (o) $S=1$ ($\theta_{dq}=\pi/6$)

Table 6.3 Switching table of the direct sliding mode control

C_d	**1**	**0**	**1**	**0**
C_q	**1**	**1**	**0**	**0**
$S = 0$	$\underline{V}^{1}_{conv}$	$\underline{V}^{3}_{conv}$	$\underline{V}^{1}_{conv}$	$\underline{V}^{5}_{conv}$
$S = 1$	$\underline{V}^{2}_{conv}$	$\underline{V}^{4}_{conv}$	$\underline{V}^{1}_{conv}$	$\underline{V}^{6}_{conv}$
$S = 2$	$\underline{V}^{2}_{conv}$	$\underline{V}^{4}_{conv}$	$\underline{V}^{2}_{conv}$	$\underline{V}^{6}_{conv}$
$S = 3$	$\underline{V}^{3}_{conv}$	$\underline{V}^{5}_{conv}$	$\underline{V}^{2}_{conv}$	$\underline{V}^{1}_{conv}$
$S = 4$	$\underline{V}^{3}_{conv}$	$\underline{V}^{5}_{conv}$	$\underline{V}^{3}_{conv}$	$\underline{V}^{1}_{conv}$
$S = 5$	$\underline{V}^{4}_{conv}$	$\underline{V}^{6}_{conv}$	$\underline{V}^{3}_{conv}$	$\underline{V}^{2}_{conv}$
$S = 6$	$\underline{V}^{4}_{conv}$	$\underline{V}^{6}_{conv}$	$\underline{V}^{4}_{conv}$	$\underline{V}^{2}_{conv}$
$S = 7$	$\underline{V}^{5}_{conv}$	$\underline{V}^{1}_{conv}$	$\underline{V}^{4}_{conv}$	$\underline{V}^{3}_{conv}$
$S = 8$	$\underline{V}^{5}_{conv}$	$\underline{V}^{1}_{conv}$	$\underline{V}^{5}_{conv}$	$\underline{V}^{3}_{conv}$
$S = 9$	$\underline{V}^{6}_{conv}$	$\underline{V}^{2}_{conv}$	$\underline{V}^{5}_{conv}$	$\underline{V}^{4}_{conv}$
$S = 10$	$\underline{V}^{6}_{conv}$	$\underline{V}^{2}_{conv}$	$\underline{V}^{6}_{conv}$	$\underline{V}^{4}_{conv}$
$S = 11$	$\underline{V}^{1}_{conv}$	$\underline{V}^{3}_{conv}$	$\underline{V}^{6}_{conv}$	$\underline{V}^{5}_{conv}$

The structure of the sliding mode direct power control for three-phase PWM rectifier is presented in Figure 6.17(a). Two hysteresis controllers are used to compute the logical level of C_d and C_q according to the switching functions on the axes d ($S(i_{convd}) = i^*_{convq} - i_{convd}$) and q ($S(i_{convq}) = i^*_{convq} - i_{convq}$). The position θ_{dq} of the grid voltage vector is estimated using a grid synchronization module [12]–[13]. This position is used to compute the sector S in which is located the grid voltage vector and to compute i_{convd} and i_{convq} currents through an *abc-to-dq* coordinate transformation. An external control loop based on a PI controller is used to control the dc-link voltage V_{dc}.

As for the case of single-phase PWM rectifier, a direct sliding mode control based on delta modulation technique can be also achieved by substituting the hysteresis controller by sign comparators and by sampling the switching signals $S_{(a,b,c)}$ at a fixed rate as shown in Figure 6.17(b). For that case, the switching frequency is variable and limited to the half of the sampling frequency F_s. Also, it should be noted that the magnitude of current ripples is not constant, but depends on the L filter parameters, the dc-link voltage and the used sampling frequency.

6.4.4 FPGA-based controller

Figure 6.18 shows the hardware architecture for the FPGA implementation of the direct sliding mode control for a three-phase PWM rectifier. The architecture is divided into two main IP modules, which are managed by a global control unit. The first one is the *A/D control* module that controls the A/D conversion process. The second one is the *direct sliding mode control* module, which is managed by its local control unit and includes six other IP modules:

- A *PI module* that computes the reference current i^*_{convq}.
- A *Grid synchronization* module that computes the grid position θ_{dq} from the measured grid voltages $V_{g(a,b,c)}$. This module is based on the *COordinate*

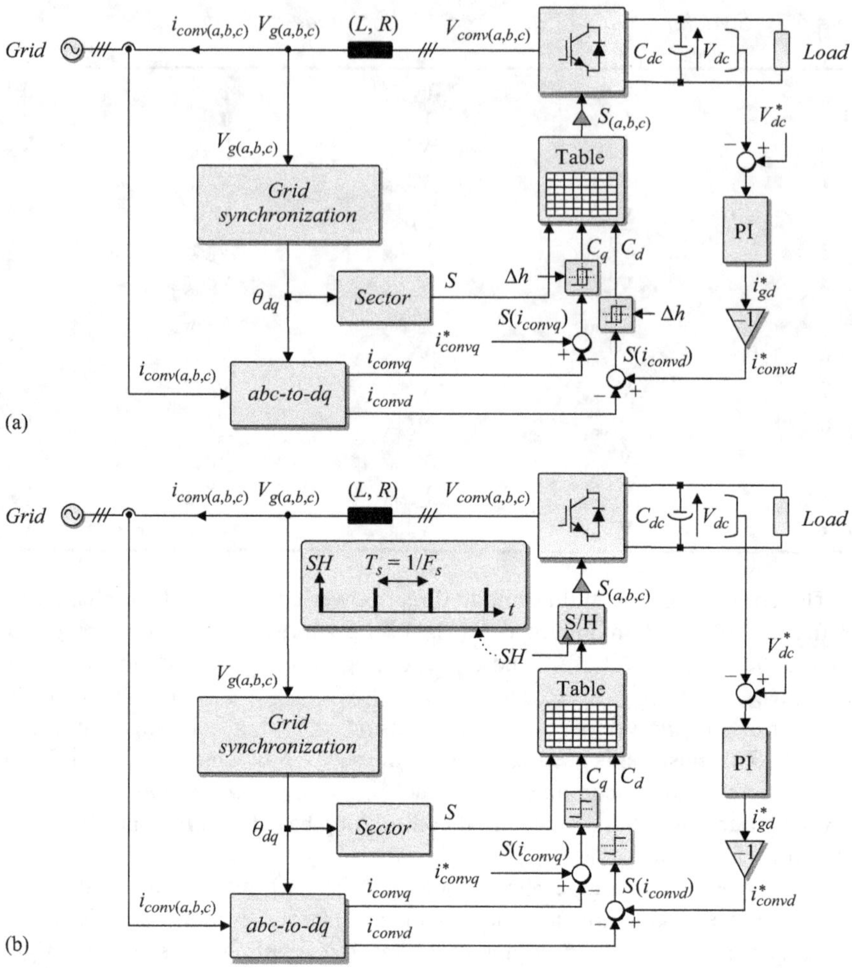

Figure 6.17 Direct sliding mode for a three-phase PWM rectifier based on (a) hysteresis controller (b) delta modulation technique

Rotation DIgital Computer (CORDIC) algorithm [14]. It should be noted that Phase Locked-Loops (PLL) can also be used for grid synchronization [12], [13].

- An *abc-to-dq* module that computes $i_{conv(d,q)}$ currents based on the measured currents $i_{conv(a,b,c)}$ and the computed θ_{dq} position.
- A *Sector* module that computes the sector number *S* based on the θ_{dq} position.
- An *Hysteresis* module that includes two hysteresis controllers and computes the $C_{(d,q)}$ Boolean variables.
- A *Table* module that includes the switching table of Table 6.3 and which is addressed through $C_{(d,q)}$ Boolean variables and *S* sector value in order to compute the switching signals $S_{(a,b,c)}$.

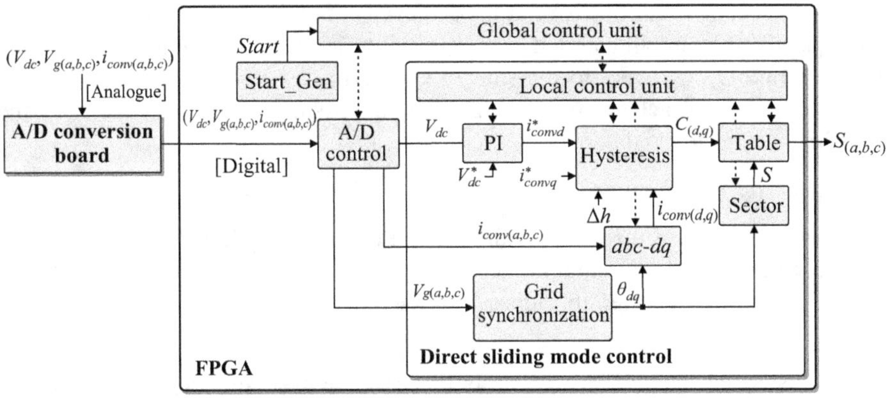

Figure 6.18 FPGA-based hardware architecture

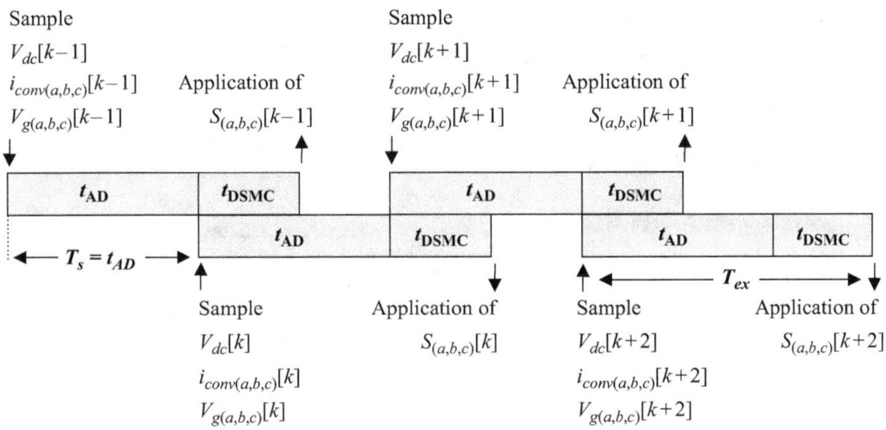

Figure 6.19 Timing diagram of the first operation mode

The *Start* signal that activates the global control unit is generated by a *Start_Gen* module and two operation modes can be achieved:

- *First operation mode*

Figure 6.19 shows the timing diagram corresponding to the first FPGA implementation method, which is aimed to set the ripples of $i_{conv(d,q)}$ currents inside an hysteresis band equal to Δh. Such timing sequences can be easily done by forcing the *Start* signal of the global control unit at an active level in order to activate each computation cycle immediately after the achievement of the previous one. Each computation cycle, the global control unit activates in parallel the *A/D control* module and the *direct sliding mode control* module. The computation time of the *direct sliding mode control* module (t_{DSMC}) is generally lower than the A/D conversion time t_{AD}. For this reason, the global control unit restarts a new computation cycle, immediately after the achievement of the A/D conversion process. Consequently, the sampling period T_s is equal to the A/D conversion time t_{AD} and the execution time T_{ex} is equal to the sum of t_{AD} and t_{DSMC}. So, the use of fast A/D converters allows the achievement of high sampling frequency and the hysteresis band exceeding is detected and corrected in real time as shown in Figure 6.20.

- *Second operation mode*

The realization of the direct sliding mode control of a three-phase PWM rectifier based on the delta modulation technique can be easily performed based on the same architecture presented in Figure 6.18. Only a few modifications are required:

- A pulse *Start* signal with a sampling period equal to T_s is used.
- The hysteresis band is set to zero. In this case the hysteresis controller works as a sign comparator.
- The global control unit activates sequentially the A/D control module and the direct sliding mode control module.

This leads to the sequential timing diagram depicted in Figure 6.21, which characterizes the FPGA implementation of a direct sliding mode control for a

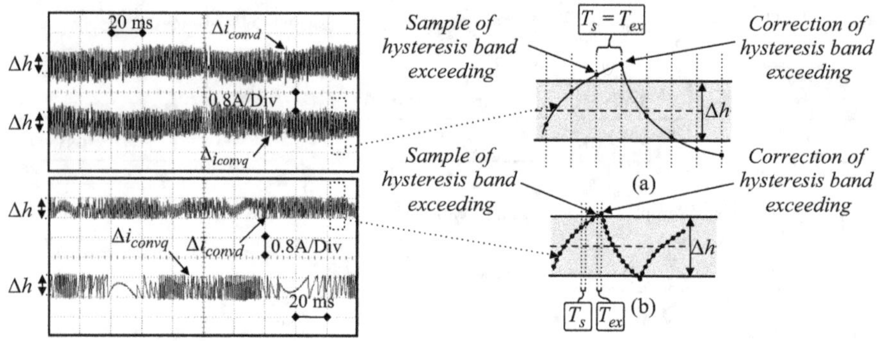

Figure 6.20 Hysteresis-based control with (a) low sampling frequency and $T_s = T_{ex}$ (b) high sampling frequency and $T_s = t_{AD}$

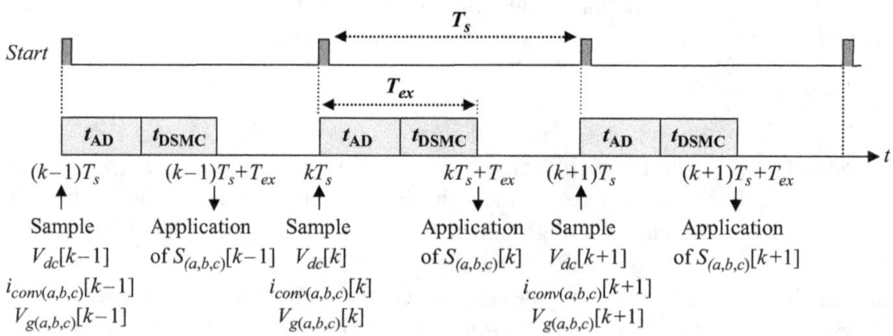

Figure 6.21 Timing diagram of the second operation mode

Table 6.4 Main data and parameters of the experimental tests

Line voltage	122 V_{RMS}
Voltage frequency	50 Hz
Filter inductor L	40 mH
Filter resistor R	0.3 Ω
DC-link capacitor C	1100 μF
Load	190 Ω
Initial dc-link voltage V_{dc}^{init}	300 V
DC-link reference voltage V_{dc}^{*}	400 V
PI gains (proportional, integral)	(0.1, 4.35)
(t_{AD}, t_{DSMC}, T_{ex}, T_s)	(1.34 μs, 0.84 μs, 2.18 μs, 25 μs)

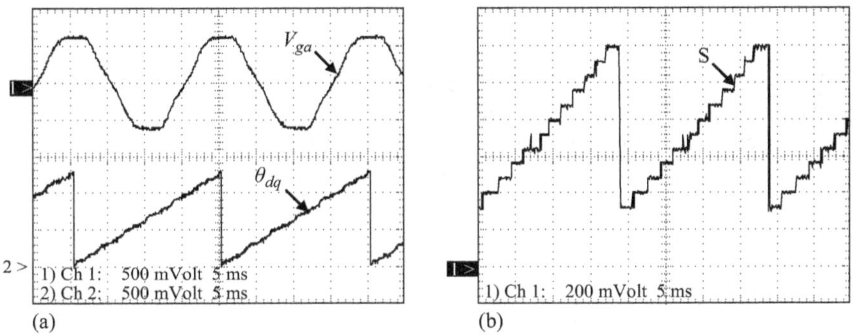

Figure 6.22 (a) θd_q position and grid voltage V_{ga} (b) sector S

three-phase PWM rectifier based on the delta modulation technique. As it can be noted in this figure, the computed logical levels of the switching signals $S_{(a,b,c)}$ are updated every sampling period T_s. Consequently, the switching frequency will be variable without exceeding the half of the sampling frequency F_s.

As for the case of a single-phase PWM rectifier, the test of the implemented hardware architecture can be done in three steps. First, the switching states of the PWM rectifier are all set to zero. In this case, the three-phase PWM rectifier works as a diode rectifier and the dc-link capacitor is charged to its initial voltage value V_{dc}^{init}. During this step, the load is disconnected from the dc-link. Then, the hardware architecture is activated and the switching signals $S_{(a,b,c)}$ are applied to the power converter. The main data and parameters of the used experimental prototype are summarized in Table 6.4.

Figure 6.22(a) and 6.22(b) present respectively the estimated θ_{dq} position with regard to the grid voltage V_{ga} and the estimated sector S.

Figure 6.23(a) shows the response of the dc-link voltage V_{dc}, which is well controlled and becomes equal to its reference V_{dc}^{*} during steady state. Finally, in the third step, the load is connected to the dc-link. As shown in Figure 6.23(b), the dc-link voltage V_{dc} decreases slightly when the load is connected, then the control compensates the load disturbance. Compared to the experimental results reported in

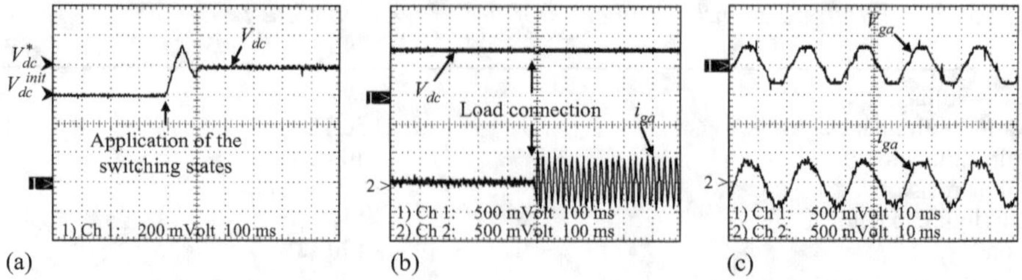

Figure 6.23 Response to step reference of V^*_{dc} and load connection (with direct sliding mode control based on delta modulation technique) (a) dc-link voltage V_{dc} response (b) dc-link voltage V_{dc} and grid current i_{ga} after load connection (c) grid current i_{ga} and grid voltage V_{ga}

Figure 6.12 for the single-phase topology, the dc-link voltage is characterized by a better dynamic response. This is mainly due to the fact that higher PI proportional gain was used for the three-phase topology. It can be noted also that the magnitude of the grid current i_{ga} increases after load connection. Figure 6.23(c) presents grid current i_{ga} ($i_{ga} = -i_{conva}$) and grid voltage V_{ga} during steady-state operation. This figure shows that a unitary power factor operation is achieved.

6.5 Conclusion

This chapter introduced the use of sliding mode theory for the control of power converters. Based on this theory, the controller design approach comprises two steps:

- The first step is aimed to define switching functions based on the system state vector.
- The second step is aimed to synthesize the control law so that the trajectories of the defined switching functions are directed toward their sliding surfaces.

Relying on this design approach, this chapter presented the synthesis of direct sliding mode control for both single-phase and three-phase PWM rectifiers. Due to the fact that power converters are naturally discontinuous and are constrained to operate with limited switching frequency to avoid power losses, two kinds of direct sliding mode controllers were considered. The first one is based on hysteresis controllers, while the second one is based on the delta modulation technique.

The FPGA implementation of each control type was also discussed in this chapter. Thanks to the high computation capabilities of FPGAs and their inherent parallelism, high sampling frequency limited to the A/D conversion frequency can be used for the implementation of hysteresis-based direct sliding mode controllers. In this case, the hysteresis band exceeding is detected and corrected in real time. Also, the FPGA-based digital solutions are well suited for the implementation of direct sliding mode controllers based on delta modulation technique. In this case, the switching frequency is variable but limited to the half of the sampling frequency.

References

[1] F. Blaadjerg, K. Ma, "Future on power electronics for wind turbine systems," *IEEE Journal of Emerging and Selected Topics in Power Electronics*, vol. 1, no. 3, pp. 139–152, September 2013.

[2] R. Teodorescu, M. Liserre, P. Rodriguez, "Grid converters for photovoltaic and wind power systems," Editeur: Wiley IEEE Press Ebook, 2011.

[3] "IEEE Recommended practice and requirements for harmonic control in electric power systems," *IEEE Std* 519-2014 (*revision of IEEE Std* 519-1992), pp. 1–29, June 2014.

[4] M.P. Kazmierkowski, M. Jasinski, G. Wrona,"DSP-based control of grid connected power converters under grid distorsions," *IEEE Transactions on Industrial Informatics*, vol. 7, no. 2, pp. 204–211, May 2011.

[5] Riesgo, Y. Torroja, E. de la Torre, "Design methodologies based on hardware description languages," *IEEE. Trans. Ind. Electron.*, vol. 46, no. 1, pp. 3–12, February 1999.

[6] C. Buccella, C. Cecati, H. Latafat, "Digital control of power converters – A survey," *IEEE Transactions on Industrial Informatics*, vol. 8, no. 3, pp 437–447, August 2012.

[7] M.W. Naouar, E. Monmasson, A.A. Naassani, I. Slama-Belkhodja, N. Patin "FPGA-based current controllers for AC machine drives – a review," *IEEE Trans. Ind. Electron.*, vol. 54, no. 4, pp. 1907–1925, August 2007.

[8] V. Utkin, "Variable structure systems with sliding modes," *IEEE Trans. on Automatic Control*, vol. 22, no. 2, pp. 212–222, April 1977.

[9] A. Sabanovic, "Variable structure systems with sliding modes in motion control – a survey," *IEEE. Trans. Ind. Inform.*, vol. 7, no. 2, pp. 212–223, May 2011.

[10] M.-P. Kazmierkowski, L. Malesani, "Current control techniques for three-phase voltage-source PWM converters: a survey," *IEEE Trans. Ind. Electron.*, vol. 45, no. 5, October 1998.

[11] E. Monmasson, M. Cirstea, "FPGA design methodology for industrial control systems – a review," *IEEE Trans. Ind. Electron.*, vol. 54, no. 4, pp. 1824–1842, August 2007.

[12] P. Rodriguez, A. Luna, R.S. Munoz-Aguilar, I. Etxeberria-Otadui, R. Teodorescu, F. Blaabjerg, "A stationary reference frame grid synchronization system for three-phase grid connected power converters under adverse grid conditions," *IEEE Transactions on Power Electronics*, vol. 27, no. 1, pp. 99–112, January 2012.

[13] L. Vachhani, K. Sridharan, P.K. Meher, "Efficient FPGA realization of CORDIC with application to robotic exploration," *IEEE Trans. on Ind. Electron.*, vol. 56, no. 12, pp. 4915–4929, December 2009.

[14] A. Timbus, M. Liserre, R. Teodorescu, F. Blaabjerg, "Synchronisation methods for three phase distributed power generation systems – an overview and evaluation," *IEEE PESC Conf. Proceedings*, pp. 2474–2481, Recife, Brazil, June 2005.

References

Chapter 7

DSP controllers for three-phase unity-power-factor rectifiers

Marco Rivera[1], Javier Muñoz[1] and Carlos Baier[1]

7.1 Introduction

This chapter starts with an overview of the available digital signal processor (DSP) boards suitable for power converters control, indicating the main features for each case, such as analogue to digital conversion, communication channels, timers, PWM outputs, interruptions, among others. The second topic of this chapter consists in a review of the main topologies used for three-phase unity-power-factor (UPF) rectifiers, where classic six-switches topologies will be presented (Voltage and Current Source Converters), as well as novel configurations as Y- or Δ-switch rectifiers and VIENNA rectifier. As a third subject, the implementation of Phase-Locked Loop (PLL) algorithms will be analyzed in this chapter, remarking the importance of this stage to achieve the UPF operation of the rectifiers. The presented PLL algorithms will be classified as Fixed or Varying Sampling Time implementation. Finally, a couple of control algorithms will be shown, with special emphasis on *dq* frame-based controllers, *pq* theory-based controllers and predictive control. The focus in this last part will lie in how to cope with some practical issues in the implementation of the aforementioned control strategies.

7.2 DSP boards for power converters control

The digital signal processing and the use of DSPs have been increasingly necessary for systems control. At the beginning they were focused on digital filtering, speech vocoding, fast Fourier transforms, image processing, digital audio, but nowadays they are a crucial stage for power converters control. All of these applications present some issues in common such as mathematically intensive algorithms because they can handle lot of computations in a short time, operation in real time without creating any noticeable delay, implementation by using sample data and system flexibility to allow improvements thanks to the technological advances.

[1]Department of Industrial Technologies, Universidad de Talca, Merced, Curicó, Chile

At difference of the classical microcontrollers, the DSPs are two or three times faster because they are optimized for specific tasks.

Based on a recent literature review, the most used DSPs for power converter control correspond to the TMS320 family of Texas Instruments which emerged in the 1980s. Among them, the most used in power converter control are the devices C32, C28346, 6A8168, C6414, C6678, C6713, F240, F28035, F28069, F28335, F2812, and VC33.

As established earlier, the DSPs allow fast arithmetic operations and they can handle intensive algorithms in real time. These issues are accomplished by the TMS320 family by considering hardware architecture for speed and flexibility. At the same time, TMS320 family uses pipelining in order to reduce the instruction cycle time to the minimum, which allow processing from two to four instructions in parallel. Additionally, in the TMS320 family, multiplication is a single cycle instruction as a result of a dedicated hardware multiplier. Table 7.1 shows the main features of some DSP devices from the TMS320 family.

The basic elements in a DSP are: input and output converters, data memory, program memory, multipliers and accumulators, arithmetic logic unit (ALU), registers, phase-locked loop (PLL), and pulse width modulation (PWM). Figure 7.1 shows a general block diagram of one of the most used DSP for the control of power converters, the DSP C6713. The general characteristics of this DSP are as follows:

- The DSP C6713 works at 225 MHz.
- This is a floating point DSP with 32 bits.
- A JTAG support is included via USB.
- Audio codec (AIC23) of high fidelity up to 32 bits and 96 kHz.
- Four audio inputs of 3.5 mm for microphone, input/output of stereo line and headphones.
- Four LEDs and DIP switches available to the user.
- Flash memory of 512 kB and 16 MB DRAM.
- Expansion port for plug-in modules.
- IEEE JTAG interface included.
- Universal source port of +5 V.

7.3 Topologies for three-phase unity-power-factor rectifiers

Several topologies can be used as UPF rectifiers; in fact References 1 and 2 show an exhaustive review of suitable configurations to achieve unitary power factor. In these papers, a classification into passive, hybrid and active systems is proposed. It is not the aim of this chapter to present a thorough description of the available topologies; then, just the most widespread configurations will be depicted, as well as the emerging structures used to implement UPF rectifiers. Particularly, classic three-phase voltage (VSR) and current (CSR) source rectifiers are presented, and among the emerging topologies, Y- and Δ-switch rectifiers and VIENNA rectifier are studied.

Table 7.1 Features of some DSP devices from TMS320 family

	C32	**C28346**	**6A8168**	**C6414**	**C6678**	**C6713**	**F28069**	**F28335**
Instruction cycle (ns)	33	3.3	–	2	0.714	4.4, 5, 6	11.11	6.67
Clock frequency	–	300 MHz	1.5 GHz	–	1.4 GHz	225–200 MHz	90 MHz	150 MHz
MOPS	330	–	–	–	–	–	–	–
MFPOPS	60	–	–	–	–	1350/1200/1000	–	–
MIPS	30	–	–	4000	–	1800/1600/1336	–	–
CPU	32 bits	32 bits	DSP + ARM Processors	32 bits	32 bits	–	32 bits	32 bits
Memory	$2 \times 256 \times 32$ bits single cycle, dual access on chip RAM blocks	258 kB × 16 SARAM 8 kB × 16 Boot ROM		1056 kB	88832 kB	264 kB	1056 kB	256 kB Flash 34 kB RAM 1 kB ROM

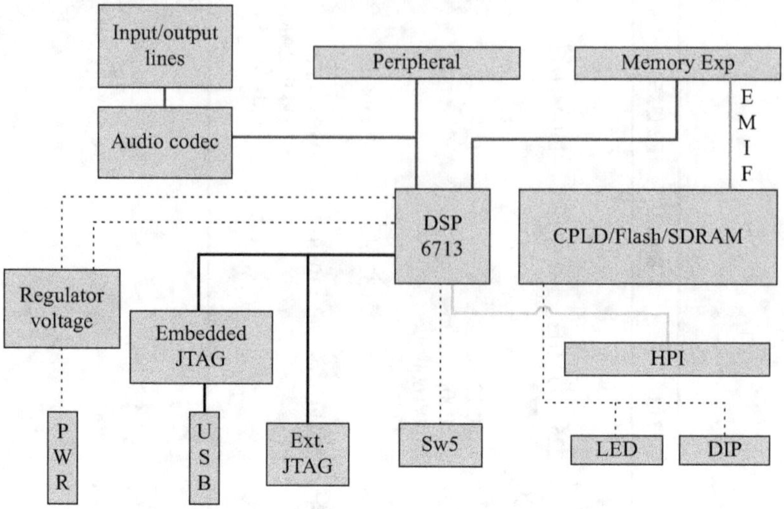

Figure 7.1 Block diagram of the DSP C6713

7.3.1 Three-phase rectifiers: VSR and CSR

The diagrams of the VSR and CSR are shown in Figure 7.2. In both cases, VSR and CSR, the gating patterns are adjusted in order to follow the reference either in the *abc*, $\alpha\beta$, or *dq* frame [3], [4]. Usually, the systems' outputs are the *DC*-side variable (voltage or current) and the input current [5], [6]. This last variable allows controlling the rectifiers in order to obtain the unitary power factor operation.

The digital implementation of the carrier-based PWM technique in the VSR is quite trivial, as it is the direct comparison among the modulating signals and the triangular wave. However, for the CSR there is a logic circuit that must be implemented in order to generate the gating patterns [7]. This circuit can be directly implemented in a digital platform or some other practical approaches can be proposed; for instance, a look-up table is suitable to map the equivalent states of the CSR, which is especially useful for the space vector modulation (SVM) technique. The equivalence between the active states of the VSR and the CSR is shown in Table 7.2. It is important to highlight that the resultant input current of the CSR will be displaced 30° compared to the VSR modulator reference waveforms due to the phase shift in the state phasors used by the CSR [8]. This phenomenon can be easily compensated by phase shifting the reference inputs to the modulator.

Another way to generate the active states of the CSR, which is more suitable for carrier-based PWM, is to calculate the pulses with the value of the comparison with the triangular wave. In fact, the code shown in Figure 7.3 can generate the active vectors of the CSR, using the comparison with the triangular wave. Regarding the zero states, it is not straightforward to obtain the null vectors because there are many available combinations to use them. In fact, the previous

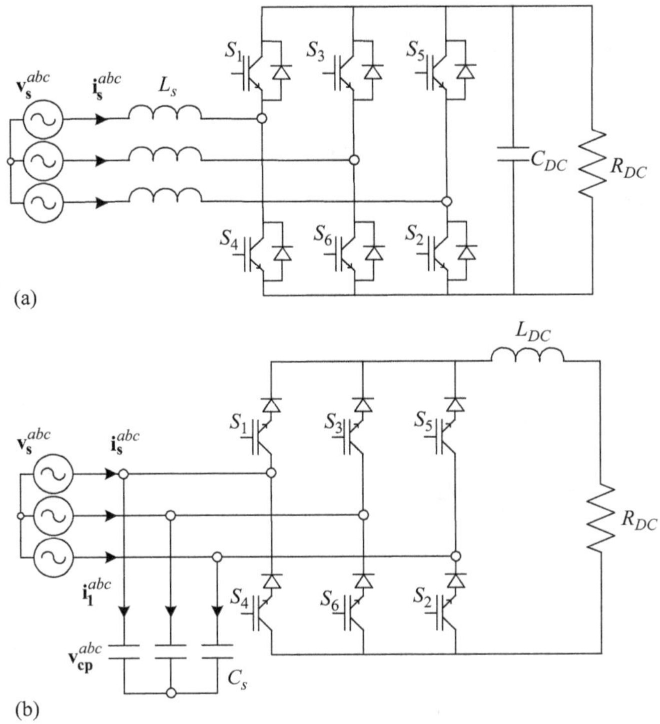

Figure 7.2 (a) Voltage source rectifier and (b) current source rectifier

Table 7.2 Active states equivalence between VSR and CSR

	VSR						CSR						AC side		
	S_1	S_2	S_3	S_4	S_5	S_6	S_1	S_2	S_3	S_4	S_5	S_6	v_{ab}/i_a	v_{bc}/i_b	v_{ca}/i_c
1	1	1	0	0	0	1	1	1	0	0	0	0	v_{dc}/i_{dc}	0	$-v_{dc}/i_{dc}$
2	1	1	1	0	0	0	0	1	1	0	0	0	0	v_{dc}/i_{dc}	$-v_{dc}/i_{dc}$
3	0	1	1	1	0	0	0	0	1	1	0	0	$-v_{dc}/i_{dc}$	v_{dc}/i_{dc}	0
4	0	0	1	1	1	0	0	0	0	1	1	0	$-v_{dc}/i_{dc}$	0	v_{dc}/i_{dc}
5	0	0	0	1	1	1	0	0	0	0	1	1	0	$-v_{dc}/i_{dc}$	v_{dc}/i_{dc}
6	1	0	0	0	1	1	1	0	0	0	0	1	v_{dc}/i_{dc}	$-v_{dc}/i_{dc}$	0

code is incomplete, because it does not ensure that two switches will be always on; therefore it must be complemented with the code lines shown in Figure 7.4.

In the case of the SVM, it is possible to generate the zero states considering an equivalent switching behavior of the saw tooth carrier, which in fact achieves the minimum switching frequency. Table 7.3 summarizes the zero states for this case. Further details about the modulation can be found in Reference 8.

```
// Comparison with carrier
// All the states are initially in zero
        sa = 0.0;
        sb = 0.0;
        sc = 0.0;
// And they are changed to one, only with the comparison
        if (carrier < ma) sa = 1.0;
        if (carrier < mb) sb = 1.0;
        if (carrier < mc) sc = 1.0;
// Finally, each switch state is calculated
        s1 = sa * (1. - sb);
        s2 = sa * (1. - sc);
        s3 = sb * (1. - sc);
        s4 = sb * (1. - sa);
        s5 = sc * (1. - sa);
        s6 = sc * (1. - sb);
```

Figure 7.3 Generation of active states in the CSR topology using carried-based PWM

```
// Total number of switches turned on
Ts = s1 +s2 + s3 + s4 + s5 + s6;
// If only one switch is on
        if (Ts < 2) {
                if (ma >= mb & ma >= mc)
                        // First zero state is applied
                        s1 = 1; s2 = 0; s3 = 0; s4 = 1; s5 = 0; s6 = 0;
                if (mb >= ma & mb >= mc)
                        // Second zero state is applied
                        s1 = 0; s2 = 1; s3 = 1; s4 = 0; s5 = 0; s6 = 0;
                if (mc >= ma & mc >= mb)
                        // Third zero state is applied
                        s1 = 0; s2 = 0; s3 = 0; s4 = 1; s5 = 1; s6 = 0;
                        }
```

Figure 7.4 Generation of zero states in the CSR topology using carried-based PWM

Table 7.3 Generation of zero states in the CSR topology using SVM

Sextant	**Last active state #1**	**Last active state #2**	**NULL**					
			S_1	S_2	S_3	S_4	S_5	S_6
$-\pi/6\rightarrow\pi/6$	6	1	0	0	1	0	0	1
$\pi/6\rightarrow\pi/2$	1	2	0	1	0	0	1	0
$\pi/2\rightarrow 5\pi/6$	2	3	1	0	0	1	0	0
$5\pi/6\rightarrow 7\pi/6$	3	4	0	0	1	0	0	1
$7\pi/6\rightarrow 3\pi/2$	4	5	0	1	0	0	1	0
$3\pi/2\rightarrow -\pi/6$	5	6	1	0	0	1	0	0

Regarding the practical implementation of the PWM techniques, particularly in the VSR the switching is made using dead times, which can be implemented by hardware and/or software. Some DSPs has some embedded functions to include the dead times or it can be externally implemented using Field programmable gate arrays (FPGAs). Also there are some integrated circuits that have internal circuitry to implement the dead times. On the other hand, the CSR needs to consider overlap times in order to ensure the safe commutation of the semiconductor devices. Similar to VSR, these times can be implemented by hardware and/or software, but latest trends privilege the use of software tools in order to reduce space and cost of the units.

7.3.2 Novel topologies: Y- or Δ-switch rectifier and VIENNA rectifier

As an attempt to reduce the semiconductor stress, the electromagnetic interference (EMI) and the reliability of the active rectifiers, some novel topologies can be found in technical literature [1], [2]. Although some of these topologies were proposed during early nineties [9], they have become popular in present years due to the necessity to improve the power quality in almost every electric system. In this context, two similar topologies were proposed [9] as a suitable solution to achieve unitary power factor in three-phase systems: Y- and Δ-switch rectifiers, which are shown in Figures 7.5 and 7.6, respectively. Both topologies behave as a boost converter – just like the VSR – and therefore the output *DC* voltage is higher than

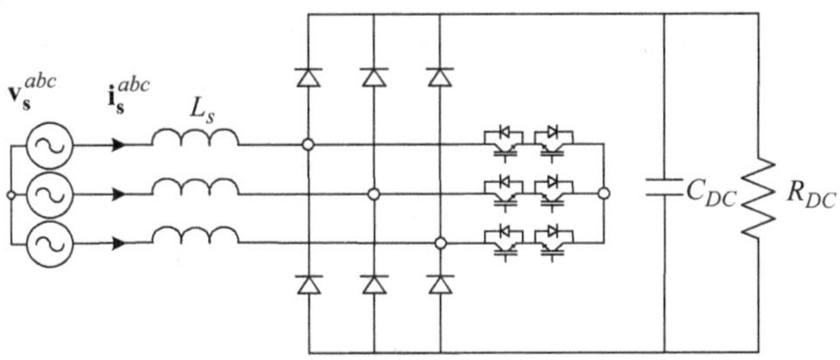

Figure 7.5 Y-switch rectifier

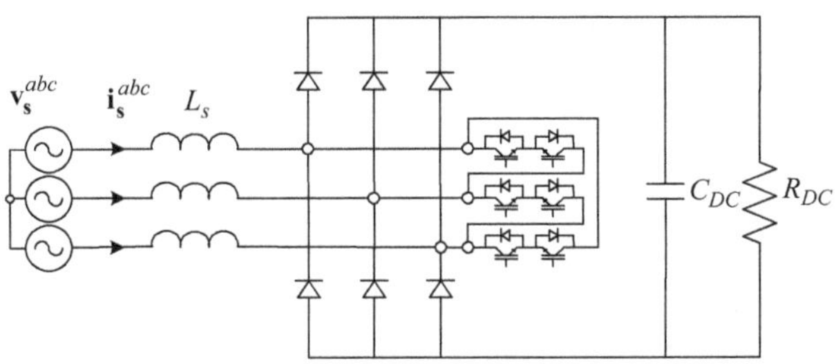

Figure 7.6 Δ-switch rectifier

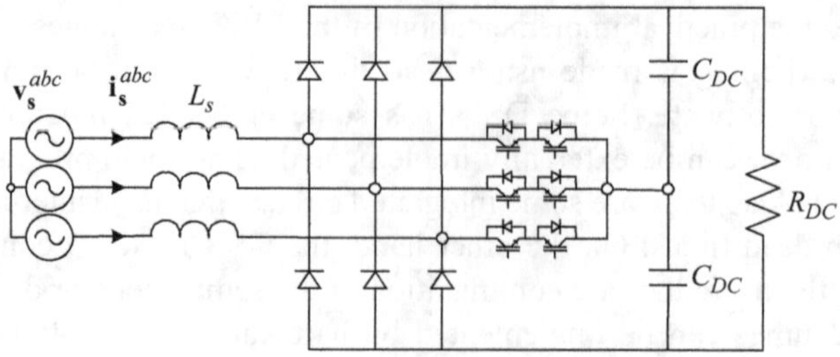

Figure 7.7 VIENNA rectifier

the input *AC* mains. The active switches handle a smaller amount of power than their VSR counterparts, because they are used to shape the current as a sinusoidal waveform by injecting current from the other phases [1]. These topologies provide a natural protection against shoot-through, in contrast to the six switch PWM rectifier, where a failure can cause a short circuit of the *DC* voltage [10].

A restriction of these converters is that they can only operate with unidirectional power flow due to the inherent behavior of a diode bridge. However, thanks to the switches commutation, they can have a displacement angle between −30° and +30° [1]. This power flow restriction may discard these topologies for several applications; nevertheless in other cases, this is a desirable condition, e.g. in aircrafts electric systems [10]. It is important to remark that these topologies require bidirectional switches, as the current needs to flow in both directions in order to obtain sinusoidal waveforms in all the three phases. In this sense, the Y-switch rectifier has more losses because the current flows through two switches (four semiconductors), in contrast to the Δ-switch rectifier, where the current goes through only one switch to travel from one phase to another. These rectifiers are known as two level converters because their phase input is connected either to the positive or negative point of the *DC* bus. However, if the *DC* capacitor is split up in two and the central point is connected to the midpoint of the Y-switch rectifier, a resultant three level converter is obtained, as now its phase inputs can be connected to the positive, negative and the central point of the *DC* bus. This converter is known as the VIENNA rectifier and its structure can be appreciated in Figure 7.7. The advantage of the three-level feature of the VIENNA rectifier is that the active semiconductors must handle only half of the voltage compared to the Y- and Δ-switch rectifiers. On the other hand, they have in common that the power flow is unidirectional and the displacement angle is in between −30° and +30° [1].

Other realizations of the VIENNA converter have been found in literature, such as the presented in Figure 7.8. Particularly, this configuration is advantageous concerning pre-charging the output capacitor at start-up [11], [12]. Other realizations can be found in technical literature, aimed to reduce losses, lessen the number of transistors and/or lower the EMI [13], [14].

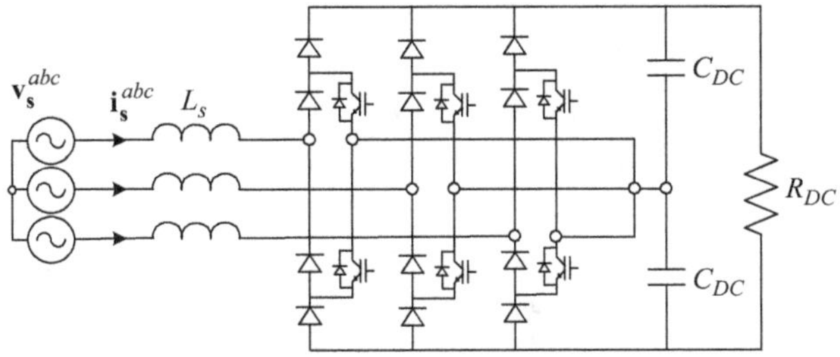

Figure 7.8 Alternative VIENNA rectifier

7.4 Phase-locked loops algorithms

The VSR and CSR rectifiers need to be synchronized with the *AC* grid to control the *DC* voltage or *DC* current as well as the power factor at the point of common coupling (PCC). A simple way to achieve the synchronization with the voltage at the PCC is by detecting the zero crossing of this signal. However, this method is not the most appropriate for industrial systems, where noise and distortion are common.

For PWM rectifiers, the synchronization with the *AC* voltage signal is nowadays performed through digital PLL algorithms. A digital PLL basically takes samples of the PCC voltage signal, which could be distorted and with noise. Internally in the DSP where the PLL has been programmed, a sinusoidal signal is reconstructed in phase or with a phase shift of 90°, at the same fundamental frequency than the original signal.

Owing to the sine and cosine signals delivered by the PLL algorithm, it is possible to not only synchronize the switching of different types of rectifiers but also give rise to different control strategies based on orthogonal signals, as the strategies in the *dq0* framework.

For single-phase and three-phase rectifiers, there are two types of PLL algorithms that stand out and can be implemented in digital processing systems: (a) the PLL algorithms with fixed sampling time, and (b) the PLL algorithms with varying sampling time.

This section reviews some of the most used PLL algorithms for the synchronization of PWM rectifiers. It is expected to give the readers a guide that lets them decide what is the best algorithm to be used at the moment of implementing a rectifier.

PLL algorithms generally have a structure that follows a classic design that is known since the 1930s in the past century [15]. This structure can be seen in Figure 7.9. The PLLs include a phase detector element, a low-pass filter, and a voltage controlled oscillator, which today can be digitally implemented in different ways.

In active rectifier circuits, the synchronization with the *AC* grid is a key to achieve the objective of controlling the *DC* variable and the power flow between

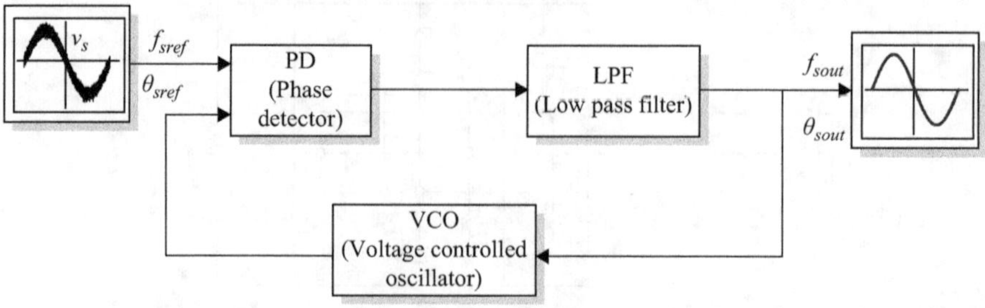

Figure 7.9 Basic phase-locked loop structure

the *AC* grid and the load. Synchronization alternatives based on PLLs have been used since rectifiers were built from mercury vapor valves [16]; however, the use of a PLL was required primarily to perform a correct zero crossing technique or to achieve a good phase control in thyristor rectifiers [17].

In the late 1970s, the digital PLLs were already known in a wide range of applications [18]. However, it was not until the mid-1990s that the implementation of a PLL in a digital system (mainly DSP) became common for PWM rectification systems [19], [20].

7.4.1 Implementation of PLL algorithms with fixed sampling time in three-phase systems

It is a known fact that a PLL algorithm that takes advantage of all phases of a three-phase system converges faster and is more robust to distortion or noise that algorithms for single-phase systems [21].

Many PLL algorithms consider fixed sampling time in acquiring the signals of interest (usually voltages v_a, v_b, v_c). Among them, those based on the *pq* instantaneous power theory developed by Akagi *et al.* in Reference 22 are the most used in three-phase systems. These algorithms are based on the Clarke and Park transformations for detecting the phase of the positive sequence of the three-phase voltage. Some studies classified these algorithms as *p*PLL and *q*PLL [23]–[26]. A *p*PLL algorithm considers the direct component of the Park transformation as the output in the phase detector, as shown in Figure 7.10(a), whereas a *q*PLL algorithm considers the quadrature component as the output in the phase detector, as it can be seen in Figure 7.10(b).

The Park transformation used as phase detector receives the output phase $\theta_o(k)$, delivered by the *p*PLL. Therefore, the instantaneous direct component *d*(*k*) of the transformation can be written as

$$d(k) = \frac{2\hat{V}}{3}\left(\sin(\theta_i(k))\cos(\theta_o(k)) + \sin\left(\theta_i(k) - \frac{2\pi}{3}\right)\cos\left(\theta_o(k) - \frac{2\pi}{3}\right) + \sin\left(\theta_i(k) + \frac{2\pi}{3}\right)\cos\left(\theta_o(k) + \frac{2\pi}{3}\right)\right) \tag{7.1}$$

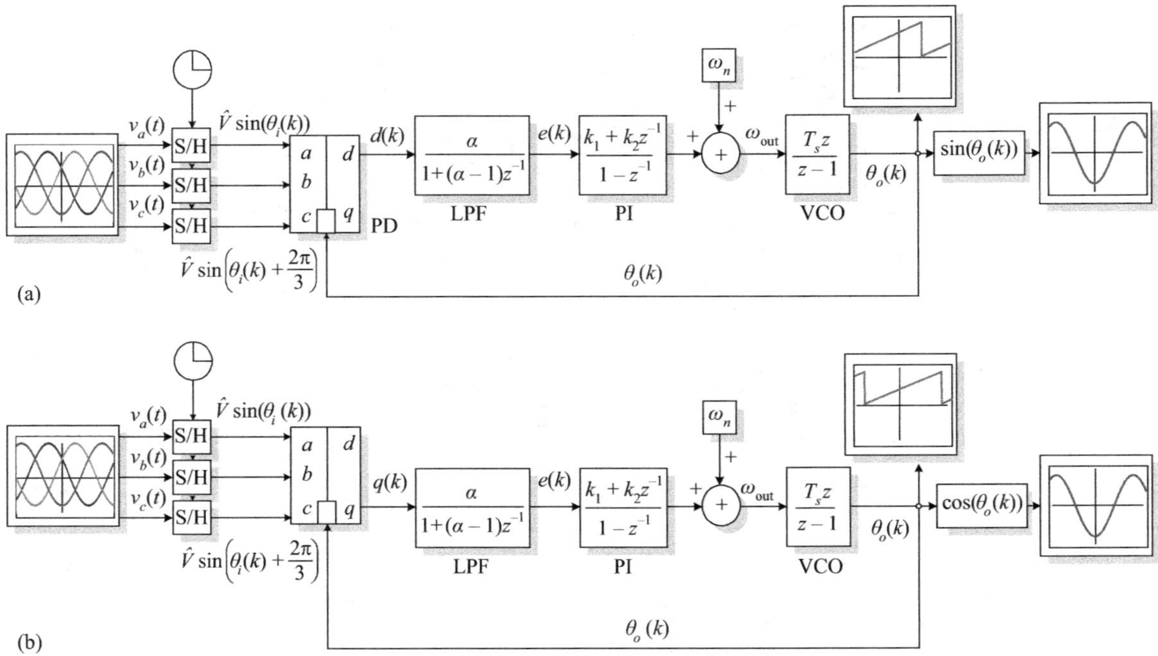

Figure 7.10 Basic three-phase fixed sampling time PLL algorithms (a) pPLL (b) qPLL

where $\hat{V}$ is the amplitude of the input signal, $\theta_i(k)$ is the phase of the positive sequence of the input signal, and $\theta_o(k)$ is the estimated phase of the positive sequence given by the *p*PLL. Working with the expression (7.1), from trigonometric identities, it is possible to find that

$$d(k) = \hat{V}(\sin(\theta_i(k) - \theta_o(k))). \tag{7.2}$$

If a small difference between angular positions is considered, the above equation can be expressed as

$$d(k) \approx \hat{V}(\theta_i(k) - \theta_o(k)) \tag{7.3}$$

Therefore the signal $d(k)$ is an error signal between the phase of the input signal and the estimated phase angle given by the PLL. A similar analysis can be made for the signal $q(k)$ in the *q*PLL. Following the classic design for a PLL (see Figure 7.9) and considering industrial conditions (where the input signals may have noise and distortion), the programming of both algorithms in a DSP requires of a filter. This stage can be a recursive first-order low-pass filter, located at the output of the phase detector, as it can be seen in Figure 7.10(a) and (b). The output of the digital filter delivers the filtered error signal to the PI controller. This controller delivers a frequency signal, which allows the convergence to zero of the filtered error. This PI output signal is summed with a feed-forward frequency value (ω_n) in order to increase the dynamic performance of the PLL. The sum of the PI output signal and the feed-forward frequency value passes through an integrator, which allows to find the estimated phase of the positive sequence signal. This estimated phase is feedback to the Park transformation block, which operates as a phase detector. These algorithms are easily implemented in the timer interrupt routine of a DSP. An interrupt routine of a *p*PLL is provided as an example in Figure 7.11.

One advantage of the PLL algorithms with fixed sampling time is that they can be easily simulated. To demonstrate this, a simulation was performed herein to show how quickly a *q*PLL algorithm can converge during the starting stage or whether is undergone to an abrupt phase shift in its input voltages (see time 0.025 s). Table 7.4 summarizes the *q*PLL parameters considered in the simulations and Figure 7.12 shows the simulation results (phase *a* of the three-phase voltage, the cosine output, and error signal).

7.4.2 *Implementation of single-phase PLLs with fixed sampling time*

For single-phase systems, it is possible to take the same idea presented earlier. However, for getting a direct component $d(k)$ (as done in a *p*PLL), it is necessary to consider a system with at least two phases (that does not exist in a single-phase system). For this reason, the PLL should digitally generate a signal orthogonal to the input, so the DSP can obtain the $\alpha\beta$ signals to be used in the PLL. Although this is not a simple task, it can be performed through the use of direct and inverse Park transformations as it can be seen in Figure 7.13.

```
interrupt void cpu_timer0(void)
{
        // A/D conversion sample of the phase voltage "a"
        Va = ( (float) ( (AdcRegs.ADCRESULT0) >> 4) - Voffset) * Gsensor;
        // A/D conversion sample of the phase voltage "b"
        Vb = ( (float) ( (AdcRegs.ADCRESULT1) >> 4) - Voffset2) * Gsensor;
        // A/D conversion sample of the phase voltage "c"
        Vc = ( (float) ( (AdcRegs.ADCRESULT2) >> 4) - Voffset3) * Gsensor;
        //d voltage obtained  in the case of pPLL
        // COS() is a function that interpolate cosine values to reduce the CPU burden.
        // outVCO is the output angle of PLL
        vd = (Va * COS(outVCO) + Vb * COS(outVCO - 2 * pi / 3) + Vc * COS(outVCO + 2 * pi/3) );
        // first order digital filter (alpha value must be less than 1)
        ek = ek + alpha * (vd - ek);
        // PI control, where Kp and Ti are parameters of the PI, and Ts is the sampling time
        oPI = oPI + ek * Kp * (1 + Ts/Ti ) - ek_1 * Kp;
        ek_1= ek; // the error value is stored to be used in next execution of timer interrupt
        // an integrator is used as VCO, here a feed-forward of the expected frequency (wn) is considered
        outVCO = outVCO + Ts * (oPI + wn);
        if (outVCO >= 2 * pi) //  the output is limited each cycle
                outVCO = outVCO - 2 * pi;
                ...
}
```

Figure 7.11 Interrupt routine of a three-phase pPLL algorithm with fixed sampling time

Table 7.4 Parameters considered for simulation of a qPLL

Parameters	Description	Value
Kp	Proportional gain of PI	25
Ti	Integral time constant of PI	1
α	Gain of low-pass filter	0.09
Ts	Sampling time A/D conversion	100 μs

In this scheme, a pair of poles must be incorporated between transformations, which allow ensuring a proper $\alpha\beta$ set of variables as well as an adequate bandwidth.

For the DSP implementation in the timer interrupt routine, it can proceed as shown in Figure 7.14.

As expected, a single-phase PLL (based on Clarke and Park transformations) is a bit slower than a three-phase PLL. A simulation of the single-phase PLL scheme shows the response in the starting stage and undergone an abrupt phase shift. Table 7.5 summarizes the PLL parameters considered in the simulations and Figure 7.15 shows the simulated results.

Other single-phase PLL alternatives differ only in the way to obtain the $\alpha\beta$ signals inside the DSP. For example, to get a 90° phase-shifted signal in a DSP, it is possible to use a digital filter based on the Hilbert Transformation (HT) [27], [28],

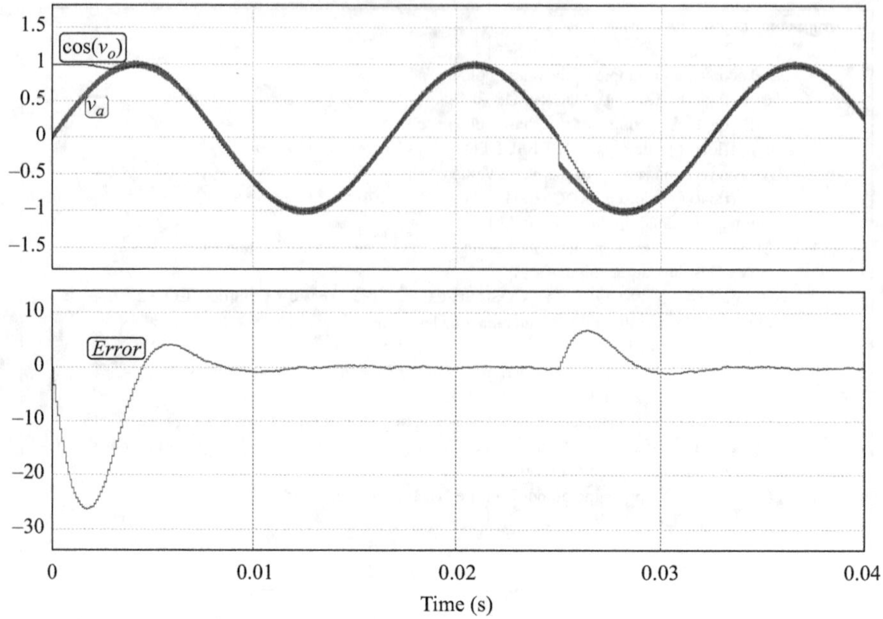

Figure 7.12 Key waveforms of a simulated qPLL

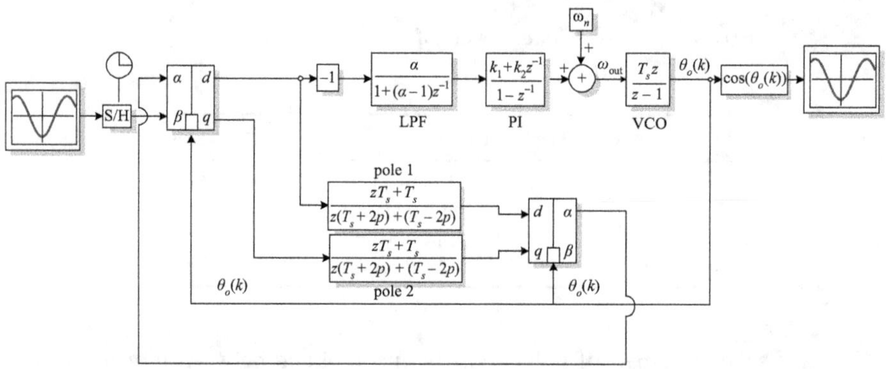

Figure 7.13 Single-phase fixed sampling time PLL algorithm with poles to generate phase-shifts

as shown in Figure 7.16(a). The digital filter needed to implement the HT is a finite impulse response (FIR) filter, whose coefficients $h(k)$ are the following:

$$h(k) = \begin{cases} \dfrac{1 - \cos[(k - 0.5N)\pi]}{(k - 0.5N)\pi} & \text{for } k \neq 0.5N \\ 0 & \text{for } k \neq 0.5N \end{cases} \tag{7.4}$$

```
interrupt void cpu_timer0_isr(void)
{
        ...
        // A/D conversion sample of the phase voltage "a"
        beta = ( (float) ( (AdcRegs.ADCRESULT0) >> 4) - Voffset) * Gsensor;
        vq = -alpha * SINp (outVCO) + beta* COSp (outVCO);
        vd = alpha * COSp (outVCO) + beta* SINp (outVCO);
        // COSp() is a function that interpolate cosine values to reduce the CPU burden.
        // SINp() is a function that interpolate sine values to reduce the CPU burden.
        // ------TF poles to help generate phase shifts -------
        yd = (vd * b0d + vdm1* b1d + a1d * ydm1) / a0d;
        yq = (vq * b0q + vqm1 * b1q +a1q* yqm1) / a0q;
        ydm1 = yd;
        // --- storing output values of TF poles
        yqm1 = yq;
        // dq/alfa-beta transformation
        Alpha = COSp (outVCO) * yd - SINp (outVCO) * yq;
        // first order digital filter (gain value must be less than 1)
        ek = ek + gain * (-vd - ek);
        // PI control, where Kp and Ti are parameters of the PI, and Ts is the sampling time
        oPI = oPI + ek * Kp * (1 + Ts/Ti) - vdfm1 * Kp;
        // the previous filtered "d" voltage value is stored
        ek_1 = ek;
        // the previous "d" voltage value is stored
        vd_1 = vd;
        // the previous "q" voltage value is stored
        vq_1 = vq;
        // an integrator as VCO, here a feedforward of the expected frequency is considered.
        outVCO = outVCO + Ts * (oPI+wes);
        //  the output is limited each cycle
        If (outVCO >= 2 * pi)
        outVCO = outVCO - 2 * pi;
        ...
}
```

Figure 7.14 Interrupt routine of a single-phase pPLL algorithm with fixed sampling time

Table 7.5 Parameters considered for simulation of a single-phase pPLL

Parameters	Description	Value
Kp	Proportional gain of PI	80
Ti	Integral time constant of PI	10
α	Gain of low-pass filter	0.035
Ts	Sampling time A/D conversion	100 μs

where N is the filter order and k is the coefficient index. As a second example, a simpler method uses a delay of one quarter cycle ($T/4$) for the input signal, thus obtaining the β signal, as shown in Figure 7.16(b) [29], [30]. However, this method is not suitable for rectifiers if the frequency of the AC grid is continuously changing.

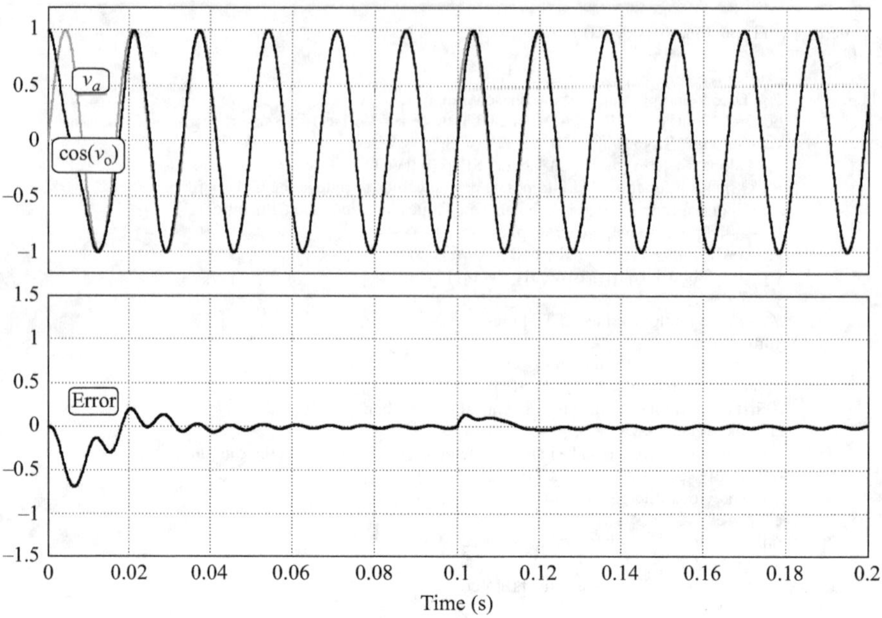

Figure 7.15 Key waveforms of a simulated single-phase PLL

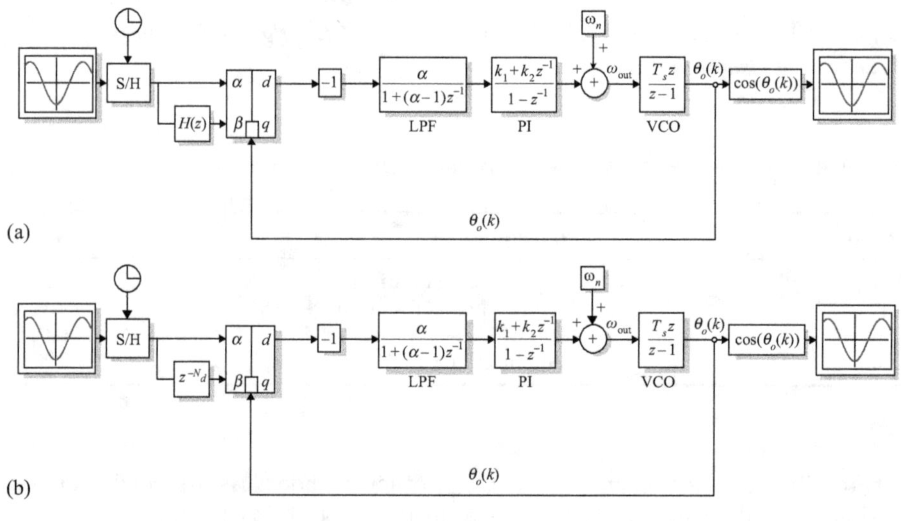

Figure 7.16 Single-phase fixed sampling time PLL algorithms. (a) Based on the Hilbert Transformation, (b) using a delay of one quarter cycle to generate phase shift

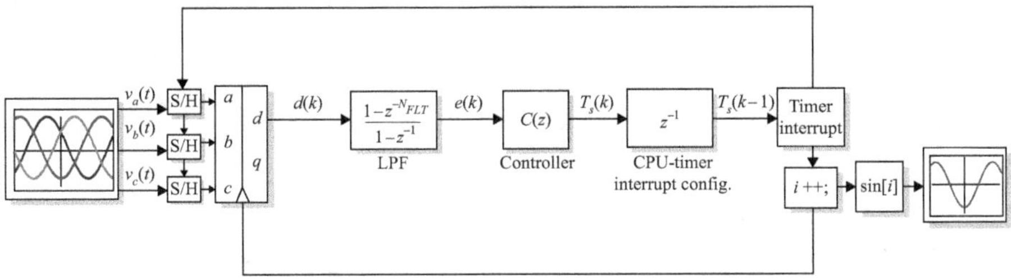

Figure 7.17 Block diagram of a three-phase varying sampling time PLL algorithm

7.4.3 Implementation of PLL algorithms with varying sampling time in three-phase systems

A discrete time-varying PLL can be very useful because it always works with an exact number of N samples per cycle. These algorithms allow, for example, store one cycle of a sine signal on N points in the memory of the DSP, and adjust this sine signal exactly at the frequency and phase of the input signal, without the need to calculate, interpolate, or approximate the value of the sinusoidal signal in the time interrupt of the DSP. This characteristic can help to reduce the burden on the processor, in contrast with the PLL with fixed sampling time.

With a discrete time-varying PLL, if one wants to generate an internal signal, phase shifted 120° from the input wave, it is possible to use the signal stored in N samples inside a look-up table, and offset its recurrence index $2N/3$ times, since the number of samples N is proportional to two times "pi" number.

A three-phase PLL algorithm with varying sampling time requires a phase detector that, as in the case of the fixed sampling time algorithm, may be the Park transformation. Since the d component of this transformation has enough information to detect the phase of the positive sequence signal, this signal must be filtered to reduce any variation caused by noise or distortion. Because the number of samples per cycle N is known beforehand, a windowed low-pass filter is a good choice to be implemented as a part of the scheme in the PLL.

As shown in Figure 7.17 and just like the fixed sampling time algorithms, the output of the filter is interpreted as an error signal, which must pass through a controller $C(z)$ to assure that it is equal to zero in steady state.

Once again, the controller $C(z)$ can be a simple PI. However, and for this case, the output of the control must give, as a result, a new period $T(k+1)$ for the timer interruption and therefore for the sampling time [31]. This new period is set into the DSP timer interruption configuration, and after setting it, the execution of the next timer interruption is expected again. Once at the next timer interruption, the input signals are sampled, the index that modifies the phase is updated and the loop returns to the start.

Figure 7.18 shows an example of the code for this PLL, programmed in the timer interrupt routine of a DSP.

```
    interrupt void cpu_timer0_isr(void)
{
    CpuTimer0.InterruptCount++;
    // ----ADC input -----//
    //Synchronization voltage "a"
    Vpcc_a = (float) ( (AdcRegs.ADCRESULT0) >> 4) - Vprom.;
    //Synchronization voltage "b"
    Vpcc_b = (float) ( (AdcRegs.ADCRESULT2) >> 4) - Vprom.;
    Vpcc_c = -Vpcc_a -Vpcc_b;
    //---- PLL Control----//
    sin0 = sine [indexPLLs0]; sin120 = sine [indexPLLs120]; sin240 = sine [indexPLLs240];
    cos0 = sine [indexPLLc0]; cos120 = sine [indexPLLc120]; cos240 = sine [indexPLLc240];
    dk = Vpcc_a * cos0 + Vpcc_b * cos120 + Vpcc_c * cos240;
    // moving window filter
    V = Vec [indexPLLs0];
    // window filter
    ek = ek_1 + (dk -V);
    // the signal is stored into the window for the next execution of the filter
    Vec [indexPLLs0] = dk;
    //PI controller
    Ts = Ts - kp * ek + ki * kp * ek_1;
    ek_1 = ek;
    /// sine index ///
    indexPLLs0++;
    if (indexPLLs0 == N)
            indexPLLs0 = 0;
    indexPLLs120++;
    if (indexPLLs120 == N)
            indexPLLs120 = 0;
    indexPLLs240++;
    if (indexPLLs240==N)
            indexPLLs240 = 0;
    /// cosine index ///
    indexPLLc0++;
    if (indexPLLc0 == N)
            indexPLLc0 = 0;
    indexPLLc120++;
    if (indexPLLc120 == N)
```

Figure 7.18 Interrupt routine of a three-phase PLL algorithm with varying sampling time

7.4.4 Implementation of single-phase PLLs with varying sampling time

A single-phase PLL with varying sampling time works similarly to the three-phase algorithm, except for the way in which the input and output phases are compared. To compare both phases, it is necessary to multiply the input signal by the cosine of the output phase resulting from the PLL.

If the expression of this multiplication is written as in (7.5), then it is possible to see that the product of the trigonometric functions will lead to two superimposed sinusoidal signals.

```
                indexPLLc120 = 0;
        indexPLLc240++;
        if (indexPLLc240 == N)
                indexPLLc240 = 0;
        /// End PLL Control ///
        If (Ts < Ts_min)
                // Min sample time
                Ts = Ts_min;
        if (Ts > Ts_max)
                //Max sample time
                Ts = Ts_max;
        TsInt=(int)Ts;
        ConfigCpuTimer(&CpuTimer0, 5, TsInt);
        // Use write-only instruction to set TSS bit = 0
        CpuTimer0Regs.TCR.all = 0x4001;
        IER |= M_INT1;
        // Enable TINT0 in the PIE: Group 1 interrupt 7
        PieCtrlRegs.PIEIER1.bit.INTx7 = 1;
        // Enable ADCINT in PIE
        // Enable global Interrupts and higher priority real-time debug events:
        // Enable Global interrupt INTM
        EINT;
        // Enable Global realtime interrupt DBGM
        ERTM;
}
```

Figure 7.18 Continued

$$U(k) = v_a \sin(\theta_i(k))\cos(\theta_o(k)) = \frac{v_a}{2}(\sin(\theta_i(k) + \theta_o(k)) - \sin((\theta_i(k) - \theta_o(k))) \tag{7.5}$$

Taking into account that the low-pass filter removes higher frequencies and considering small differences between the two phases, the filter output can be written as

$$e(k) = -\frac{v_a}{2}\sin(\theta_i(k) - \theta_o(k)) \approx \frac{v_a}{2}[\theta_i(k) - \theta_o(k)] \tag{7.6}$$

Thus, multiplication in (7.5) is directly related to the error between the phase of the input signal and the output phase of the PLL. This error signal is then fed to the controller, which gives as a result a new period for the timer interruption associated to the signal acquisition, such that the signal (sine or cosine) to be rebuilt, always use the same number of samples per cycle. The new period set in the timer for the interruption is not executed until the current count is completed in the timer (just like happens in the three-phase algorithm), therefore, there is an associated delay, as shown in Figure 7.19. Once executed, the interruption modifies the *k* index in the cosine function, and triggers the signal acquisition.

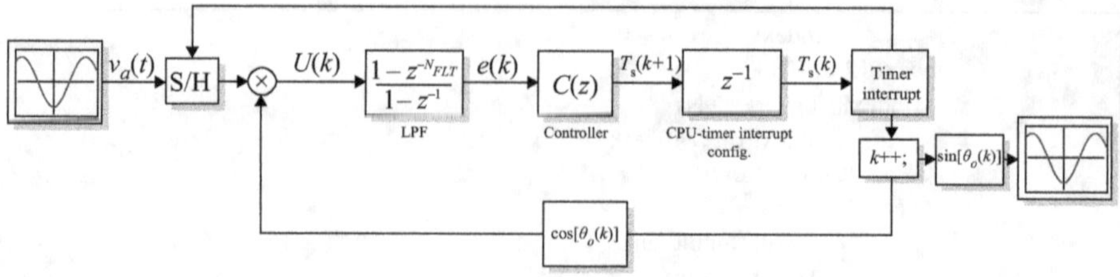

Figure 7.19 Block diagram of a single-phase varying sampling time PLL algorithm

In this scheme, different window filters and different controllers can be used [32], [33]. However, a filter that achieves excellent results considers the number of samples for the window N_{FLT} equal to half the total number of samples ($N/2$).

For the controller, a simply PI can be used, as done in Reference 32. However, better results are obtained if a gain, two zeros and two poles, is considered in the controller [33], as follows:

$$C(z) = K\frac{(z-a)^2}{z(z-1)} \tag{7.7}$$

It is important to note that the output phase of the PLL is exactly 90° phase shifted from the phase of the input signal. Figure 7.20 shows a code example that considers the controller described by (7.7).

To demonstrate that this algorithm can quickly achieve the tracking of a sinusoidal signal v_a, this strategy has been implemented in a TMS320F28335 DSP. The parameters used in the PLL implementation are given in Table 7.6. The challenge of the proposed experience is that the input signal to be tracked by the PLL suffers a delay jump of 20° in $t = 12$ ms. As shown in Figure 7.21, after the disturbance, a significant error between phases of the input signal and the PLL output is noticed (in time $t = 24$ ms). This error begins to be reduced until a synchronized signal is achieved again in almost two cycles ($t = 48$ ms).

7.4.5 Comments

For rectifiers, the use of a PLL not only allows a successful synchronization with the *AC* grid but rather it also helps in generating useful variables for different control strategies. Today, it is possible to program a digital PLL in a DSP or other programmable devices. As shown in this chapter, we might consider two types of PLL algorithms, with fixed sampling period and variable sampling period. Both kind of algorithms are easy to implement, nevertheless, if what we need is to create an internal sinusoidal signal in phase or with some shifting from the input signal phase, may be more convenient to implement a PLL with variable sampling period, because the PLL with fixed sampling period possess a greater burden on the

```
interrupt void cpu_timer0_isr(void)
{
        // The input signal is normalized with the gain (Ngain)
        Va = ( (float) ( (AdcRegs.ADCRESULT0) >> 4) - Vprom) * Ngain;
        // moving window filter
        V = Vec[indexPLL];
        // Input signal is multiplied by cosine of the output.
        Uk = Va * sine [index_c0];
        // Window filter
        ek = ek_1 + (Uk - V);
        // The signal is stored to the window for the next execution of the filter.
        Vec [indexPLL] = Uk;
        // control is applied
        Ts = Ts + K * ek - 2 * a * K * ek_1 + K* a * a * ek_2;
        // e(k-2) is stored for the next execution.
        ek_2 = ek_1;
        //  e(k-1) is stored for the next execution.
        ek_1 = ek;
        /// the sampling time is limited below and above ///
        If (Ts > 122)
                Ts = 122;
        If (Ts < 110)
                Ts = 110;
        /// sine index ///
        index_s0++;
        if (index_s0 == N)
                  index_s0 = 0;
        /// cosine index ///
        index_c0++;
        if (index_c0 == N)
                index_c0 = 0;
        /// PLL index ///
        indexPLL++;
        if (indexPLL == N / 2)
                indexPLL = 0;
        ConfigCpuTimer (&CpuTimer0 , 150 , (int) Ts);
        // Use write-only instruction to set TSS bit = 0
        CpuTimer0Regs.TCR.all = 0x4001;
        // Enable CPU INT1 which is connected to CPU-Timer 0:
        IER |= M_INT1;
        // Enable TINT0 in the PIE: Group 1 interrupt 7
        PieCtrlRegs.PIEIER1.bit.INTx7 = 1;

        // Enable global Interrupts and higher priority real-time debug events:
        // Enable Global interrupt INTM
        EINT;
        // Enable Global realtime interrupt DBGM
        ERTM;
}
```

Figure 7.20 Interrupt routine of a single-phase PLL algorithm with varying sampling time

Table 7.6 Parameters in experimental results of a single-phase varying sampling time PLL

Parameters	Description	Value
K	Proportional gain of controller	25
a	Double zero value of controller	0.98
N	Number of samples per cycle	168

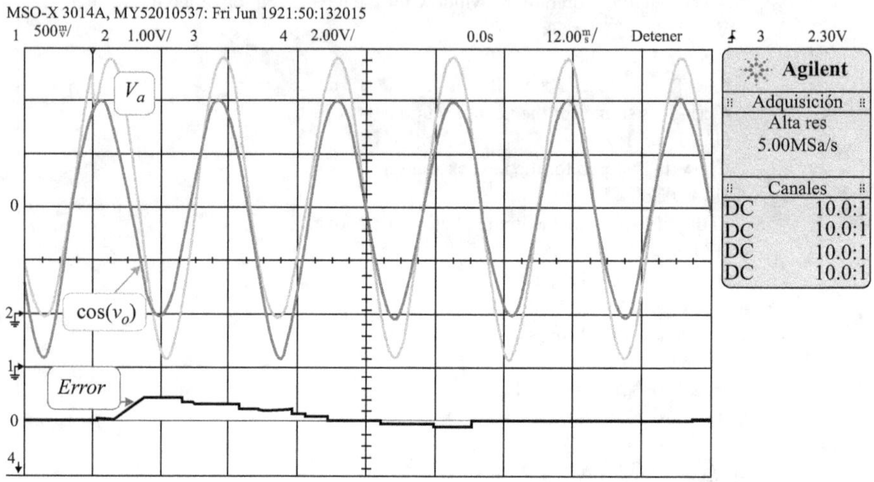

Figure 7.21 Experimental results of a single-phase varying sampling time PLL algorithm subjected to a phase-shift in t = 12 ms

processor, since it requires calculating or interpolation sinusoidal values in a recursive way. In contrast, an algorithm with variable sampling period allows adjusting the frequency and phase of a stored sinusoidal signal with the input signal, without the need to recursively calculate or interpolate the sine values.

7.5 Control algorithms for UPF rectifiers

Several control algorithms are presented in the following text, with special emphasis in *dq* frame-based controllers, *pq* theory-based controllers, and predictive control. The focus in this section lies in how to cope with some practical issues in the implementation of the aforementioned control strategies.

7.5.1 dq frame-based control

The *dq* frame-based control is one of the most widespread techniques in power converters. In fact, it was derived from motor control applications [34]. For these control strategies, it is necessary to map currents and voltages to the synchronous reference frame (SRF), using the invariant power Park transformation which can be defined as

$$\begin{bmatrix} x_d \\ x_q \end{bmatrix} = \sqrt{\frac{2}{3}} \begin{bmatrix} \cos(\omega t + \theta) & \cos\left(\omega t - \frac{2\pi}{3} + \theta\right) & \cos\left(\omega t + \frac{2\pi}{3} + \theta\right) \\ \sin(\omega t + \theta) & \sin\left(\omega t - \frac{2\pi}{3} + \theta\right) & \sin\left(\omega t + \frac{2\pi}{3} + \theta\right) \end{bmatrix} \begin{bmatrix} x_a \\ x_b \\ x_c \end{bmatrix} \tag{7.8}$$

and the inverse Park transformation defined as

$$\begin{bmatrix} x_a \\ x_b \\ x_c \end{bmatrix} = \sqrt{\frac{2}{3}} \begin{bmatrix} \cos(\omega t + \theta) & \sin(\omega t + \theta) \\ \cos\left(\omega t - \frac{2\pi}{3} + \theta\right) & \sin\left(\omega t - \frac{2\pi}{3} + \theta\right) \\ \cos\left(\omega t + \frac{2\pi}{3} + \theta\right) & \sin\left(\omega t + \frac{2\pi}{3} + \theta\right) \end{bmatrix} \begin{bmatrix} x_d \\ x_q \end{bmatrix} \tag{7.9}$$

In both cases θ corresponds to the synchronization angle that must be provided by a PLL algorithm.

In order to illustrate a practical application of the *dq* frame-based control, a typical scheme suitable for VSR topologies is shown in Figure 7.22. In this method, a PLL algorithm must be included to calculate the Park transformation of the source currents and voltages, and therefore obtain their direct and quadrature components.

From technical literature [35], [36], it is possible to state that the *AC* side of the VSR can be modeled in the *dq* frame as

$$\frac{d}{dt}\mathbf{i}_{\mathbf{s}}^{dq} = \frac{R_s}{L_s}\mathbf{i}_{\mathbf{s}}^{dq} - \mathbf{W}\mathbf{i}_{\mathbf{s}}^{dq} - \frac{G_{AC}\mathbf{m}_{\mathbf{c}}^{dq} v_{DC}}{L_s} + \frac{\mathbf{v}_{\mathbf{s}}^{abc}}{L_s}, \tag{7.10}$$

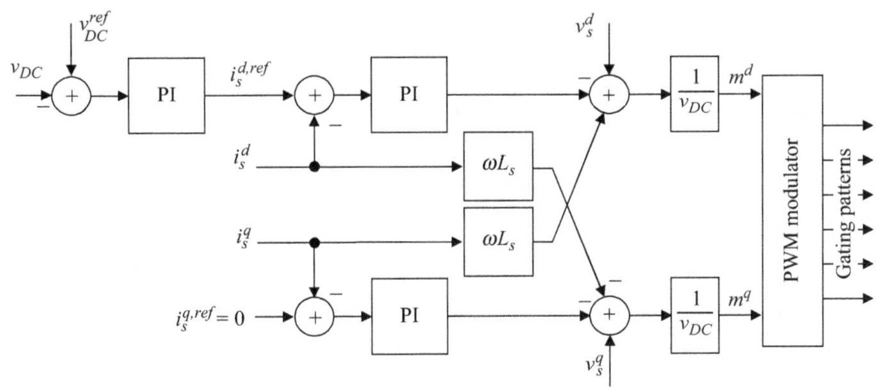

Figure 7.22 dq control schematic

where the **W** matrix produces a coupled behavior between the direct and quadrature components [36]. Therefore, in order to compensate the coupled terms, a decoupling stage must be included in the control strategy, as shown in Figure 7.22.

Once the equivalent model of the system is decoupled, simple PI controllers can be included to track the *dq* components of the source current. If the Park transformation is synchronized with the source voltage, the quadrature current is associated with the reactive power, and therefore it can be used to control the power factor. In fact, to obtain the unitary power factor operation, the reference of the quadrature source current $i_s^{q,ref}$, must be set to zero, as shown in Figure 7.22.

On the other hand, as the direct current is related with the active power, its reference is generated with the *DC* voltage control block, which also considers a PI controller, as Figure 7.22 illustrates.

The advantage of using PI controllers in the *dq* frame is that they are very well-known and easy to program in any digital platform, such as a DSP. Also, they can be tuned using classical tools as Root Locus, Bode Diagram and/or Nyquist Diagram. A sample code that implements the *dq* control strategy is presented in Figure 7.23.

The first part of the code calculates the sinusoidal functions required to implement both the direct and inverse Park Transformations. Depending on the implemented PLL algorithm (fixed or varying sampling time) and the features of the DSP, this part of the code can be modified using look-up tables, polynomial approximations, quarter-wave symmetry, or any other method that reduces the computational burden.

With the sinusoidal function already generated, the source currents and voltages are mapped to the *dq* frame with the Park transformation. The weighting factor of this transformation is pre-stored in memory in order to reduce the number of calculations of the DSP. Moreover, the invariant amplitude Park transformation is implemented in this code in order to avoid an extra multiplication in the inverse transform (at the end of the code).

Then, the master *DC* voltage controller is included in the code, which output *udc* serves as reference for the slave controller for the direct current. The quadrature current PI controller receives its independent reference $i_s^{q,ref}$, which must be equal to zero to ensure unitary power factor operation. As claimed earlier, the digital implementation of the PI controller is quite simple because it can be programmed with a few lines. The value of the parameters can be obtained using the discretization methods reported in literature [37].

After PI controllers, the decoupling stage uses the *dq* components, the *DC* voltage and the controllers' outputs to generate the modulating signals. For practical reasons it is preferred to use the reference instead of the actual value of the *DC* voltage, particularly during the start-up, where the *DC* voltage is near to zero. Finally, the modulating signals in the *abc* frame are calculated using the inverse Park transformation.

The simulated behavior of the VSR rectifier controlled with the presented code is shown in Figure 7.24, where a 100% load impact is produced at 20 ms and

```
// sine and cosine functions
s0 = sin (w * t);
c0 = cos (w * t);
s120 = sin (w * t - 2 * pi / 3);
c120 = cos (w * t - 2 * pi / 3);
s240 = sin (w * t + 2 * pi / 3);
c240 = cos (w * t + 2 * pi / 3);

//Source current in the dq frame
Isd = frac2_3 * (s0 * Isa + s120 * Isb + s240 * Isc);
Isq = frac2_3 * (c0 * Isa + c120 * Isb + c240 * Isc);
//Source voltage in the dq frame
Vsd = frac2_3 * (s0 * Vsa + s120 * Vsb + s240 * Vsc);
Vsq = frac2_3 * (c0 * Vsa + c120 * Vsb + c240 * Vsc);

// DC Voltage discrete PI controller (Master)
edc = Vref - Vdc;
udc = udc1 + q0dc * edc + q1dc * edc1;
udc1 = udc;
edc1 = edc;

// Discrete PI controller d-axis (Slave)
ed = udc - Isd;
ud = ud1 + q0 * ed + q1 * ed1;
ud1 = ud;
ed1 = ed;

//  Discrete PI controller q-axis
eq = Isqref - Isq;
uq = uq1 + q0 * eq + q1 * eq1;
uq1 = uq;
eq1 = eq;

// Decoupling Stage
md = (- ud + w * Ls * Isq + Vsd) / Vref;
mq = (- uq - w * Ls * Isd + Vsq) / Vref;

// Modulating Signals
ma = md * s0 + mq * c0;
mb = md * s120 + mq * c120;
mc = md * s240 + mq * c240;
```

Figure 7.23 Routine of a dq control strategy

a DC voltage reference step-up change is considered at 80 ms. As it can be seen, the unitary power factor operation is achieved even during transient conditions.

7.5.2 pq theory-based control

To implement control schemes that apply the *pq* theory, the Clarke transformation is required to map the three-phase instantaneous voltages and currents in the

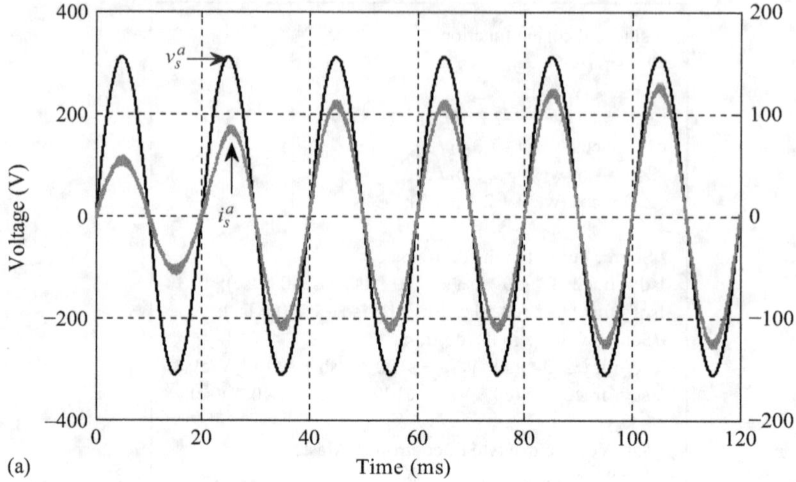

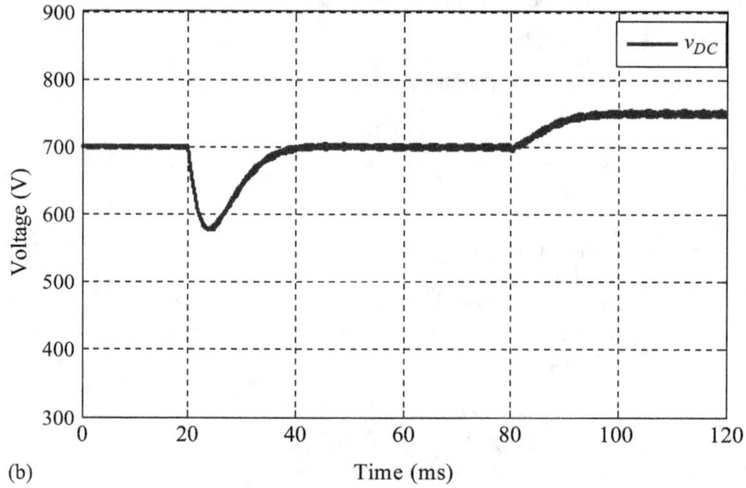

Figure 7.24 dq control performance (a) input voltage and current, (b) DC voltage

abc frame into the instantaneous quantities in the $\alpha\beta$ frame. The invariant power Clarke transformation for an arbitrary variable *x* (voltage or current) is defined as

$$\begin{bmatrix} x_\alpha \\ x_\beta \end{bmatrix} = \sqrt{\frac{2}{3}} \begin{bmatrix} 1 & -\frac{1}{2} & -\frac{1}{2} \\ 0 & \frac{\sqrt{3}}{2} & -\frac{\sqrt{3}}{2} \end{bmatrix} \begin{bmatrix} x_a \\ x_b \\ x_c \end{bmatrix} \tag{7.11}$$

and the inverse Clarke transformation is

$$\begin{bmatrix} x_a \\ x_b \\ x_c \end{bmatrix} = \sqrt{\frac{2}{3}} \begin{bmatrix} 1 & 0 \\ -\frac{1}{2} & \frac{\sqrt{3}}{2} \\ -\frac{1}{2} & -\frac{\sqrt{3}}{2} \end{bmatrix} \begin{bmatrix} x_\alpha \\ x_\beta \end{bmatrix} \tag{7.12}$$

Control strategies based on the *pq* theory use the calculation of instantaneous active and reactive power in the electrical system, which leads to the definition of the instantaneous complex power as

$$s = p + jq = \left(v_\alpha i_\alpha + v_\beta i_\beta\right) + j\left(v_\beta i_\alpha - v_\alpha i_\beta\right) \tag{7.13}$$

where p and q are the instantaneous active and reactive power, respectively [22]. This last equation can be also written as

$$\begin{bmatrix} p \\ q \end{bmatrix} = \begin{bmatrix} v^\alpha & v^\beta \\ v^\beta & -v^\alpha \end{bmatrix} \begin{bmatrix} i^\alpha \\ i^\beta \end{bmatrix} \tag{7.14}$$

which is used to implement the control strategies.

By inverting the matrix in (7.14), the currents can be calculated as function of voltages and powers as

$$\begin{bmatrix} i^\alpha \\ i^\beta \end{bmatrix} = \frac{1}{(v^\alpha)^2 + (v^\beta)^2} \begin{bmatrix} v^\alpha & v^\beta \\ v^\beta & -v^\alpha \end{bmatrix} \begin{bmatrix} p \\ q \end{bmatrix} \tag{7.15}$$

A practical application of the *pq* theory-based control is shown in Figure 7.25, where a simple hysteresis modulator is used to control the input currents of the Vienna converter of Figure 7.7. Different to the algorithm in the *dq* frame, the PLL is not required because the synchronization angle is not necessary to calculate the Clarke transformation and therefore neither to obtain the instantaneous active and reactive power. In this case, the unitary power factor operation is achieved when the reactive power q is zero. The active power is set according to the *DC* voltage, which is controlled with a master PI loop, as shown in Figure 7.25.

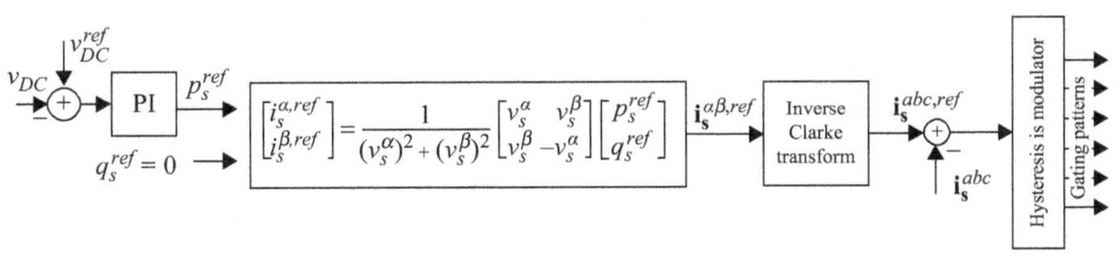

Figure 7.25 pq control schematic

Once the power references are set, (7.15) is used to calculate the current references in the $\alpha\beta$ frame, which are then mapped back to the *abc* frame where a suitable controller must be implemented. In the presented case a hysteresis modulator is included that directly commutates the switches when the error value goes outside the hysteresis bands [38]. The advantage of implementing a hysteresis modulator lies in its simplicity; however, in order to obtain satisfactory results, the sampling frequency must be high. If fixed switching frequency is required, other control techniques can be considered in the *abc* frame, such as the resonant controller which achieves zero steady state error for sinusoidal references [36].

Figure 7.26 shows the sample code that implements the proposed control algorithm for a Vienna converter. First of all, the source voltages in the $\alpha\beta$ frame are calculated with the Clarke transformation. The square root factors are pre-stored in order to reduce the calculation time of the DSP. Then, the PI controller for the *DC* voltage is implemented. The output of this stage corresponds to the active power reference. As claimed earlier, the reactive power reference is set to zero in order to achieve the unitary power factor operation. With the references of active and reactive power already generated, the current references are obtained in the $\alpha\beta$ frame, and using the inverse Clarke transformation, they are converted to the *abc* frame. A simple hysteresis modulator allows implementing the current tracking, just by comparing the instantaneous error with a pre-fixed threshold value. If the error goes beyond the hysteresis bands, the switches change their state. The advantage of this technique is that it does not require any PLL algorithm, as it does not use the sine and cosine functions required in *dq* controllers. It makes this method very easy to implement in a DSP. The performance of the proposed controller is shown in Figure 7.27.

The same test presented in Figure 7.24 is performed over the Vienna converter using the *pq*-based controller. As it can be seen, a similar behavior can be obtained, as Figure 7.27 illustrates. Notice that the unitary power factor is obtained even during transient conditions.

7.5.3 Predictive control

Thanks to technological advances and the emergence of faster microcontrollers that are capable of more powerful calculations, predictive control has emerged as an alternative modulation method for power converter applications [39]. This technique is a very intuitive concept that is easy to implement and performs well considering numerous restrictions. The technique can compensate for downtime or nonlinearities in the system, offers a flexible control method, and is easily extendible for different applications [39]–[43]. As an example, in this chapter the predictive current control applied to a CSR is explained.

The CSR consists of an array of six unidirectional power semiconductor switches which feed a DC-load, as shown in Figure 7.28. The converter operates by connecting at any given time one switch from the high side and one switch from the low side of the rectifier to the load. This constraint limits the rectifier to nine feasible switching states [44]. The input currents are given as a function of the

```
//Source Voltage in the alpha-beta frame
Val  = sqrt2_3 * (Vsa -Vsb / 2 - Vsc / 2) ;
Vbe = sqrt2 * (Vsb / 2 -Vsc / 2);

// DC Voltage discrete PI controller
edc = Vref - Vdc;
udc = udc1 + q0dc * edc + q1dc * edc1;
udc1 = udc;
edc1 = edc;

// Active and reactive power reference
ps = udc;
qs = 0;

// Denominator of fraction to calculate the current
D = 1 / (Val * Val + Vbe * Vbe + 1e-3);

// Source current references in the alpha-beta frame
Ial = D * (Val * ps + Vbe * qs);
Ibe = D * (Vbe * ps - Val * qs);

// Source current references in the abc frame
Isaref = sqrt2_3 * Ial;
Isbref = sqrt2_3 * (- Ial / 2 + sqrt3_2 * Ibe);
Iscref = sqrt2_3 * (- Ial / 2 - sqrt3_2 * Ibe);

// Current control with Hysteresis Modulator
// Current errors
esa = Isaref - Isa;
esb = Isbref - Isb;
esc = Iscref - Isc;

// Error comparison with the threshold value (TH)
If (esa > TH) SWa = 1;
if (esa <- TH) SWa = 0;
if (esb > TH) SWb = 1;
if (esb <- TH) SWb = 0;
if (esc > TH) SWc = 1;
if (esc <- TH) SWc =0;
```

Figure 7.26 Routine of a pq control strategy

switches and DC-side current, and the DC-side voltage are synthesized by the switches and input voltages as follows:

$$\mathbf{i}_{\mathbf{i}} = \mathbf{S}_{\mathbf{r}} i_{dc} \tag{7.16}$$

$$v_{DC} = \mathbf{S}_{\mathbf{r}}^{T} \mathbf{v}_{\mathbf{i}} \tag{7.17}$$

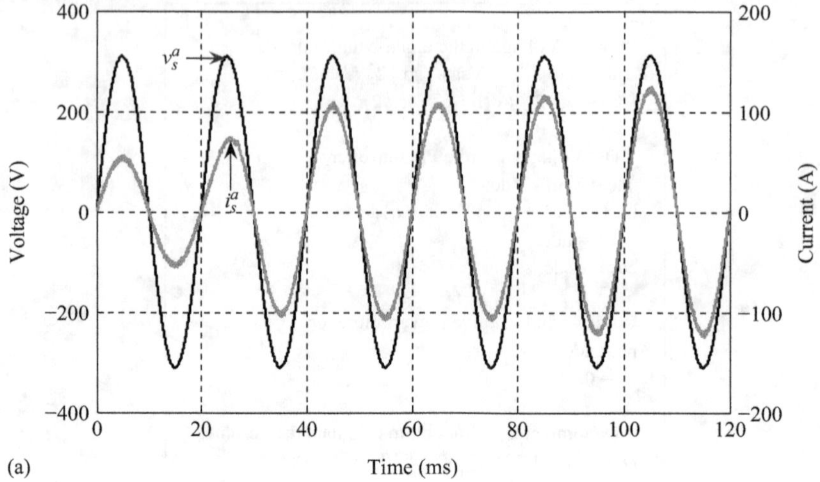

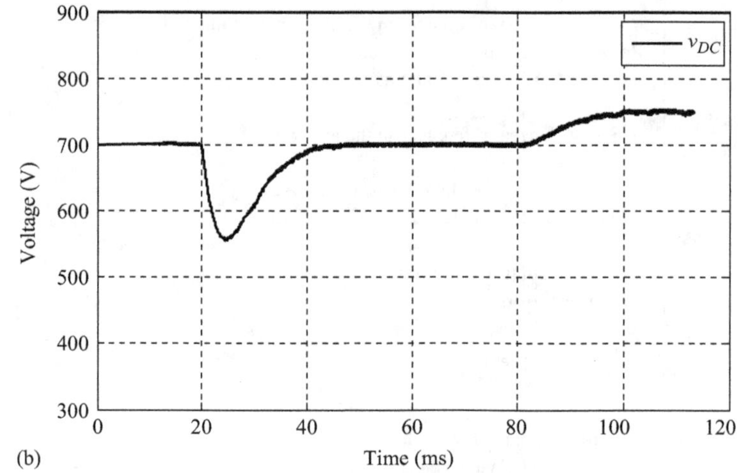

Figure 7.27 pq control performance: (a) input voltage and current, (b) DC voltage

where $\mathbf{S_r}$ corresponds to the switches matrix of the rectifier as follows:

$$\mathbf{S_r} = \begin{bmatrix} S_1 - S_4 \\ S_3 - S_6 \\ S_5 - S_2 \end{bmatrix} \tag{7.18}$$

In addition, an input filter is necessary to avoid over-voltages during the switching transitions and to avoid high $d\mathbf{i_s}/dt$. Its model is given by

$$\mathbf{v_s} = R_f \mathbf{i_s} + L_f \frac{d\mathbf{i_s}}{dt} + \mathbf{v_i} \tag{7.19}$$

$$\mathbf{i_s} = C_f \frac{d\mathbf{v_i}}{dt} + \mathbf{i_i} \tag{7.20}$$

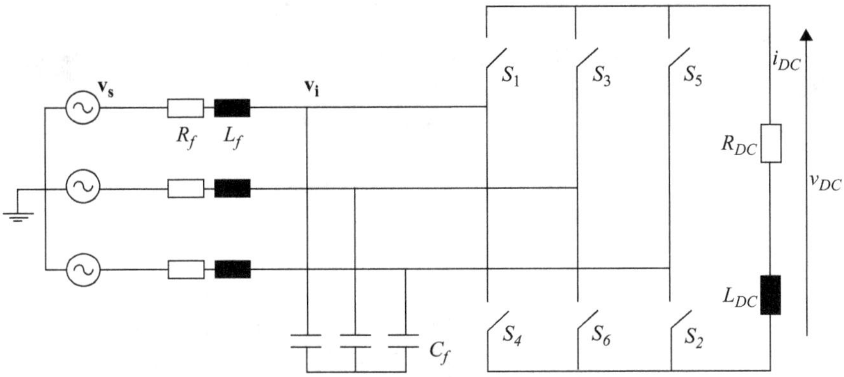

Figure 7.28 General topology of the current source converter

On the load side, the mathematical model is given as follows:

$$v_{DC} = L_{DC}\frac{di_{DC}}{dt} + R_{DC}i_{DC} \tag{7.21}$$

In order to minimize the computational burden, the $\alpha\beta$ linear transformation is applied to all three-phase current and voltage vectors, defined as

$$\begin{bmatrix} u_\alpha \\ u_\beta \end{bmatrix} = \begin{bmatrix} \frac{2}{3} & -\frac{1}{3} & -\frac{1}{3} \\ 0 & \frac{\sqrt{3}}{3} & -\frac{\sqrt{3}}{3} \end{bmatrix} \begin{bmatrix} u_a \\ u_b \\ u_c \end{bmatrix} \tag{7.22}$$

where the vector $[u_a \quad u_b \quad u_c]^T$ is the natural frame current or voltage vector and $[u_\alpha \quad u_\beta]^T$ is the stationary frame vector.

7.5.3.1 Predictive control strategy

A simplified block diagram of the predictive control scheme is presented in Figure 7.29, where superscript p corresponds to $k+1$ sampling time. The predictive control scheme considers the discrete change of switching states at equidistant points with a constant sampling period. This state is selected out of the nine available possibilities in order to meet two requirements: first, the line side of the converter must deliver mainly active power, and second, on the load side, the current must follow the reference. The first condition is fulfilled by minimizing the instantaneous reactive power, which is indicated as follows:

$$\Delta q_s(k+1) = |v_s^\alpha(k+1)i_s^\beta(k+1) - v_s^\beta(k+1)i_s^\alpha(k+1)| \tag{7.23}$$

where k is the sample instant. On the other side, the second condition requires a minimum error between the load current and its reference as

$$\Delta i_{DC}(k+1) = |i_{DC}^*(k+1) - i_{DC}(k+1)| \tag{7.24}$$

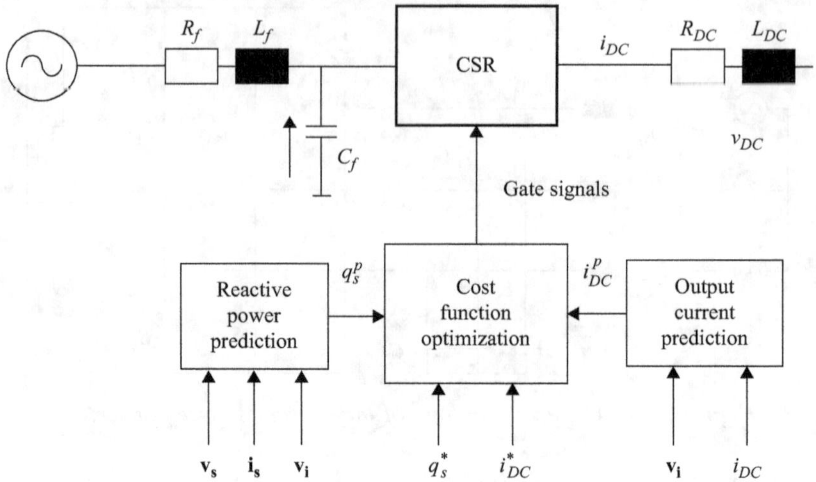

Figure 7.29 Predictive control scheme for the current source rectifier

where i_{DC} is the load current and i_{DC}^* corresponds to the DC-side reference current. Both requirements are merged in a unique so-called cost function given by

$$g(k+1) = \Delta i_{DC}(k+1) + \lambda_q \Delta q_s(k+1) \tag{7.25}$$

The control scheme works as follows: at each sampling time, all possible switching states are used to calculate the predicted values of the load and input currents, which are then used to calculate the cost function g. Next, the switching state yielding the minimum value of g will be applied to the converter [45], [46]. The weighting factor λ_q decides whether priority is given to those states which minimize the error in the load current or those which improve the power factor. Basically, λ_q is adjusted empirically, such that the output current presents no noticeable deviations with respect to the reference.

The mathematical model introduced from (7.16) to (7.21) provides the basis for predicting the input and output currents which will be evaluated by the cost function.

By using (7.19) and (7.20), the prediction of the input current is computed from a first-order difference equation, as described in Reference 45.

$$\mathbf{i_s}(k+1) = c_1 \mathbf{v_s}(k) + c_2 \mathbf{v_i}(k) + c_3 \mathbf{i_s}(k) + c_4 \mathbf{i_i}(k) \tag{7.26}$$

The coefficients c_1, c_2, c_3, and c_4 are defined so that the same values for the currents of the equivalent continuous-time system are obtained for all sampling instants. This is carried out through a state-space system with state variables $\mathbf{i_s}$ and $\mathbf{v_i}$ and the input variables $\mathbf{i_i}$ and $\mathbf{v_s}$ as

$$\begin{bmatrix} \dot{\mathbf{v}}_\mathbf{i} \\ \dot{\mathbf{i}}_\mathbf{s} \end{bmatrix} = \mathbf{A} \begin{bmatrix} \mathbf{v_i} \\ \mathbf{i_s} \end{bmatrix} + \mathbf{B} \begin{bmatrix} \mathbf{v_s} \\ \mathbf{i_i} \end{bmatrix} \tag{7.27}$$

where

$$\mathbf{A} = \begin{bmatrix} 0 & \frac{1}{C_f} \\ -\frac{1}{L_f} & -\frac{R_f}{L_f} \end{bmatrix} \quad \mathbf{B} = \begin{bmatrix} 0 & \frac{1}{C_f} \\ -\frac{1}{L_f} & -\frac{R_f}{L_f} \end{bmatrix} \tag{7.28}$$

and assuming that

$$\begin{aligned} \mathbf{v_s} &\cong \mathbf{v_s}(kT_s) \cong \mathbf{v_s}(k) \\ \mathbf{i_i} &\cong \mathbf{i_i}(kT_s) \cong \mathbf{i_i}(k) \end{aligned} \tag{7.29}$$

For $kT_s \leq t \leq (k+1)T_s$, with T_s being the sampling time, the discrete-time space model is determined as [46]

$$\begin{bmatrix} \mathbf{v_i}(k+1) \\ \mathbf{i_s}(k+1) \end{bmatrix} = \mathbf{\Phi} \begin{bmatrix} \mathbf{v_i}(k) \\ \mathbf{i_s}(k) \end{bmatrix} + \mathbf{\Gamma} \begin{bmatrix} \mathbf{v_s}(k) \\ \mathbf{i_i}(k) \end{bmatrix} \tag{7.30}$$

where

$$\mathbf{\Phi} = e^{\mathbf{A}T_s} \quad \mathbf{\Gamma} = \mathbf{A}^{-1}(\mathbf{\Phi} - \mathbf{I}_{2\times 2})\mathbf{B} \tag{7.31}$$

Solving (7.31), the coefficients are determined as

$$c_1 = \sqrt{\frac{C_f}{L_f}} \sin(T_s\omega_r) \quad c_2 = -c_1 \quad c_3 = \cos(T_s\omega_r) \quad c_4 = 1 - c_3 \tag{7.32}$$

The term w_r corresponds to the resonance frequency, which is given by $w_r = 1/\sqrt{C_f L_f}$. On the load side, the output current prediction can be obtained using a forward Euler approximation given in (7.21), such that

$$i_{DC}(k+1) = d_1 v_{DC}(k) + d_2 i_{DC}(k) \tag{7.33}$$

where $d_1 = T_s/L_{DC}$ and $d_2 = 1 - R_{DC}T_s/L_{DC}$ are constants which depend on load parameters and the sampling time T_s [42]. Note that the voltage $v_{dc}(k)$ and the current $i_{DC}(k)$ depend upon $S_r(k)$ through (7.16) and (7.17).

7.5.3.2 Simulations results

Simulation results have been obtained using MATLAB/Simulink. The controller operates with a sample time of $T_s = 20\,\mu\text{s}$. Thanks to technological advances in switch devices (such as Silicon Carbide, SiC) and fast and powerful processors, this sampling time is not a problem nowadays. The control strategy is evaluated considering the cost function indicated in (7.25) and with a weighting factor λ_q equal to $\lambda_q = 1.7 \times 10^{-7}$ which has been empirically adjusted as explained in Reference 46. First it was established in a value equal to zero in order to prioritize the control of the output current and later it was slowly increased aiming to obtain a minimal THD in the source current.

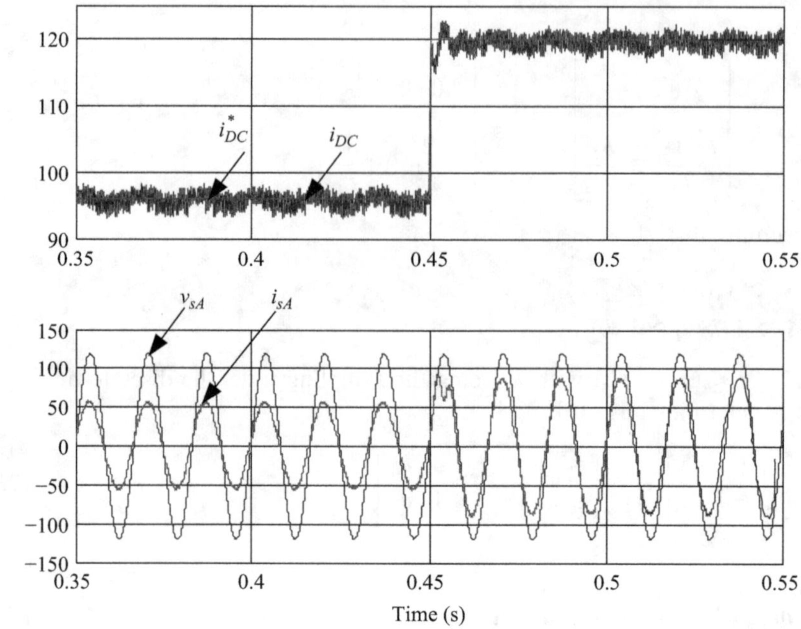

Figure 7.30 Simulation results of the predictive current control with instantaneous reactive power minimization: (top) output i_{DC} [A] current and its reference i^*_{DC} [A]; (bottom) source voltage v_{sA} [V/20] and phase current i_{sA} [A]

Figure 7.30 shows the output current i_{DC} and its reference i^*_{DC}, the source voltage v_{sA} and the source input current i_{sA} for steady and transient operation. In the first 0.45 s, a load reference of $I^*_{DC} = 120$A is applied, observing a THD of 4.30% in current i_{sA}. A step change is given at $t = 0.45$ s where the load reference I^*_{DC} is increased from $I^*_{DC} = 96$ A to $I^*_{DC} = 120$ A.

As shown in Figure 7.30, a good tracking of the load current with respect to its reference can be obtained along with a good dynamic performance. On the input side, the source current is sinusoidal and in phase with its respective source voltage. Of course, due to the relation between the load current and input current, an amplitude change in source current i_{sA} is observed. At low output current a major distortion of the source currents is observed.

7.6 Conclusions

In this chapter an overview of the main available DSP boards used in power electronics has been discussed, providing the structure and their main characteristics. Different topologies of three-phase UPF rectifiers have been depicted, where basic implementation issues were described, such as zero states generation in a CSR. Moreover, some implementations of PLL algorithms were described, giving emphasis in three-phase and single-phase algorithms for fixed and variable sampling times. This chapter finalizes with the analysis of different digital control methods for UPF rectifiers. Particularly, the control of a VSR is performed with the *dq* frame transformation; a VIENNA rectifier is commanded using the power theory and a hysteresis modulator, and the predictive control of a CSR is demonstrated as a suitable alternative for the control of power converters. Therefore, commercially available devices can successfully cope with the actual challenges in UPF rectifier control.

References

[1] Kolar, J.W., Friedli, T., "The essence of three-phase PFC rectifier systems—part I," *Power Electronics, IEEE Transactions on*, vol. 28, no. 1, pp. 176–198, Jan. 2013.

[2] Friedli, T., Hartmann, M., Kolar, J.W., "The essence of three-phase PFC rectifier systems—part II," *Power Electronics, IEEE Transactions on*, vol. 29, no. 2, pp. 543–560, Feb. 2014.

[3] Zargari, N.R., Joos, G., "Performance investigation of a current-controlled voltage-regulated PWM rectifier in rotating and stationary frames," *Industrial Electronics, Control, and Instrumentation, 1993. Proceedings of the IECON '93, International Conference on*, vol. 2, pp. 1193–1198, Nov. 1993.

[4] Richter, S.A., Bader, B., De Doncker, R.W., "Control of a high power PWM current source rectifier," *Power Electronics Conference (IPEC), 2010 International*, pp. 1287–1292, Jun. 2010.

[5] Jiaming Chen, Wenjie Zhang, Dakun Duan, Dianguo Xu, "An optimized strategy to control DC bus voltage decline for three-phase VSR," *Power Electronics and Motion Control Conference (IPEMC), 2012 7th International*, vol. 4, pp. 2769–2773, Jun. 2012.

[6] Gangquan Si, Weili Zhao, Lixin Jia, Yanbin Zhang, "Robust sliding mode control of three-phase voltage source PWM rectifier with uncertainties and disturbances," *Control and Decision Conference (2014 CCDC), The 26th Chinese*, pp. 542–546, 31 May 2014–2 Jun. 2014.

[7] Espinoza, J.R., Joos, G., "Current-source converter on-line pattern generator switching frequency minimization," *Industrial Electronics, IEEE Transactions on*, vol. 44, no. 2, pp. 198–206, Apr. 1997.

[8] Holmes, D., Lipo, T., *Pulse Width Modulation for Power Converters: Principles and Practice*, Hoboken, NJ: John Wiley & Sons, 2003.

[9] Salmon, J.C., "Reliable 3-phase PWM boost rectifiers employing a series-connected dual boost converter sub-topology," *Industry Applications Society Annual Meeting, 1994. Conference Record of the 1994 IEEE*, pp. 781–788, Oct. 1994.

[10] Hartmann, M., Miniboeck, J., Kolar, J.W., "A three-phase delta switch rectifier for more electric aircraft applications employing a novel PWM current control concept," *Applied Power Electronics Conference and Exposition, 2009. APEC 2009. Twenty-Fourth Annual IEEE*, pp. 1633–1640, Feb. 2009.

[11] Kolar J.W. and Zach F.C., "A novel three-phase three-switch three-level PWM rectifier," in *Proceedings of 28th Power Conversion Conference*, pp. 125–138, 28–30 Jun. 1994.
[12] Yifan Zhao, Yue Li, Lipo, T.A., "Force commutated three level boost type rectifier," *Industry Applications, IEEE Transactions on*, vol. 31, no. 1, pp. 155–161, Jan./Feb. 1995.
[13] Heldwein, M.L., Mussa, S.A., Barbi, I., "Three-phase multilevel PWM rectifiers based on conventional bidirectional converters," *Power Electronics, IEEE Transactions on*, vol. 25, no. 3, pp. 545–549, Mar. 2010.
[14] Stogerer, F., Minibock, J., Kolar, J.W., "Implementation of a novel control concept for reliable operation of a VIENNA rectifier under heavily unbalanced mains voltage conditions," *Power Electronics Specialists Conference, 2001. PESC. 2001 IEEE 32nd Annual*, vol. 3, pp. 1333–1338, 2001.
[15] Henri de Bellescize, *La réceptionsynchrone*. Paris, France: Ed. E. Chiron, 1932.
[16] Ainsworth, J.D., "The Phase-Locked Oscillator – A New Control System for Controlled Static Convertors," *IEEE Transactions on Power Apparatus and Systems*, vol. PAS-87, no. 3, pp. 859–865, Mar. 1968.
[17] El-Amawy, A.A., Mirbod, A., "An efficient software-controlled PLL for low-frequency applications," *Industrial Electronics, IEEE Transactions on*, vol. 35, no. 2, pp. 341–344, May 1988.
[18] Lindsey, W.C., Chak Ming Chie, "A survey of digital phase-locked loops," *Proceedings of the IEEE*, vol. 69, no. 4, pp. 410–431, Apr. 1981.
[19] Kaura, V., Blasko, V., "Operation of a phase locked loop system under distorted utility conditions," *IEEE Transactions on Industry Applications*, vol. 33, no. 1, pp. 58–63, Jan./Feb. 1997.
[20] Song, E.-H., Bong-Hwan Kwon, "A direct digital control for the phase-controlled rectifier," *Industrial Electronics, IEEE Transactions on*, vol. 38, no. 5, pp. 337–343, Oct. 1991.
[21] Freijedo, F.D., Doval-Gandoy, J., Lopez, O., *et al.*, "Grid-synchronization methods for power converters," *Industrial Electronics, 2009. IECON '09. 35th Annual Conference of IEEE*, pp. 522–529, 3–5 Nov. 2009.
[22] Akagi, H., Watanabe, E., Aredes, M., *Instantaneous Power Theory and Applications to Power Conditioning*, Hoboken, NJ: Wiley-IEEE Press, Apr. 2007.
[23] Liccardo, F., Marino, P., Schiano, C., Visciano, N., "A new robust phase tracking system for asymmetrical and distorted three phase networks," *Harmonics and Quality of Power, 2004. 11th International Conference on*, pp. 525–530, 12–15 Sept. 2004.
[24] Liccardo, F., Marino, P., Raimondo, G., "Robust and fast three-phase PLL tracking system," *Industrial Electronics, IEEE Transactions on*, vol. 58, no. 1, pp. 221–231, Jan. 2011.
[25] Singh, B., Dwivedi, S., Hussain, I., Verma, A.K., "Grid integration of solar PV power generating system using QPLL based control algorithm," *Power*

India International Conference (PIICON), 2014 6th IEEE, pp. 1–6, 5–7 Dec. 2014.

[26] Lopez Gonzalez, O., Buja, G., "Novel PLL scheme for grid connection of three-phase power converters," *Diagnostics for Electric Machines, IEEE International Symposium on Power Electronics & Drives (SDEMPED), 2011*, pp. 372–377, 5–8 Sept. 2011.

[27] Saitou M., Matsui N., Shimizu T., "A control strategy of single-phase active filter using a novel d-q transformation," *38th IAS Annual Meeting Conference Record of the Industry Applications Conference*, pp. 1222–1227, Oct. 2003.

[28] Silva, S.M., Lopes, B.M., Filho, B.J.C., Campana, R.P., Bosventura, W.C., "Performance evaluation of PLL algorithms for single-phase grid-connected systems," *39th IAS Annual Meeting. Conference Record of the Industry Applications Conference, 2004*, pp. 2259–2263, Oct. 2004.

[29] Ali, E., Yilmaz, S., Malik, E., "A robust and efficient PLL algorithm for single-phase grid-connected renewable energy sources," *Twenty-Eighth Annual IEEE Applied Power Electronics Conference and Exposition (APEC)*, pp. 2940–2946, 17–21 Mar. 2013.

[30] Ciobotaru, M., Teodorescu R., Blaabjerg, F., "Improved PLL structures for single-phase grid inverters," *Proceedings of PELINCEC 2005*, 16–19 Oct. 2005.

[31] Carugati, I., Maestri, S., Donato, P.G., Carrica, D., Benedetti, M., "Variable sampling period filter PLL for distorted three-phase systems," *Power Electronics, IEEE Transactions on*, vol. 27, no. 1, pp. 321–330, Jan. 2012.

[32] Perez, M.A., Espinoza, J.R., Moran, L.A., Torres, M.A., Araya, E.A., "A robust phase-locked loop algorithm to synchronize static-power converters with polluted AC systems," *Industrial Electronics, IEEE Transactions on*, vol. 55, no. 5, pp. 2185–2192, May 2008.

[33] Carugati, I., Donato, P., Maestri, S., Carrica, D., Benedetti, M., "Frequency adaptive PLL for polluted single-phase grids," *Power Electronics, IEEE Transactions on*, vol. 27, no. 5, pp. 2396–2404, May 2012.

[34] Onel, I.Y., El Hachemi Benbouzid, M., "Induction motor bearing failure detection and diagnosis: park and concordia transform approaches comparative study," *Mechatronics, IEEE/ASME Transactions on*, vol. 13, no. 2, pp. 257–262, Apr. 2008.

[35] Yajing Gao, Junqiang Li, Haifeng Liang, "The simulation of three-phase voltage source PWM rectifier," *Power and Energy Engineering Conference (APPEEC), 2012 Asia-Pacific*, pp. 1–4, 27–29 Mar. 2012.

[36] Hwang, J.G., Lehn, P.W., Winkelnkemper, M., "A generalized class of stationary frame-current controllers for grid-connected AC–DC converters," *Power Delivery, IEEE Transactions on*, vol. 25, no. 4, pp. 2742–2751, Oct. 2010.

[37] Ogata, K., *Discrete-Time Control Systems*, Englewood Cliffs, NJ: Prentice-Hall, 1995.

[38] Foureaux, N.C., Oliveira, J.J.H., de Oliveira, F.D., De J Cardoso Filho, B., de Faria, R.S., "Command generation for wide range operation of hysteresis

controlled Vienna rectifiers," *Energy Conversion Congress and Exposition (ECCE), 2013 IEEE*, pp. 1379–1384, 15–19 Sept. 2013.

[39] Rodriguez, J., Kazmierkowski, M.P., Espinoza, J.R., *et al.*, "State of the art of finite control set model predictive control in power electronics," *Industrial Informatics, IEEE Transactions on*, vol. 9, no. 2, pp. 1003–1016, May 2013.

[40] Rivera, M., Rodriguez, J., Espinoza, J.R., *et al.*, "Imposed sinusoidal source and load currents for an indirect matrix converter," *Industrial Electronics, IEEE Transactions on*, vol. 59, no. 9, pp. 3427–3435, Sept. 2012.

[41] Perez, M.A., Fuentes, R.L., Rodriguez, J., "Predictive control of DC-link voltage in an active-front-end rectifier," *Industrial Electronics (ISIE), 2011 IEEE International Symposium on*, pp. 1811–1816, 27–30 Jun. 2011.

[42] Cortes, P., Vattuone, L., Rodriguez, J., "A comparative study of predictive current control for three-phase voltage source inverters based on switching frequency and current error," *Power Electronics and Applications (EPE 2011), Proceedings of the 2011-14th European Conference on*, pp. 1–8, 30 Aug. 2011–1 Sept. 2011.

[43] Rivera, M., Rodriguez, J., Yaramasu, V., Wu, B., "Predictive load voltage and capacitor balancing control for a four-leg NPC inverter," *Power Electronics and Motion Control Conference (EPE/PEMC), 2012 15th International*, pp. DS3c.8-1–DS3c.8-5, 4–6 Sept. 2012.

[44] Wu, B., *High Power Converters and AC Drives*, 1st ed. ser. Wiley-IEEE Press. John Wiley & Sons, Inc. 2006.

[45] Correa, P., Rodriguez, J., Lizama, I., Andler, D., "A predictive control scheme for current-source rectifiers," *Industrial Electronics, IEEE Transactions on*, vol. 56, no. 5, pp. 1813–1815, May 2009.

[46] Cortes, P., Kouro, S., La Rocca, B., *et al.*, "Guidelines for weighting factors design in Model Predictive Control of power converters and drives," *Industrial Technology, 2009. ICIT 2009. IEEE International Conference on*, pp. 1–7, 10–13 Feb. 2009.

Chapter 8

DSP controllers for grid-connected three-phase voltage-sourced inverters

Gabriel Garcerá[1], Emilio Figueres[1], Raúl González-Medina[1] and Iván Patrao[1]

This chapter deals with the DSP control of three-phase voltage source inverters. A study on a 10-kW grid-connected photovoltaic inverter with two control options, namely, the α-β stationary reference frame and the d-q rotating reference frame, is shown. Experimental results are provided in order to validate the performance and limitations of the control strategies under study.

8.1 Introduction

Grid-tied inverters play a key role in the renewable energy conversion scenario, in particular for the integration of wind and photovoltaic systems into the distribution grids [1, 2]. The relevance of such systems is growing substantially all around the world. For instance, the cumulative installed photovoltaic capacity has risen from 1.23 GW in 2000 to about 140 GW in 2013, with a forecast between 320 GW and 430 GW for 2018 [3].

The performance of grid-connected inverters, especially in terms of power quality, strongly depends on the design of the inverter control loops. Current control loops [4] are responsible for the low distortion of the current waveforms that are injected into the grid. Grid synchronization techniques are needed to properly manage the active and reactive power generation of the inverter [5]. In some cases, DC voltage control allows to implement Maximum Power Point Tracking (MPPT) algorithms needed to optimally deal with the photovoltaic energy source [6].

The grid-connected PV inverter presented in this chapter is a three-phase 10-kW PV inverter connected to the grid by means of an LCL filter and a delta-star (Δ-Y) transformer as it is depicted in Figure 8.1. This is a common topology for medium and high power PV inverters.

[1]Grupo de Sistemas Electrónicos Industriales del Departamento de Ingeniería Electrónica, Universitat Politécnica de Valencia, Valencia, España

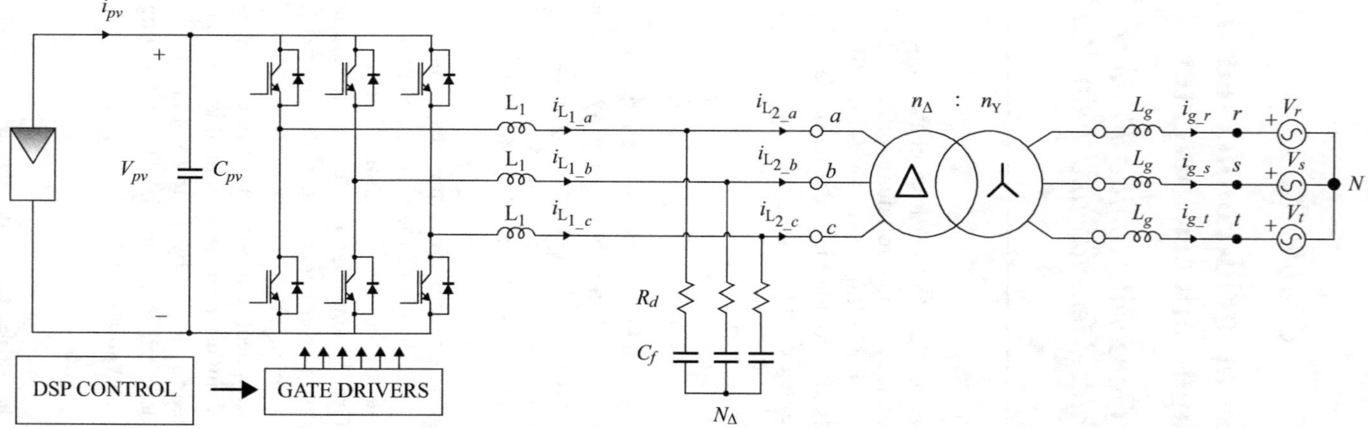

Figure 8.1 *Power stage of a three-phase PV inverter connected to the grid by means of an LCL filter and a Δ-Y transformer*

8.2 Modeling and control structures of grid-connected three-phase voltage-sourced inverters (VSIs)

There are two typical control strategies for implementing the DSP control of the inverter in orthogonal reference frames: the stationary α-β reference frame (StatRF), based on the Clarke transform, and the rotating d-q reference frame, based on the Park transform [2]. The d-q reference frame is also called the synchronous reference frame (SRF), because its orthogonal axes rotate at the angular frequency of the grid voltage.

8.2.1 Modeling in an orthogonal stationary reference frame (StatRF)

Given a sinusoidal three-phase system in *a-b-c* coordinates, the Clarke transformation consists of a vector with three coordinates α-β-γ as shown by (8.1). Note that when the sum of the *a, b, c* components of an electrical magnitude (voltage or current) is zero, the γ component of that magnitude is null and that magnitude can be solely represented by its α-β components. The γ component is known as the zero-sequence or homopolar component.

Note that the matrix, $T_{\alpha\beta\gamma\to abc}$, for transforming the α-β-γ components of a three-phase system into its *a-b-c* components is the inverse one of $\mathrm{T}_{abc\to\alpha\beta\gamma}$, as it is shown in (8.3).

The version of the Clarke transformation matrix (8.2) preserves the instantaneous power of a three-phase system, which can be expressed by (8.4), where v_α, v_β and v_γ denote the α-β-γ components of the phase voltages and i_α, i_β and i_γ are the α-β-γ components of the phase currents. Obviously, in a three-phase three-wire system it results either $i_\gamma = 0$ (Y-connected voltage source) or $v_\gamma = 0$ (Δ-connected voltage source), so that the instantaneous power only depends on the α and β components of the phase voltages and phase currents.

Equation (8.5) shows the Clarke transformation of a balanced and positive-sequence three-phase voltage system. Taking into account that $v_\gamma = 0$, the system can be represented as a complex vector, as it is expressed by (8.6). In the version of the Clarke transformation (8.1) the amplitude of that vector is $\sqrt{\frac{3}{2}}$ times the amplitude of the three-phase voltage in *a-b-c* coordinates. The resulting vector rotates at the angular speed of the three-phase system, ω, with respect to the static α-β axes, as it is depicted in Figure 8.2. In this case the resulting α and β components, v_α and v_β, are sinusoidal magnitudes with an angular frequency ω. When the homopolar component of a three-phase system is null, it can be represented as an orthogonal two-phase system applying the Clarke transformation. The resulting system has two orthogonal components, namely the α and the β coordinates. In this transformation the orthogonal α-β axes are static, conforming a stationary reference frame (StatRF).

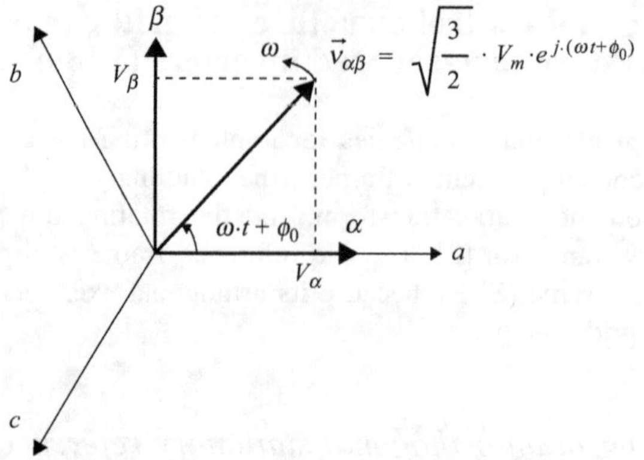

Figure 8.2 Complex vector resulting from the Clarke transformation of a balanced and positive-sequence three-phase voltage system

$$[v_{abc}] = \begin{bmatrix} v_a \\ v_b \\ v_c \end{bmatrix} \Rightarrow [v_{\alpha\beta\gamma}] = T_{abc\rightarrow\alpha\beta\gamma} \cdot [v_{abc}] = \sqrt{\frac{2}{3}} \cdot \begin{bmatrix} 1 & -\frac{1}{2} & -\frac{1}{2} \\ 0 & \frac{\sqrt{3}}{2} & -\frac{\sqrt{3}}{2} \\ \frac{1}{\sqrt{2}} & \frac{1}{\sqrt{2}} & \frac{1}{\sqrt{2}} \end{bmatrix} \cdot \begin{bmatrix} v_a \\ v_b \\ v_c \end{bmatrix}$$

$$[v_{\alpha\beta\gamma}] = \begin{bmatrix} v_\alpha \\ v_\beta \\ v_\gamma \end{bmatrix} = \sqrt{\frac{2}{3}} \cdot \begin{bmatrix} 1 & -\frac{1}{2} & -\frac{1}{2} \\ 0 & \frac{\sqrt{3}}{2} & -\frac{\sqrt{3}}{2} \\ \frac{1}{\sqrt{2}} & \frac{1}{\sqrt{2}} & \frac{1}{\sqrt{2}} \end{bmatrix} \cdot \begin{bmatrix} v_a \\ v_b \\ v_c \end{bmatrix} \tag{8.1}$$

$$T_{abc\rightarrow\alpha\beta\gamma} = \sqrt{\frac{2}{3}} \cdot \begin{bmatrix} 1 & -\frac{1}{2} & -\frac{1}{2} \\ 0 & \frac{\sqrt{3}}{2} & -\frac{\sqrt{3}}{2} \\ \frac{1}{\sqrt{2}} & \frac{1}{\sqrt{2}} & \frac{1}{\sqrt{2}} \end{bmatrix} \tag{8.2}$$

$$[v_{abc}] = T_{\alpha\beta\gamma\rightarrow abc} \cdot [v_{\alpha\beta\gamma}] = \sqrt{\frac{2}{3}} \cdot \begin{bmatrix} 1 & 0 & \frac{1}{\sqrt{2}} \\ -\frac{1}{2} & \frac{\sqrt{3}}{2} & \frac{1}{\sqrt{2}} \\ -\frac{1}{2} & -\frac{\sqrt{3}}{2} & \frac{1}{\sqrt{2}} \end{bmatrix} \cdot \begin{bmatrix} v_\alpha \\ v_\beta \\ v_\gamma \end{bmatrix}$$

$$T_{\alpha\beta\gamma\to abc} = \left[T_{abc\to\alpha\beta\gamma}\right]^{-1} = \left[T_{abc\to\alpha\beta\gamma}\right]^{T} = \sqrt{\frac{2}{3}}\cdot\begin{bmatrix} 1 & 0 & \frac{1}{\sqrt{2}} \\ -\frac{1}{2} & \frac{\sqrt{3}}{2} & \frac{1}{\sqrt{2}} \\ -\frac{1}{2} & -\frac{\sqrt{3}}{2} & \frac{1}{\sqrt{2}} \end{bmatrix} \tag{8.3}$$

$$p(t) = v_a \cdot i_a + v_b \cdot i_b + v_c \cdot i_c = v_\alpha \cdot i_\alpha + v_\beta \cdot i_\beta + v_\gamma \cdot i_\gamma \tag{8.4}$$

$$\begin{aligned}\left[v_{abc}\right] = \begin{bmatrix} v_a \\ v_b \\ v_c \end{bmatrix} &= V_m \cdot \begin{bmatrix} \cos(\omega\cdot t + \phi_0) \\ \cos\left(\omega\cdot t + \phi_0 - \frac{2\pi}{3}\right) \\ \cos\left(\omega\cdot t + \phi_0 + \frac{2\pi}{3}\right) \end{bmatrix} \Rightarrow \left[v_{\alpha\beta\gamma}\right] = \begin{bmatrix} v_\alpha \\ v_\beta \\ v_\gamma \end{bmatrix} \\ &= \sqrt{\frac{3}{2}}\cdot V_m \cdot \begin{bmatrix} \cos(\omega\cdot t + \phi_0) \\ \sin(\omega\cdot t + \phi_0) \\ 0 \end{bmatrix}\end{aligned} \tag{8.5}$$

$$\begin{aligned}\vec{v}_{\alpha\beta} &= v_\alpha + j\cdot v_\beta \\ &= \underbrace{\sqrt{\frac{3}{2}}\cdot V_m \cdot \cos(\omega\cdot t + \phi_0)}_{v_\alpha} + j\cdot \underbrace{\sqrt{\frac{3}{2}}\cdot V_m \cdot \sin(\omega\cdot t + \phi_0)}_{v_\beta} = \sqrt{\frac{3}{2}}\cdot V_m \cdot e^{j\cdot(\omega\cdot t + \phi_0)}\end{aligned} \tag{8.6}$$

When modeling a grid-connected PV inverter like that in Figure 8.1, it's useful to reflect the grid voltage to the inverter side, i.e. to consider the grid voltage seen from the Δ-side of the grid connection transformer, as depicted in Figure 8.3, where

- L_1: inductance placed at the output of each inverter leg
- L_2: sum of the leakage inductance of one phase of the transformer and the grid inductance, seen from the Δ-side (inverter side)
- R_d, C_f: values of damping resistance and capacitance of the capacitive branch of the LCL filter
- V_{grid}: RMS value of the of the grid voltage (phase to neutral) seen from the Δ-side
- $v_{g\Delta\_a}$, $v_{g\Delta\_b}$, $v_{g\_\Delta c}$: a, b, c phase voltages of the grid seen from the Δ-side.

Figure 8.4 summarizes the operation point of the inverter by means of a single phase equivalent phasor diagram, neglecting the capacitive branch of the LCL filter. The phasors run at the fundamental of the grid frequency, ω. In Figure 8.4 it has been assumed that the phasor of the grid voltage seen from the inverter side, $\vec{V}_{grid}$, is on the phase origin. The inverter generates a phasor, $\vec{V}_{inv}$, with a positive angle, $\phi_{inv} > 0$, in inverter operation. The inverter closed loop control has to generate the phase

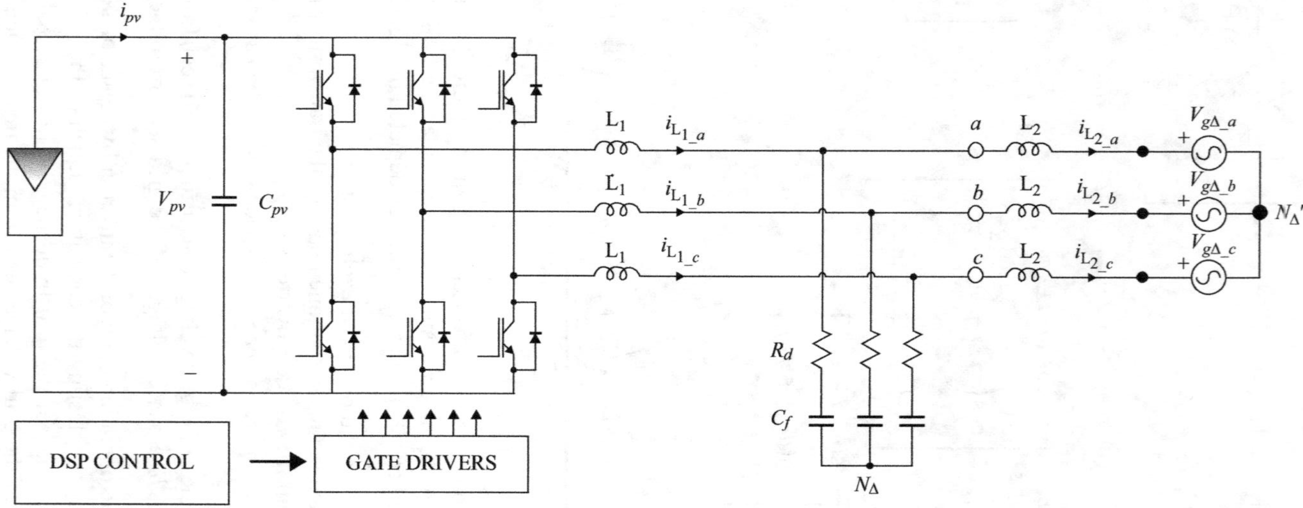

Figure 8.3 Equivalent circuit of the grid-connected PV inverter with LCL filter seen from the inverter side (Δ-side of the transformer)

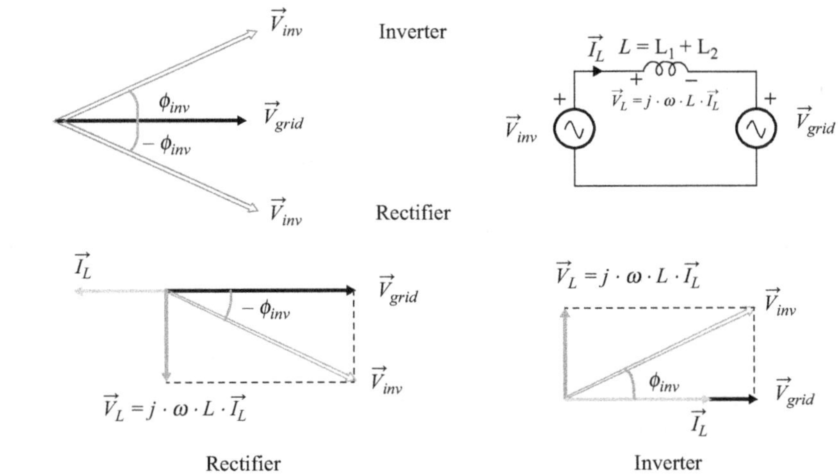

Figure 8.4 Equivalent phasors (fundamental) of a grid-connected inverter: inverter and rectifier operation modes

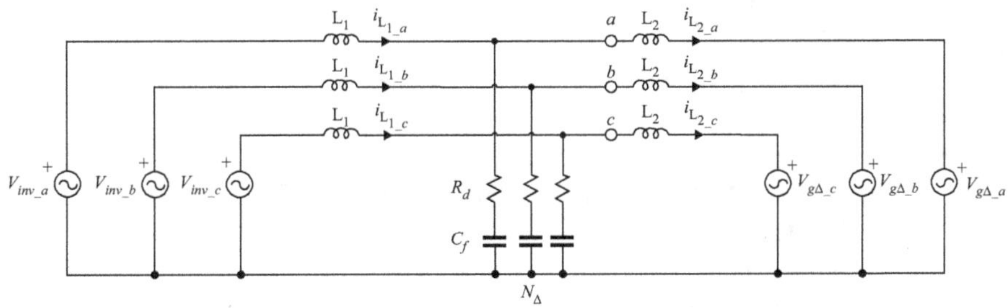

Figure 8.5 Equivalent circuit of a three-phase grid-connected inverter with LCL filter in a, b, c coordinates

voltages in each inverter leg (v_{inv_a}, v_{inv_b} and v_{inv_c}) so that each grid injected current ($i_{L_{2\_a}}$, $i_{L_{2\_b}}$ and $i_{L_{2\_c}}$) is in phase with its corresponding phase voltage and has the suitable RMS value to inject the PV array power to the grid. Figure 8.5 shows the equivalent circuit of the three-phase grid-connected inverter in *a*, *b*, *c* coordinates.

The inverter phase voltages can be transformed using the Clarke transform into two components, v_{inv_α} and v_{inv_β}, leading to the equivalent circuit of the three-phase grid-connected PV inverter in α-β coordinates depicted in Figure 8.6. The values of v_{inv_α} and v_{inv_β} are expressed by (8.7). In Figure 8.6 and in (8.7) the following nomenclature is used:

- d_α and d_β: α-β coordinates of the duty cycles of the inverter legs, after applying the Clarke transform to the duty cycles in *a-b-c* coordinates, d_a, d_b and d_c.
- v_{grid_α} and v_{grid_β}: α-β coordinates of the grid voltage seen from the inverter side, after applying the Clarke transform to the grid voltages in *a-b-c* coordinates, $v_{g\Delta\_a}$, $v_{g\Delta\_b}$ and $v_{g\Delta\_c}$.

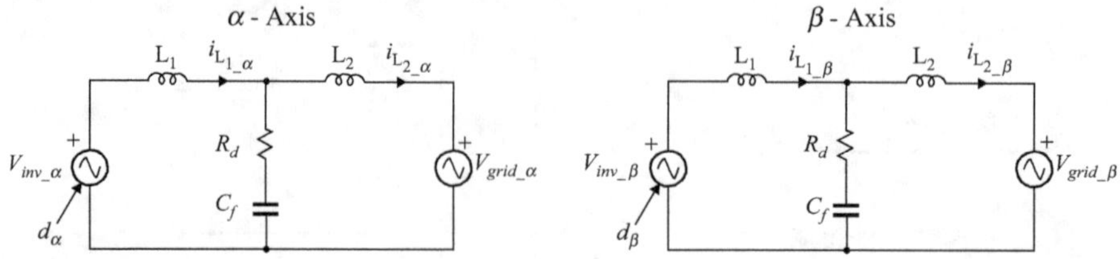

Figure 8.6 Equivalent circuit of a three-phase grid-connected inverter with LCL filter in α-β coordinates

- v_{PV}: PV array voltage.
- The gain of the inverter, K_{inv}, which depends on the PWM modulation being used. Equation (8.8) provides the values of K_{inv} for carrier-based sinusoidal PWM and for Space Vector Modulation (SVM).

$$\begin{aligned} v_{inv\_\alpha}(t) &= K_{inv} \cdot v_{PV} \cdot d_\alpha(t) \\ v_{inv\_\beta}(t) &= K_{inv} \cdot v_{PV} \cdot d_\beta(t) \end{aligned} \tag{8.7}$$

$$K_{inv} = \begin{cases} \dfrac{\sqrt{3}}{2 \cdot \sqrt{2}} \approx 0.61237 & \textit{sinusoidal PWM} \\ \dfrac{1}{\sqrt{2}} \approx 0.70711 & \textit{SVM} \end{cases} \tag{8.8}$$

8.2.1.1 Operation point of the grid-connected VSI in α-β coordinates

If the phase origin is placed at $v_{g\Delta\_a}$, $v_{g\Delta\_b}$ and $v_{g\Delta\_c}$, so that the grid voltage seen from the inverter side can be expressed as (8.9), the inverter voltages in *a*, *b*, *c* and in α-β coordinates are given by (8.10), where ϕ_{inv} is the phase shift of the voltage generated by the VSI and $\hat{V}_{inv}$ is the amplitude of that voltage. The former equations describe the operation point (OP) of the VSI, where the values of ϕ_{inv} and $\hat{V}_{inv}$ can be derived from the phasor circuit of Figure 8.4 by neglecting the capacitor branch of the LCL filter, yielding to (8.11) and (8.12). This assumption is valid for the common design values of the LCL filter because the effects of the capacitor branch impedance at the grid frequency can be neglected. Note that V_{grid} and V_{inv} are the RMS values of the grid voltage and of the inverter voltage, respectively, seen from the inverter side of the transformer. *P* stands for the power injected to the grid.

$$\begin{bmatrix} v_{g\Delta\_a} \\ v_{g\Delta\_b} \\ v_{g\Delta\_c} \end{bmatrix} = \hat{V}_{grid} \cdot \begin{bmatrix} \cos(\omega \cdot t) \\ \cos\left(\omega \cdot t - \dfrac{2\pi}{3}\right) \\ \cos\left(\omega \cdot t + \dfrac{2\pi}{3}\right) \end{bmatrix} \Rightarrow \begin{bmatrix} v_{grid\_\alpha} \\ v_{grid\_\beta} \\ v_{grid\_\gamma} \end{bmatrix} = \sqrt{\frac{3}{2}} \cdot \hat{V}_{grid} \cdot \begin{bmatrix} \cos(\omega \cdot t) \\ \sin(\omega \cdot t) \\ 0 \end{bmatrix} \tag{8.9}$$

$$\begin{bmatrix} v_{inv\_a} \\ v_{inv\_b} \\ v_{inv\_c} \end{bmatrix} = \hat{V}_{inv} \cdot \begin{bmatrix} \cos(\omega \cdot t + \phi_{inv}) \\ \cos\left(\omega \cdot t + \phi_{inv} - \frac{2\pi}{3}\right) \\ \cos\left(\omega \cdot t + \phi_{inv} + \frac{2\pi}{3}\right) \end{bmatrix} \Rightarrow \begin{bmatrix} v_{inv\_\alpha} \\ v_{inv\_\beta} \\ v_{inv\_\gamma} \end{bmatrix}$$

$$= \sqrt{\frac{3}{2}} \cdot \hat{V}_{inv} \cdot \begin{bmatrix} \cos(\omega \cdot t + \phi_{inv}) \\ \sin(\omega \cdot t + \phi_{inv}) \\ 0 \end{bmatrix} \tag{8.10}$$

$$\phi_{inv} \approx \tan^{-1}\left[\frac{\omega \cdot (\mathrm{L}_1 + \mathrm{L}_2) \cdot P}{3 \cdot V_{grid}^2}\right] \tag{8.11}$$

$$\hat{V}_{inv} = \sqrt{2} \cdot V_{inv} \approx \sqrt{2} \cdot \sqrt{V_{grid}^2 + \left[\frac{\omega \cdot (\mathrm{L}_1 + \mathrm{L}_2) \cdot P}{3 \cdot V_{grid}}\right]^2} \tag{8.12}$$

By comparing (8.7) and (8.10) it can be derived the expression (8.13) of the duty cycles d_α and d_β at the OP. Note that the amplitudes of the duty cycles in α and β coordinates agree at the OP, yielding (8.14). The amplitude of the grid injected current seen from the inverter side at the OP in α-β coordinates can be expressed as (8.15), so that the OP values in α-β coordinates of the grid injected current seen from the inverter side are given by (8.16)

$$\begin{aligned} d_\alpha(t) &= \frac{\sqrt{\frac{3}{2}} \cdot \hat{V}_{inv}}{K_{inv} \cdot v_{PV}} \cdot \cos(\omega \cdot t + \phi_{inv}) = \hat{D}_\alpha \cdot \cos(\omega \cdot t + \phi_{inv}) \\ d_\beta(t) &= \frac{\sqrt{\frac{3}{2}} \cdot \hat{V}_{inv}}{K_{inv} \cdot v_{PV}} \cdot \sin(\omega \cdot t + \phi_{inv}) = \hat{D}_\beta \cdot \sin(\omega \cdot t + \phi_{inv}) \end{aligned} \tag{8.13}$$

$$\hat{D}_{\alpha\beta} \equiv \hat{D}_\alpha = \hat{D}_\beta = \frac{\sqrt{\frac{3}{2}} \cdot \hat{V}_{inv}}{K_{inv} \cdot v_{PV}} \approx \frac{\sqrt{3} \cdot \sqrt{V_{grid}^2 + \left[\frac{\omega \cdot (\mathrm{L}_1 + \mathrm{L}_2) \cdot P}{3 \cdot V_{grid}}\right]^2}}{K_{inv} \cdot v_{PV}} \tag{8.14}$$

$$\hat{I}_{\mathrm{L}_2\_\alpha} = \hat{I}_{\mathrm{L}_2\_\beta} = \frac{P}{\sqrt{3} \cdot V_{grid}} \equiv \hat{I}_{\mathrm{L}_2\_\alpha\beta} \tag{8.15}$$

$$\begin{aligned} i_{\mathrm{L}_2\_\alpha}(t) &= \hat{I}_{\mathrm{L}_2\_\alpha} \cdot \cos(\omega \cdot t) = \hat{I}_{\mathrm{L}_2\_\alpha\beta} \cdot \cos(\omega \cdot t) \\ i_{\mathrm{L}_2\_\beta}(t) &= \hat{I}_{\mathrm{L}_2\_\beta} \cdot \sin(\omega \cdot t) = \hat{I}_{\mathrm{L}_2\_\alpha\beta} \cdot \sin(\omega \cdot t) \end{aligned} \tag{8.16}$$

8.2.1.2 Dynamic model of the grid-connected VSI in α-β coordinates

Any variable x of the dynamic model can be decomposed as the sum of an OP term, x_{OP}, and a small-signal term, $\tilde{x}$, so that: $x = x_{OP} + \tilde{x}(s)$, where s is the Laplace variable of the small-signal term. If the voltages generated by the VSI in α-β coordinates shown in (8.7) are decomposed as the sum of the OP value plus the dynamic small-signal value, as illustrated by (8.17), a dynamic model for the closed loop control of the VSI can be obtained. It is worth pointing out that in α-β coordinates many of the OP values vary sinusoidally at the grid frequency as it has been described in Section 8.2.1.1. In order to obtain the small-signal dynamics, the linearization around the OP described by (8.18) can be performed. From (8.11) to (8.14) the OP values of the duty cycles in α-β coordinates can be derived, yielding (8.19). Those OP duty cycles are used in (8.18) and have a sinusoidal variation with time.

$$\begin{aligned} v_{inv\_\alpha} &= K_{inv} \cdot v_{PV} \cdot d_{\alpha} = \underbrace{v_{inv\_\alpha\_OP}}_{OP\ value} + \underbrace{\tilde{v}_{inv\_\alpha}}_{small-signal\ dynamics} \\ v_{inv\_\beta} &= K_{inv} \cdot v_{PV} \cdot d_{\beta} = \underbrace{v_{inv\_\beta\_OP}}_{OP\ value} + \underbrace{\tilde{v}_{inv\_\beta}}_{small-signal\ dynamics} \end{aligned} \tag{8.17}$$

$$\begin{aligned} \tilde{v}_{inv\_\alpha}(s) &= \left.\frac{\partial v_{inv\_\alpha}}{\partial d_{\alpha}}\right|_{OP} \cdot \tilde{d}_{\alpha}(s) + \left.\frac{\partial v_{inv_{\alpha}}}{\partial v_{PV}}\right|_{OP} \cdot \tilde{v}_{PV}(s) \\ &= K_{inv} \cdot v_{PV\_OP} \cdot \tilde{d}_{\alpha}(s) + K_{inv} \cdot d_{\alpha\_OP} \cdot \tilde{v}_{PV}(s) \\ \tilde{v}_{inv\_\beta}(s) &= \left.\frac{\partial v_{inv\_\beta}}{\partial d_{\beta}}\right|_{OP} \cdot \tilde{d}_{\beta}(s) + \left.\frac{\partial v_{inv_{\beta}}}{\partial v_{PV}}\right|_{OP} \cdot \tilde{v}_{PV}(s) \\ &= K_{inv} \cdot v_{PV\_OP} \cdot \tilde{d}_{\beta}(s) + K_{inv} \cdot d_{\beta\_OP} \cdot \tilde{v}_{PV}(s) \end{aligned} \tag{8.18}$$

$$\begin{aligned} d_{\alpha\_OP} &= \hat{D}_{\alpha\beta\_OP} \cdot \cos\left(\omega \cdot t + \phi_{inv\_OP}\right) \\ &= \underbrace{\frac{\sqrt{3} \cdot \sqrt{V_{grid\_OP}^2 + \left[\dfrac{\omega \cdot (\mathrm{L}_1 + \mathrm{L}_2) \cdot P}{3 \cdot V_{grid\_OP}}\right]^2}}{K_{inv} \cdot v_{PV\_OP}}}_{\hat{D}_{\alpha\beta\_OP}} \cdot \cos\left(\omega \cdot t + \phi_{inv\_OP}\right) \\ d_{\beta\_OP} &= \hat{D}_{\alpha\beta\_OP} \cdot \sin\left(\omega \cdot t + \phi_{inv\_OP}\right) \\ &= \underbrace{\frac{\sqrt{3} \cdot \sqrt{V_{grid\_OP}^2 + \left[\dfrac{\omega \cdot (\mathrm{L}_1 + \mathrm{L}_2) \cdot P}{3 \cdot V_{grid\_OP}}\right]^2}}{K_{inv} \cdot v_{PV\_OP}}}_{\hat{D}_{\alpha\beta\_OP}} \cdot \sin\left(\omega \cdot t + \phi_{inv\_OP}\right) \end{aligned} \tag{8.19}$$

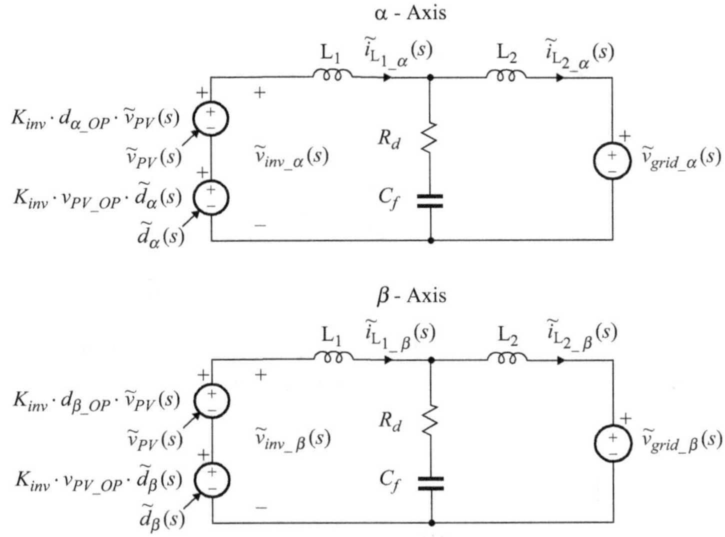

Figure 8.7 Small-signal dynamic model of the VSI in α-β coordinates

Equation (8.18) provides the dynamic model of the voltages generated by the VSI in α-β coordinates. These voltages are applied to one of the ports of the LCL filter, whereas the grid voltage disturbances excite the other port of the LCL filter, resulting in the dynamic model of the VSI depicted in Figure 8.7. Note that $\tilde{v}_{PV}(s)$ stands for the disturbances in the PV array voltage, e.g. low frequency ripples, whereas $\tilde{v}_{grid\_\alpha}(s)$ and $\tilde{v}_{grid\_\beta}(s)$ denote disturbances of the grid voltage, like low frequency harmonics. It is worth pointing out that the small-signal dynamic model is valid for frequencies lower than $f_{SW}/2$, where f_{SW} is the switching frequency of the VSI.

8.2.2 Modeling in an orthogonal synchronous reference frame (SRF)

A rotating vector $\vec{v}_{\alpha\beta}$ in stationary α-β coordinates can be transformed into a stationary vector in rotating d-q coordinates. That's the idea underlying the Park transform. The vector operation to transform the $\vec{v}_{\alpha\beta}$ vector into a stationary vector, $\vec{v}_{dq}$, is shown in (8.20), where $\theta(t)$ is the angle of the Park transformation, expressed by (8.21). An arbitrary initial value of θ, $\theta(0)$, can be chosen. In this case the new d-q axes are rotating at an angular speed ω, but the d-q components of the vector, v_d and v_q, have constant, i.e. DC, values at the OP of the three-phase system (constant ω), as it can be observed from (8.22). The d-q axis conform an SRF, because it's always synchronized with the angle $\theta(t)$. Figure 8.8 depicts the relationship between the vectors $\vec{v}_{\alpha\beta}$ and $\vec{v}_{dq}$ and their respective reference frames, in the case of a balanced and positive-sequence three-phase voltage system.

$$\vec{v}_{dq} = v_d + j \cdot v_q = \vec{v}_{\alpha\beta} \cdot e^{-j\cdot\theta(t)} = \sqrt{\frac{3}{2}} \cdot V_m \cdot e^{j\cdot(\omega\cdot t+\phi_0)} \cdot e^{-j\cdot\theta(t)} \qquad (8.20)$$

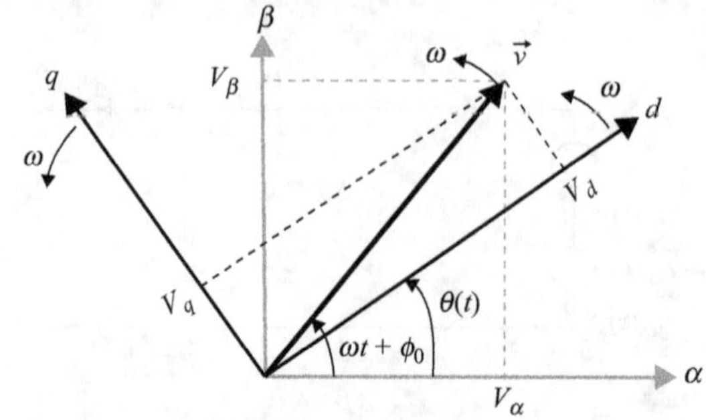

Figure 8.8 Complex vector resulting from the Clarke and Park transformations of a balanced and positive-sequence three-phase voltage system

$$\theta(t) = \theta(0) + \int_0^t \omega(\tau)d\tau \tag{8.21}$$

$$\text{if } \omega = \text{constant} \rightarrow \theta(t) = \theta(0) + \omega \cdot t \rightarrow \vec{v}_{dq}$$

$$= \vec{v}_{\alpha\beta} \cdot e^{-j\cdot\theta(t)} = \sqrt{\frac{3}{2}} \cdot V_m \cdot e^{j\cdot(\omega\cdot t+\phi_0)} \cdot e^{-j\cdot[\theta(0)+\omega\cdot t]}$$

$$\rightarrow \vec{v}_{dq} = \sqrt{\frac{3}{2}} \cdot V_m \cdot e^{j\cdot(\phi_0-\theta(0))} = \underbrace{\sqrt{\frac{3}{2}} \cdot V_m \cdot \cos(\phi_0 - \theta(0))}_{v_d} \tag{8.22}$$

$$+j \cdot \underbrace{\sqrt{\frac{3}{2}} \cdot V_m \cdot \sin(\phi_0 - \theta(0))}_{v_q} = v_d + j \cdot v_q$$

The transformation matrix which transforms the α-β-γ components of a three-phase system to its the *d*-*q*-*o* components is shown in (8.23). Note that the *o*-component is identical to the γ-component. The matrix used for transforming from the *d*-*q*-*o* components to the α-β-γ components is shown in (8.24). In order to calculate the *d*-*q*-*o* components starting from the *a*-*b*-*c* components, the transformation shown in (8.25) can be used. The inverse transformation is described by (8.26).

$$[v_{dqo}] = \begin{bmatrix} v_d \\ v_q \\ v_o \end{bmatrix} = T_{\alpha\beta\gamma\rightarrow dqo} \cdot \begin{bmatrix} v_\alpha \\ v_\beta \\ v_\gamma \end{bmatrix}$$

$$T_{\alpha\beta\gamma\rightarrow dqo} = \begin{bmatrix} \cos\theta & \sin\theta & 0 \\ -\sin\theta & \cos\theta & 0 \\ 0 & 0 & 1 \end{bmatrix} \tag{8.23}$$

$$[v_{\alpha\beta\gamma}] = T_{dqo\to\alpha\beta\gamma}\cdot\begin{bmatrix} v_d \\ v_q \\ v_o \end{bmatrix}$$

$$T_{dqo\to\alpha\beta\gamma} = [T_{\alpha\beta\gamma\to dqo}]^{-1} = [T_{\alpha\beta\gamma\to dqo}]^T = \begin{bmatrix} \cos\theta & -\sin\theta & 0 \\ \sin\theta & \cos\theta & 0 \\ 0 & 0 & 1 \end{bmatrix} \quad (8.24)$$

$$[v_{dqo}] = \begin{bmatrix} v_d \\ v_q \\ v_o \end{bmatrix} = T_{abc\to dqo}\cdot\begin{bmatrix} v_a \\ v_b \\ v_c \end{bmatrix};$$

$$T_{abc\to dqo} = \sqrt{\frac{2}{3}}\cdot\begin{bmatrix} \cos\theta & \cos\left(\theta-\frac{2\pi}{3}\right) & \cos\left(\theta+\frac{2\pi}{3}\right) \\ -\sin\theta & -\sin\left(\theta-\frac{2\pi}{3}\right) & -\sin\left(\theta+\frac{2\pi}{3}\right) \\ \frac{1}{\sqrt{2}} & \frac{1}{\sqrt{2}} & \frac{1}{\sqrt{2}} \end{bmatrix}$$

$$T_{abc\to dqo} = T_{\alpha\beta\gamma\to dqo}\cdot T_{abc\to\alpha\beta\gamma} = \begin{bmatrix} \cos\theta & \sin\theta & 0 \\ -\sin\theta & \cos\theta & 0 \\ 0 & 0 & 1 \end{bmatrix}\cdot\sqrt{\frac{2}{3}}\cdot\begin{bmatrix} 1 & -\frac{1}{2} & -\frac{1}{2} \\ 0 & \frac{\sqrt{3}}{2} & -\frac{\sqrt{3}}{2} \\ \frac{1}{\sqrt{2}} & \frac{1}{\sqrt{2}} & \frac{1}{\sqrt{2}} \end{bmatrix} \quad (8.25)$$

$$[v_{abc}] = T_{dqo\to abc}\cdot[v_{dqo}]$$

$$= \sqrt{\frac{2}{3}}\cdot\begin{bmatrix} \cos\theta & -\sin\theta & \frac{1}{\sqrt{2}} \\ \cos\left(\theta-\frac{2\pi}{3}\right) & -\sin\left(\theta-\frac{2\pi}{3}\right) & \frac{1}{\sqrt{2}} \\ \cos\left(\theta+\frac{2\pi}{3}\right) & -\sin\left(\theta+\frac{2\pi}{3}\right) & \frac{1}{\sqrt{2}} \end{bmatrix}\cdot\begin{bmatrix} v_d \\ v_q \\ v_o \end{bmatrix}$$

$$T_{dqo\to abc} = [T_{abc\to dqo}]^{-1} = [T_{abc\to dqo}]^T$$

$$= \sqrt{\frac{2}{3}}\cdot\begin{bmatrix} \cos\theta & -\sin\theta & \frac{1}{\sqrt{2}} \\ \cos\left(\theta-\frac{2\pi}{3}\right) & -\sin\left(\theta-\frac{2\pi}{3}\right) & \frac{1}{\sqrt{2}} \\ \cos\left(\theta+\frac{2\pi}{3}\right) & -\sin\left(\theta+\frac{2\pi}{3}\right) & \frac{1}{\sqrt{2}} \end{bmatrix}\cdot\begin{bmatrix} v_d \\ v_q \\ v_o \end{bmatrix} \quad (8.26)$$

The version of the Park transformation described in this section preserves the instantaneous power of a three-phase system, so that the instantaneous power can be expressed by (8.27).

$$p(t) = v_a \cdot i_a + v_b \cdot i_b + v_c \cdot i_c = v_d \cdot i_d + v_q \cdot i_q + v_o \cdot i_o \tag{8.27}$$

8.2.2.1 Operation point of the grid-connected VSI in *d-q* coordinates

The inverter voltages in *d-q* coordinates can be calculated by applying the vectorial transformation (8.20) or the matrix transformation (8.23) to the inverter voltages in α-β coordinates, resulting in (8.28). The angle chosen for the Park transformation is $\theta = \omega \cdot t$, i.e. the angle of the phase voltage '*a*' from the inverter side, $v_{g\Delta\_a}$. A Phase-Locked-Loop (PLL) [7], [8] is necessary for that synchronization.

The duty cycles of the inverter in *d-q* coordinates can be obtained by applying the Park transform to (8.13), resulting in (8.29). From (8.28), (8.29) and (8.14) the alternative expression (8.30) for the VSI voltages in *d-q* coordinates can be derived.

$$\begin{aligned}
\vec{v}_{inv_dq} &= v_{inv_d} + j \cdot v_{inv_q} = \vec{v}_{inv_{\alpha\beta}} \cdot e^{-j \cdot \theta(t)} = \sqrt{\frac{3}{2}} \cdot \hat{V}_{inv} \cdot e^{j \cdot (\omega \cdot t + \phi_{inv})} \cdot e^{-j \cdot \theta(t)} \\
\text{if } \theta(t) = \omega \cdot t &\rightarrow \vec{v}_{inv_{dq}} = \sqrt{\frac{3}{2}} \cdot \hat{V}_{inv} \cdot e^{j \cdot \phi_{inv}} \\
&= \underbrace{\sqrt{\frac{3}{2}} \cdot \hat{V}_{inv} \cdot \cos\phi_{inv}}_{v_{inv_d}} + j \cdot \underbrace{\sqrt{\frac{3}{2}} \cdot \hat{V}_{inv} \cdot \sin\phi_{inv}}_{v_{inv_q}}
\end{aligned} \tag{8.28}$$

$$\begin{aligned}
\hat{d}_{dq} &= d_d + j \cdot d_q = \vec{d}_{\alpha\beta} \cdot e^{-j \cdot \theta(t)} = \hat{D}_{\alpha\beta} \cdot e^{j \cdot (\omega \cdot t + \phi_{inv})} \cdot e^{-j \cdot \theta(t)} \\
\text{if } \theta(t) = \omega \cdot t &\rightarrow \hat{d}_{dq} = \hat{D}_{\alpha\beta} \cdot e^{j \cdot \phi_{inv}} = \underbrace{\hat{D}_{\alpha\beta} \cdot \cos\phi_{inv}}_{d_d} + j \cdot \underbrace{\hat{D}_{\alpha\beta} \cdot \sin\phi_{inv}}_{d_q}
\end{aligned} \tag{8.29}$$

$$\begin{aligned}
v_{inv_d} &= \sqrt{\frac{3}{2}} \cdot \hat{V}_{inv} \cdot \cos\phi_{inv} = K_{inv} \cdot v_{PV} \cdot d_d \\
v_{inv_q} &= \sqrt{\frac{3}{2}} \cdot \hat{V}_{inv} \cdot \sin\phi_{inv} = K_{inv} \cdot v_{PV} \cdot d_q
\end{aligned} \tag{8.30}$$

The grid voltages seen from the inverter side in *d-q* coordinates can be calculated by applying (8.20) to the grid voltages in α-β coordinates (8.9), resulting in (8.31). Note that $v_{grid\_q} = 0$ at the OP due to the angle $\theta = \omega \cdot t$ chosen for the transformation, i.e. the synchronization angle. By applying (8.20) to the grid injected currents in α-β coordinates (8.16), the grid injected currents in *d-q* coordinates

(8.32) are obtained. Again, it is observed that the q component, in this case of the grid injected current, is null at the OP: $i_{\mathrm{L}_{2\_q}} = 0$.

$$\vec{v}_{grid\_dq} = v_{grid\_d} + j \cdot v_{grid\_q} = \vec{v}_{grid\_\alpha\beta} \cdot e^{-j \cdot \theta(t)} = \sqrt{\frac{3}{2}} \cdot \hat{V}_{grid} \cdot e^{j \cdot \omega \cdot t} \cdot e^{-j \cdot \theta(t)}$$

$$\text{if } \theta(t) = \omega \cdot t \rightarrow \vec{v}_{grid\_dq} = \sqrt{\frac{3}{2}} \cdot \hat{V}_{grid} \cdot e^{j \cdot 0}$$

$$= \underbrace{\sqrt{\frac{3}{2}} \cdot \hat{V}_{grid}}_{v_{grid\_d}} + j \cdot \underbrace{0}_{v_{grid\_q}} \rightarrow \begin{cases} v_{grid\_d} = \sqrt{\frac{3}{2}} \cdot \hat{V}_{grid} = \sqrt{3} \cdot V_{grid} \\ v_{grid\_q} = 0 \end{cases} \tag{8.31}$$

$$\vec{i}_{\mathrm{L}_{2\_dq}} = i_{\mathrm{L}_{2\_d}} + j \cdot i_{\mathrm{L}_{2\_q}} = \vec{i}_{\mathrm{L}_{2\_\alpha\beta}} \cdot e^{-j \cdot \theta(t)} = \hat{I}_{\mathrm{L}_{2\_\alpha\beta}} \cdot e^{j \cdot \omega \cdot t} \cdot e^{-j \cdot \theta(t)}$$

$$\text{if } \theta(t) = \omega \cdot t \rightarrow \vec{i}_{\mathrm{L}_{2\_dq}} = \hat{I}_{\mathrm{L}_{2\_\alpha\beta}} \cdot e^{j \cdot 0} = \underbrace{\hat{I}_{\mathrm{L}_{2\_\alpha\beta}}}_{i_{\mathrm{L}_{2\_d}}} + j \cdot \underbrace{0}_{i_{\mathrm{L}_{2\_q}}} \rightarrow \begin{cases} i_{\mathrm{L}_{2\_d}} = \hat{I}_{\mathrm{L}_{2\_\alpha\beta}} \\ i_{\mathrm{L}_{2\_q}} = 0 \end{cases} \tag{8.32}$$

Taking into account that in the SRF the OP magnitudes are DC, the OP values in d-q coordinates will be represented with capital letters throughout the text (DC values). The OP value $I_{\mathrm{L}_{2\_d}}$ can be represented as a function of the grid injected power, P, by particularizing (8.27) at the OP, yielding (8.33).

$$P = V_{grid\_d} \cdot I_{\mathrm{L}_{2_d}} + \overbrace{V_{grid_q} \cdot I_{\mathrm{L}_{2_q}}}^{0} + \overbrace{V_{grid_o} \cdot I_{\mathrm{L}_{2_o}}}^{0} = V_{PV} \cdot I_{PV} \rightarrow I_{\mathrm{L}_{2_d}}$$

$$= \frac{P}{V_{grid_d}} = \frac{V_{PV} \cdot I_{PV}}{V_{grid_d}} = \frac{V_{PV} \cdot I_{PV}}{\sqrt{3} \cdot V_{grid}} \tag{8.33}$$

A complete equivalent circuit for the grid-connected VSI including the LCL filter in d-q coordinates can be found in [9]. The part which is interesting for the control of the grid injected current is depicted in Figure 8.9a.

Figure 8.9b depicts the equivalent circuit of the VSI at the OP, and it has been obtained by short-circuiting the inductors and leaving the capacitors in open circuit in Figure 8.9a. The exact expressions of the OP values of all variables of interest can be derived from that circuit, taking into account (8.33) and (8.34).

$$V_{grid\_d} = \sqrt{\frac{3}{2}} \cdot \hat{V}_{grid} = \sqrt{3} \cdot V_{grid}; \;\; V_{grid_q} = 0; \;\; I_{\mathrm{L}_{2_d}} = \frac{P}{V_{grid_d}} = \frac{V_{PV} \cdot I_{PV}}{V_{grid_d}}; \;\; I_{\mathrm{L}_{2_q}} = 0 \tag{8.34}$$

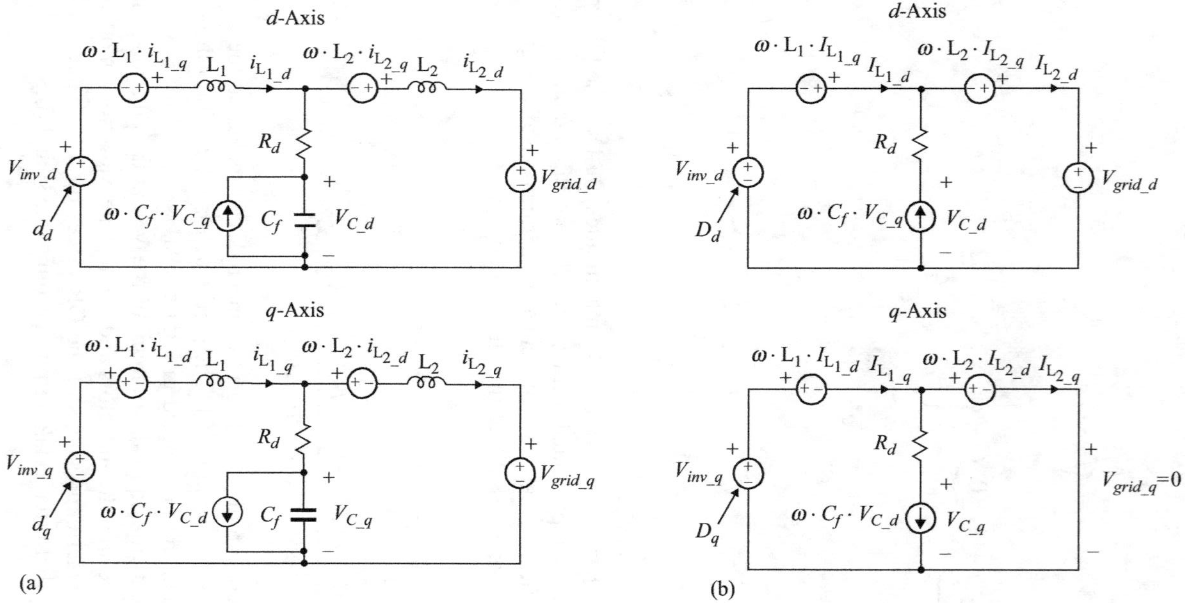

Figure 8.9 Equivalent circuit of a three-phase grid-connected VSI with LCL filter in d-q coordinates. (a) Complete circuit. (b) Equivalent circuit at the operation point

8.2.2.2 Dynamic model of the grid-connected VSI in *d-q* coordinates

Any variable x of the dynamic model in *d-q* coordinates can be decomposed as the sum of a DC OP term, X, and a small-signal term, $\tilde{x}$, so that $x = X + \tilde{x}(s)$, where s is the Laplace variable. If the voltages generated by the VSI in *d-q* coordinates shown in (8.30) are decomposed as illustrated by (8.31) and (8.32), a dynamic small-signal dynamic model for the closed loop control of the VSI can be obtained.

$$\begin{aligned} v_{inv\_d} &= K_{inv} \cdot v_{PV} \cdot d_d = \underbrace{V_{inv\_d}}_{OP\ value} + \underbrace{\tilde{v}_{inv\_d}}_{small-signal\ dynamics} \\ v_{inv\_q} &= K_{inv} \cdot v_{PV} \cdot d_q = \underbrace{V_{inv\_d}}_{OP\ value} + \underbrace{\tilde{v}_{inv\_q}}_{small-signal\ dynamics} \end{aligned} \tag{8.31}$$

$$\begin{aligned} \tilde{v}_{inv\_d}(s) &= \left.\frac{\partial v_{inv\_d}}{\partial d_d}\right|_{OP} \cdot \tilde{d}_d(s) + \left.\frac{\partial v_{inv_d}}{\partial v_{PV}}\right|_{OP} \cdot \tilde{v}_{PV}(s) \\ &= K_{inv} \cdot V_{PV} \cdot \tilde{d}_d(s) + K_{inv} \cdot D_d \cdot \tilde{v}_{PV}(s) \\ \tilde{v}_{inv\_q}(s) &= \left.\frac{\partial v_{inv\_q}}{\partial d_q}\right|_{OP} \cdot \tilde{d}_q(s) + \left.\frac{\partial v_{inv_q}}{\partial v_{PV}}\right|_{OP} \cdot \tilde{v}_{PV}(s) \\ &= K_{inv} \cdot V_{PV} \cdot \tilde{d}_q(s) + K_{inv} \cdot D_q \cdot \tilde{v}_{PV}(s) \end{aligned} \tag{8.32}$$

Equation (8.32) provides the small-signal dynamic model of the voltages generated by the VSI in *d-q* coordinates. These voltages are applied to the inverter-connected ports of the LCL filter transformed to *d-q* coordinates, whereas the grid voltage disturbances excite the grid-connected ports of the *d-q* transformed LCL filter, resulting in the dynamic model of the VSI depicted in Figure 8.10. Note that $\tilde{v}_{PV}(s)$ stands for the disturbances of the PV array voltage, whereas $\tilde{v}_{grid\_d}(s)$ and $\tilde{v}_{grid\_q}(s)$ denote disturbances of the grid voltage. Couplings between the d and q circuits of Figure 8.10 can be observed. As it holds true for the dynamic model in α-β coordinates, the small-signal dynamic model in *d-q* coordinates is valid for frequencies lower than $f_{SW}/2$.

8.2.3 Control of a grid-connected PV inverter with LCL filter

In this section the control structure in the StatRF and in the SRF of the grid-connected inverter of Figures 8.1 and 8.3 will be described. The VSI is connected to a 50 Hz, 400 V_{RMS} grid (nominal value of the RMS line voltage) through a Δ-Y transformer with turns ratio: $n_\Delta / n_Y = 1$, so that the line voltage at the inverter side (Δ-side) is approximately 230 V_{RMS}, corresponding to a phase voltage around 133 V_{RMS}. The values of the inverter under study seen from the inverter side of the transformer (Δ-side) are summarized by Table 8.1. Its digital control has been implemented by means of a floating point DSP TMS320LF28335 from Texas Instruments. The power transistors of the 2-level VSI are the BSM50GB120DN2 IGBTs from Infineon. SVM is used for generating the driving signals of the IGBTs.

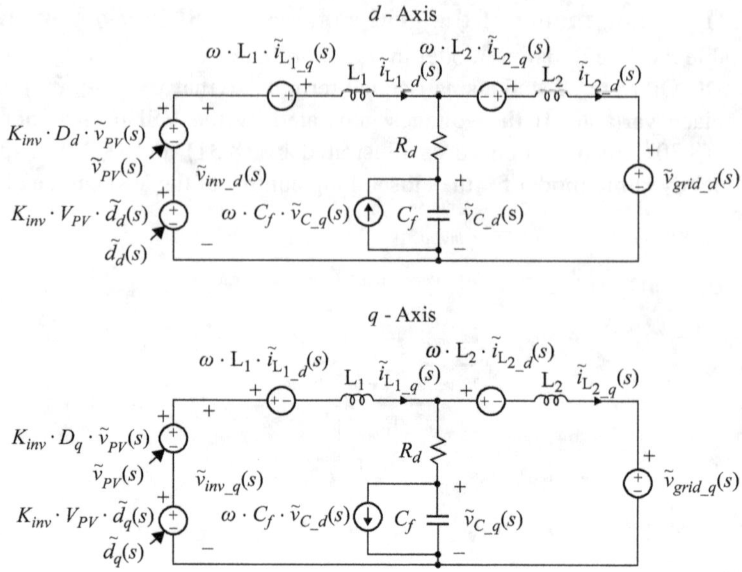

Figure 8.10 Small-signal dynamic model of the VSI in d-q coordinates

Table 8.1 Values of the grid-connected PV inverter with LCL filter under study. All values are referred to the inverter side (Δ-side) of the transformer

	Parameter name	Value
Nominal power	P_{nom}	10 kW
Switching frequency	f_{SW}	10 kHz
Sampling frequency	f_{samp}	20 kHz
PV panel voltage at the maximum power point	V_{PV_MPP}	450 ... 900 V
Inductance connected to the inverter	L_1	1.2 mH
Inductance connected to the grid	L_2	267 ... 800 μH
Filter capacitance	C_f	30 μF
Damping resistors	R_d	3 Ω
DC-link capacitance	C_{PV}	1.36 mF
Nominal grid frequency	f_{nom}	50 Hz
RMS phase voltage	V_{grid}	$\frac{230\text{ V}}{\sqrt{3}} \approx 133$ V
Inverter gain (SVM)	K_{inv}	$\frac{1}{\sqrt{2}}$
MPPT refreshing frequency	f_{MPPT}	10 Hz
MPPT step sixe	ΔV^*_{PV}	8 V

It can be observed from Table 8.1 that damping resistors with a value $R_d = 3\ \Omega$ are used in series with the LCL filter capacitors. Their goal is to damp the resonance of the LCL filter which can produce an instability of the current loops [10, 11], especially when those loops work from the grid injected currents. On the one hand, to minimize the distortion of the grid injected currents the best option is to close the current loops starting from the measurement of the grid injected currents. On the other hand, a high crossover frequency of the current loops, f_{Ci}, is necessary to get a low value of the current harmonics placed at the frequencies of the grid voltage harmonics (5th, 7th, 11th and 13th harmonics). If f_{Ci} approaches the resonance frequency of the LCL filter and the resonance has a low damping value, the stability of the current loop will be seriously compromised. This issue is worsened when the grid inductance has a wide variation. In the 10-kW inverter under study the value of the damping resistors has been chosen intentionally high to cope with those problems. The maximum power dissipation in each damping resistor is 12 W.

Note that the sampling frequency is twice the switching frequency. That's a common feature offered by current microcontrollers and DSPs, which allows for lower sampling delays.

8.2.3.1 Control structure in a stationary reference frame

Figure 8.11 shows a control structure of the VSI in the StatRF, based on sensing the grid injected currents, i.e. of the currents through inductors L_2. This is the preferred option for the minimization of the THDi of the grid injected currents, being a good option when the value of L_2, which is influenced by the grid inductance, doesn't undergo a great variation. As it was reported in [10], another possibility is to control the currents through inductors L_1. In that case the current control is more robust to variations of L_2, but a worse THDi of the grid injected currents is obtained because those currents are not directly sensed.

The control structure of the VSI is based on an inner current loop in α-β coordinates, i.e. two current loops, which gets the reference currents, i_α^* and i_β^*, from an outer voltage loop, whose task is to control the PV panel voltage to achieve MPPT of the PV source [12, 13]. The current controllers, $G_i(s)$, of both channels are identical. The usual choice for $G_i(s)$ is a Proportional + Resonant (PR) Controllers [14–16], with a high gain at the fundamental of the grid frequency and of its more important harmonics. The resonant parts of the PR controller can be implemented by means of Second-Order Generalized Integrators (SOGIs) [17] which provide automatic tuning of the central frequencies with the grid frequency or its harmonics.

An interesting improvement of the current loop consists of the multipliers at the output of the current controllers observed in Figure 8.11. As it will be explained in Section 8.3.1, the open loop gain of the current loop is proportional to V_{PV}, so that in the case of a wide variation of the PV voltage, the adjustment of $G_i(s)$ can be complicated. The action of multiplying the output of the current regulator by the factor $\frac{V_{PV\_max}}{v_{PV}}$ or $\frac{V_{PV\_max}}{v_{PV}^*}$ eliminates the influence of V_{PV} on the current loop. In this case the current loop gain becomes proportional to a constant value V_{PV_max} (the maximum attainable value of V_{PV} in the PV system) instead, so that the adjustment can be performed for this constant value, achieving robustness to PV voltage variations. This technique can be called 'dc-link voltage feedforward in the current loop' [18]. The reason of choosing v_{PV}^* instead of v_{PV} for this

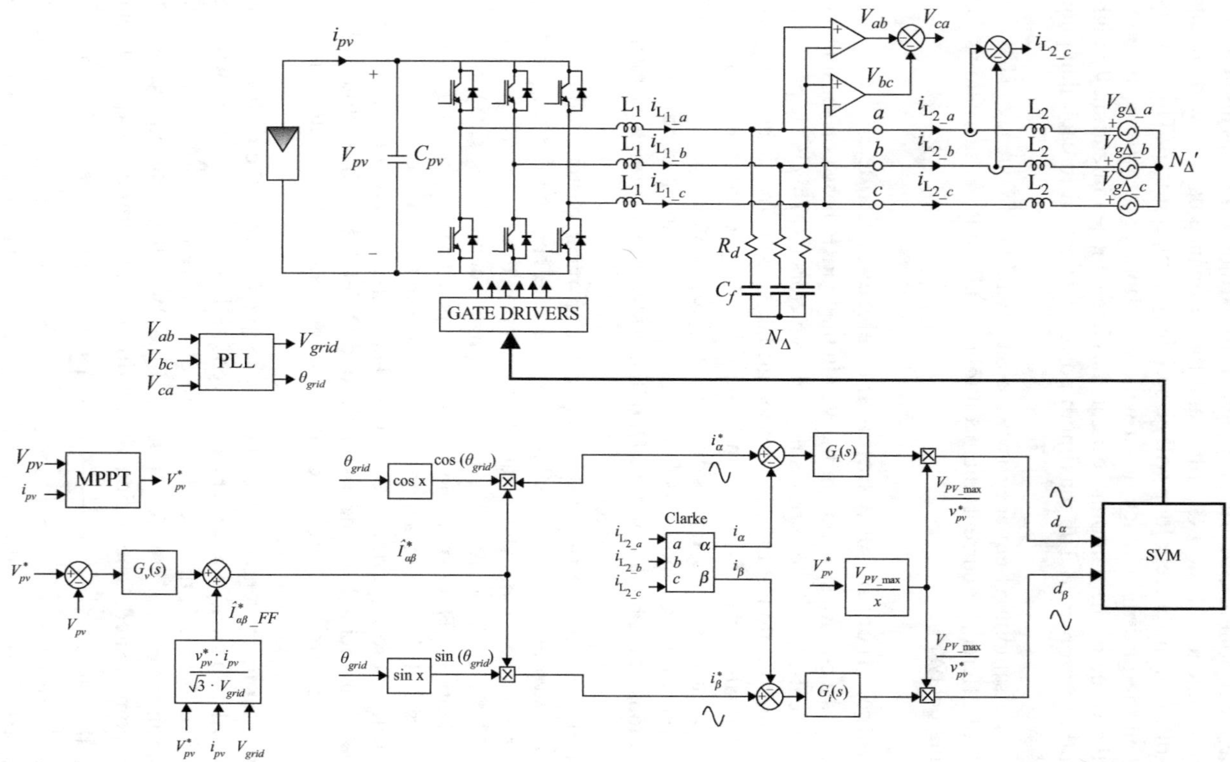

Figure 8.11 Control structure in the StatRF. The current control is performed on the grid injected currents

feedforward is to avoid a possible distortion on the duty cycles d_α and d_β caused by the low frequency ripple at the PV voltage, which would result in a higher THDi of the grid injected currents. That ripple may be important with low values of C_{PV} in the presence of grid voltage unbalance.

It is observed in Figure 8.11 that the voltage controller, $G_V(s)$, works from the reference v^*_{PV}, which is provided by an MPPT algorithm. The most common algorithms are Perturb & Observe, Incremental Conductance and all their variations [6, 12, 13, 19]. The PV voltage reference varies in steps at a low frequency, several Hz or tens of Hz, which is the refreshing frequency of the MPPT algorithm. In order to speed-up the response of the slow voltage controller, a feedforward value, $\hat{I}^*_{\alpha\beta\_FF}$, may be added to its output. This value provides the OP point of the amplitude of the grid injected currents in α-β coordinates, as it was expressed by (8.15). It is worth pointing out that the voltage controller of Figure 8.11 is a proportional-integral (PI) one with a negative sign because v_{PV} decreases when the grid injected current increases, i.e. $G_V(s)$ 'sees' a dynamics with a negative sign.

For transforming the signal $\hat{I}^*_{\alpha\beta}$ to references i^*_α and i^*_β, multiplication by $\sin(\theta_{grid})$ and $\cos(\theta_{grid})$ is performed. The grid phase, θ_{grid}, is provided by a three-phase PLL which extracts the phase and the frequency of the fundamental of the grid voltage, which might be highly distorted. The classical solution is the SRF PLL (SRF-PLL) [7] and its improvements, like adaptive PLLs [20]. Advanced synchronization techniques capable of a good performance under grid voltage unbalance and transient grid faults are the Decoupled Double SRF PLL (DDSRF-PLL) [21] and the Double Second-Order Generalized Integrator Frequency-Locked-Loop (DSOGI-FLL) [22].

8.2.3.2 Control structure in a synchronous reference frame

Figure 8.12 shows a control structure of the VSI in the SRF, based on sensing the grid injected currents. This structure is analog to that of Figure 8.11, but the following differences deserve to be stressed:

- In the current loops in d-q coordinates the Park transformation and its inverse are used.
- Decoupling terms with gain $\frac{\omega\cdot(L_1+L_2)}{K_{inv}\cdot v^*_{PV}}$ are used in the current loops. Their goal is to eliminate the effect of the current i_q on the d-axis current loop and the effect of the current i_d on the q-axis current loop.
- The references i^*_d and i^*_q are synthesized in a different way. As it was expressed by (8.34), $i^*_q = 0$. Reference i^*_d is provided by the voltage controller and a feedforward term.
- Signals i^*_d, i^*_q, i_d, i_q, d_d and d_q are DC at the OP. Note that their homologues in α-β coordinates are sinusoids working at the grid frequency at the OP.
- Using PI current controllers, $G_i(s)$, in d-q coordinates is equivalent to provide an infinite gain at the grid frequency in α-β coordinates [15, 23].

It is worth pointing out that the dynamics 'seen' by the voltage controller in d-q coordinates is the same as in α-β ones, so that the same PI voltage controller can be used.

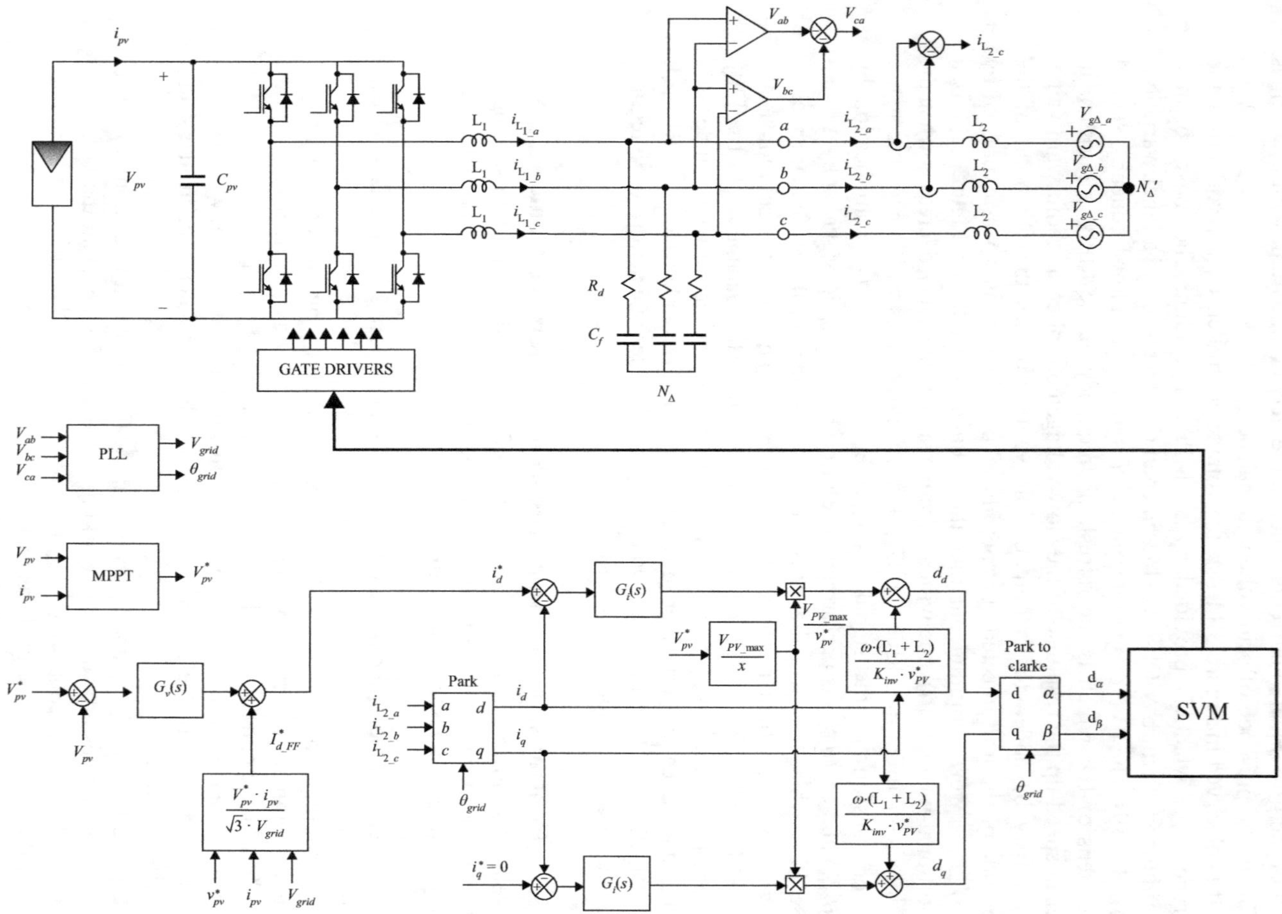

Figure 8.12 Control structure in the SRF. The current control is performed on the grid injected currents

8.3 DSP control of a grid-connected PV inverter with LCL filter in the StatRF

In this section guidelines about the design and programming of the current and voltage control loops in α-β coordinates are provided. The small-signal models explained in Section 8.2.1 will be used.

8.3.1 Design and programming of the current loops in the StatRF

If the small-signal disturbance of the PV voltage in the circuit of Figure 8.7 is cancelled, i.e. $\tilde{v}_{PV}(s) = 0$, the simplified circuits of Figure 8.13 result, which describe the dynamics of the currents in the StatRF once the current loops have been closed. The cancellation of the PV voltage disturbances from the point of view of the dynamics of the current loops is possible because the current loop is designed to be much faster than the voltage loop, being v_{PV} almost static 'seen' from the current loops. In this way the design of the current controllers is notably simplified.

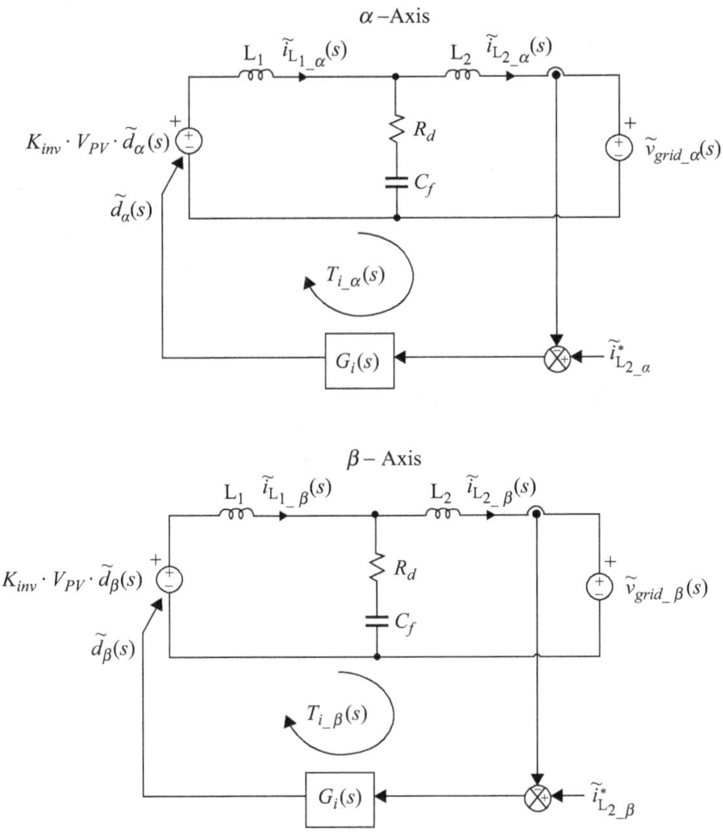

Figure 8.13 Equivalent circuits of the current loops in α-β coordinates (StatRF) after cancelling the PV voltage disturbance

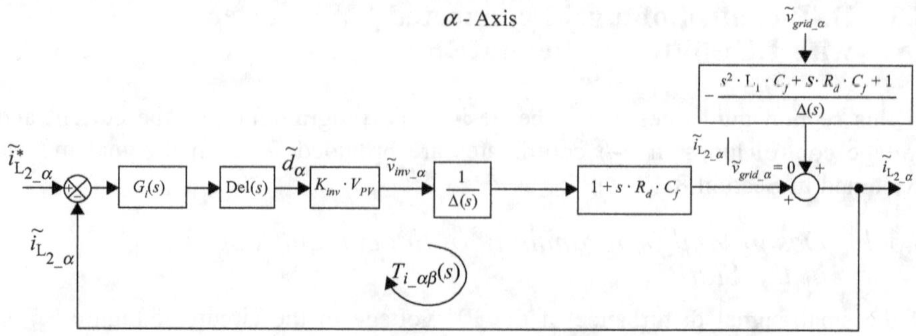

Figure 8.14 Equivalent block diagram of the current loop (α-channel). The current control is performed on the grid injected currents

In Figure 8.13 the current loop has been closed using the same current controller, $G_i(s)$, in both channels. Note that due to the total symmetry of both channels, the open loop gains of both current loops agree, i.e. $T_{i\_\alpha}(s) = T_{i\_\beta}(s) \equiv T_{i\_\alpha\beta}(s)$. Therefore, for design purposes it's enough to study one of the channels, e.g. the α-channel. In Figure 8.13 disturbances of the grid voltage ($\tilde{v}_{grid\_\alpha}$ and $\tilde{v}_{grid\_\beta}$) are considered.

Figure 8.14 depicts the equivalent block diagram of the current loop (α-channel) resulting from the study of Figure 8.13, where several transfer functions related to the LCL filter and the delay, Del(s), of the SVM implementation inside the digital controller appear. The characteristic polynomial of the LCL filter, $\Delta(s)$, is expressed by (8.33). Equation (8.34) shows the expression of the delay of the digital implementation and its second-order Padé implementation, where T_d is the delay of the digital controller. In most DSPs or microcontrollers this delay is usually either a switching period ($T_d = 1/f_{SW}$) or half a switching period ($T_d = 0.5/f_{SW}$) [24]. Equation (8.35) expresses the loop gain of the current loops, $T_{i\_\alpha\beta}(s)$. Note that $T_{i\_\alpha\beta}(s)$ is proportional to the PV voltage at the OP, V_{PV}. When using the 'voltage-feedforward in the current loop' explained in Section 8.2.3.1 by means of the multipliers of Figure 8.11, $T_{i\_\alpha\beta}(s)$ is made proportional to a constant value V_{PV_max}, i.e. the maximum attainable PV voltage, as it is expressed by (8.36)

$$\Delta(s) = (\mathrm{L}_1 + \mathrm{L}_2) \cdot s \cdot \left(s^2 \cdot L_P \cdot C_f + s \cdot R_d \cdot C_f + 1\right); \quad L_P = \frac{\mathrm{L}_1 \cdot \mathrm{L}_2}{\mathrm{L}_1 + \mathrm{L}_2} \tag{8.33}$$

$$Del(s) = e^{-T_d \cdot s} \approx \frac{1 - \frac{T_d}{2} \cdot s + \frac{T_d^2}{12} \cdot s^2}{1 + \frac{T_d}{2} \cdot s + \frac{T_d^2}{12} \cdot s^2} \tag{8.34}$$

$$T_{i\_\alpha\beta}(s) = G_i(s) \cdot Del(s) \cdot K_{inv} \cdot V_{PV} \cdot \frac{1 + s \cdot R_d \cdot C_f}{\Delta(s)} \tag{8.35}$$

$$T_{i\_\alpha\beta}(s)\Big|_{voltage\_FF} = G_i(s) \cdot Del(s) \cdot K_{inv} \cdot V_{PV\_max} \cdot \frac{1 + s \cdot R_d \cdot C_f}{\Delta(s)} \tag{8.36}$$

The usual choice for the current controller, $G_i(s)$, is a PR controller [Teodorescu 06], whose general transfer function is expressed by (8.37), being ω the grid angular frequency (in rad/s). The proportional gain, K_{Pi}, is adjusted for obtaining the desired crossover frequency of the current loop, ω_{Ci}, (in rad/s). If 'voltage-feedforward' is performed and a maximum crossover frequency of the current loop, ω_{Ci_max}, is desired, the expression of the resulting K_{Pi} value is given by (8.38), where L_{2_min} is the minimum expected value of L_2. A double limitation has to be taken into account when choosing ω_{Ci_max}, as it is expressed by (8.39). On the one hand, ω_{Ci_max} is limited by the VSI switching frequency. The main cause of this limitation is the phase lag of the digital delay transfer function, $Del(s)$. On the other hand, ω_{Ci_max} must be below the minimum possible value of the resonance frequency of the LCL filter, ω_P, expressed by (8.40), to avoid its phase lag which would produce stability problems.

The damped resonant controllers with gains K_{Rk} and bandwidths B_k (in rad/s) achieve a good tracking of the current reference. They are commonly placed at the grid frequency ($k = 1$), at the third harmonic ($k = 3$) and at the odd non-tripplen harmonics ($k = 5, 7, 11, 13, \ldots$) in order to minimize the THDi of the grid injected currents. It is worth pointing out that a small third harmonic of both sequences (positive and negative) appears in the grid injected currents when the grid voltage is unbalanced, so that placing resonant controllers at the third harmonic slightly improves the THDi. A possible criterion for the choice of B_k is to set the same settling time for all the resonant controllers, described by (8.41). The values of the gains K_{Rk} are chosen for giving a high gain at the center frequencies of the resonant controllers. Special care must be taken with the gains of the resonant controllers with high order center frequencies, which may approach ω_{Ci}. It is usually necessary to reduce those gains in order to get a stable current loop. Reference 25 gives a different design criterion for the resonant controllers, applied to single phase PV inverters.

$$G_i(s) = K_{Pi} + \sum_{k=1,3,5,7,11,13} \frac{K_{Rk} \cdot B_k \cdot s}{s^2 + B_k \cdot s + (k \cdot \omega)^2} \tag{8.37}$$

$$K_{Pi} \approx \frac{\left(L_1 + L_{2\_min}\right) \cdot \omega_{Ci\_max}}{K_{inv} \cdot V_{PV\_max}} \tag{8.38}$$

$$\omega_{Ci\_max} < MIN\left(\frac{2\pi \cdot f_{SW}}{10}, \omega_{P\_min}\right) \tag{8.39}$$

$$\omega_P = \frac{1}{\sqrt{L_P \cdot C_f}}; \quad \omega_{P\_min} = \frac{1}{\sqrt{L_{P\_max} \cdot C_f}} \tag{8.40}$$

$$t_{settl} = \frac{9.2}{B_k} \tag{8.41}$$

The expression of the current controller chosen for the VSI under study is given by (8.42). Note that all the resonant controllers have the same bandwidth (0.1 Hz), so that their settling times agree. The lower the bandwidth of the resonant controllers is, the more stable the current loop is. The gain of the high order resonant controllers ($K_{R11} = K_{R13} = 4$) has been chosen intentionally lower than that of the rest ($K_{R1} = K_{R3} = K_{R5} = K_{R7} = 8$) in order to get a stable current loop. It is worth pointing out that a high crossover frequency of the current loops has been chosen, near the resonance frequency of the LCL filter, so that a low THDi can be obtained. That implies a careful selection of the higher order resonant controllers. The Bode plots of the chosen current controller for both axis are shown by Figure 8.15a, whereas the Bode plots of the resulting loop gain, $T_{i\_\alpha\beta}(j\omega)$, are depicted in

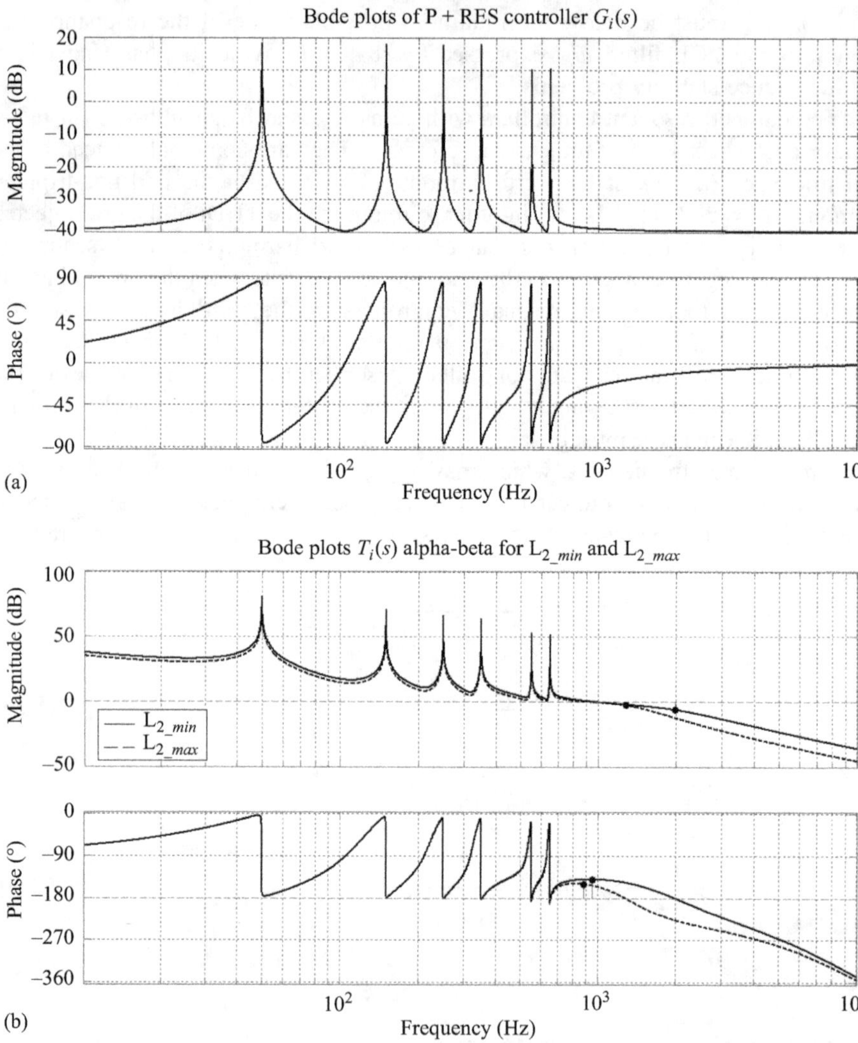

Figure 8.15 Bode plots of (a) $G_i(j\omega)$, (b) $T_{i\_\alpha\beta}(j\omega)$ for L_{2_min} *(solid line) and* L_{2_max} *(dashed line)*

Table 8.2 Stability margins and crossover frequencies of $T_{i\_\alpha\beta}(j\omega)$ for L_{2_min} and L_{2_max}

	$L_{2\_min} = 267\ \mu H$	$L_{2\_max} = 800\ \mu H$
Parallel resonance frequency of LCL filter, $f_P = \omega_P/2\pi$	1.966 kHz	1.326 kHz
0 dB crossover frequency, $f_{Ci} = \omega_{Ci}/2\pi$	952 Hz	882 Hz
Phase margin, PM	≈40°	≈30°
−180° crossover frequency	1.99 kHz	1.28 kHz
Gain margin, GM	6.3 dB	2.82 dB

Figure 8.15b for the minimum and maximum values of L_2, that can be found in Table 8.1. The digital delay considered for the calculation of the loop gain is $T_d = 1/f_{samp} = 0.5/f_{SW}$. The stability margins and crossover frequencies for L_{1_min} and L_{2_max} can be read from Table 8.2, along with the corresponding values of $f_P = \omega_P/2\pi$. It is worth pointing out that $G_i(s)$ has been designed for having: $f_{Ci\_max} = \omega_{Ci\_max}/2\pi \approx 1$ kHz. From Table 8.2 it can be observed that both f_{Ci} and f_P decrease with the value of L_2, being necessary for stability that $f_{Ci} < f_P$, so that the phase lag of the LCL filter doesn't degrade the phase margin (PM) too much. The most critical situation, i.e. minimum stability margins, occurs when $L_2 = L_{2\_max}$, because in that case f_P is more close to f_{Ci} than for lower L_2 values.

$$G_i(s) = 0.01 + \sum_{k=1,3,5,7} \frac{8 \cdot (0,1 \cdot 2 \cdot \pi) \cdot s}{s^2 + (0,1 \cdot 2 \cdot \pi) \cdot s + (k \cdot \omega)^2} + \sum_{k=11,13} \frac{4 \cdot (0,1 \cdot 2 \cdot \pi) \cdot s}{s^2 + (0,1 \cdot 2 \cdot \pi) \cdot s + (k \cdot \omega)^2} \tag{8.42}$$

Figure 8.16 depicts the Bode plots of the transfer function from the current reference to the grid injected current, $i_{L_{2\_\alpha\beta}}/i^*_{L\alpha\beta}(j\omega)$, and of the grid voltage to the grid injected current. The expressions of those transfer functions can be derived from Figure 8.14 and can be expressed by (8.43) and (8.44), respectively. Note that $|T_{i\_\alpha\beta}| \gg 1$ at the central the frequencies of the resonant controllers, so that at those frequencies the grid injected current perfectly follows its reference ($|i_{L_2}/i^*_L| = 1$) and the susceptibility of the grid injected currents to the grid voltage disturbances is very small ($|i_{L_{2\_\alpha\beta}}/v_{grid\_\alpha\beta}| \ll 1$).

$$\left.\frac{\tilde{i}_{L_{2\_\alpha\beta}}(s)}{\tilde{i}^*_{L_{2\_\alpha\beta}}(s)}\right|_{\tilde{v}_{grid\_\alpha\beta}=\tilde{v}_{PV}=0} = \frac{1}{1 + T_{i\_\alpha\beta}(s)} \tag{8.43}$$

$$\left.\frac{\tilde{i}_{L_{2\_\alpha\beta}}(s)}{\tilde{v}_{grid\_\alpha\beta}(s)}\right|_{\tilde{i}^*_{L_{2\_\alpha\beta}}(s)=\tilde{v}_{PV}=0} = \frac{-\dfrac{s^2 + s \cdot L_1 \cdot C_f + 1}{\Delta(s)}}{1 + T_{i\_\alpha\beta}(s)} \tag{8.44}$$

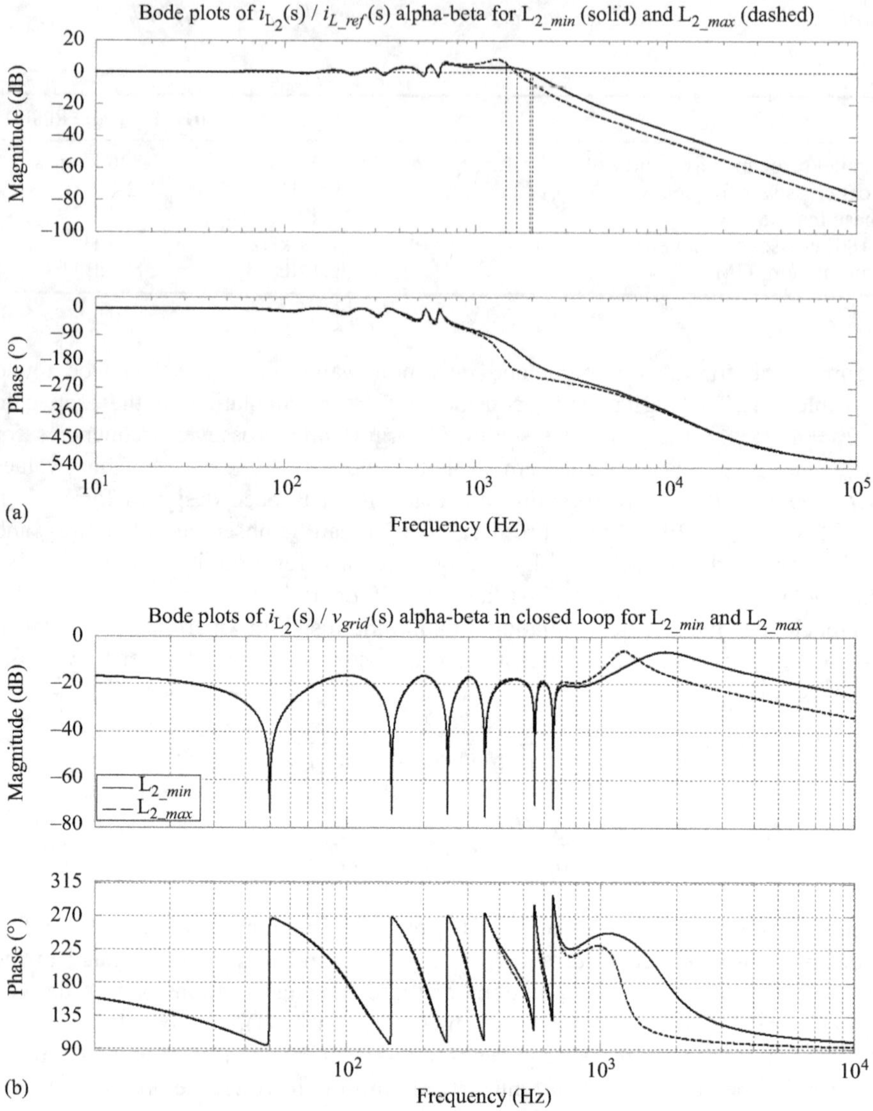

Figure 8.16 Bode plots of (a) $i_{L_2}/i_L^*(j\omega)$ and (b) $i_{L_2}/v_{grid}(j\omega)$. L_{2_min} (solid line) and L_{2_max} (dashed line)

Figure 8.17 depicts the control structure of the PR current controller in α-β coordinates described by (8.37) and (8.42), where the resonant part is implemented by means of unity gain SOGI resonant filters multiplied by the corresponding resonant gains K_{Rk}. Each SOGI is tuned at an angular frequency (ω, $3 \cdot \omega$, $5 \cdot \omega$, ...) provided by the PLL implemented on the DSP. Note that v_{PV} feedforward and a limiter

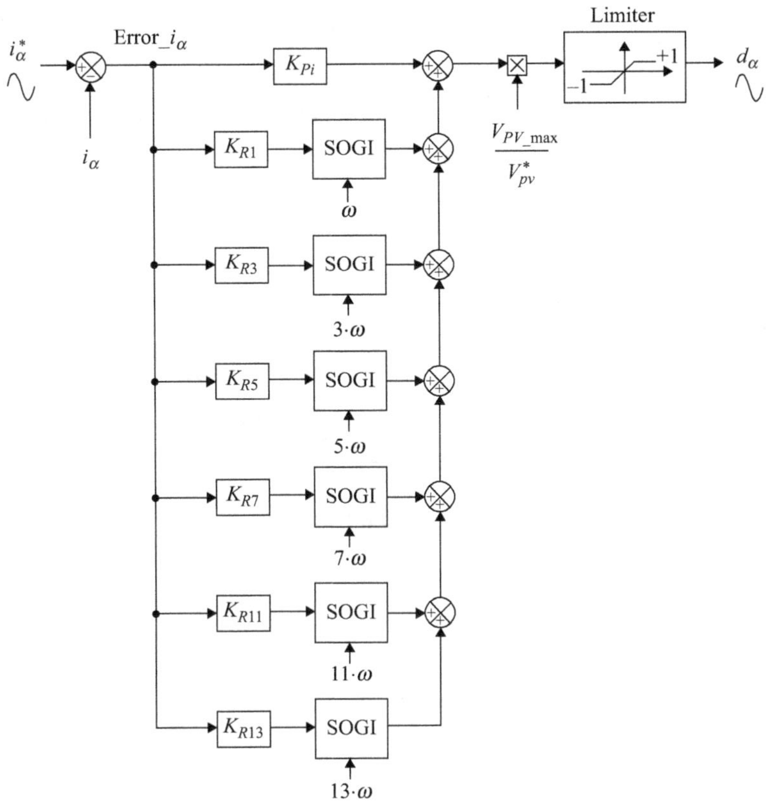

Figure 8.17 Control structure of a SOGI-based PR current controller in α-β coordinates

at the SOGI output have been included. Figure 8.18 shows the C++ code for the implementation of the SOGI-based PR current controller. Figure 8.19a shows a possible programming code of the function 'Run_SOGI_RES' to run a SOGI resonant filter of unity gain, where 'v_in' is the current error, 'w_in' is the tuning angular frequency and 'v_out' is the SOGI output. The digital implementation of the programmed SOGI-based unity gain resonant filter is depicted in Figure 8.19b. It can be observed in Figure 8.19a that the coefficients a_1, a_2, b_0 and b_2 of the digital implementation depend on B, T_s (sampling period) and ω_{in}, being ω_{in} the angular frequency at which the resonant filter is tuned. The digital implementation is the result of the discretization by means of the Tustin method of the continuous-time transfer function of the SOGI-based unity gain resonant filter, shown by (8.45).

$$G_{SOGI}(s) = \frac{V_{out}(s)}{V_{in}(s)} = \frac{B \cdot s}{s^2 + B \cdot s + \omega_{in}^2} \tag{8.45}$$

```
#define KPI 0.01 //Definition of Kp of current controller
#define KR 8 //Definition of the Kr of the current controller for the fundamental and harmonics 3, 5 and 7
#define KR_11_13 4 // Definition of the Kr of the current controller for harmonics 11 and 13

void Control_Iac( float Iref_pk)
{
        float IL_alfa = CLARKE_P_al(Inv1_Ia.value,Inv1_Ib.value,-(Inv1_Ia.value+Inv1_Ib.value)); //Alpha coordinate of the //Clarke transform of the sensed currents
        float IL_beta = CLARKE_P_be(Inv1_Ia.value,Inv1_Ib.value,-(Inv1_Ia.value+Inv1_Ib.value)); //Beta coordinate of the //Clarke transform  of the sensed currents.
        //Inv1_Ia and Inv1_Ib are C++ objects of the class Canal_ADC. They get the values of the sensed currents ia and ib
        float Error_Ialfa = Iref_pk * (cos (PLL.theta))-IL_alfa; //Alpha coordinate of the current error
        float Error_Ibeta = Iref_pk * (sin (PLL.theta))-IL_beta; //Beta coordinate of the current error
        // PLL.theta is the value of the grid angle produced by the SRF-PLL. PLL is a C++ object of the class SRF_PLL_3F
        float d_alfa=0, d_beta=0; //Duty cycles in alpha and beta coordinates initialized to zero.

        //P+RES CURRENT CONTROLLER IN THE ALPHA-AXIS
        d_alfa =KPI*Error_Ialfa;
        d_alfa+= KR*Reg_I_alfa_1.Run_SOGI_RES(Error_Ialfa,PLL.w);
        d_alfa+= KR*Reg_I_alfa_3.Run_SOGI_RES(Error_Ialfa,3*PLL.w);
        d_alfa+= KR*Reg_I_alfa_5.Run_SOGI_RES(Error_Ialfa,5*PLL.w);
        d_alfa+= KR*Reg_I_alfa_7.Run_SOGI_RES(Error_Ialfa,7*PLL.w);
        d_alfa+= KR_11_13*Reg_I_alfa_11.Run_SOGI_RES(Error_Ialfa,11*PLL.w);
        d_alfa+= KR_11_13*Reg_I_alfa_13.Run_SOGI_RES(Error_Ialfa,13*PLL.w);
        //d_alfa is the alpha axis value of the duty cycle calculated by the P+RES current controller in alpha-beta coordinates
        //Reg_I_alfa_x are C++ objects of the class SOGI_FLL which use the function Run_SOGI_RES to run a SOGI resonant //controller of unity gain.
        //Each unity gain SOGI is multiplied by the corresponding gain Kr. PLL.w is the w value calculated by the C++ object //called PLL of the class SRF_PLL_3F

        d_alfa = d_alfa * (VDC_MAX/Inv1_Vdc.value); //Vpv voltage feedforward in the alpha axis of //the current loop
        if (d_alfa> 1) d_alfa= 1;
        if (d_alfa<-1) d_alfa=-1; //Saturation of the alpha-axis duty cycle

        //P+RES CURRENT CONTROLLER IN THE BETA-AXIS
        d_beta =KPI*Error_Ibeta;
        d_beta+= KR*Reg_I_beta_1.Run_SOGI_RES(Error_Ibeta,PLL.w);
        d_beta+= KR*Reg_I_beta_3.Run_SOGI_RES(Error_Ibeta,3*PLL.w);
        d_beta+= KR*Reg_I_beta_5.Run_SOGI_RES(Error_Ibeta,5*PLL.w);
        d_beta+= KR*Reg_I_beta_7.Run_SOGI_RES(Error_Ibeta,7*PLL.w);
        d_beta+= KR_11_13*Reg_I_beta_11.Run_SOGI_RES(Error_Ibeta,11*PLL.w);
        d_beta+= KR_11_13*Reg_I_beta_13.Run_SOGI_RES(Error_Ibeta,13*PLL.w);

        d_beta = d_beta * (VDC_MAX/Inv1_Vdc.value); //Vpv voltage feedforward in the beta axis of //the current loop

        if (d_beta> 1) d_beta= 1;
        if (d_beta<-1) d_beta=-1;
        //The values d_alfa and d-beta are the inputs of the code implementing the Space Vector //Modulation
```

Figure 8.18 C++ code for the implementation of the SOGI-based PR current controller in α-β coordinates

8.3.2 Design and programming of the voltage loop in the StatRF

In [2] it can be found a nice derivation of the small–signal model of the voltage control loop in *d-q* coordinates (SRF). Following a similar procedure, it can be demonstrated that the small-signal transfer function from the peak value of the current injected to the grid in α or β coordinates to the PV voltage can be expressed by (8.46), so that the dynamic model of the voltage loop can be depicted by Figure 8.20. Note that the current loop operating in closed loop has been modeled as a first-order low-pass transfer function with a pole at ω_{Ci}. This transfer function is an approximation of (8.43) in the low frequency range ($\omega < \omega_{Ci}$). Taking into account that the crossover frequency of the voltage loop, ω_{Cv}, is designed well below ω_{Ci}, the approximation (8.47) is valid for the design of the voltage controller, $G_V(s)$, which is usually a PI controller. The main difficulty for the adjustment of $G_V(s)$ is that it has to compensate a first-order low-pass transfer function (8.46) with

```
//CODE OF THE UNITY GAIN RESONANT SOGI FILTER
//Define constants
float pi=3.141592654;
float Ts=0.00005; //sampling period
float B=2*pi*0.1; //Bandwidth of the resonant SOGI (rad/s)
//Define and initialize global variables.
//In the C++ code these variables are defined in each
//C++ object of class 'SOGI_FLL'
float h_1=0, h_2=0;
float SOGI_FLL::Run_SOGI_RES (float v_in, float w_in)
{
        //Define and initialize local variables
        float den=2*B*Ts+4+w_in*w_in*Ts*Ts;
        float bo=2*B*Ts/den;
        float b2=-bo;
        float a1=(2*w_in*w_in*Ts*Ts-8)/den;
        float a2=(-2*B*Ts+4+w_in*w_in*Ts*Ts)/den;
        float h=v_in-a1*h_1-a2*h_2;
        float v_out=bo*h+b2*h_2;
        h_2=h_1; //update register for the next iteration
        h_1=h;//update register for the next iteration
        return v_out;
}
```

(a)

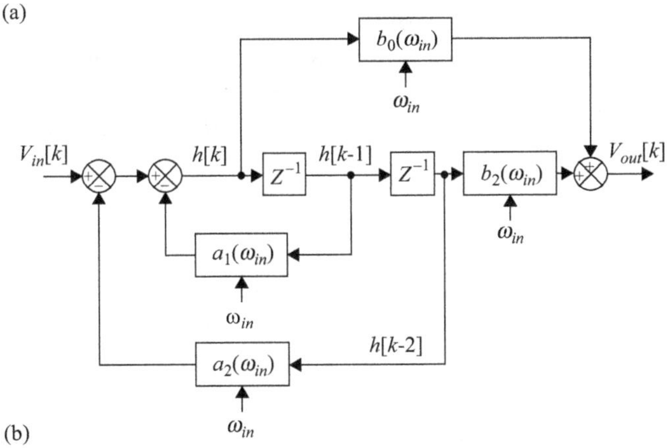

(b)

Figure 8.19 (a) A programming code for the implementation of a unity gain SOGI-based resonant filter, (b) Structure of the digital implementation

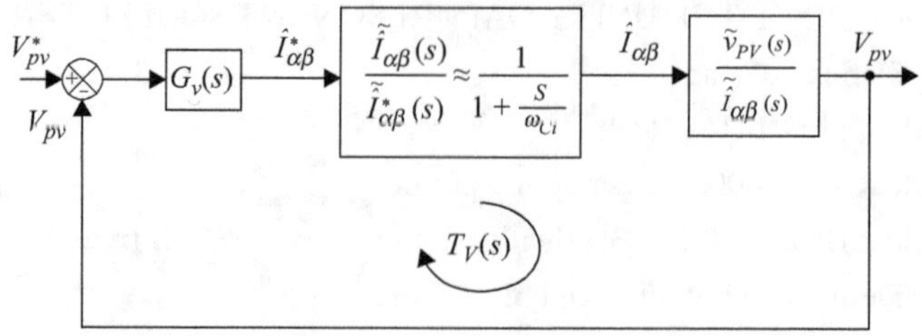

Figure 8.20 Dynamic model of the voltage loop

an open loop unstable pole, ω_0. As a consequence, for the stabilization of that pole the value of ω_{Cv} must be much higher than its maximum possible value, $\omega_{0\_max} = I_{PV\_max}/(V_{PV\_min} \cdot C_{PV})$, as it's expressed by (8.48). Note that (8.48) will hold true easily if the value of C_{PV} is high enough, i.e. if the DC-link works as an effective energy buffer between the PV array and the inverter.

$$\frac{\tilde{v}_{PV}(s)}{\tilde{\hat{i}}_{\alpha}(s)} = \frac{\tilde{v}_{PV}(s)}{\tilde{\hat{i}}_{\beta}(s)} = \frac{\tilde{v}_{PV}(s)}{\tilde{\hat{i}}_{\alpha\beta}(s)} = \frac{\frac{\sqrt{3}\cdot V_{grid\_OP}}{I_{PV}}}{1-\frac{s}{\omega_0}}; \ \omega_0 = \frac{I_{PV}}{V_{PV}\cdot C_{PV}} \tag{8.46}$$

$$\frac{\tilde{\hat{i}}_{\alpha\beta}(s)}{\tilde{\hat{i}}^*_{\alpha\beta}(s)} \approx \frac{1}{1+\frac{s}{\omega_{Ci}}}; \quad If\, \omega_{Ci} \gg \omega_{Cv} \rightarrow \frac{1}{1+\frac{s}{\omega_{Ci}}} \approx 1 \tag{8.47}$$

$$\omega_{Cv} \gg \omega_{0\_max} = \frac{I_{PV\_max}}{V_{PV\_min}\cdot C_{PV}} \tag{8.48}$$

Assuming the condition (8.48), expression (8.46) is simplified into (8.49) in the frequency range near the desired value of ω_{Cv}, so that the design of a PI voltage controller with a negative sign (8.50) is greatly simplified. The final expression of the loop gain of the voltage loop is expressed by (8.51). For the VSI under study, the expression of the chosen voltage controller is provided by (8.52), whereas the resulting Bode plots of $T_V(s)$ at different OPs defined by the power, $P_{PV} = V_{PV} \cdot I_{PV}$, and V_{PV} are depicted in Figure 8.21, for the minimum and maximum values of L_2. The crossover frequency of the voltage loop ranges between 50Hz and 100Hz, with a phase margin close to 90º. No important influence of the value of L_2 in the voltage loop is observed. Figure 8.22 shows the Bode plots of the closed loop transfer function of the voltage loop, expressed by (8.53). It is observed that the voltage loop, once closed, has a dominant first-order pole around 50 Hz and 100 Hz.

$$\left.\frac{\tilde{v}_{PV}(s)}{\tilde{\hat{i}}_{\alpha\beta}(s)}\right|_{near\ \omega_{Cv}} \approx \frac{\frac{\sqrt{3}\cdot V_{grid_{OP}}}{I_{PV}}}{-\frac{s}{\omega_0}} = -\frac{\frac{\sqrt{3}\cdot V_{grid_{OP}}}{V_{PV}\cdot C_{PV}}}{s} \tag{8.49}$$

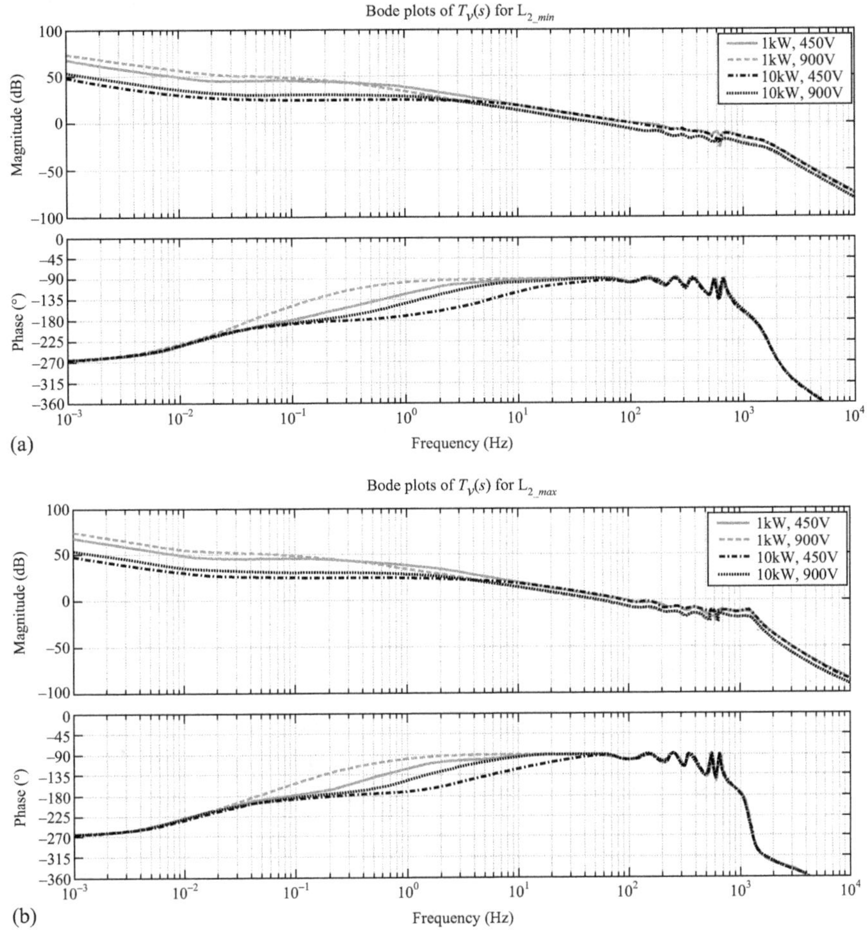

Figure 8.21 Bode plots of $T_V(j\omega)$ as a function of P_{PV} and V_{PV}. (a) L_{2_min}, (b) L_{2_max}.

$$G_V(s) = -\left(K_{Pv} + \frac{K_{iv}}{s}\right) \tag{8.50}$$

$$T_V(s) = G_V(s) \cdot \frac{\tilde{\hat{I}}_{\alpha\beta}(s)}{\tilde{\hat{I}}^*_{\alpha\beta}(s)} \cdot \frac{\tilde{v}_{PV}(s)}{\tilde{\hat{I}}_{\alpha\beta}(s)} \tag{8.51}$$

$$G_V(s) = -\left(1.67 + \frac{0.15}{s}\right) \tag{8.52}$$

$$\frac{\tilde{v}_{PV}(s)}{\tilde{v}^*_{PV}(s)} = \frac{T_V(s)}{1 + T_V(s)} \tag{8.53}$$

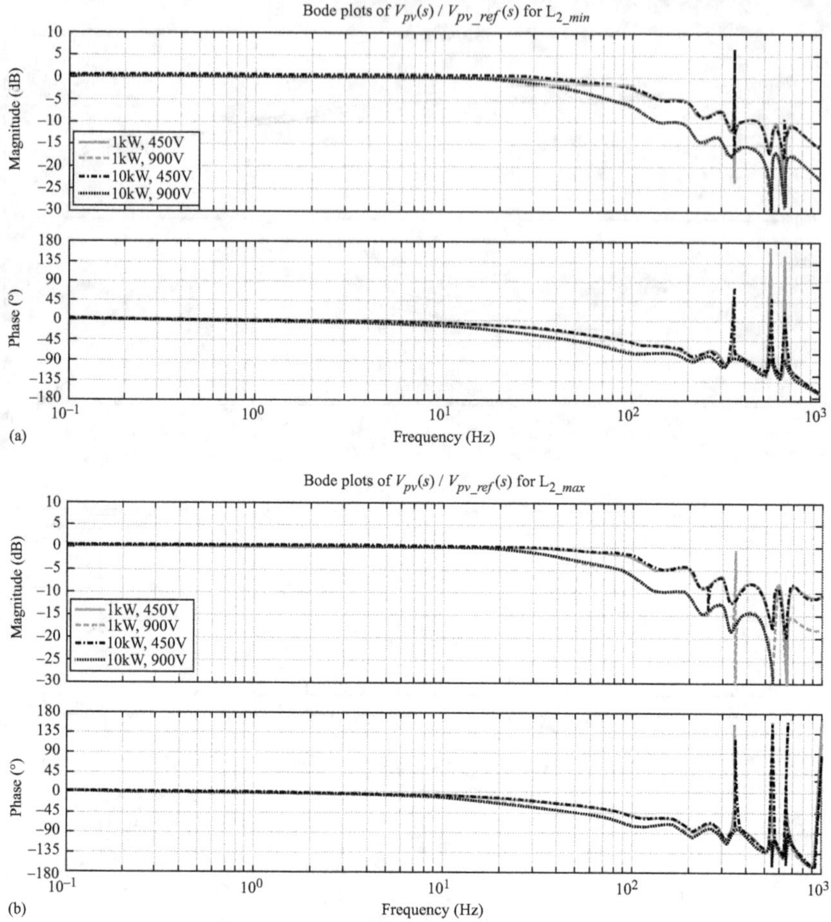

Figure 8.22 *Bode plots of $v_{PV}/v^*_{PV}(j\omega)$ as a function of P_{PV} and V_{PV}. (a)* L_{2_min}, *(b)* L_{2_max}

8.4 DSP control of a grid-connected PV inverter with LCL filter in the SRF

In this section guidelines about the design and programming of the current and voltage control loops in *d-q* coordinates are provided. The small-signal models explained in Section 8.2.2 will be used.

8.4.1 Design and programming of the current loops in the SRF

From the small-signal circuit of the current loops in the SRF shown in Figure 8.10, two important differences are observed when compared with its homologue in the StatRF (Figure 8.7):

- The effect of the PV voltage disturbance in the d and q channels, $K_{inv} \cdot D_d \cdot \tilde{v}_{PV}$ and $K_{inv} \cdot D_q \cdot \tilde{v}_{PV}$, may be very different in both channels because D_d and D_q

usually take very different values at the OP, as it can be observed from (8.29). This feature causes low frequency differences between the loop gain of both current loops. Those differences were explained in [9], but they dissapear near the crossover frequency of the current loops.

- The currents of a channel produce coupling voltage sources in the other one. The voltages at the capacitors are also coupled from one channel to the other one, producing coupling current sources. The main effect is produced by the coupling of the currents (coupling voltage sources) but it can be solved by means of a proper decoupling, as it will be explained further.

If the small-signal disturbance of the PV voltage in the circuit of Figure 8.10 is cancelled, i.e. $\tilde{v}_{PV} = 0$, the simplified circuits of Figure 8.23 result, which describe the dynamics of the currents in the SRF once the current loops have been closed.

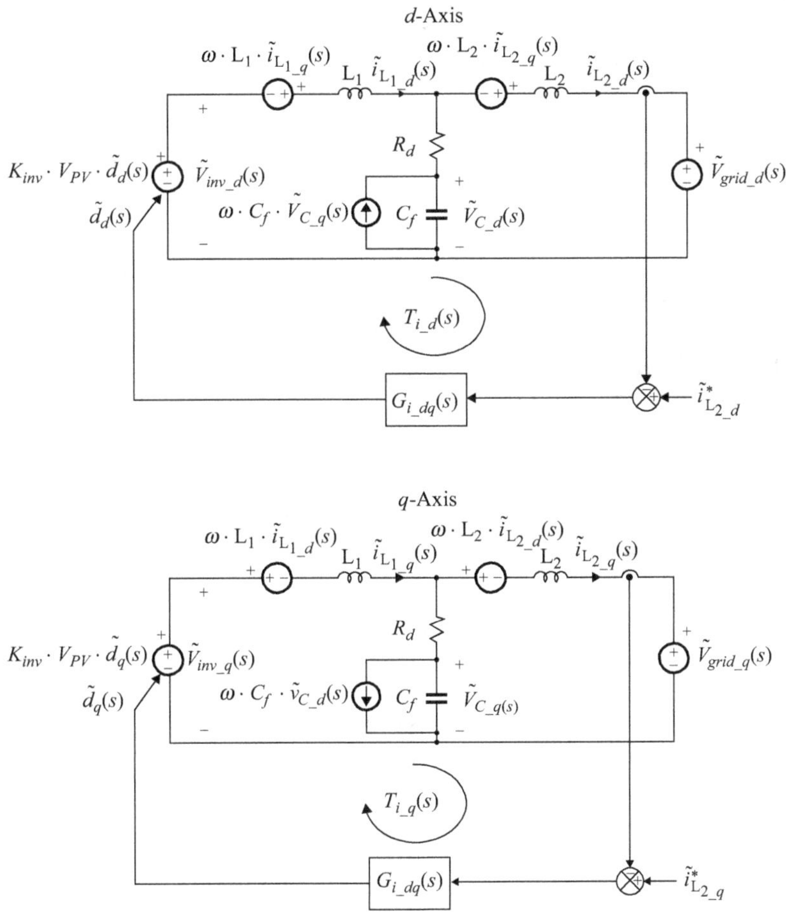

Figure 8.23 Equivalent circuits of the current loops in d-q coordinates (SRF) after cancelling the PV voltage disturbance ($\tilde{v}_{PV}(s) = 0$)

Note that in Figure 8.23 the current loops have been closed using the same current controller, $G_{i\_dq}(s)$, in both channels. Although the *d*- and *q*-axis circuits are not completely symmetric due to the coupling elements, the transfer functions 'seen' by the current controllers of each axis, defined by (8.54) and (8.55), are practically identical as it can be observed in the Bode plots of both transfer functions, depicted in Figure 8.24.

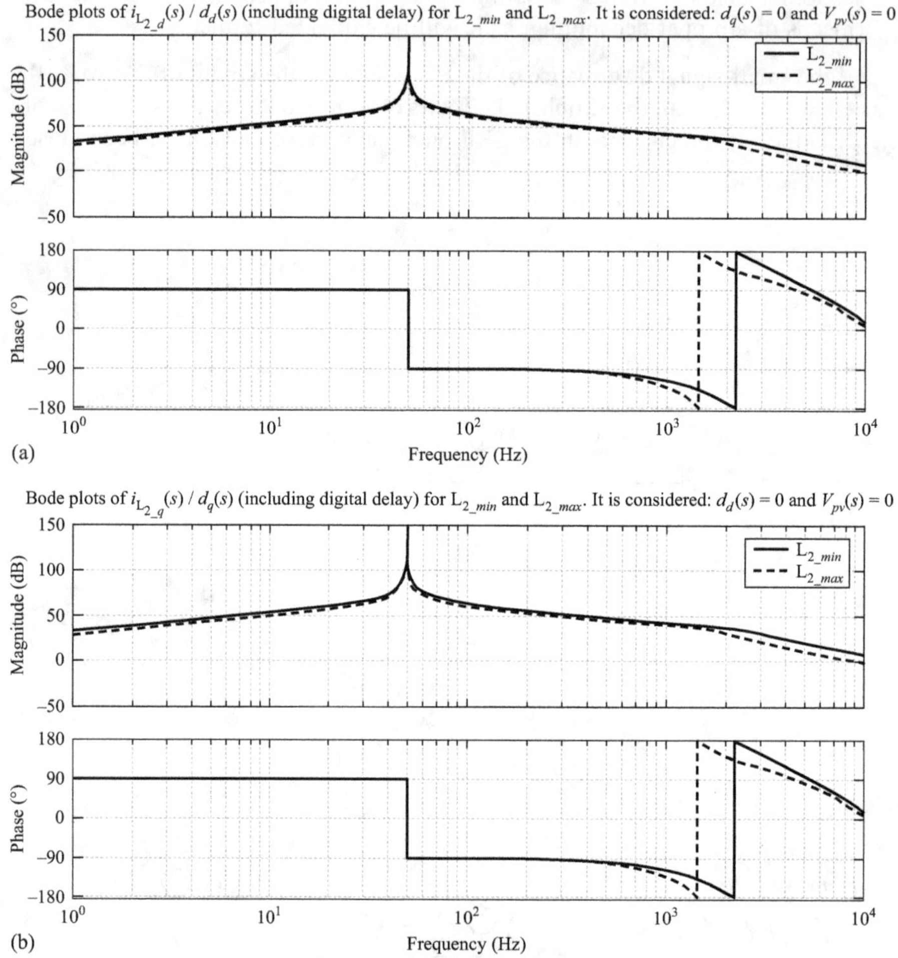

Figure 8.24 Bode plots of the transfer functions 'seen' by the current controllers in both axis of the SRF, cancelling the PV voltage disturbance and the grid voltage disturbance ($\tilde{v}_{PV}(s) = \tilde{v}_{grid}(s) = 0$), as a function of L_2 *for* V_{PV_max}. *(a)* $\left.\frac{\tilde{i}_{L_{2\_d}}(s)}{\tilde{d}_d(s)}\right|_{\substack{\tilde{d}_q=0 \\ \tilde{v}_{grid}=\tilde{v}_{PV}=0}}$, *(b)* $\left.\frac{\tilde{i}_{L_{2\_q}}(s)}{\tilde{d}_q(s)}\right|_{\substack{\tilde{d}_d=0 \\ \tilde{v}_{grid}=\tilde{v}_{PV}=0}}$

$$\left.\frac{\tilde{i}_{L_{2\_d}}(s)}{\tilde{d}_d(s)}\right|_{\substack{\tilde{d}_q=0\\ \tilde{v}_{grid}=\tilde{v}_{PV}=0}} \tag{8.54}$$

$$\left.\frac{\tilde{i}_{L_{2\_q}}(s)}{\tilde{d}_q(s)}\right|_{\substack{\tilde{d}_d=0\\ \tilde{v}_{grid}=\tilde{v}_{PV}=0}} \tag{8.55}$$

Therefore, the open loop gains of both current loops agree, i.e. $T_{i\_d}(s) \approx T_{i\_q}(s) \equiv T_{i\_dq}(s)$. Therefore, for design purposes it's enough to study one of the channels, e.g. the d-channel. In Figure 8.24 the second-order Padé approximation of the digital delay of the SVM implementation has been considered. Note that an important resonance at the grid frequency (50 Hz) appears in Figure 8.24, which is produced by the couplings in the SRF.

The choice of the current controller in the SRF can be derived from the current controller (8.37) and (8.42) designed in the StatRF by considering the equivalence pointed out by [15] and [26]. The general expression of the chosen $G_{i\text{-}dq}(s)$ is given by (8.56), whereas its expression as a function of the values of the StatFR controller is shown by (8.57). The value of the chosen current controller is provided by (8.58). Note that the integral term in (8.56)–(8.58) gives an infinite gain to the DC component of the input of the controller (current error signal) in d-q coordinates, which is equivalent to an infinite gain to the grid frequency (ω) positive sequence component of the current error in α-β coordinates. The resonant controller centered at $2 \cdot \omega$ in the SRF gives a high gain to both the negative sequence ω-frequency and to the positive sequence $3 \cdot \omega$-frequency components of the current error in the StatRF. The resonant controller centered at $4 \cdot \omega$ in the SRF gives a high gain to both the negative sequence $3 \cdot \omega$-frequency and to the positive sequence $5 \cdot \omega$-frequency components of the current error in the StatRF. Similar conclusions can be found for the resonant controllers in the SRF centered at $6 \cdot \omega$ and $12 \cdot \omega$: one resonant term in the SRF gives a high gain to one negative sequence component and to a positive sequence component in the StatRF. The SRF controller is a PI+Resonant one if it is derived from a PR StatRF controller.

$$G_{i\_dq}(s) = K_{Pi\_dq} + \frac{K_{i\_dq}}{s} + \sum_{k=2,4,6,12} \frac{K_{Rk\_dq} \cdot B_{k\_dq} \cdot s}{s^2 + B_k \cdot s + (k \cdot \omega)^2} \tag{8.56}$$

$$\begin{aligned} G_{i\_dq}(s) = K_{Pi} + \frac{K_{R1} \cdot \frac{B_1}{2}}{s} + \sum_{k=2,4} \frac{K_{R3} \cdot B_3 \cdot s}{s^2 + B_3 \cdot s + (k \cdot \omega)^2} \\ + \sum_{k=6,12} \frac{2 \cdot K_{R(k\pm 1)} \cdot B_{k\pm 1} \cdot s}{s^2 + B_{k\pm 1} \cdot s + (k \cdot \omega)^2} \end{aligned} \tag{8.57}$$

$$G_{i\_dq}(s) = 0.01 + \frac{8 \cdot \frac{2\pi \cdot 0.1}{2}}{s} + \sum_{k=2,4} \frac{8 \cdot 2\pi \cdot 0.1 \cdot s}{s^2 + 2\pi \cdot 0.1 \cdot s + (k \cdot \omega)^2}$$

$$+ \frac{16 \cdot 2\pi \cdot 0.1 \cdot s}{s^2 + 2\pi \cdot 0.1 \cdot s + (6 \cdot \omega)^2} + \frac{8 \cdot 2\pi \cdot 0.1 \cdot s}{s^2 + 2\pi \cdot 0.1 \cdot s + (12 \cdot \omega)^2} \quad (8.58)$$

Figure 8.25 shows the Bode plots of the designed current controller in the SRF and of the loop gain of the current loop (d-axis), $T_{i\_d}(s)$, with no decoupling terms with

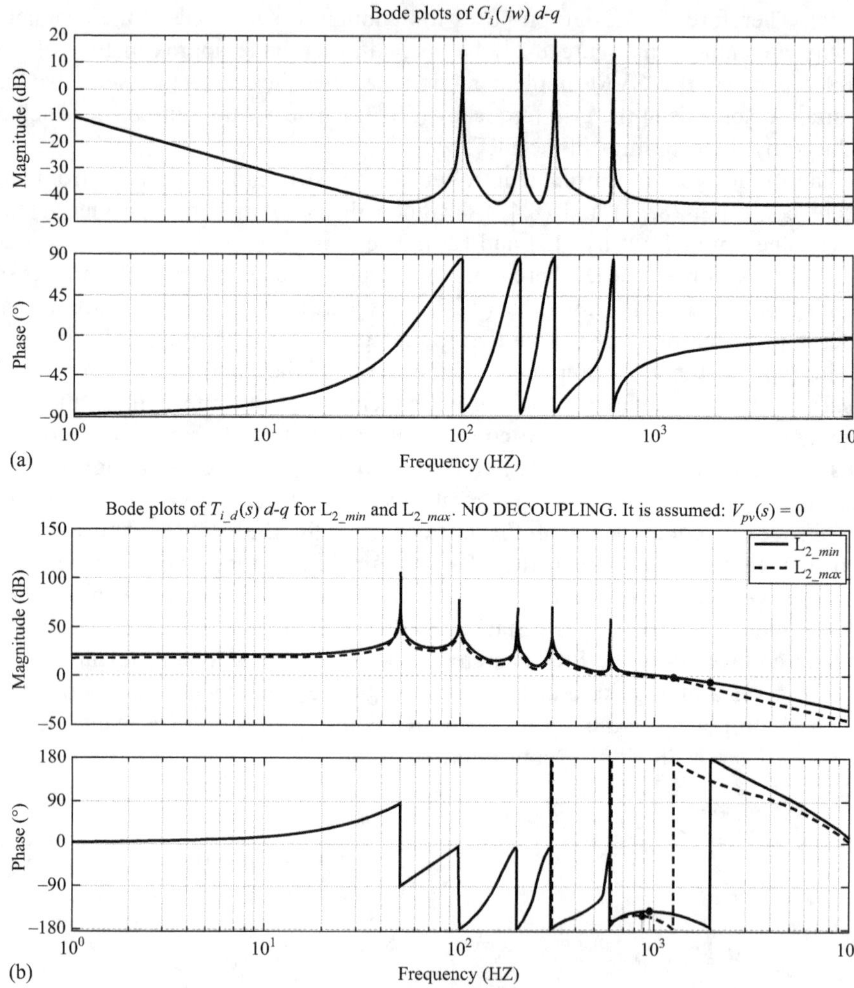

Figure 8.25 Bode plots of (a) $G_{i\_dq}(j\omega)$, (b) $T_{i\_d}(j\omega)$ for L_{2_min} (solid line) and L_{2_max} (dashed line) and for V_{PV_max}. Both the PV voltage disturbance and the grid voltage disturbance are cancelled ($\tilde{v}_{PV}(s) = \tilde{v}_{grid}(s) = 0$). No decoupling in the current loops has been considered

Table 8.3 Stability margins and crossover frequencies of $T_{i\_dq}(j\omega)$ for L_{2_min} and L_{2_max}. No decoupling

	$\mathbf{L_{2\_min} = 267\ \mu H}$	$\mathbf{L_{2\_max} = 800\ \mu H}$
0 dB crossover frequency, $f_{Ci} = \omega_{Ci}/2\pi$	954 Hz	885 Hz
Phase margin, PM	≈ 40°	≈ 30°
−180° crossover frequency	1.96 kHz	1.26 kHz
Gain margin, GM	6.26 dB	2.76 dB

gain $\frac{\omega\cdot(L_1+L_2)}{K_{inv}\cdot v^*_{PV}}$ in the current loops (see Figure 8.12). The grid frequency resonance in the transfer function (8.54) can be observed in the Bode plots $T_{i\_d}(s)$, along with the four resonant terms of $G_{i\_dq}(s)$. Note that without decoupling the low frequency gain of $T_{i\_d}(s)$ is not very high in spite of the integrator in $G_{i\_dq}(s)$, which is translated into some tracking error of the positive sequence grid frequency component of current reference in the StatRF (DC component in the SRF). The stability margins and crossover frequencies of $T_{i\_d}(s)$ are summarized in Table 8.3. All values are very similar to those shown in Table 8.2, corresponding to the loop gain of the current loop in the StatRF.

Figure 8.26 shows the equivalent dynamic circuits of the current loops in the SRF after applying a perfect decoupling by means of the gain blocks of gain $\frac{\omega\cdot(L_1+L_2)}{K_{inv}\cdot v_{PV}}$ in the current loops (see Figure 8.12). If the current loops are considered at frequencies much lower than the corner frequency of the LCL filter, where the capacitor branches can be removed from the circuits of Figure 8.26, it holds: $\tilde{i}_{L_{1\_d}} = \tilde{i}_{L_{2\_d}} \equiv \tilde{i}_d$ and $\tilde{i}_{L_{1\_q}} = \tilde{i}_{L_{2\_q}} \equiv \tilde{i}_q$. In the d-axis the decoupling terms $\frac{\omega\cdot(L_1+L_2)}{K_{inv}\cdot v_{PV}}\cdot\tilde{i}_{L_{2\_q}} = \frac{\omega\cdot(L_1+L_2)}{K_{inv}\cdot v_{PV}}\cdot\tilde{i}_q$ produce that a part of $\tilde{v}_{inv\_d}(s)$ eliminates the voltage drops produced by the q-axis currents ($\omega\cdot L_1\cdot\tilde{i}_q$ and $\omega\cdot L_2\cdot\tilde{i}_q$). A similar conclusion can be drawn for the q-axis decoupling. The proposed decoupling is highly effective in the low frequency range, where $f \ll f_P$.

Figure 8.27 shows the Bode plot of $T_{i\_d}(s)$ and of the closed loop gain of the current loop in the d-axis (8.59), both with the decoupling depicted by Figure 8.26. When comparing Figures 8.25b and 8.27a, it is observed that the low frequency gain of $T_{i\_d}(s)$ with decoupling is much higher than without it. The decoupling in the current loop hardly affects the crossover frequencies and stability margins shown in Table 8.3, because its effect occurs at low frequency, far away from the crossover frequency.

In order to illustrate the effect of the current in one channel on the current on the other channel, Figure 8.28 shows the Bode plots of the cross-coupling transfer function (8.60) after closing the current loop with and without decoupling. The symbolic expression of the cross-coupling transfer function (8.60) can be derived from the analysis of Figure 8.26, but due to the coupling between the d and q channels it results a cumbersome task. In order to obtain the Bode plots of Figure 8.28, a Matlab® LTI array has been programmed, having used the state-space representation of the VSI

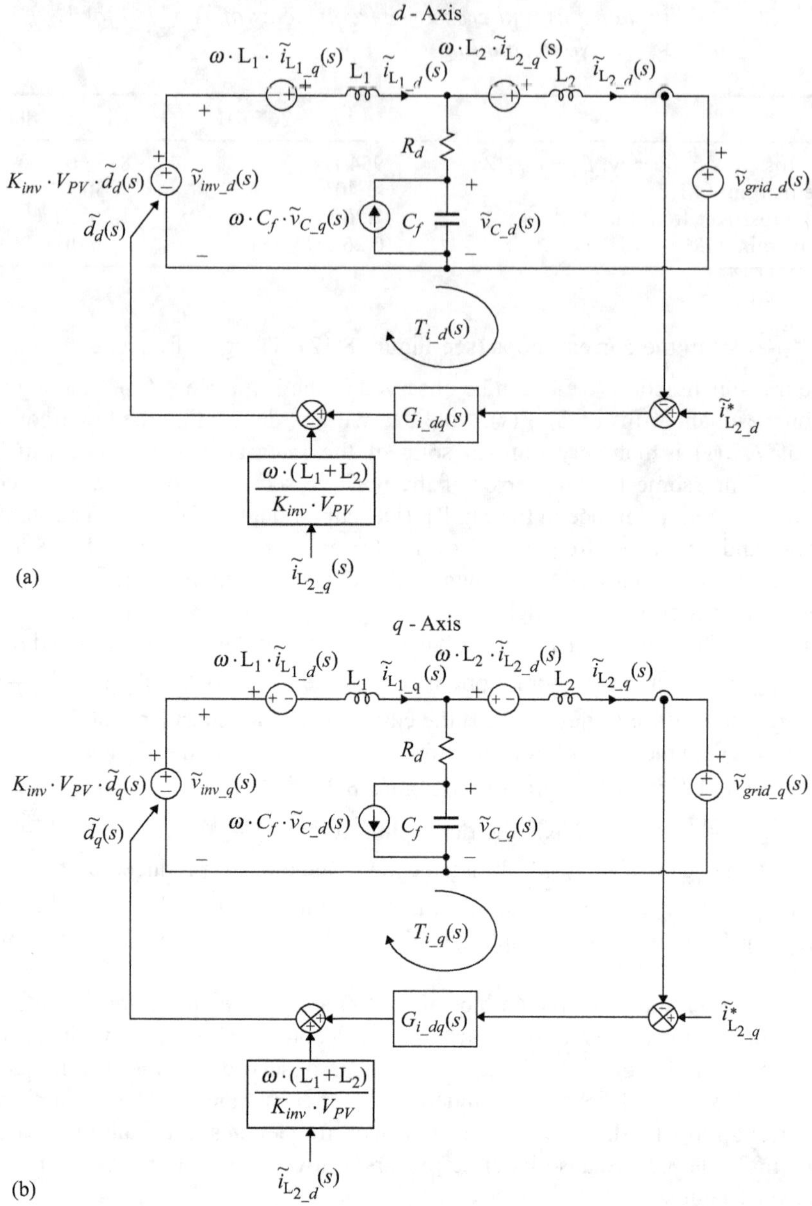

Figure 8.26 (a) and (b) Equivalent circuits of the current loops in d-q coordinates (SRF) with decoupling. The PV voltage disturbance is cancelled ($\tilde{v}_{PV}(s) = 0$)

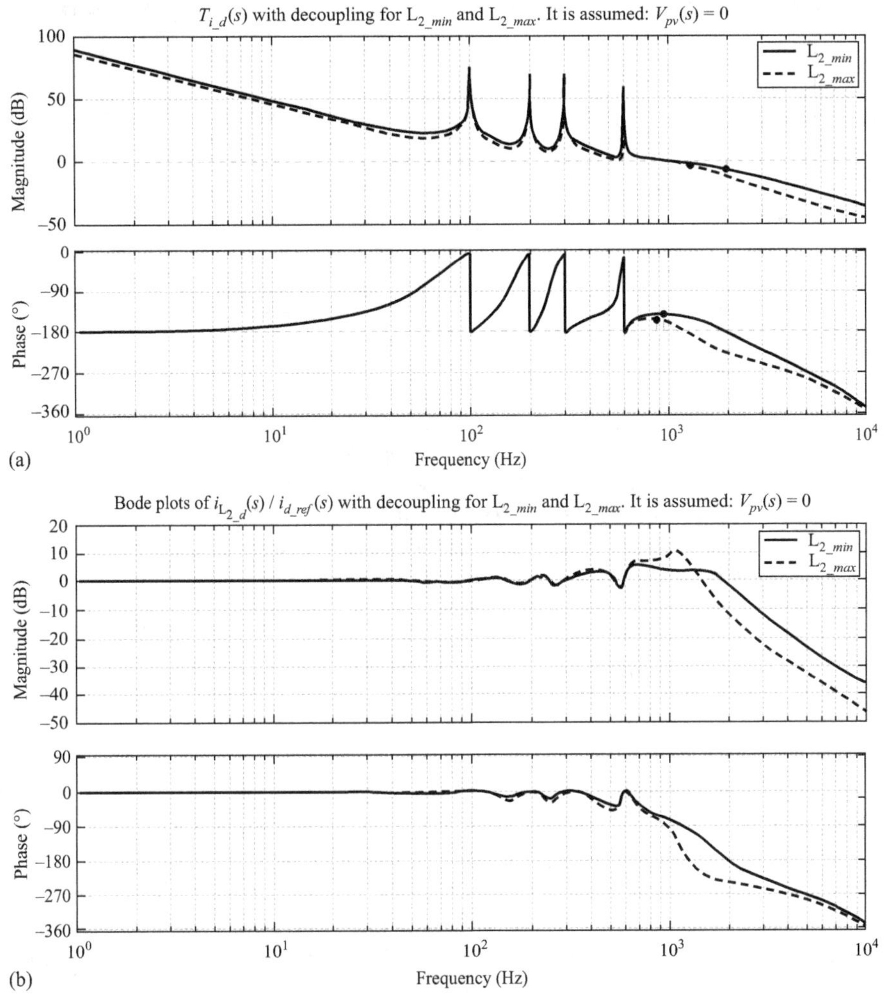

Figure 8.27 Bode plots of the d-axis current loop transfer functions with decoupling. $V_{PV} = V_{PV\_max}$. Both the PV voltage disturbance and the grid voltage disturbance are cancelled ($\tilde{v}_{PV}(s) = \tilde{v}_{grid}(s) = 0$). (a) $T_{i\_d}(j\omega)$ for L_{2_min} (solid line) and L_{2_max} (dashed line). (b) $\left.\frac{\tilde{i}_{L_{2\_d}}(s)}{\tilde{i}^*_{L_{2\_d}}(s)}\right|_{\tilde{v}_{grid}=\tilde{v}_{PV}=0}$ for L_{2_min} (solid line) and L_{2_max} (dashed line)

shown in [9]. A very important low frequency improvement of 30 dB or higher achieved by decoupling is observed.

$$\left.\frac{\tilde{i}_{L_{2\_d(q)}}(s)}{\tilde{i}^*_{L_{2\_d(q)}}(s)}\right|_{\tilde{v}_{grid}=\tilde{v}_{PV}=0} = \frac{T_{i\_d(q)}(s)}{1 + T_{i\_d(q)}(s)} \tag{8.59}$$

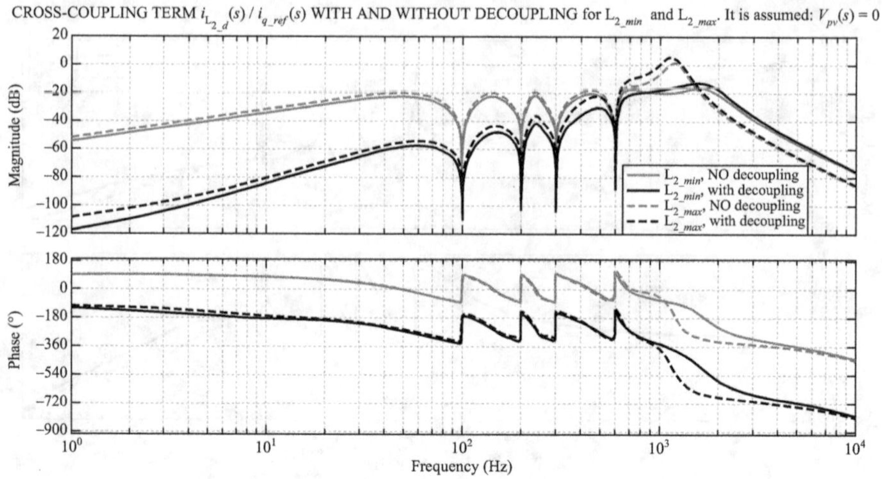

Figure 8.28 Bode plots of the d-axis current loop cross-coupling transfer functions in closed loop with and without decoupling. $V_{PV} = V_{PV\_max}$. PV voltage and grid voltage disturbances are cancelled $(\tilde{v}_{PV}(s) = \tilde{v}_{grid}(s) = 0$. $\left.\frac{\tilde{i}_{L_{2\_d}}(s)}{\tilde{i}^*_{L_{2\_q}}(s)}\right|_{\tilde{v}_{grid}=\tilde{v}_{PV}=0}$ for L_{2_min} (solid lines) and L_{2_max} (dashed lines)

$$\left.\frac{\tilde{i}_{L_{2\_d(q)}}(s)}{\tilde{i}^*_{L_{2\_q(d)}}(s)}\right|_{\tilde{v}_{grid}=\tilde{v}_{PV}=0} \tag{8.60}$$

8.4.2 Design and programming of the voltage loop in the SRF

It can be demonstrated that the same voltage controller which was designed in the StatRF in Section 8.3.2 is suitable for the SRF. Reference 2 provides design hints for the design of the voltage controller in the SRF.

8.5 Experimental results

This section provides experimental results of the control scheme in the StatRF shown in Figure 8.11, corresponding to the grid-connected 10-kW VSI whose parameters were described in Section 8.2.3 (see Table 8.1). The adjustment of the controllers was described in Section 8.3.

The inverter is connected to a highly distorted grid defined by (8.61), in which voltage unbalance factors VUF_b and VUF_c have been introduced. The value of the fundamental of the phase voltage seen from the grid side (Y of the grid transformer) is 230V_{RMS}. The nominal grid frequency is 50 Hz. Table 8.4 summarizes the values of the harmonics and of the unbalance factors of the grid voltage. The resulting distortion of the grid voltage is $THDv = 13.4\%$. The distorted grid voltage has been obtained from a 12 kVA AC power source (Pacific Power 360-AMX).

Table 8.4 Experimental grid voltage. Fundamental, harmonics and unbalance factors seen from the grid side

Parameter	**Value**
V_1	$\sqrt{2} \cdot 230\,\mathrm{V}$
V_5	$0.1 \cdot \mathrm{V_1}$
V_7	$0.07 \cdot \mathrm{V_1}$
V_{11}	$0.05 \cdot \mathrm{V_1}$
V_{13}	$0.04 \cdot \mathrm{V_1}$
VUF_b	−0.1
VUF_c	0.3

$$\vec{v}_g = \begin{bmatrix} v_a \\ v_b \\ v_c \end{bmatrix}$$

$$= \begin{bmatrix} V_1 \cdot \cos(\omega t) - V_5 \cdot \cos(5\omega t) + V_7 \cdot \cos(7\omega t) \\ - V_{11} \cdot \cos(11\omega t) + V_{13} \cdot \cos(13\omega t) \\ (1 + VUF_b) \cdot \left[V_1 \cdot \cos\left(\omega t - \frac{2\pi}{3}\right) - V_5 \cdot \cos\left(5 \cdot \left(\omega t - \frac{2\pi}{3}\right)\right)\right. \\ \left. + V_7 \cdot \cos\left(7 \cdot \left(\omega t - \frac{2\pi}{3}\right)\right) + \cdots \right] \\ (1 + VUF_c) \cdot \left[V_1 \cdot \cos\left(\omega t + \frac{2\pi}{3}\right) - V_5 \cdot \cos\left(5 \cdot \left(\omega t + \frac{2\pi}{3}\right)\right)\right. \\ \left. + V_7 \cdot \cos\left(7 \cdot \left(\omega t + \frac{2\pi}{3}\right)\right) + \cdots \right] \end{bmatrix} \tag{8.61}$$

Figure 8.29 shows the waveforms at the inverter side (Δ of the grid connection transformer) with the distorted grid at $P \approx 10$ kW and $V_{PV} = 760$ V, but without unbalance ($VUF_b = VUF_c = 0$). The distortion of the grid currents is $THDi \approx 0.8\%$. Figure 8.30 shows the waveforms at the inverter side in the same conditions as Figure 8.29, but at $P \approx 2$ kW. The distortion of the grid currents is $THDi \approx 2.8\%$. Figure 8.31 shows the waveforms at the inverter at $P \approx 10$ kW and $V_{PV} = 760$ V with the distortion and unbalance shown in Table 8.4 ($VUF_b = -0.1$, $VUF_c = 0.3$). The distortion of the grid currents is THDi ≈ 1.4%.

Figure 8.32 shows the waveforms at $P \approx 7$ kW with a P&O MPPT algorithm working with: $f_{MPPT} = 10$ Hz, $\Delta V^*_{PV} = 8$ V. It is observed that the OP of V_{PV}

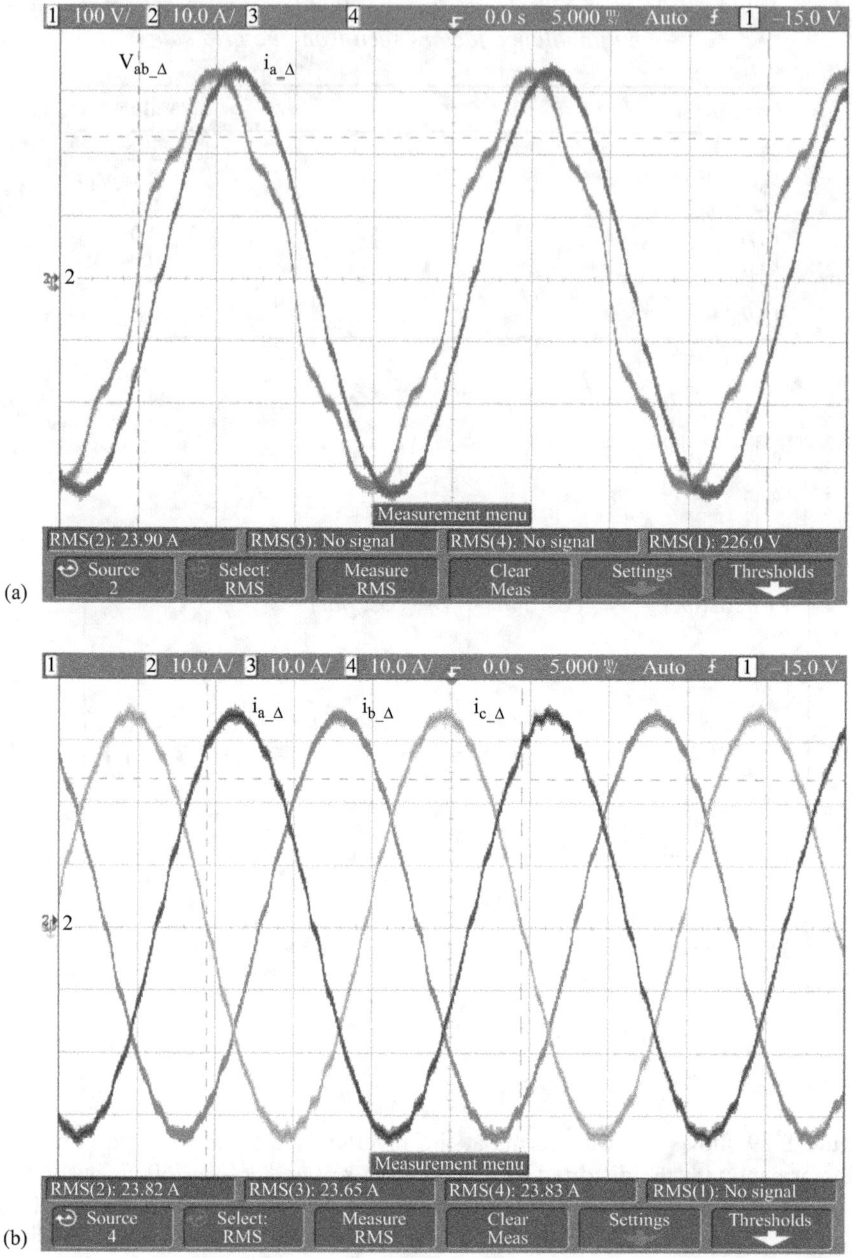

Figure 8.29 $P \approx 10$ kW at $V_{PV} = 760$ V. Balanced grid voltage with a distortion of THDv = 13.4%. (a) V_{ab_Δ}, i_{a_Δ}. (b) i_{a_Δ}, i_{b_Δ}, i_{c_Δ}. THDi = 0.8%

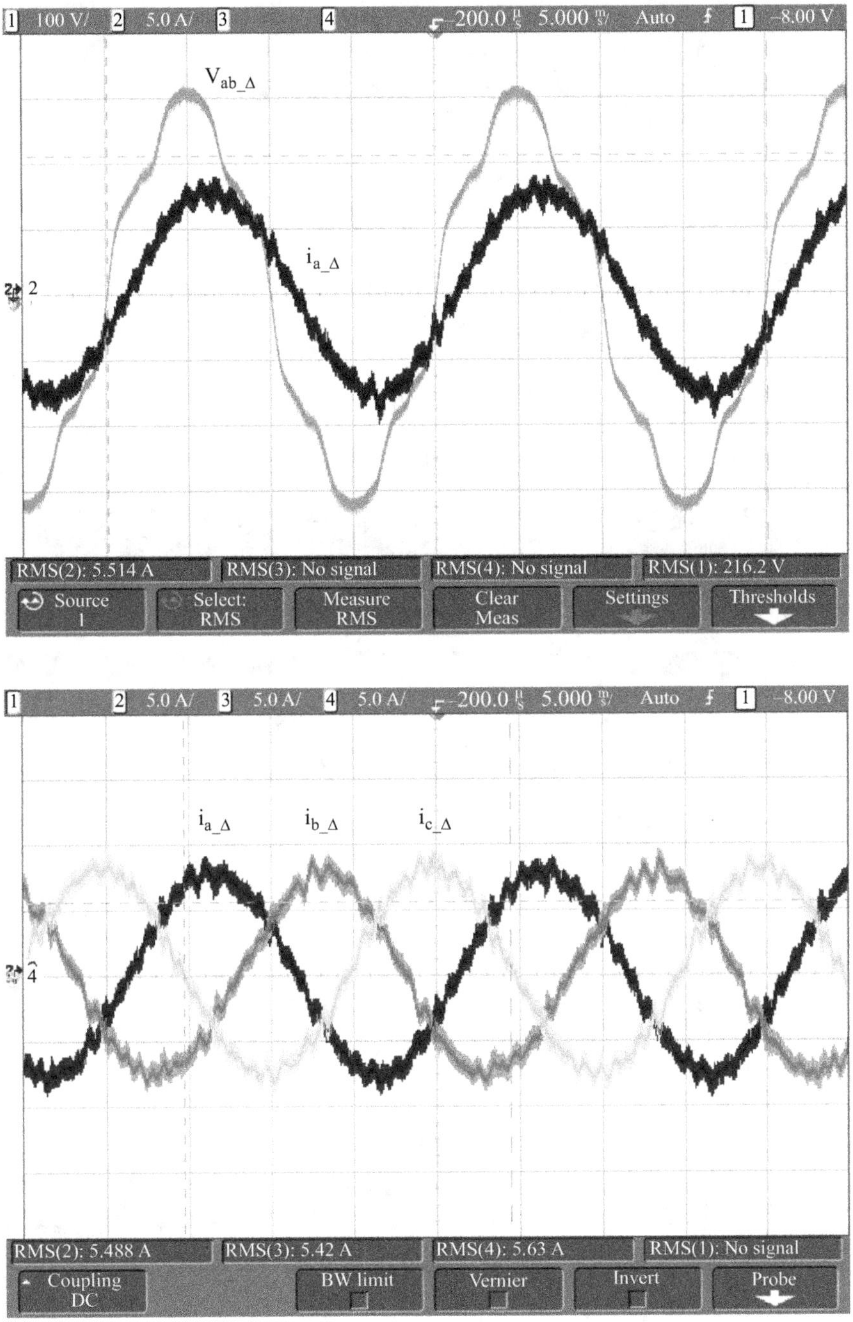

Figure 8.30 $P \approx 2$ kW at $V_{PV} = 760$ V. Balanced grid voltage with a distortion of THDv = 13.4%. (a) V_{ab_Δ}, i_{a_Δ}. (b) i_{a_Δ}, i_{b_Δ}, i_{c_Δ}. THDi = 2.8%

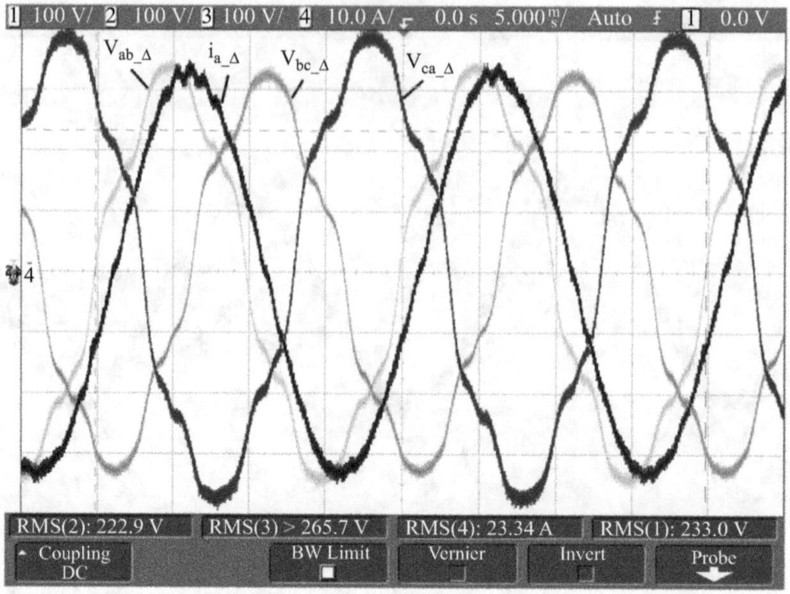

(a)

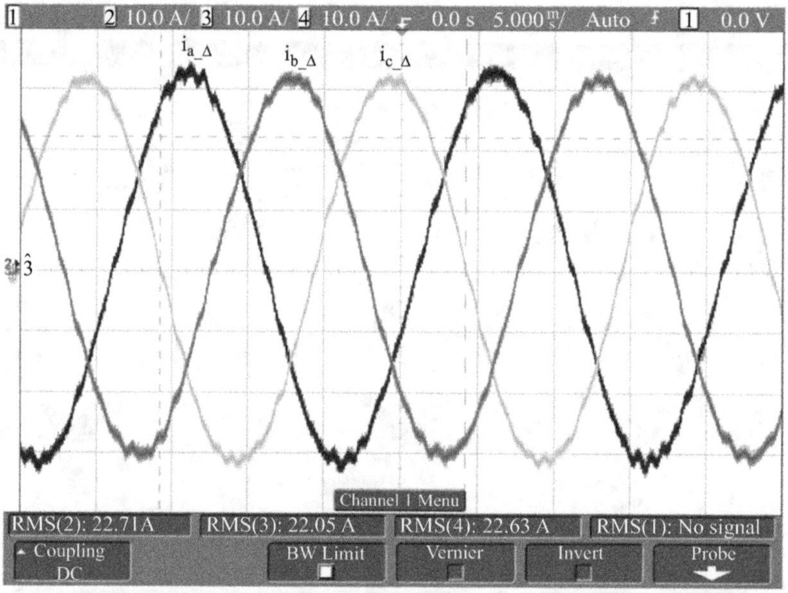

(b)

Figure 8.31 $P \approx 10$ kW at $V_{PV} = 760$ V. Unbalanced grid voltage with a distortion of THDv = 13.4%. (a) V_{ab_Δ}, V_{bc_Δ}, V_{ca_Δ}, i_{a_Δ}. (b) i_{a_Δ}, i_{b_Δ}, i_{c_Δ}. THDi = 1.4%

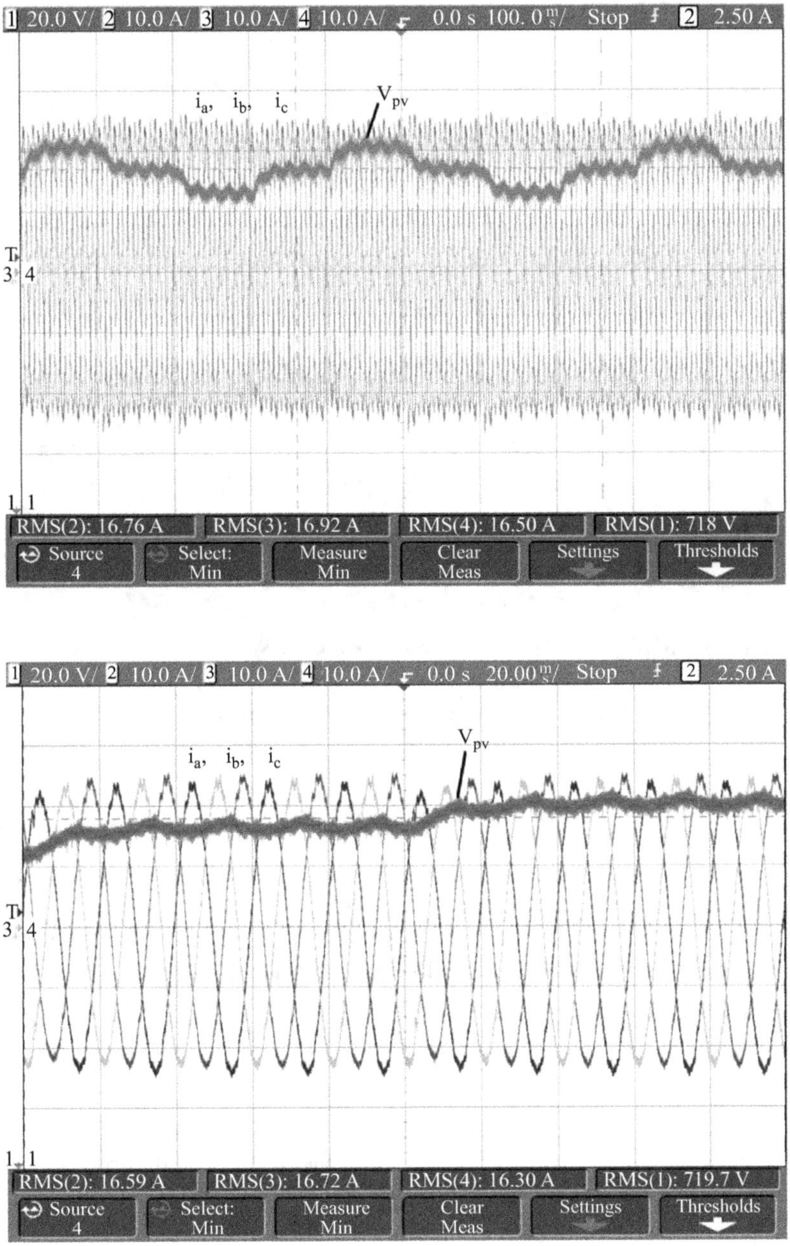

Figure 8.32 Waveforms at $P \approx 7$ kW with a P&O MPPT algorithm working with: $f_{MPPT} = 10$ Hz, $\Delta V^{*}_{PV} = 8$ V. Balanced grid voltage with a distortion of THDv = 13.4%. (a) V_{PV} and i_{a_Δ}, i_{b_Δ}, i_{c_Δ}. (b) Detail

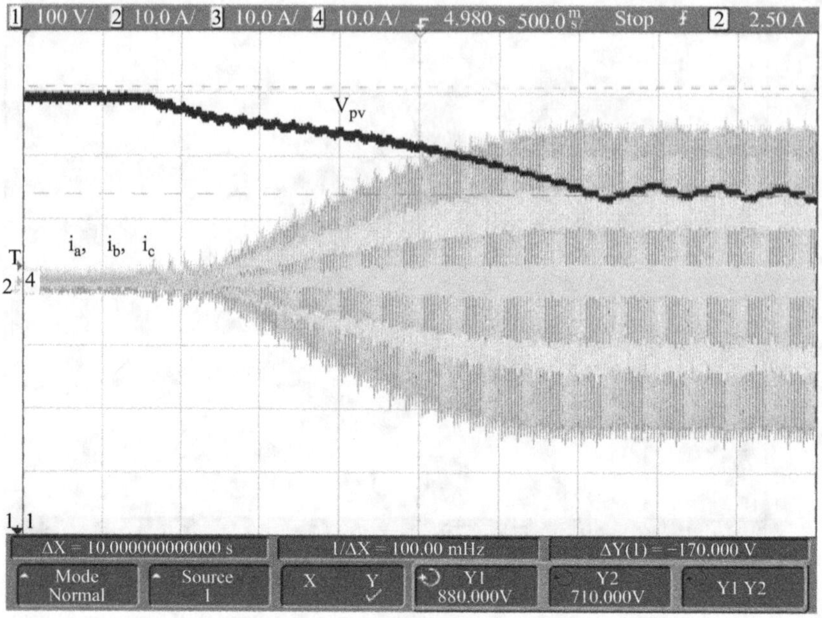

Figure 8.33 Waveforms during the start-up of the VSI until $P \approx 7$ kW is reached with a P&O MPPT algorithm working with: $f_{MPPT} = 10Hz$, $\Delta V^*_{PV} = 8$ V. Balanced grid voltage with a distortion of THDv = 13.4%. V_{PV} and i_{a_Δ}, i_{b_Δ}, i_{c_Δ}

undergoes the typical steps of the P&O algorithm (8 V steps in this case) at a refreshing frequency of 10 Hz (the PV voltage changes each 100 ms). In this case the VSI is working from a 10-kW TerraSAS PV array emulator from AMREL. The PV array being emulated has the following characteristics:

- Open circuit voltage. $V_{OC} = 902$ V
- Short circuit current. $I_{SC} = 9.8$ A
- Current at the maximum power point: $I_{MPP} = 9.2$ A
- Voltage at the maximum power point: $V_{MPP} = 760$ V

Figure 8.33 shows the PV panel voltage and the grid injected currents (seen from the inverter side) during the start-up of the inverter until $P \approx 7$ kW is reached with the P&O algorithm under operation. A very smooth transient of both the currents and the voltage is observed.

Figure 8.34 shows the grid injected currents seen from the inverter side at $P \approx 7$ kW and $V_{PV} = 760$ V with a grid frequency step from 45 Hz to 55 Hz in $\Delta t = 40$ms. It is observed that the adaptation mechanism of the SOGIs in the current loop works with a high stability.

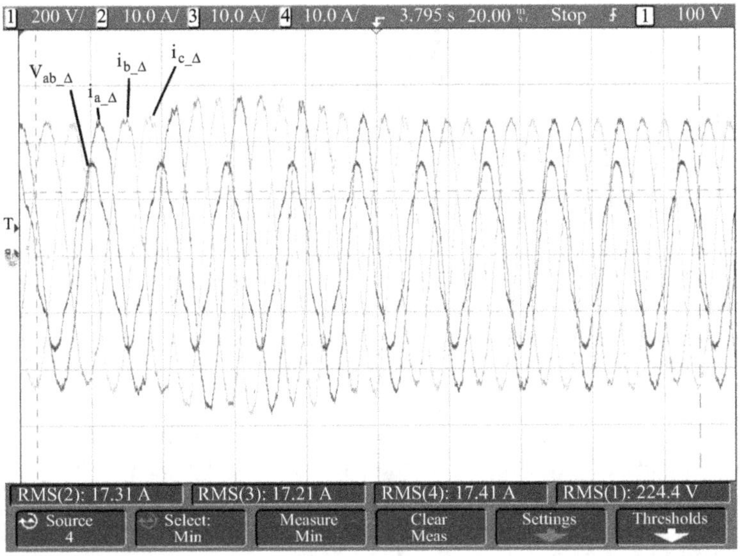

Figure 8.34 *Waveforms at $P \approx 7$ kW with a grid frequency step from 45 Hz to 55 Hz in $\Delta t = 40$ ms at $V_{PV} = 760$ V. Balanced grid voltage with a distortion of THDv = 13.4%. V_{ab_Δ} and i_{a_Δ}, i_{b_Δ}, i_{c_Δ}*

8.6 Conclusions

This chapter has dealt with the design and implementation on a floating point DSP of the control loops of a three-phase 10-kW grid-connected photovoltaic inverter both in a stationary reference frame (StatRF) and in an SRF. SOGI-based resonant controllers for the current loops have been used for obtaining a low THDi of the grid injected currents in spite of a high distortion and unbalance of the grid voltage. The design of the current controllers is challenging taking into account the high crossover frequency which is needed for the current loops, slightly lower than the resonance frequency of the LCL filter. A thorough robustness study of both the current and the voltage loops has been shown, having taken into account the variations of the grid inductance, the injected power and the PV array voltage. Experimental results have been shown, including the behavior of the MPPT algorithm and sudden variations of the grid frequency.

Acknowledgment

This work is supported by the Spanish Ministry of Economy and Competitiveness (MINECO) and the European Regional Development Fund (ERDF) under Grants ENE2012-37667-C02-01 and ENE2015-64087-C02-02.

The AC power source TerraSAS from AMREL used to carry out the experimental results was financed by the Generalitat Valenciana under 'Línea T0104000 del programa presupuestario 542.50 innovación, transferencia tecnológica y apoyo a infraestructuras', which received funds from the ERDF.

References

[1] J.M. Carrasco, L.G. Franquelo, J.T. Bialasiewicz, *et al.*, 'Power-electronic systems for the grid integration of renewable energy sources: a survey', *IEEE Transactions on Industrial Electronics*, 2006, 53 (4), pp. 1002–1016.

[2] R. Teodorescu, M. Liserre, P. Rodriguez, *'Grid converters for photovoltaic and wind power systems'*, Wiley-IEEE, 2011.

[3] M. Gaëtan, O. Sinead, R. Manoël, *'Global market for photovoltaics 2014–2018'*, European Photovoltaic Industry Association, 2014.

[4] A. Timbus, M. Liserre, R. Teodorescu, P. Rodriguez, F. Blaabjerg, 'Evaluation of current controllers for distributed power generation systems', *IEEE Transactions on Power Electronics*, 2009, 24 (3), pp. 654–664.

[5] F. Blaabjerg, R. Teodorescu, M. Liserre, *et al.*, 'Overview of control and grid synchronization for distributed power generation systems', *IEEE Transactions on Industrial Electronics*, 2006, 53 (5), pp. 1398–1409.

[6] S. Bidyadhar, P. Raseswari, 'A comparative study on maximum power point tracking techniques for photovoltaic power systems', *IEEE Transactions on Sustainable Energy*, 2013, 4 (1).

[7] V. Kaura V, V. Blasko, 'Operation of a phase locked loop system under distorted utility conditions', *IEEE Transactions on Industry Applications*, 1997, 33 (1), pp. 58–63.

[8] M. Karimi-Ghartemani, M.R. Iravani, 'Method for synchronization of power electronic converters in polluted and variable-frequency environments', *IEEE Transactions on Power Systems*, 2004, 19 (3), pp. 1263–1270.

[9] E. Figueres, G. Garcerá, S. Jesús, F. González-Espín, J. Calvo Rubio, 'Sensitivity study of the dynamics of three-phase photovoltaic inverters with an LCL grid filter', *IEEE Transactions on Industrial Electronics*, 2009, 56 (3), pp. 706–717.

[10] E. Figueres, G. Garcerá, S. Jesús, F. González-Espín, J. Calvo, M. Vales, 'Dynamic analysis of three-phase photovoltaic inverters with a high order grid filter', International Symposium on Industrial Electronics, 2007, pp. 691–696.

[11] M. Liserre, F. Blaabjerg, R. Teodorescu, 'Stability of photovoltaic and wind turbine grid-connected inverters for a large set of grid impedance values', *IEEE Transactions on Power Electronics*, 2006, 21 (1), pp. 263–272.

[12] D.P. Hohm, M.E. Ropp, 'Comparative study of maximum power point tracking algorithms', *Progress in Photovoltaics: Research and Applications*, 2003, 11 (1), pp. 47–62.

[13] N. Femia, G. Petrone, G. Spagnuolo, M. Vitelli, 'Optimization of perturb and observe maximum power point tracking method', *IEEE Transactions on Power Electronics*, 2005, 20 (4), pp. 963–973.

[14] R. Teodorescu, F. Blaabjerg, M. Liserre, P.C. Loh, 'Proportional-resonant controllers and filters for grid-connected voltage-source converters', *IEEE Proceedings Electric Power Applications*, 2006, 153 (5), pp. 750–762.

[15] D.N. Zmood, D.G. Holmes, G. Bode, Frequency domain analysis of three-phase linear current regulator', *IEEE Transactions on Industry Applications*, 2001, 37, pp. 601–610.

[16] D.N. Zmood, D.G. Holmes, 'Stationary frame current regulation of PWM inverters with zero steady-state error', *IEEE Transactions on Power Electronics*, 2003, 18, pp. 814–822.

[17] X. Yuan, W. Merk, H. Stemmler, J. Allmeling 'Stationary frame generalized integrators for current control of active power filters with zero steady-state error for current harmonics of concern under unbalanced and distorted operating conditions', *IEEE Transactions on Industry Applications*, 2002, 38, pp. 523–532.

[18] M. Ciobotaru, R. Teodorescu, F. Blaabjerg, 'Control of single-stage single-phase PV inverter'. European Power Electronics Conference, EPE 2005, pp. P.1–P.10.

[19] M.A.S. Masoum, H. Dehbonei, E.F. Fuchs, 'Theoretical and experimental analysis of photovoltaic systems with voltage and current-based maximum power point tracking', *IEEE Transactions on Energy Conversion*, 2002, 17 (4), pp. 514–522.

[20] F. Gonzalez-Espin, E. Figueres, G. Garcera, 'An adaptive synchronous-reference-frame phase-locked loop for power quality improvement in a polluted utility grid', *IEEE Transactions on Industrial Electronics*, 2012, 59 (6), pp. 2718–2731.

[21] P. Rodriguez, J. Pou, J. Bergas, *et al.* 'Decoupled double synchronous reference frame PLL for power converters control', *IEEE Transactions on Power Electronics*, 2007, 22 (2), pp. 584–592.

[22] P. Rodriguez, A. Luna, I. Candela, R. Mujal, R. Teodorescu, F. Blaabjerg, 'Multiresonant frequency-locked loop for grid synchronization of power converters under distorted grid conditions', *IEEE Transactions on Industrial Electronics*, 2011, 58 (1), pp. 127–138.

[23] P. Mattavelli, 'Synchronous-frame harmonic control for high performance AC power supplies', *IEEE Transactions on Industry Applications*, 2001, 37, pp. 864–872.

[24] S. Buso, P. Mattavelli, *Digital Control in Power Electronics* (Morgan & Claypool Publishers, 2006).

[25] M. Castilla, J. Miret, J. Matas, Luis García de Vicuña, J.M. Guerrero, 'Control design guidelines for single-phase grid-connected photovoltaic inverters with damped resonant harmonic compensators', *IEEE Transactions on Industrial Electronics*, 2009, 56 (11), pp. 4492–4500.

[26] M. Liserre, R. Teodorescu, F. Blaabjerg 'Double harmonic control for three-phase systems with the use of resonant current controllers in a rotating frame', *IEEE Transactions on Power Electronics*, 2006, 21, pp. 836–841.

Chapter 9

FPGA-DSP controllers for DC-DC converters in renewable energy applications

Yihua Hu[1], Wenping Cao[2] and Chushan Li[3]

9.1 Introduction

Solar energy is the main source of renewable energy. As the cost of photovoltaic (PV) panels continues to reduce, PV-based power generation becomes popular across the world and they can be either grid-connected or stand-alone systems. Currently, the global installation is over 40 GW and has been growing at 50% per year since 2005 [1]. Unfortunately, the output voltage of the single PV cell is quite low (about 0.7 V) and thus the output voltage of the PV module is limited to usually 35 V. In order to connect to the power grid, the PV bus voltage should be higher than 350 V for a 220V/50Hz AC grid. Therefore, a high-voltage gain is needed to convert the low PV cell voltage into high voltage for grid connection.

With the advent of power electronics converter technologies, such as modular multi-level converter and matrix converter, the use of a large number of PWM channels can help achieve the voltage conversion but conventional DSPs lack sufficient resources to achieve a reliable and flexible control scheme. In this chapter, a combination of FPGA and DSP-based multi-functional digital controller has been developed to control a new high-voltage-gain DC-DC converter for renewable energy applications.

9.2 FPGA and DSP-based multi-functional digital controller

FPGA and DSP-based multi-functional digital controller is the controller platform for energy conversion system.

9.2.1 Controller platform

A FPGA EP3C16 [2] and a floating-point DSP TMS320F28335 [3] are linked together to compose a powerful controller platform. External memory interface (EMIF) bus is employed to realize high-speed communication between the two

[1]Department of Electrical Engineering and Electronics, University of Liverpool, Liverpool, UK
[2]School of Engineering and Applied Science, Aston University, Birmingham, UK
[3]Department of Electrical Engineering, Zhejiang University, Hangzhou, China

digital controllers. The DSP+FPGA controller platform supports up to 60 PWM channels, 36 I/O channels, and 32 ADC channels, as well as CAN, RS485, and RS232 communication interfaces.

a) Design objectives

For modern power electronic converters, their control systems become complex as their switching frequency and power rate increase. These lead to increased hardware size and costs, especially for the widely used multi-level power converters where one controller (e.g. DSP) alone cannot handle all control tasks and the PWM channels are not sufficient. As a result, a control platform with multi-controllers linked together is required to achieve fast calculation speed and required power circuit interface connection. However, there is no need for power electronic engineers to develop their own platform. A general-purpose control platform will support their development of high-level control algorithms.

This chapter develops a powerful control platform with the following five design objectives:

- Using the DSP and FPGA as the co-processors to increase PWM output channels and algorithm realization speed.
- The FPGA deals with analogue-to-digital converter (ADC) and digital-to-analogue converter (DAC).
- FPGA-only control mode, DSP-only control mode, and DSP+FPGA control mode are provided.
- A touch panel is used as the user interface (UI) to realize on-line monitoring and debugging.
- Several common communication interfaces (e.g. CAN, RS232, RS485) are included.

b) Hardware development

The controller platform is realized by using both floating-point DSP TMS320F28335 and FPGA EP3C16. A photograph of the control platform (including a control board, a touch panel, and a lower signal conditioning board) is shown in Figure 9.1. The block diagram of this platform is given in Figure 9.2.

The platform includes the following function blocks:

DSP part:

- 12-channel DSP PWM and 12 DSP I/O Ports: A voltage-level shift function is added to all the PWM and I/O ports connected to DSP. Thus the DSP is compatible with 3.3-V level and 5-V level standard. A 10-bit level shifting bus switch IC SN74CBTD3384 [4] is used in this platform.
- 16-channel DSP ADC Input Ports: Every ADC input is connected to an over-voltage protection circuit to keep the input voltage within 0–3.3 V.
- DSP Communication Ports: DSP is designed to communicate with touch panel using dedicate RS232 interface. Communication with up-level host devices is based on either CAN interface or RS485 interface. By applying isolated transceivers with integrated isolated DC-to-DC converter such as ADM3053 [5] and ADM2587 [6], the interface circuits can be simplified significantly.

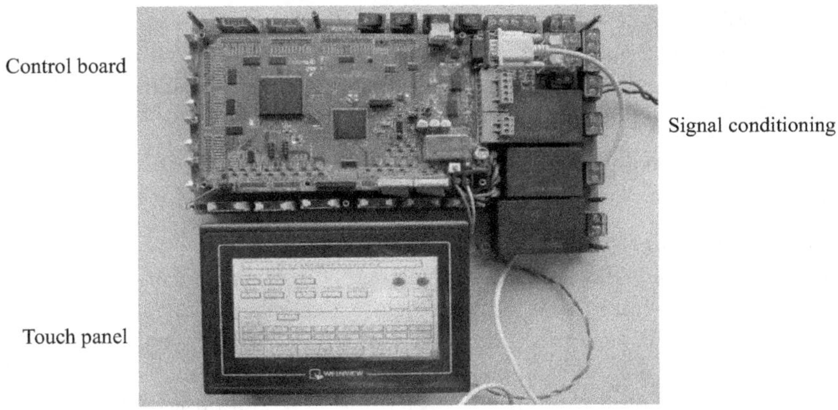

Figure 9.1 Photograph of the control platform

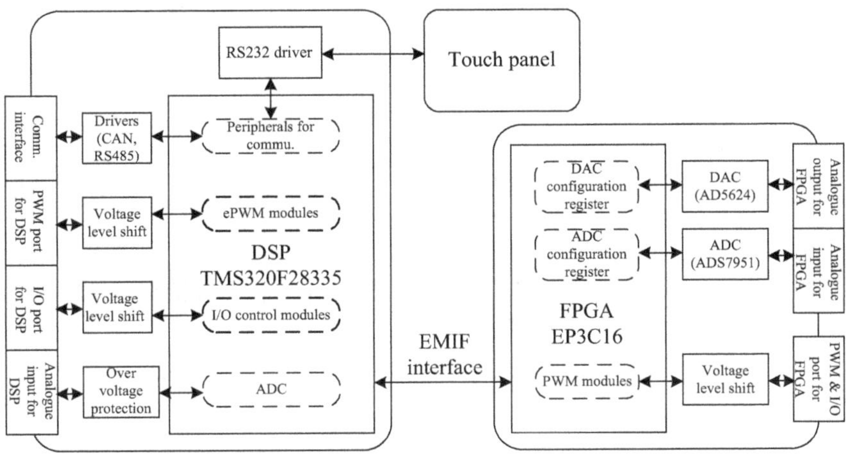

Figure 9.2 Block diagram of the controller platform

FPGA part:

- 60-channel PWM Output Ports: The voltage level of these ports is shifted from 3.3 to 5 V. SN74ALVC164245 [7] is used as the interface IC.
- 24-channel FPGA PWM and I/O Ports: The voltage-level shift function is also added to all the PWM and I/O ports. Thus, it is compatible with 3.3-V level and 5-V level standard. The 10-bit level shifting bus switch IC SN74CBTD3384 [4] is used in this part.
- 16-channel extended ADC Input Ports: ADS7953 is a 16-channel ADC IC [8], which is used here to be the external ADC for FPGA. The data exchange is based on SPI bus. Also, ADC input is connected with the same over-voltage protection circuit.

- 4-channel extended DAC Output Ports: AD5624 is a 4-channel DAC IC [9], which is implemented to be the external DAC for FPGA. As no DAC is provided in DSP part, this is unique and can provide useful display functions for both DSP and FPGA part by data exchanging.

At the same time, this control board is enabling single power supply. The input voltage range can be from 9 to 36 V. The maximum power loss is about 10 W.

c) Software implementation

In order to accelerate the development process for users and to facilitate code transplantation, the control platform offers a series of well-developed functions blocks. DSP programs include a code section which enables data exchange between DSP and FPGA, and communication between DSP and touch panel. FPGA function blocks include an external memory interface EMIF module used for DSP–FPGA data exchange, a PWM generator module, an I/O port control module, and ADC and DAC control modules.

9.2.2 DSP–FPGA synchronization

The programs for DSP–FPGA data exchange demand high speed in real-time control. The code structure and the operational details of the DSP and FPGA are as follows.

A block diagram of the design implemented in the FPGA device and its data exchange method with DSP is shown in Figure 9.3. The data exchange uses the asynchronous EMIF bus. In the code section, the configuration registers of all the

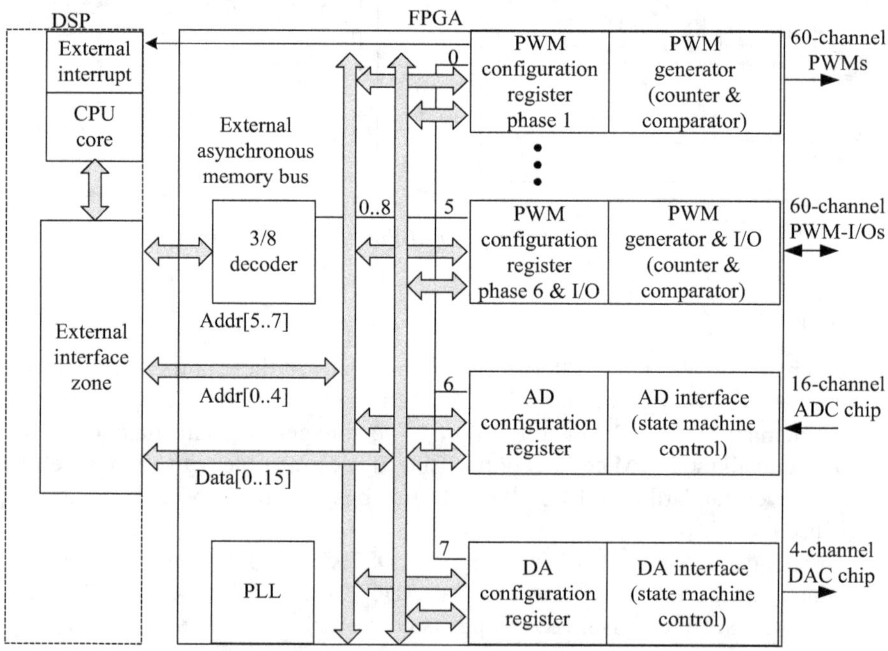

Figure 9.3 Block diagram of the design implemented in the FPGA device

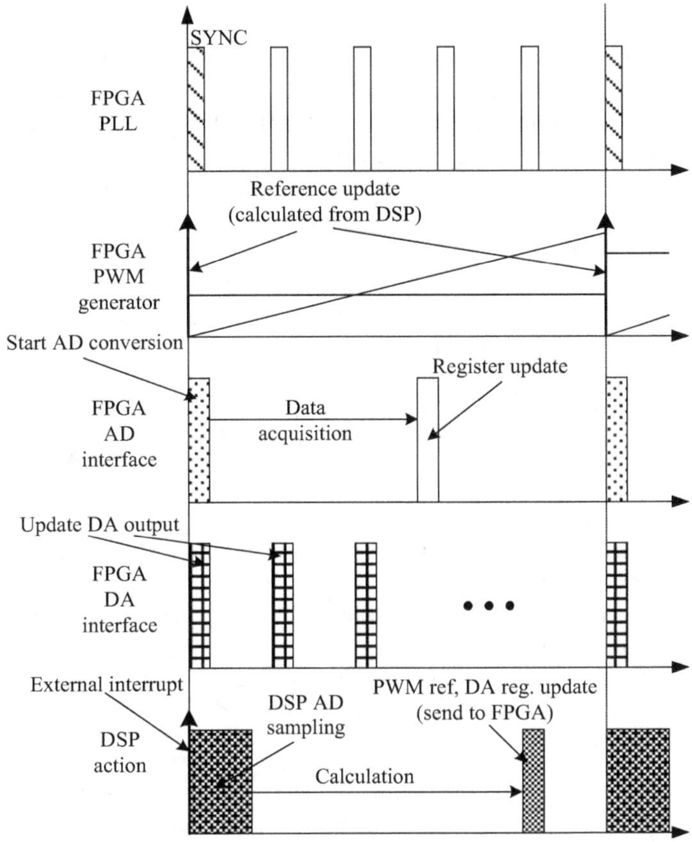

Figure 9.4 Timing of the synchronization between the DSP and FPGA

function blocks in FPGA are mapped into the external memory address of DSP. The data exchange is realized by DSP writing to or reading these sections.

Since in most cases the PWM signals are generated by FPGA, the FPGA device should be the master device of the power electronic control system. As is shown in Figure 9.4, a synchronization module in FPGA generates a periodic pulse (SYNC). The period is the same as the switching frequency and the sampling frequency (e.g. 100 μs). This SYNC pulse is also used to periodically interrupt the DSP to start sampling and calculation for the next control period. The new results will be sent to FPGA before the end of the period. Also, the duty cycle of the PWM generators is updated. The reference used in this control cycle is calculated in the last cycle by DSP or FPGA. The external ADC converter is triggered at this instant. The same sampling speed with DSP ADC enables a sampling of up to 32 channels in 1 switching cycle.

The operating mechanism of the external DAC is different since its conversion rate can be up to 250 kHz for the four DAC channels operation. This rate is much higher than the switching frequency of high-power converters. As a result, it can operate continuously to update its output. Either DSP or FPGA can modify the registers of the DAC interface at any time, which enables a real-time update of DAC output.

9.2.3 Explanation of function blocks in the FPGA device

The FPGA part of programs consists of eight function blocks, which are five identical PWM generator modules, one PWM and I/O multiplexing module, one external DAC interface module, and one external ADC interface module.

As shown in Figure 9.3, the external address bus Addr is of 8 bits and the external data bus data is of 16 bits. The high 3-bit signals of the address bus Addr are connected to the 3/8 decoder modules in FPGA. Eight chip selection signals are generated based on the high 3-bit address signals. Each function block uses one chip selection signal and the low 5-bit address signals. That is, every module has 32 16-bit registers. There registers are accessible with DSP. The DSP can write or read them to modify the operation of these function blocks.

9.2.4 Implementation of a touch panel

A touch panel is implemented as the UI for the general-purpose control platform. Compared to the commonly used LCD output and keyboard input, the touchpad offers both input and output functions and user-friendliness. Programming on touch panels is also easy. Some software (e.g. EasyBuilder 8000) can provide a large number of developed internal object resources. The users only need to familiarize with the functions of all the internal objects and drag them to the desktop of the touch panel to construct the interface.

Importantly, the touch panel is designed to a control and monitor system for power electronics converters. The main page of the UI is shown in Figure 9.5. Several functions are realized:

- Observation and setting of control registers: aiming to provide key control parameters such as PI coefficient, current and voltage reference to realize

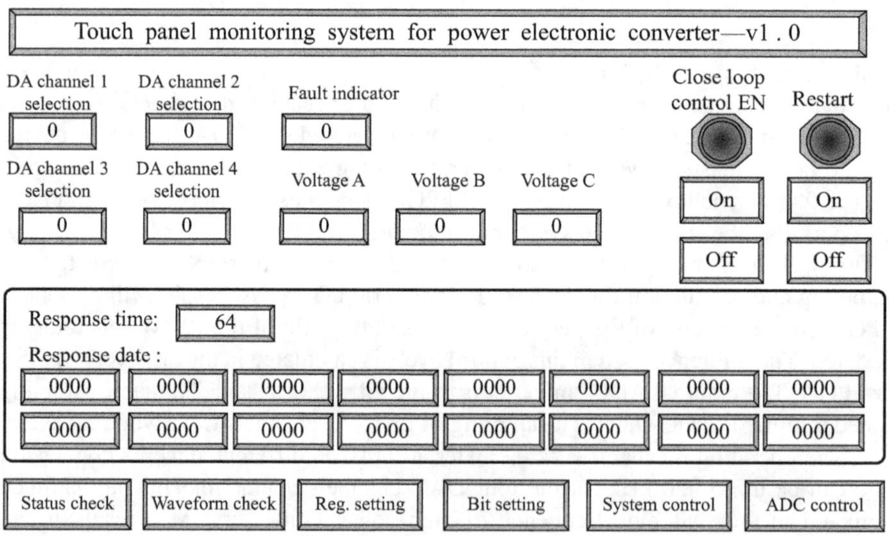

Figure 9.5 Main page of the touch panel user interface

online debugging. Users can check and modify the value of the registers. This value is then sent to the corresponding address in the DSP memory. Thus the parameter can be modified online.

- Observation of status registers: a series of registers are designed for read-only. This function is used to check the variables such as current and voltage sampling, PI output etc. The value from corresponding address in the DSP memory is sent to the touch panel. Users can check them online.
- Observation of bit registers and switches control: bit registers are used to modify the system state such as converter on/off control, closed loop enabling. These states are also shown on the touch panel.
- Waveform display: the waveform display is realized by plotting a sequence of data in the DSP memories. This function can be used to continuously monitor the sampling data and calculation output.

Figure 9.6 shows the overall structure of the data exchange between the DSP and the touch panel. The RS232 protocol is applied to the Physical Layer, and the simplified Modebus protocol is applied to the Data Link Layer.

On the application layer, the display interface of the touch panel is built up by predefined components provided by EasyBuilder 8000. Actions can be defined inside the components to access the internal registers in the touch panel. On the data link layer, the Modebus protocol is realized by using Macro Instructions defined in EasyBuilder 8000. These instructions can access the data in internal registers and trigger the RS232 communication on physical layer, thus the data exchange between the three layers is enabled.

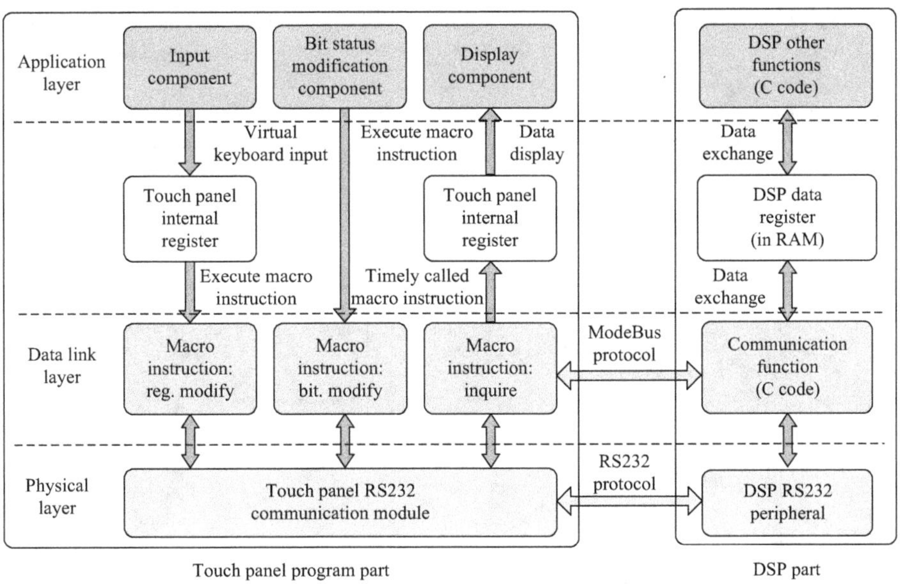

Figure 9.6 Overall structure of data exchange between the DSP and touch panel

In the DSP part, the programs on the data link layer only receive data from RS232 and save them to a specified memory buffer. As a result, after redefining the registers in the touch panel and their respective data registers in DSP, the users can embed this monitor and control system into their own design. Other DSP functions on application layer only need to access the memory buffer for the converter control.

9.3 Development of new topologies and control schemes for DC-DC converters

9.3.1 High step-up passive clamp circuits

High step-up and high-power-density non-electrical-isolated DC-DC converters are widely used in distributed generation, such as fuel cells, PVs, and uninterruptible power systems (UPS) [10]–[12]. Low-voltage stress, high step-up capability, and less switching devices are three key requirements for DC-DC converters [13]–[15].

Figure 9.7 shows the proposed passive clamp circuit, which is deduced from the Luo converter. Figure 9.8(a) shows the combination of a traditional Flyback converter and the proposed passive clamp circuit. Figure 9.8(b) illustrates the improved Flyback converter. With the proposed passive clamp circuit, the leakage energy can be transferred back to the load, and the voltage spike of the main switches is limited. The high step-up interleaved Flyback–Forward converter is first proposed in Reference 16, as shown in Figure 9.9(a). The active clamp circuit is composed of switches S_{c1} and S_{c2} and clamp capacitors C_{C1} and C_{C2}. These are utilized to recover the leakage energy and suppress the turnoff voltage spikes on the main switches. As shown in Figure 9.9(b), the active clamp circuit is replaced by the proposed passive clamp circuit, which consists of two diodes and one capacitor. Figure 9.9(c) shows the proposed topology that the passive clamp circuits are employed to clamp main switch voltage spikes and transfer the leakage energy into the output. There are two coupled inductors in the proposed high step-up converter: L_1 and L_2. The primary inductors L_{1a} and L_{2a} with n_1 turns are coupled with secondary inductors L_{1b} and L_{2b}

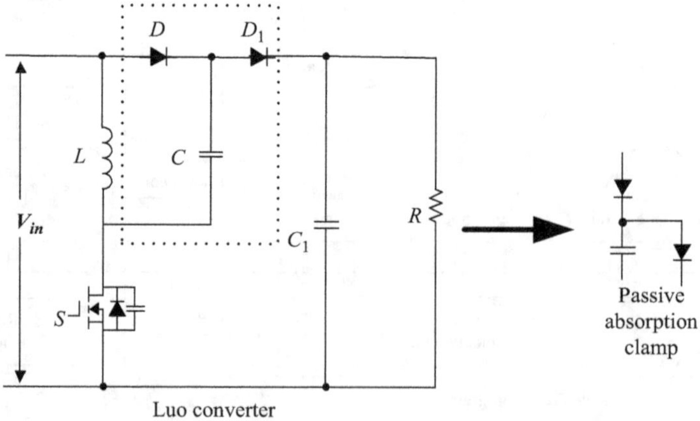

Figure 9.7 The proposed passive clamp circuit

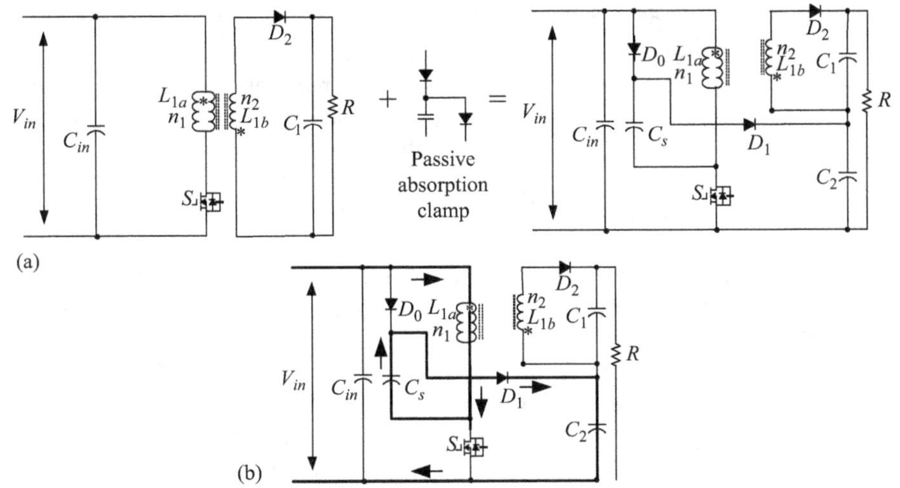

Figure 9.8 Combined Flyback converter with the proposed passive clamp circuit. (a) Circuit deduction and (b) the current path for recycling leakage energy

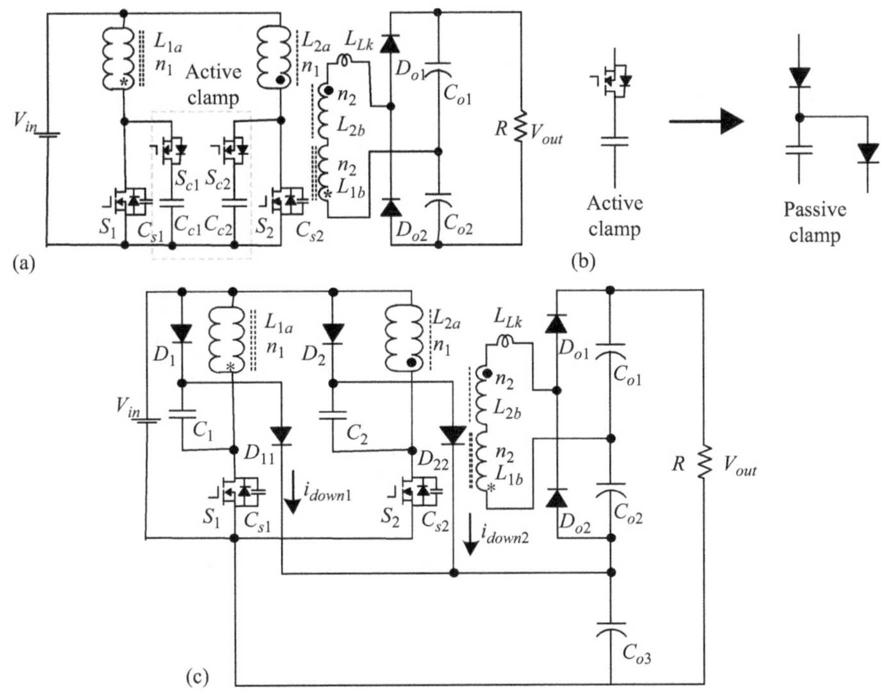

Figure 9.9 Passive clamp high step-up dc-dc converter. (a) Flyback–forward DC-DC converter, (b) clamp circuit, and (c) proposed high step-up DC-DC converter

with n_2 turns. The coupling references are remarked as "*" and ".". The secondary windings of L_1 and L_2 are connected in series. L_{lk} includes the leakage inductances of the two coupled inductors. D_1, D_{11}, D_2, and D_{22} are rectifier diodes, C_1 and C_2 are self-lift capacitors, C_{o1}, C_{o2}, C_{o3} are output capacitors. D_{o1} and D_{o2} are output rectifier diodes. Capacitors C_{S1} and C_{S2} are the parasitic capacitors of S_1 and S_2 respectively. D_1, D_{11}, and C_1 compose the passive clamp circuit, thus the leakage energy can be recycled and the spike voltage of main switch S_1 can be clamped.

When the leakage inductance is low, the working waveforms are as follows. Two 180° out-of-phase gate signals with the same duty cycle are applied to S_1 and S_2. Based on the steady state, there are 12 operational modes in a switching period. Due to the symmetry of the circuit, the six modes are analyzed briefly. The typical waveforms of the proposed converter at steady state are shown in Figure 9.10, in which i_{down1} and i_{down2} are the current in D_{11} and D_{22}, respectively.

Mode 1 [t_0, t_1]: During this stage, the main switches S_1 and S_2 are turned on. The capacitor C_1 is initially charged to V_{in} in a short time. The output rectified diode D_{o1} is reverse-biased while D_{o2} is positive-biased. Due to the voltage of V_{co2}, the current decreases in a linear manner.

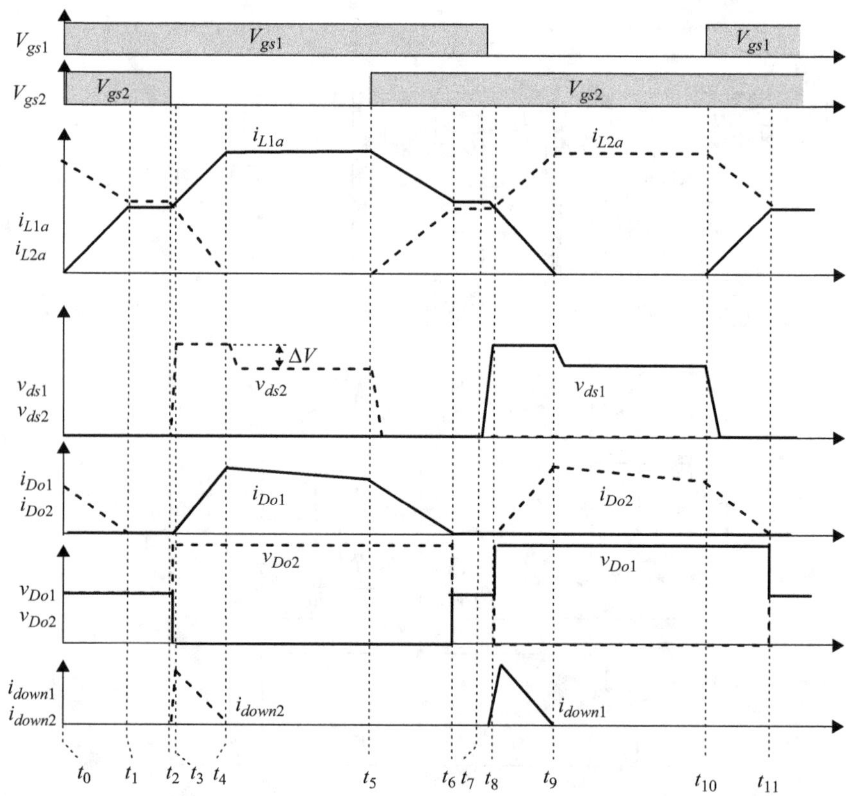

Figure 9.10 Waveforms of the typical passive clamp high step-up converter

Mode 2 $[t_1, t_2]$: At t_1, the current in D_{o2} decreases to 0. Both D_{o1} and D_{o2} are reverse-biased. The two coupled inductors operate in the Flyback state to store energy. The energy to the load is sustained by the output capacitors C_{o1}, C_{o2}, and C_{o3}.

Mode 3 $[t_2, t_3]$: At t_2, the main switch S_2 is turned off, which increases its drain-source voltage nearly linearly due to the parallel capacitor C_{S2}. Because of the large winding current and small parallel capacitor, this interval is short. During the interval $t_2\_t_3$, the diode D_{o1} conducts.

Mode 4 $[t_3, t_4]$: At t_3, the voltage on C_{S2} increases to its climax, and the diode D_2 conducts. The current in L_{2a} decreases in a linear way. Both the currents in L_{1a} and D_{o1} increase linearly. During this stage, the coupled inductor L_1 operates in the forward mode and L_2 in the Flyback mode to transfer energy to the load.

Mode 5 $[t_4, t_5]$: At t_4, the current in L_{2a} decreases to 0, so does the voltage across C_{S2} and the current in D_{o1}. During this stage, the coupled inductor L_1 operates in the forward mode and L_2 in the Flyback mode to transfer energy to the load.

Mode 6 $[t_5, t_6]$: At t_5, the main switch S_2 turns on with zero voltage switching (ZVS). The secondary diode D_{o1} is still in the conduction state. At t_6, the leakage current decreases to zero and the diode D_{o1} turns off with zero current switching (ZCS) operation.

The operation of the converter during the interval $t_7 \sim t_{12}$ is similar to that of $t_0 \sim t_6$. The passive clamp circuit can absorb the turnoff voltage spikes on the main switches S_1 and S_2, and recycle the leakage energy.

The voltages on the main switches are given by

$$V_{ds} = V_{Co3} - V_{in} + V_{leak} = \frac{1}{1-D} V_{in} + V_{leak} \tag{9.1}$$

where V_{leak} is the voltage caused by the coupled inductor leakage, D is defined as the duty ratio of the main switches.

Due to the voltage-second balance on the magnetizing inductor, the voltages of C_{o1} and C_{o2} can be obtained as

$$V_{co1} = V_{co2} = \frac{N \cdot V_{in}}{1-D} \tag{9.2}$$

where $N = n_2/n_1$.

The voltage on C_{o3} can be given by,

$$V_{Co3} = \frac{2-D}{1-D} V_{in} \tag{9.3}$$

The output voltage is the sum of the three output capacitors. Therefore, the voltage gain (G) of the proposed converter under ideal condition is

$$G = \frac{2(N+1) - D}{1-D} \tag{9.4}$$

In order to verify the theoretical analysis, a 500-W prototype is built and tested with the specifications shown in Table 9.1. The Koolmu magnetic core 0077109A7 is chosen as the coupled inductors. The currents through the primary inductors i_{L1a} and i_{L2a} are shown in Figure 9.11(a); they agree with the theoretical analysis.

Table 9.1 Specifications of the proposed converter

Components	Parameters
Power level (P_{out})	500 W
Input voltage (V_{in})	10 ~ 12 V
Output voltage (V_{out})	80 V
Switching frequency (f_s)	50 kHz
Main switches (S_1 and S_2)	59N30
Diodes (D_1, D_2, D_{o1}, D_{o2})	BYW99W200
Clamp capacitors (C_1 and C_2)	100 V/100 μF
Output capacitors (C_{o1} and C_{o2})	470 μF
Turns ratio ($N = n_2/n_1$)	56:28

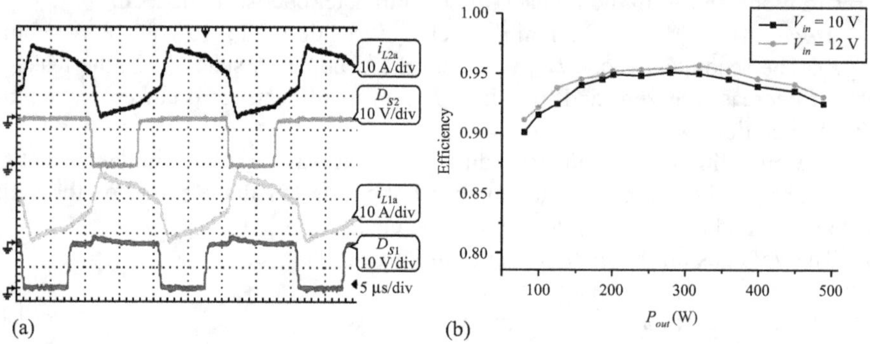

Figure 9.11 Experimental results of the proposed converter. (a) Primary current with a 66.7% duty cycle and (b) measured efficiency

For the symmetrical characteristic of the proposed converter, the amplitudes of waveforms of i_{L1a} and i_{L2a} are equal; the current sharing of the main switching device can be realized. In Figure 9.11(a), D_{s1} and D_{s2} are gate signals of the main switches S_1 and S_2. The measured efficiency at 10 V and 12 V input voltages of the proposed passive clamp Flyback–Forward converter is shown in Figure 9.11(b). When the input voltage is 12 V, the highest efficiency is 95.4% at nearly 50% load. At full load conditions, the efficiency is 92.5%. In fact, the coupled inductor in the proposed converter can deliver the energy to the load both at turn-on and turn-off stages, and the leakage energy can also be delivered to the load.

The proposed passive clamp circuit can be used to replace traditional active clamp circuits to decrease the number of switching devices, while the isolation characteristic is lost. The proposed passive clamp circuit is proper in non-isolated step-up converter application situation. The passive clamp circuit considered in the basic structure can also be extended to another high step-up circuit. An improved high step-up structure developed is presented in Figure 9.12. The passive clamp circuit is constructed by adding a diode and a capacitor to achieve a high-voltage gain.

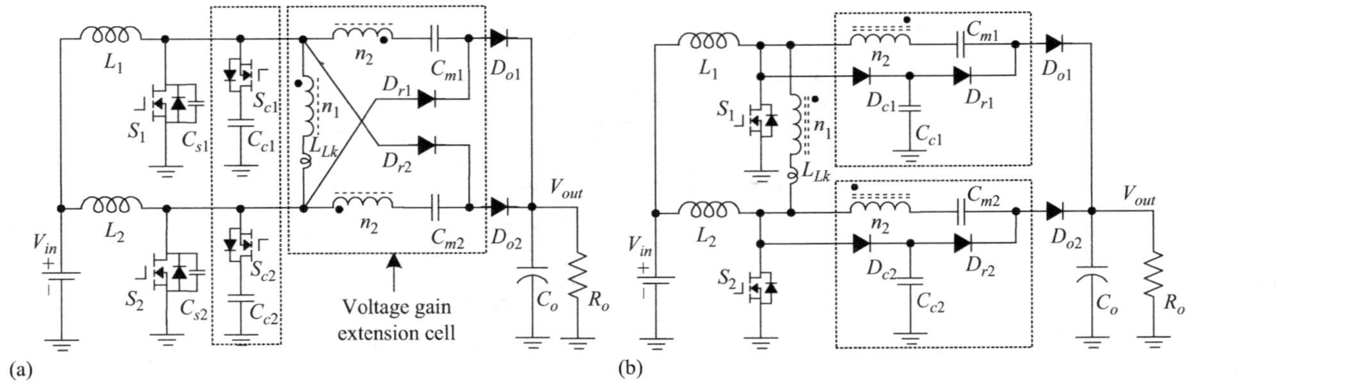

Figure 9.12 *Clamping circuit applications in the non-isolated circuit (a) high step-up converter with an active clamp circuit and (b) with a passive clamp circuit. © 2014 IEEE. Reprinted with permission from Reference 17*

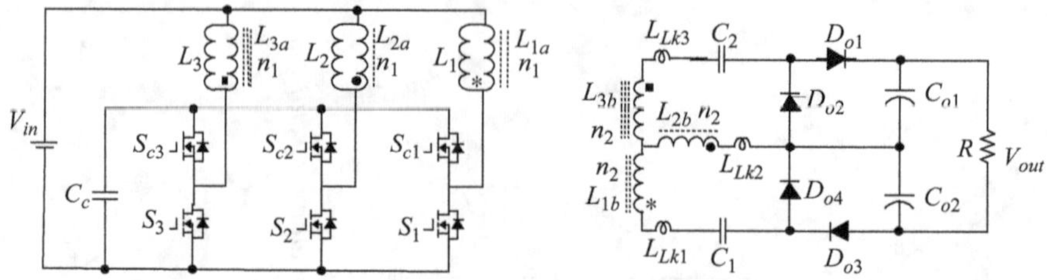

Figure 9.13 Three-phase interleaved high step-up converter

9.3.2 Three-phase interleaved high step-up converters

Due to the influence of PV module leakage current in the application of large PV array, the electrical isolation is needed to block leakage current. Therefore, high-power level, high-voltage gain, and electrical isolation DC-DC converter are the vital components. In this section, the three-phase interleaved high step-up converter is introduced.

The proposed high-power level, high step-up converter is shown in Figure 9.13, where S_1, S_2, S_3, S_{c1}, S_{c2}, and S_{c3} are the traditional three-phase bridges, and C_c is the clamp circuit. Three coupled inductors are labeled as L_1, L_2, and L_3, where the primary windings are employed as filter inductors and the secondary windings are star series connected for high output voltage gain. The primary and secondary winding turns are represented as n_1 and n_2 respectively, and the turn ratio N is derived from n_2/n_1. The coupling references are remarked as "*," "○," and "□." For future analysis, the lump sum leakage inductance is named as L_{LK1}, L_{LK2}, and L_{LK3}, corresponding to the coupled inductors of L_1, L_2, and L_3, respectively. D_{o1}, D_{o2}, D_{o3}, and D_{o4} are the output diode; C_1 and C_2 are the switched capacitor; C_{o1} and C_{o2} are the output capacitors. C_{s1}, C_{s2}, and C_{s3} are the parasitic parallel capacitors of S_1, S_2, and S_3, respectively.

The key steady-state waveforms of the proposed high step-up converter are plotted in Figure 9.14. The main switch of each coupled inductor bridge (CIB) is operated with identical duty ratio as others, but 120° phase shift among each other. The active clamp switch, $S_{C1\text{-}3}$ is conducted complementarily to the main switches. There are six main operational states in one switching period.

State 1 [t_0, t_1]: Before t_0, switch S_3 is with turn-on state. The clamp switch S_{c1} turns off at t_0. The clamp capacitor C_c is disconnected from the resonant circuit, the voltage across S_{c1} increases from zero, and the ZVS of the clamp switch S_{c1} is realized. While the voltage of the parallel capacitor C_{s1} is resonant with leakage inductance of L_1, the energy is transferred from C_{s1} to leakage inductance of L_1. During this time period, the voltage of C_{s1} decreases to zero, and anti-parallel diode of S_1 begins to conduct. The turn-on signal of the main switch S_1 is given in this period when its anti-parallel diode is in the ON state, and S_1 turns on with ZVS. At t_0, the main switch S_2 turns off, and then its parallel capacitor C_{s2} is charged by the magnetizing current of L_2. In this period, the switching voltage of the main switch S_2 reaches the voltage of the clamp capacitor C_c, and the anti-parallel diode of S_{c2} conducts. The voltage of the main switch S_2 is clamped to the voltage on C_c.

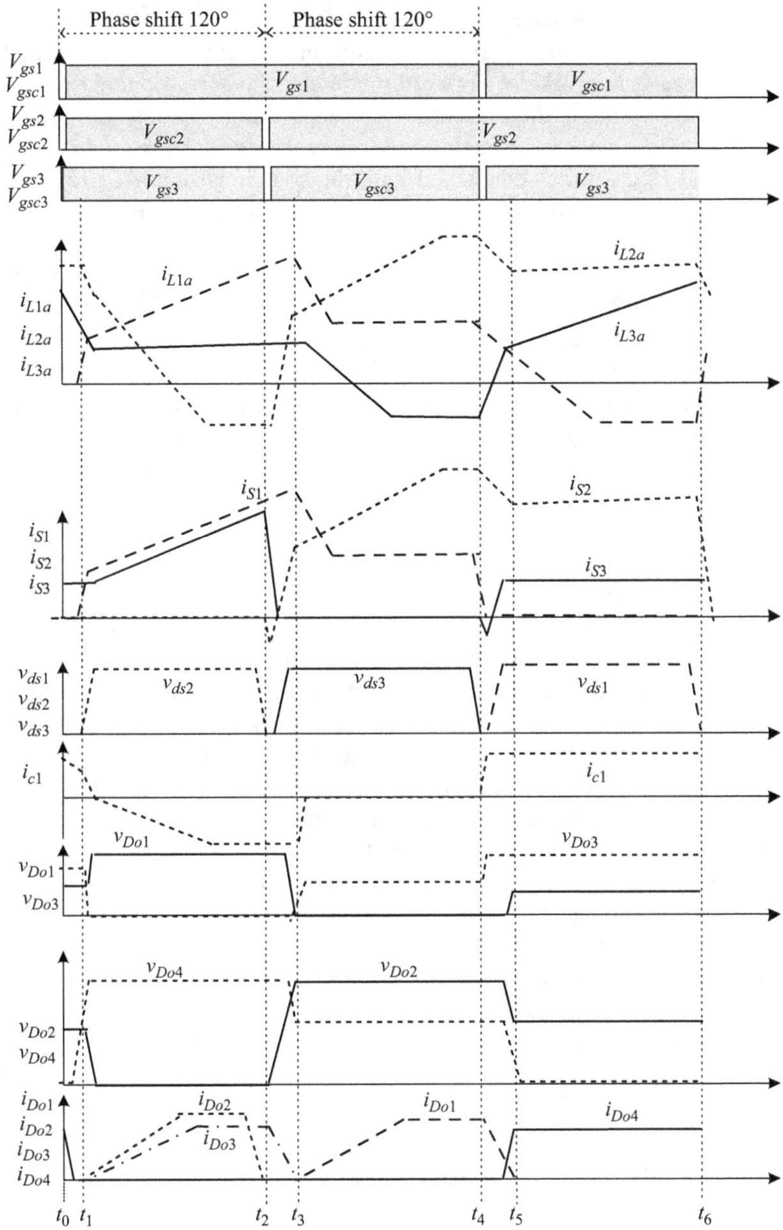

Figure 9.14 Steady-state waveforms of the proposed converter in six predefined states. © 2014 IET. Reprinted with permission from Reference 20

State 2 [t_1, t_2]: In this state, the main switches S_1 and S_3 turn on while the active clamp switch S_{c2} is at on-state. At t_1, the current of primary side of L_1 increases from zero; because L_3 works in Forward mode and the corresponding second side voltage is in low voltage, the current of L_3 keeps stable. In this sub-interval, the inductors, L_1 and L_3 work in Forward mode, L_2 works in Flyback mode, the second sides of L_1 and L_2 are series connected with the switch capacitor C_1 to charge the output capacitor C_{o2}, while the second sides of L_3 and L_2 are series connected to charge the switch capacitor C_2.

State 3 [t_2, t_3]: The clamp switch S_{c2} turns off at t_2, while the clamp capacitor C_{s2} is disconnected from the resonant circuit, the voltage across S_{c2} increases from zero, and the clamp switch S_{c2} ZVS is realized. During this time period, the voltage of C_{s2} decreases to zero, and the anti-parallel diode S_2 begins to conduct. The turn-on signal of the main switch S_2 is given in this period when its anti-parallel diode is at ON state, therefore, the S_2 turns on with ZVS. At t_2, the main switch S_3 turns off, and then its parallel capacitor C_{s3} is charged by the magnetizing current of L_3. In this period, the switching voltage of the main switch S_3 reaches the voltage of the clamp capacitor C_c, and the anti-parallel diode of S_{c3} conducts. The voltage of the main switch S_3 is clamped to the voltage across C_c. In this sub-interval, the L_1 works in Forward mode; the second sides of L_1 and L_2 are series connected with C_1 to charge the output capacitor C_{o2}.

State 4 [t_3, t_4]: The main switches S_1 and S_2 turn on while the active clamp switch S_{c3} is at on-state. At t_3, the current of the primary side of L_2 increases, while the current of the primary side of L_1 and L_3 decreases. From t_3 to t_4, the L_2 works in Forward mode, the L_3 works in Flyback mode, the second sides of L_2 and L_3 combined with the switch capacitor C_1 are series connected to charge the output capacitor C_{o1}.

State 5 [t_4, t_5]: When the clamp switch S_{c3} turns off at t_4, the clamp capacitor C_{s3} is disconnected from the resonant circuit. Since the voltage across S_{c3} increases from zero, the clamp switch S_{c3} ZVS is realized. During this time period, the voltage of C_{s3} decrease to zero, while the anti-parallel diode of S_3 begins to conduct. The S_3 turns on with ZVS since the turn-on gate signal of the main switch S_3 is given during the on-state of its anti-parallel diode. At t_4, the main switch S_1 turns off, meanwhile, its parallel capacitor C_{s1} is charged by the magnetizing current of L_1. In this state, the switching voltage of the main switch S_1 reaches the voltage of the clamp capacitor C_c, and the anti-parallel diode of S_{c1} conducts. The voltage of the main switch S_1 is clamped to the voltage across C_c. In this sub-interval, the L_2 works in Forward mode, the L_3 works in Flyback mode, the second sides of L_3 and L_2 are series connected with the switch capacitance C_2 to charge the output capacitor C_{o1}.

State 6 [t_5, t_6]: In this state, the main switches S_2 and S_3 turn on, the active clamp switch S_{c1} is with on-state. In this sub-interval, the second sides of L_1 and L_2 are series connected to charge the switch capacitance C_1. The output voltage is maintained by the capacitor C_{o1} and C_{o2}.

Due to the symmetry structure of the second side, the same charge and discharge periods are applied to the switching capacitors, C_1 and C_2. The next switching period repeats the same six operational states, as shown above.

The voltage stress on the main switches and active clamp switches is equal to the voltage on the clamp capacitors. The voltage of the clamp capacitors can be derived by the inductor voltage-second balance principle, as shown in (9.5).

$$V_{ds\_stress} = V_{Cc} \approx \frac{V_{in}}{1-D} \approx \frac{V_{out}}{4N} \tag{9.5}$$

where V_{Cc} is the voltage of active clamp switch, V_{in} is the input voltage, and D is the switching duty ratio of all main switches.

The voltage across the switch capacitors C_1 and C_2 is equal to one-quarter of the output voltage and half of the output capacitor voltage due to the series connection and doubler configuration. The peak output voltage of L_{1b}, L_{2b}, and L_{3b} is therefore one-eighth of output voltage. The current rising rate of the output diode D_{o1} is expressed as:

$$k_r = \frac{NV_{Cc} - V_{Co1}/2}{2L_k} \tag{9.6}$$

where V_{Co1} is the voltage of output capacitor, C_{o1}, k_r represents the current increasing rate of the output diodes. The current decrease rate of the output diodes can be expressed as

$$k_f = -\frac{V_{Co1}}{4L_k} \tag{9.7}$$

where k_f represents the current decreasing rate of the output diodes. According to Figure 9.14, it takes $(1-D)T_s$ for the current of the output diode D_{o1} to reach its peak value. Substituting (9.6), the peak value of the diode current can be derived as:

$$I_{Do1\_peak} = k_r(1-D)T_s = \frac{NV_{Cc} - V_{Co1}/2}{2L_k}(1-D)T_s \tag{9.8}$$

where T_s is the switching time period. Substituting (9.7), the decrement time of output diode D_{o1} current can be calculated as:

$$t_{Do1\_f} = \frac{I_{Do1\_peak}}{-k_f} = \frac{2NV_{Cc} - V_{Co1}}{V_{Co1}}(1-D)T_s \tag{9.9}$$

Due to the symmetry configuration of the second side of the coupled inductors, the voltage across C_{o1} is equal to the C_{o2} voltage, as expressed in (9.10). The average current of the output diode D_{o1} is equivalent to the average output current, which can be expressed in (9.11).

$$V_{Co1} = V_{Co2} = \frac{1}{2}V_{out} \tag{9.10}$$

$$I_{out} = \frac{V_{out}}{R} = I_{Do1} = I_{Do3} = \frac{1}{2}I_{Do1\_peak}\left[(1-D)T_s + t_{Do1\_f}\right] \times f_s \tag{9.11}$$

where f_s is the switching frequency, V_{Co2} is the voltage on C_{o2}, V_{out} is the output voltage of the converter, I_{out} is the output current, and R is the load resistance.

The voltage gain can be derived by:

$$G = \frac{V_{out}}{V_{in}} = \frac{8N}{\sqrt{(1-D)^2 + \frac{32L_k f_s}{R}} + 1 - D} \tag{9.12}$$

From (9.12), the voltage conversion ratio can be controlled by the duty cycle of main switches and affected by the parameters of N, L_k, and f_s. When the leakage inductance is completely ignored, the voltage gain can be expressed as (9.13), which represents the ideal steady state.

$$G = \frac{4N}{1-D} \tag{9.13}$$

Figure 9.15 illustrates the theoretical voltage gain of passive clamp high step-up converter and three-phase interleaved topology under $N=2$ condition. Clearly, the three-phase interleaved topology has a higher voltage gain than the passive clamp high step-up converter.

The active clamp circuit supports the function of soft switching for the main switches and clamp switches. The ZVS turn-on of the clamp switches is realized since the anti-parallel diodes of the clamp switches conduct before turn-on gate signal. To guarantee the turn-on ZVS for the main switches, the energy stored in the leakage inductor should be more than that of the parallel capacitor across the main switches. The corresponding ZVS turn-on condition for the main switches can be expressed as:

$$\frac{1}{2}\frac{L_k}{N^2} I_{in}^2 \geq \frac{1}{2} C_{s1} V_{Cs1}^2 \tag{9.14}$$

where I_{in} is the CIB average input current, and V_{Cs1} is the capacitor voltage across the main switches. The expression in (9.14) indicates that the ZVS region can be extended by increasing the values of the leakage inductance and/or the input current. However, the increment of the leakage inductance has a negative effect on the voltage conversion ratio. To compensate the voltage gain loss from the leakage

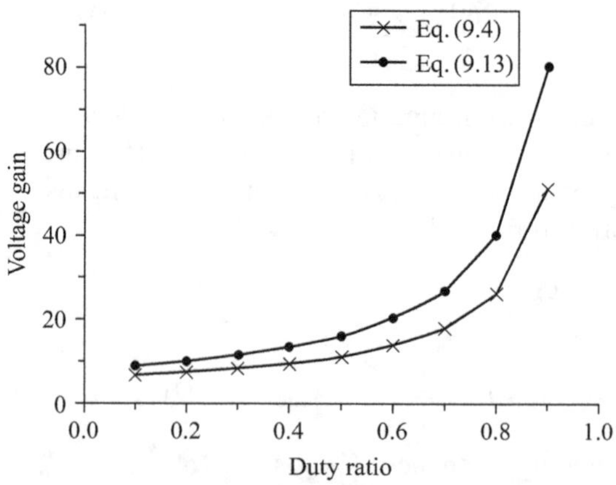

Figure 9.15 The theoretical voltage gain (N = 2)

inductance, a higher turn ratio is required, which contributes additional loss to the system due to the long wire conduction and skin effect, etc. A design tradeoff should be considered to achieve high-voltage gain, but maintain high converter efficiency.

As shown in Figure 9.13, the output voltage of the converter is the sum of the two series-connected capacitors that are connected with the two series-connected doubler rectifiers. This structure can be used as the front-end of the neutral-point-clamped three-level inverter. The fluctuation of the neutral-point voltage should be minimized in multi-level converters. Therefore, the neutral point of the proposed converter can be employed for double-source operation, such as in half-bridge inverters or three-level converters. Owing to the interleaved three-phase structure, the voltage of C_{o1} and C_{o2} can be independently controlled by the modulation of S_3 and S_1. For example, the duty ratio of S_3 determines the storage energy in L_3 to realize the active voltage balance in C_{o1} without disturbing the L_1 operation. The duty cycle of S_1 can be used to control the energy transfer through L_1 to realize the active voltage balance in C_{o2}, which is independent of L_3. Figure 9.16 shows the control and modulation diagram. The two voltage levels can be independently regulated to balance the neutral-point voltage. In the diagram, V_{down} indicates the voltage across C_{o2} while V_{up} represents the voltage of C_{o1}. Under normal operations, the voltages of V_{up} and V_{down} are regulated to generate two halves of the output voltage (i.e. $1/2V_{out}$). Classical proportional integral (PI) controllers are adopted in the regulation loop, which is expressed as (9.15) in Laplace format.

$$C_V(s) = K_p + K_i/s \tag{9.15}$$

There is a 240° phase-angle shift between the modulating waveforms of S_1 and S_3 and the energy in L_2 can influence both V_{down} and V_{up}. The duty ratio of S_2 can be determined by (9.13). In case there is a deviation from the duty ratio of S_2, the PI

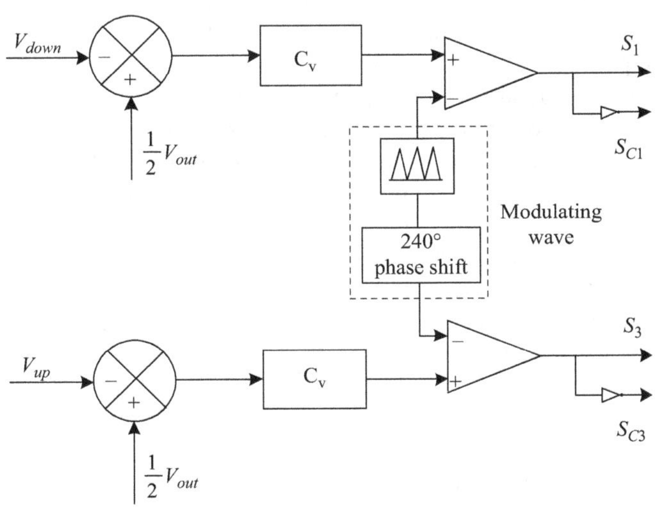

Figure 9.16 Control and modulation diagram for the neutral point. © 2014 IET. Reprinted with permission from Reference 20

regulators of S_1 and S_3 can compensate for it. To realize the three-phase operation, a 120° phase-angle shift is modulated between S_1 and S_2.

The system dynamics can be represented by two transfer functions, which show the V_{up}–D_{up} relation and V_{down}–D_{down} interaction. D_{up} and D_{down} indicate the switching duty ratios of the upper loop and the down loop, as shown in Figure 9.16. One unified transfer function can represent both leg dynamics since the V_{up} and V_{down} should be controlled to generate the output voltage equally. One assumption is made that the dynamic interaction among the three legs is insignificant due to the 120° phase shift and the middle leg operation (L_2, S_2, S_{c2}, and C_{c2}). The operation of each leg is equivalent to the modulation principle of conventional boost converters except for the turn ratio of 4N, as shown in (9.13). Therefore, the average model can be derived by neglecting the switching ripple. The small signal modeling of boost converters has been well developed [18]–[19]; the technique is used for the proposed converter, which provides decoupled control for V_{up} and V_{down}. The small-signal model is derived by:

$$\frac{\tilde{v}_{down}(s)}{\tilde{m}_1(s)} = \frac{\dfrac{2V_{in}}{L_m C}\left(-\dfrac{I_{L1}L_m}{V_{in}}s + 1\right)}{s^2 + \dfrac{1}{R_2 C}s + \dfrac{D'^2}{L_m C}} \tag{9.16}$$

where $\tilde{m}_1(s)$ indicates the small-signal perturbation in the duty cycle D_{down}, and $\tilde{v}_{down}(s)$ represents the small-signal output of V_{down}. In (9.16), the steady-state values appear in the transfer function, where D' denotes $1 - D$, R_2 is the equivalent resistance of the load across V_{down}, and I_{L1} is the average current in the excitation inductance of the coupled inductor. The small signal model for the V_{up}–D_{up} relationship can be derived in a similar manner.

The output capacitor C_{o1} is charged by the energy stored in L_3 and L_2, while capacitor C_{o2} is charged by L_1 and L_2, as illustrated in Figure 9.17(a). In the case of an asymmetrical power supply, the switching devices for coupled L_1 or L_3 can be turned off to supply asymmetrical power to C_{o1} and C_{o2}. Two fault conditions are considered to illustrate phase-deficient operation. Figure 9.17(b) and (c) show the primary-side open circuit of coupled inductor fault for L_3 and L_2, respectively. When L_3 is open-circuited, C_{o1} can only be charged by L_2 and the charging energy is halved. Because of the independent control loop in Figure 9.16, a 1:2 (V_{up}:V_{down}) can be achieved by turning off the corresponding switches on L_3. The same principle can be applied if L_1 is faulted. When L_2 is out of duty as shown in Figure 9.17(c), both C_{o1} and C_{o2} are affected. Although the system is still operational, the output voltage decreases to half of the normal output due to the symmetrical structure of the topology.

A 400-W converter prototype is designed and fabricated to verify the effectiveness of the proposed converter. The system parameters are listed in Table 9.2. Figure 9.18 presents the photo of the experimental platform, which includes the power source, accessory power supply, converter prototype, digital controller, and resistive load bank. The proposed DSP+FPGA control platform is employed as digital controller.

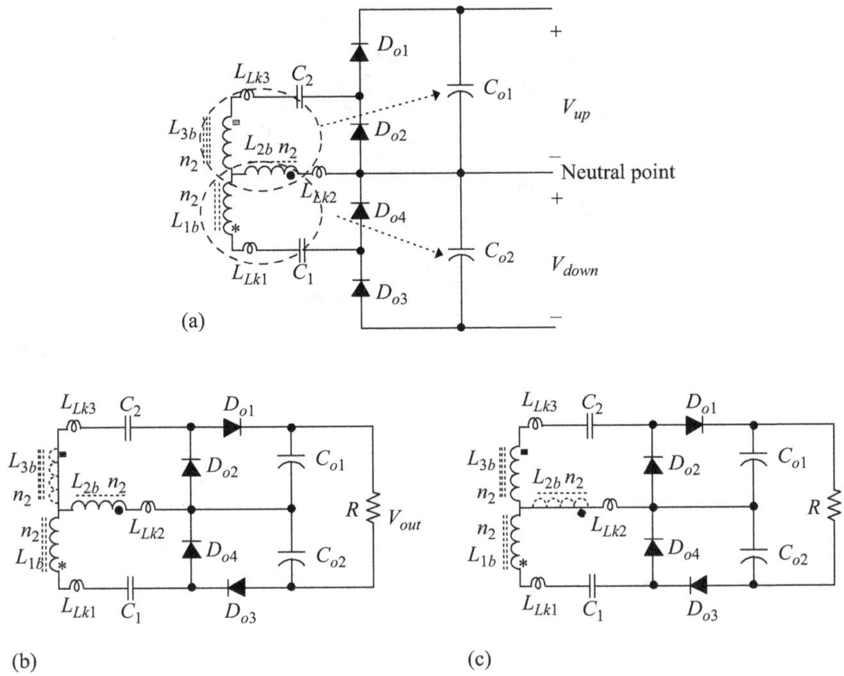

Figure 9.17 Fault deficient operation (a) charging C_{o1} and C_{o2}, (b) L_3 fault, and (c) L_2 fault. © 2014 IET. Reprinted with permission from Reference 20

Table 9.2 Specifications of proposed converter

Components	Parameters
Power level (P_{out})	400 W
Input voltage (V_{in})	10 ~ 20 V
Output voltage (V_{out})	200 V
Switching frequency (f_s)	150 kHz
Main switches (S_1, S_2, and S_3)	FDP047AN
Clamp switches (S_{c1}, S_{c2v}, and S_{c3})	FDP047AN
Diodes (D_{o1}, D_{o2}, D_{o3}, and D_{o4})	FEP30DP
Clamp capacitors (C_c)	4.7 μF
Switch capacitors (C_1 and C_2)	4.7 μF
Output capacitors (C_{o1} and C_{o2})	470 μF
Turns ratio ($N = n_2/n_1$)	56:28
Magnetic core of coupled inductor	Koolmu magnetic core 0077109A7

The main switch and clamp switch ZVS soft switching performance are presented in Figure 9.19(a) and (b). Due to the active clamp circuit, the ring problem and voltage spike are limited in main switching devices. Because of the star connection at the second side of coupled inductor, the i_{L2b} is shared by i_{L1b} and i_{L3b} to compose Flyback–Forward converter. Therefore, the primary side of the coupled

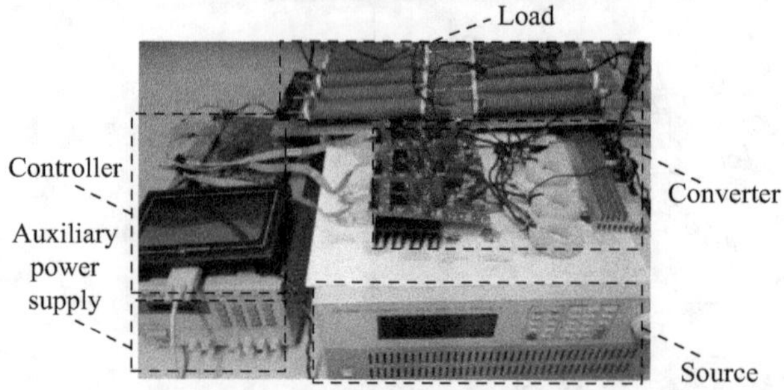

Figure 9.18 Photograph of the experimental setup. © 2014 IET. Reprinted with permission from Reference 20

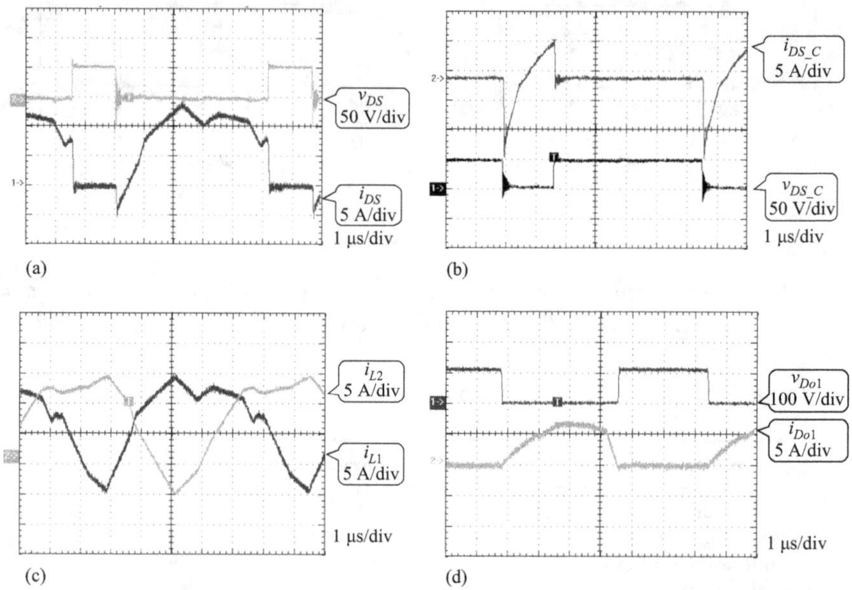

Figure 9.19 Experimental results (a) soft switching of the main switches, (b) soft switching of the clamp switches, (c) phase current, and (d) voltage and current of output diodes. © 2014 IET. Reprinted with permission from Reference 20

inductor L_2 conducts slightly different current than the current of L_1, as shown in Figure 9.19(c). Figure 9.19(d) is the waveform of the output rectifier diode voltage and current waveforms. The reverse-recovery current is limited to zero, the reverse-recovery is alleviated.

In order to verify the neutral-point control method, two resistive loads are connected in parallel with C_{o1} and C_{o2} individually. Figure 9.20 presents the experimental results with and without the neutral-point control. Without this method, the driving signals of $S_1 \sim S_3$ are generated with a single voltage controller

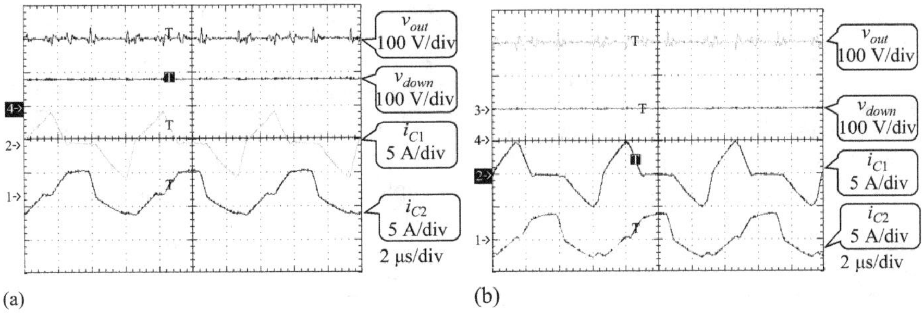

Figure 9.20 The neutral-point potential control under unbalanced load conditions (a) without the neutral-point control and (b) with the neutral-point control. © 2014 IET. Reprinted with permission from Reference 20

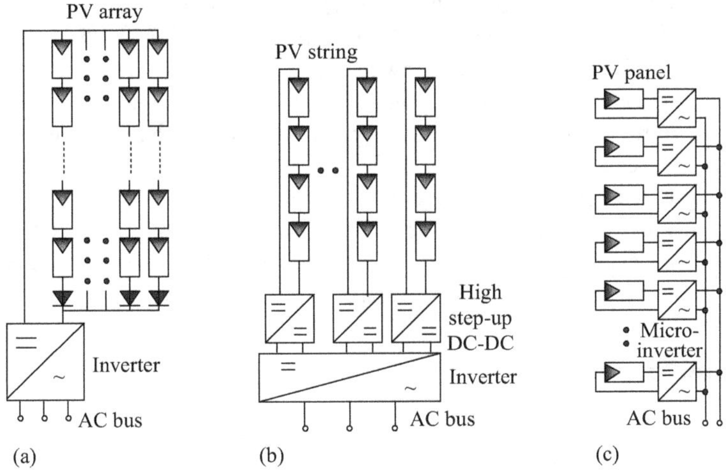

Figure 9.21 Three types of PV system configurations (a) the central type, (b) the string type and (c) the modularized type. © 2014 Elsevier. Reprinted with permission from Reference 24

by phase shifting 120° for each phase. Note that the inputs of the controller are the output voltage and its reference. The output voltage is 200 V and the distribution is unbalanced V_{down} and V_{up}. In Figure 9.20(a), $V_{up} = 115$ V and $V_{down} = 85$ V. The voltage difference is 30 V. By applying the neutral-point control, the distribution of V_{down} and V_{up} is balanced to 100 V, as shown in Figure 9.20(b).

9.4 Application of the new topologies for PV installations

In the literature, there are three types of grid-connected PV system configurations: the central type, string type, and modularized type [21]–[23], as depicted in Figure 9.21. The first systems connect PV modules in series and then in parallel so

Table 9.3 Performance comparison of grid-connected inverters

Item	Central type	String type	Modularized type
Power rating	10–250 kW	1.5–10 kW	50–500 W
Application	PV station	BIPV/PV roof	PV roof
MPPT capability	Low	High	High
Efficiency	High	Medium	Low
Reliability	Low	High	High
Cost	Low	Medium	High

as to increase their DC bus voltage and power level. They generally have high-voltage output and are suited for large-scale three-phase systems. The converters are characterized with high efficiency and low cost. However, they tend to suffer from the mismatch faults and shade effects. The second type requires a frond-end topology to achieve MPPT for each string. This structure provides high MPPT and high reliability. The cost is slightly higher than the central type. The third type of PV systems requires an individual power converter and individual MPPT for PV modules. Therefore, these systems typically have a low-voltage low-current output and require high capital investment. A performance comparison between three types is summarized in Table 9.3.

The high step-up DC-DC converters presented in this chapter are good candidates to be employed in the PV string configuration shown in Figure 9.21. In this application, the proposed FPGA and DSP-based multi-functional digital controller provides the broad range of PWM channels required by the power converters.

9.5 Conclusions

This chapter is composed of three main parts. There are three conclusions for them.

1. The DSP and FPGA are combined to provide more PWM channels; and the touch panel is employed to improve the user-friendliness of the controller.
2. In this chapter, the passive clamp circuit is proposed, which can be used to replace the traditional active clamp circuit. The passive clamp circuit not only can be used in isolated DC-DC converter, such as Flyback, Flyback–Forward, but also can be used in non-isolated DC-DC converter to decrease converter cost and improve reliability.
3. Based on the three-phase CIB, a high-power high-voltage-gain DC-DC converter is proposed. Comparing with prior studies, a significant voltage gain is achieved by the implementation of the three-phase CIB and switch capacitor voltage doubler configuration. The three-phase interleaved configuration significantly minimizes the input current ripple. Since low switch-voltage stress is achieved, the topology allows using low-voltage-rated power devices, which show low loss and cost-effectiveness. The switching loss is also minimized since the ZVS is achieved for all switches and the diode reverse-recovery problem is alleviated.

References

[1] http://www.epia.org/fileadmin/user_upload/Publications/EPIA_Global_Market_Outlook_for _Photovoltaics_2014-2018_-_Medium_Res.pdf
[2] Altera, "Cyclone III Device Handbook," www.altera.com, 2015.
[3] TI, "TMS320F28335/F28334/F28332/F28235/F28234/F28232 Digital Signal Controllers (Rev. M)," www.ti.com, 2015.
[4] TI, "SN74CBTD3384 Datasheet," www.ti.com, 2015.
[5] ADI, "ADM3053 Datasheet," www.adi.com, 2015.
[6] ADI, "ADM2582E/ADM2587E Datasheet," www.adi.com, 2015.
[7] TI, "SN74ALVC164245 Datasheet," www.ti.com, 2015.
[8] TI, "ADS7953 Datasheet," www.ti.com, 2015.
[9] ADI, "AD5624/AD5664 Datasheet," www.adi.com, 2015.
[10] K. Jin, M. Yang, X. Ruan, M. Xu, "Three-level bidirectional converter for fuel-cell/battery hybrid power system," *IEEE Trans. Ind. Electron.*, 2010, vol. 57, issue 6, pp. 1976–1986.
[11] F. Evran, M. T. Aydemir, "Z-source-based isolated high step-up converter," *IET Power Electron.*, 2013, vol. 6, issue 1, pp. 117–124.
[12] W. Li, X. He, "Zero-voltage transition interleaved high step-up converter with built-in transformer," *IET Power Electron.*, 2011, vol. 4, issue 5, pp. 523–531.
[13] R. J. Wai, R. Y. Duan, "High-efficiency DC/DC converter with high voltage gain," *IEE Proc. Electric Power Appl.*, 2005, vol. 152, issue 4, pp. 793–802.
[14] L. W. Zhou, B. X. Zhu, Q. M. Luo, "High step-up converter with capacity of multiple input," *IET Power Electron.*, 2012, vol. 5, issue 5, pp. 524–531.
[15] Y. P. Hsieh, J. F. Chen, T. J. Liang, L. S. Yang, "Analysis and implementation of a novel single-switch high step-up DC-DC converter," *IET Power Electron.*, 2012, vol. 5, issue 1, pp. 11–21.
[16] W. Li, L. Fan, Y. Zhao, X. He, D. Xu, B. Wu, "High-step-up and high-efficiency fuel-cell power-generation system with active-clamp Flyback–forward converter," *IEEE Trans. Ind. Electron.*, 2012, vol. 59, issue 1, pp. 599–610.
[17] W. Li, W. Li, X. Xiang, Y. Hu, X. He, "High step-up interleaved converter with built-in transformer voltage multiplier cells for sustainable energy applications," *IEEE Trans. Power Electron.*, 2014, vol. 29, issue 6, pp. 2829–2836.
[18] R. D. Middlebrook, S. Cuk, "A General Unified Approach to Modeling Switching-Converter Power Stages," proc. IEEE Power Electronics Specialists Conference, 1976, pp. 73–86.
[19] R. W. Erickson, D. Maksimovic, *Fundamentals of power electronics*, Springer, 2001.
[20] Y. Hu, W. Li, W. Xiao, X. He, "Three-phase interleaved high-step-up converter with coupled-inductor-based voltage quadrupler," *IET Power Electron.*, 2014, vol. 7, issue 7, pp. 1841–1849.

[21] H. Yi, F. Z. Peng, W. Jin, D. W. Yoo, "Survey of the power conditioning system for PV power generation," *The 37th IEEE Power Electronics Specialists Conference*, 2006, pp. 1–6.
[22] Q. Li, P. Wolfs, "A review of the single phase photovoltaic module integrated converter topologies with three different dc link configurations," *IEEE Trans. Power Electron.*, 2008, vol. 23, no. 3, pp. 1320–1333.
[23] M. K. Alam, F. Khan, A. M. Imtiaz, "Optimization of subcell interconnection for multijunction solar cells using switching power converters," *IEEE Trans. Sustainable Energy*, 2013, vol. 4, issue 2, pp. 340–349.
[24] Y. Hu, W. Cao, B. Ji, J. Si, X. Chen, "New multi-stage hysteresis control of DC-DC converters for grid-connected photovoltaic systems," *Renewable Energy*, 2015, vol. 74, pp. 247–254.

Chapter 10

Multilevel converters: topologies, modulation and control

Josep Pou[1], Salvador Ceballos[2] and Georgios Konstantinou[1]

10.1 Introduction

Multilevel converters are currently considered as a consolidated industrial solution to medium-voltage (MV), high-power applications. The main difference between multilevel converters and their two-level counterparts is their ability to generate voltage waveforms with three or more voltage levels. The voltage levels are produced by connecting in series several capacitors or isolated dc-sources. Figure 10.1 illustrates this concept through simplified circuit representation and the voltage waveforms generated by two-, three-, and five-level converters, respectively.

The main reasons for the growing use of multilevel converters in MV, high-power applications are, among others: (i) increased voltage and, consequently, increased power capacity achieved with standard rated semiconductors, (ii) high power quality of the voltage and current waveforms, (iii) high efficiency, (iv) low value of the voltage derivatives (dv/dt), and (v) reduced common-mode voltage. Several major manufacturers already commercialise multilevel converters in a wide range of power and voltage levels for different markets. The main markets include MV drives [2–4], regenerative applications [5–7], renewable energy sources such as wind power and solar photovoltaic plants [8–10], traction applications [11–13], high voltage direct current (HVDC) transmission systems [14–16], active power filtering, and reactive power compensation [17–19].

This chapter aims to introduce the basis of multilevel converters, including modulation and control techniques. It is especially devoted for those readers with no background in this field who want to gain a practical insight into the technology. The chapter is organised as follows. Section 10.2 describes the operation principles of the most common multilevel converter topologies. Section 10.3 focuses on the

[1]University of New South Wales, Sydney, NSW, Australia
[2]Tecnalia Research & Innovation, Parque Tecnológico de Bizkaia, Derio, Spain

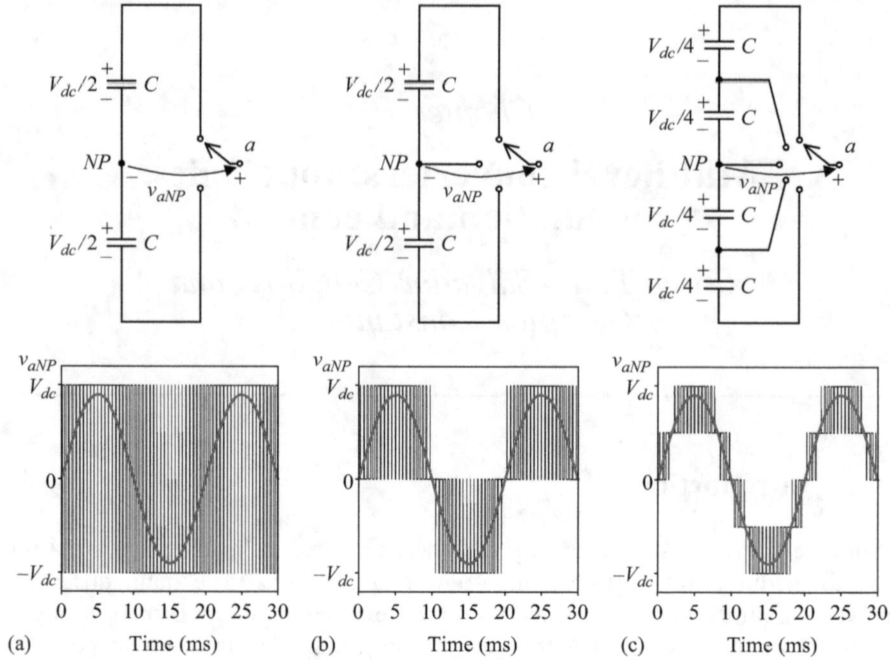

Figure 10.1 Simplified circuit representation and output voltages: (a) two-level; (b) three-level; and (c) five-level waveforms [1]

modulation strategies and includes some practical rules to facilitate implementation of some of them in a digital controller. This section also describes voltage balancing algorithms for neutral-point-clamped (NPC) and modular multilevel converters (MMCs). Finally, Section 10.4 introduces state-of-the-art digital controller implementations for multilevel converters.

10.2 Multilevel converter topologies

Several multilevel converter topologies have been proposed over the last four decades, but most of them are variations or combinations of the diode-clamped converter (DCC), flying capacitor (FC) and cascade H-bridge (CHB) multilevel converters, which are the three most common and well-established technologies. This section focuses on these three topologies. Additionally, a new technology, known as MMC, has emerged during the last decade. This topology is the state of the art in voltage-source converters (VSCs) for HVDC applications. A description of this converter is also included in this section due to its practical importance and growing research interest. An interested reader can find an overview of these and other multilevel converters in References 1 and 20–22.

10.2.1 Diode-clamped converter (DCC)

The three-level DCC, also known as NPC converter, was introduced for the first time in 1981 [23]. This converter, represented in Figure 10.2(a), can be seen as the series connection of two two-level converters [1], where the positive dc-bus rail of one of the converters is connected to the negative dc-bus rail of the other converter forming the neutral-point (NP) (labelled as *NP* in Figure 10.2(a)). The NP is connected to the legs of the converter through the clamping diodes (D_1 and D_2). These diodes are fundamental to the proper operation of the NPC converter since they participate in the generation of one of the output voltage levels and, in addition, they prevent the rest of semiconductors to be subjected to over-voltages without the use of snubbers. Assuming that the NP voltage is perfectly balanced, the clamping diodes will theoretically prevent the semiconductors from being exposed to voltages higher than $V_{dc}/2$. In practice, the voltage applied to the semiconductors will be slightly higher due to the NP voltage imbalance.

Table 10.1 shows the switching states of the semiconductors of an NPC converter and the corresponding output voltage. Using these switching states, the output

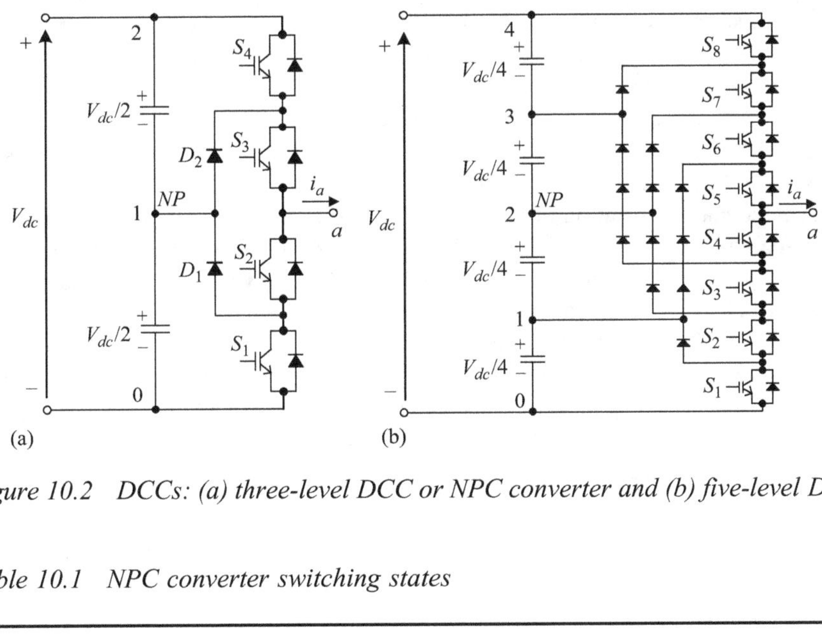

Figure 10.2 DCCs: (a) three-level DCC or NPC converter and (b) five-level DCC

Table 10.1 NPC converter switching states

S_4	S_3	S_2	S_1	Output voltage V_{a0}	Remarks
ON	ON	OFF	OFF	V_{dc}	The output phase is connected to the positive dc-bus rail through S_4 and S_3
OFF	ON	ON	OFF	$V_{c1} = V_{dc}/2$	The output phase is connected to the NP through D_2 and S_3 if i_a is positive and D_1 and S_2 of i_a is negative
OFF	OFF	ON	ON	0	The output voltage is connected to the negative dc-bus rail through S_1 and S_2

phase is connected to the positive, NP, or negative dc-bus rails (labelled as 2, 1, and 0, respectively, in Figure 10.2(a)) and, consequently, the three output voltage levels are univocally determined.

The DCC can be extended to higher voltage and power ratings by adding switches and clamping diodes to the basic topology, besides increasing the number of levels. Figure 10.2(b) displays a five-level DCC converter.

The advantages of DCCs compared to other multilevel topologies are that:

- They use a low number of capacitors. Although this topology requires some additional clamping diodes, the low number of reactive components is usually preferred from the standpoint of cost.
- They can be connected to a single dc-link voltage. The FC and MMC topologies also share this advantage, but the cascade converter does not, since that converter requires multiple isolated dc power supplies.

However, the DCC also suffers from some practical problems:

- As the number of levels increases, the clamping diodes are subject to high voltage stress equal to $V_{dc}(n-2)/(n-1)$, n being the number of levels. As a result, series connection of the diodes is required, as shown in Figure 10.2(b). This issue complicates the converter design and raises reliability and cost concerns.
- The inner switches (S_2 and S_3) and the series-connected clamping diodes might suffer from over-voltages due to the parasitic inductances. In References 24 and 25, two solutions to mitigate this problem are introduced.
- An unequal distribution of the conduction times between semiconductors takes place and causes an uneven distribution of the junction temperatures between switches. Consequently, some semiconductors have to be over-rated. The active NPC (ANPC) converter, introduced in Reference 26, helps to share the losses among the power switches.
- Maintaining stable capacitor voltages in converters with more than three levels is not possible at all operating points with conventional modulation techniques [27]. Advanced modulation strategies [28–30] can solve the problem but increase the switching losses and the high frequency harmonic distortion notably. A back-to-back connection of two DCCs also helps mitigate this problem [31].

All in all, the DCC topology is a good solution when the number of levels required is low (especially three levels) but it is unpractical for converters that require high numbers of levels.

10.2.2 Flying capacitor (FC) converter

The FC converter was introduced for the first time in 1992 [32, 33]. Figure 10.3(a) and (b) show the three- and four-level versions of the FC converter, respectively. In this converter, besides the dc-bus capacitor, there are also the so-called FCs, which are used to generate the intermediate voltage levels of the output voltage waveforms. These capacitors need to be pre-charged through external circuits to the

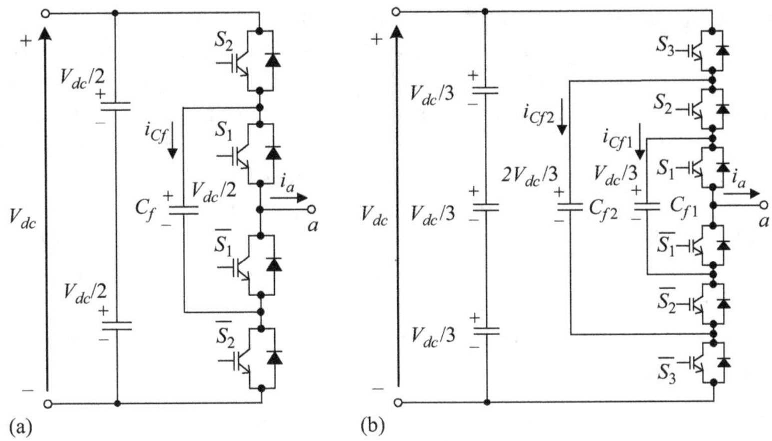

Figure 10.3 (a) Three-level and (b) four-level FC converters

Table 10.2 Switching states: (a) three-level FC converter and (b) four-level FC converter

(a)

S_2	S_1	V_{a0}	i_{Cf}
OFF	OFF	0	0
OFF	ON	$V_{dc}/2$	$-i_a$
ON	OFF	$V_{dc}/2$	i_a
ON	ON	V_{dc}	0

(b)

S_3	S_2	S_1	V_{a0}	i_{Cf2}	i_{Cf1}
OFF	OFF	OFF	0	0	0
OFF	OFF	ON	$V_{dc}/3$	0	$-i_a$
OFF	ON	OFF	$V_{dc}/3$	$-i_a$	i_a
ON	OFF	OFF	$V_{dc}/3$	i_a	0
OFF	ON	ON	$2V_{dc}/3$	$-i_a$	0
ON	OFF	ON	$2V_{dc}/3$	i_a	$-i_a$
ON	ON	OFF	$2V_{dc}/3$	0	i_a
ON	ON	ON	V_{dc}	0	0

voltage references for proper operation of the converter. For instance, the FC of the three-level converter in Figure 10.3(a) needs to be pre-charged with an initial voltage equal to $V_{dc}/2$. Similarly, capacitors C_{f2} and C_{f1} of the four-level converter shown in Figure 10.3(b) require to be pre-charged to $2V_{dc}/3$ and $V_{dc}/3$, respectively.

The different levels of the output voltage waveforms are generated by connecting a varying number of FCs in series between the dc-side and the output. This is illustrated in Table 10.2(a) and (b), where the switching states, the output voltage, and the current flowing through the FCs of three- and four-level FC converters are displayed. It can be appreciated that there are certain redundant switching states that generate the same output voltage level, but impose currents with opposite directions flowing through the FCs. This redundancy is enough to guarantee balanced voltages in the FCs for any operation conditions using a proper modulation strategy. For example, for the three-level converter, there are only two states that affect the voltage of the FC, and both of them provide the same output voltage ($V_{dc}/2$). However, the current through the FC flows in opposite directions. Hence, by choosing the suitable state according to the direction of the output current, the voltage of the FC can be controlled.

The main characteristics of the FC converters compared with other multilevel converters are that:

- They can be connected to a single dc-link voltage. As stated before, this advantage is shared with the DCC and the MMC, but not with the CHB converter.
- The voltages of the FCs can be regulated and maintained stable thanks to the degrees of freedom introduced by the redundancy of switching states.
- The structure of the FC converter is more modular than the DCC. Consequently, it is easier to increase the number of levels, besides increasing the voltage and power ratings of the converter. However, the number of FCs required increases drastically with the number of levels. Furthermore, the voltage that each FC has to withstand is different for each cell. Taking as an example the four-level converter represented in Figure 10.3(b), the FC C_{f2} has double voltage applied as compared with capacitor C_{f1}. Additionally, as the current level through all the FCs is the same, they should have the same capacitance to produce similar amplitudes in their voltage ripples. Therefore, assuming series and parallel connections based on the same elementary component, capacitor C_{f2} requires four times the number of components required by C_{f1}. Reference 34 proposes a hybrid multilevel topology to mitigate this problem.

Therefore, the FC converter is a good solution when the required power and voltage ratings, and consequently the number of levels, are higher than those achievable with the DCC. However, as with the DCC, this topology is not well suited for converters with high number of levels.

10.2.3 Cascaded H-bridge multilevel converter

The CHB converter was introduced for the first time for a single-phase plasma stabilisation application [35] and was later on extended to three-phase systems. It is formed by the series connection of several H-bridge sub-modules (SM). The basic structure of a three-phase, five-level CHB converter is shown in Figure 10.4.

The different levels of the output voltage waveform are generated inserting or bypassing the SM capacitors. This is illustrated in Table 10.3, where the switching states and the corresponding output voltages for a five-level CHB converter are displayed. It can be appreciated that, as in the case of the FC converter, there are

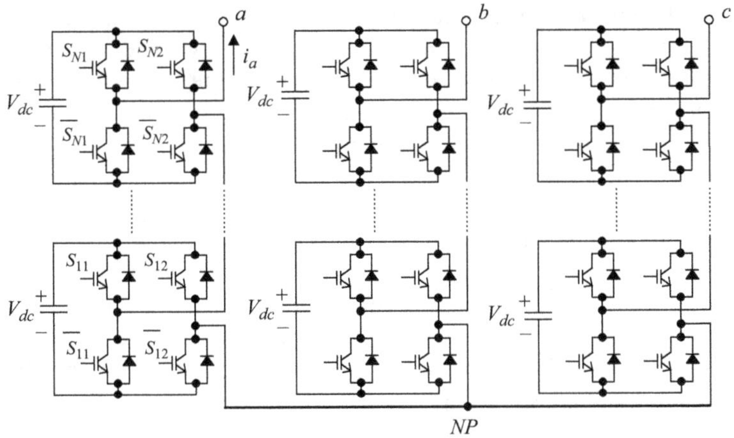

Figure 10.4 Three-phase, five-level CHB converter

Table 10.3 Switching states for a five-level CHB converter

S_{11}	S_{12}	S_{21}	S_{22}	V_{aNP}
OFF	OFF	OFF	OFF	0
OFF	OFF	OFF	ON	$-V_{dc}$
OFF	OFF	ON	OFF	V_{dc}
OFF	OFF	ON	ON	0
OFF	ON	OFF	OFF	$-V_{dc}$
OFF	ON	OFF	ON	$-2V_{dc}$
OFF	ON	ON	OFF	0
OFF	ON	ON	ON	$-V_{dc}$
ON	OFF	OFF	OFF	$+V_{dc}$
ON	OFF	OFF	ON	0
ON	OFF	ON	OFF	$2V_{dc}$
ON	OFF	ON	ON	V_{dc}
ON	ON	OFF	OFF	0
ON	ON	OFF	ON	$-V_{dc}$
ON	ON	ON	OFF	V_{dc}
ON	ON	ON	ON	0

also redundant switching states. These states introduce degrees of freedom that can be exploited by the modulation strategy to balance the power provided by each SM as well as to achieve equal switching frequency for all the devices [36, 37].

The main advantage of this topology is the high modularity, with the possibility of building converters with high power and voltage ratings. However, the fact that the SM dc-link voltages must be isolated is the major drawback for the application of these converters. In the case of processing active power, several independent dc power supplies (one per SM) are required, which can be provided either by a transformer with multiple isolated secondary windings or by several transformers. Both cases increase the size and cost of the converter. Nevertheless, this drawback is overcome when the converter is connected to fuel cells, batteries, or solar panels that already perform as isolated dc power supplies, or when it works in applications where no active power is processed such as active filtering applications and/or reactive power compensation.

Another important feature of CHB converters is that the dc-voltage of the series-connected SMs can be different. This asymmetry is used to increase the number of levels that a given CHB converter can provide. In this case, it might be interesting to use different semiconductor devices to make up each SM, so that the fastest ones, which may synthesise the SM fed by the lower voltage, will define the output voltage switching frequency [38, 39].

10.2.4 Modular multilevel converter

The MMC converter was developed in the early 2000s [40]. This converter can be considered as an evolution of the CHB converter, with the added feature of being able to process active power without the necessity of using isolated dc power supplies for the SMs. Figure 10.5 shows a circuit representation of the MMC. Each phase-leg is composed by two arms that are connected through two built-in inductances. These inductances are used to control the circulating current between the arms and to limit the fault current in the case of short-circuit in the dc-bus. Each arm is composed by N series-connected SMs. Several SM configurations have been

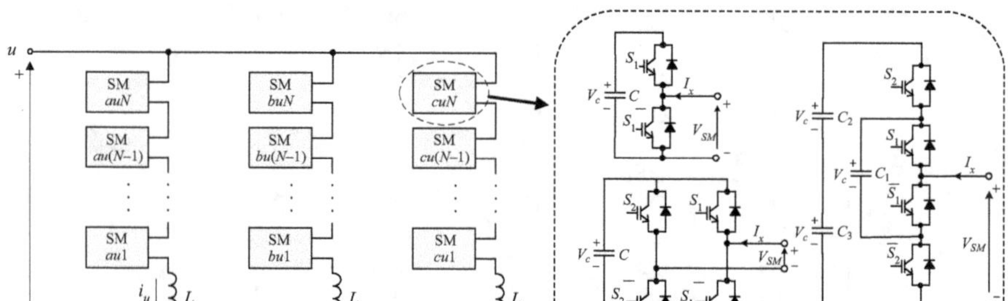

Figure 10.5 Circuit diagram of a three-phase MMC and various SM configurations

proposed in the technical literature [15, 41–45], the half-bridge topology being the most common one because of its simplicity and low conduction losses.

Each half-bridge is composed of two switches and a capacitor (see Figure 10.5). In addition, a thyristor and a bypass switch (not represented in Figure 10.5) are usually included at the output of each SM for protection purposes. The SM capacitors require to be initially pre-charged with a voltage equal to V_{dc}/N for proper operation of the converter. Once in operation, the voltage of each capacitor should be kept stable around the reference voltage level.

The two switches of the SMs are driven by complementary control signals. When the switch S_1 is in the ON state (and consequently $\overline{S}_1$ is OFF), the SM is considered activated and it provides a voltage V_c in its output. If, however, S_1 is OFF, the SM output voltage is 0 and the SM is bypassed. The different voltage levels of the arm voltage waveform are generated activating or bypassing the SMs. Figure 10.6 illustrates this concept for an MMC with three SMs per arm, generating a four-level waveform. In this example, the number of SMs connected to the phase-leg is always equal to N, however, as Section 3.4.1 suggests, control of the circulating current requires that the total number of SMs connected to the phase-leg may be different to N on an instantaneous basis.

Table 10.4 extends the concept illustrated in Figure 10.6 to converters with N series-connected SMs. It shows the number of activated SMs to generate a given arm voltage level. Table 10.4 also shows the number of redundant switching states available to generate a given voltage level. No redundant switching states are available to generate the 0 and V_{dc} voltage levels since all the SMs within the arm are bypassed or activated to generate these levels. For the rest of voltage levels the number of redundant states is calculated according to the expressions given in Table 10.4. For instance, the voltage level V_{dc}/N can be generated activating just one of the SMs. As there are N SMs within the arm, there are N possible options to generate this voltage level. Similar argumentation is applicable to the rest of voltage levels. The redundant switching states are used by the modulator to keep the voltages of the SMs balanced.

The main features of the MMC are (i) its modularity and scalability, making this topology suitable for very high voltage and power applications, (ii) the lack of a central dc-bus capacitor, (iii) the high quality of the output voltage and current waveforms, and (iv) the relatively high efficiency. These features make the MMC the preferred topology for VSC-based HVDC transmission systems [46–51]. Today state-of-the-art MMCs for HVDC applications have power and voltage ratings of 1 GW and ±320 kV, respectively. In addition, other applications such as flexible alternating current transmission systems (FACTS) [52], photovoltaic generation [53], wind turbine applications [54], motor drives [55–57], and traction [58] are also being explored nowadays as potential applications for the MMC.

10.3 Modulation techniques for multilevel converters

Extensive research has been done to extend and adapt classical pulse-width modulation (PWM) techniques to multilevel converters. Consequently, a large number of different modulation strategies for multilevel converters has been proposed,

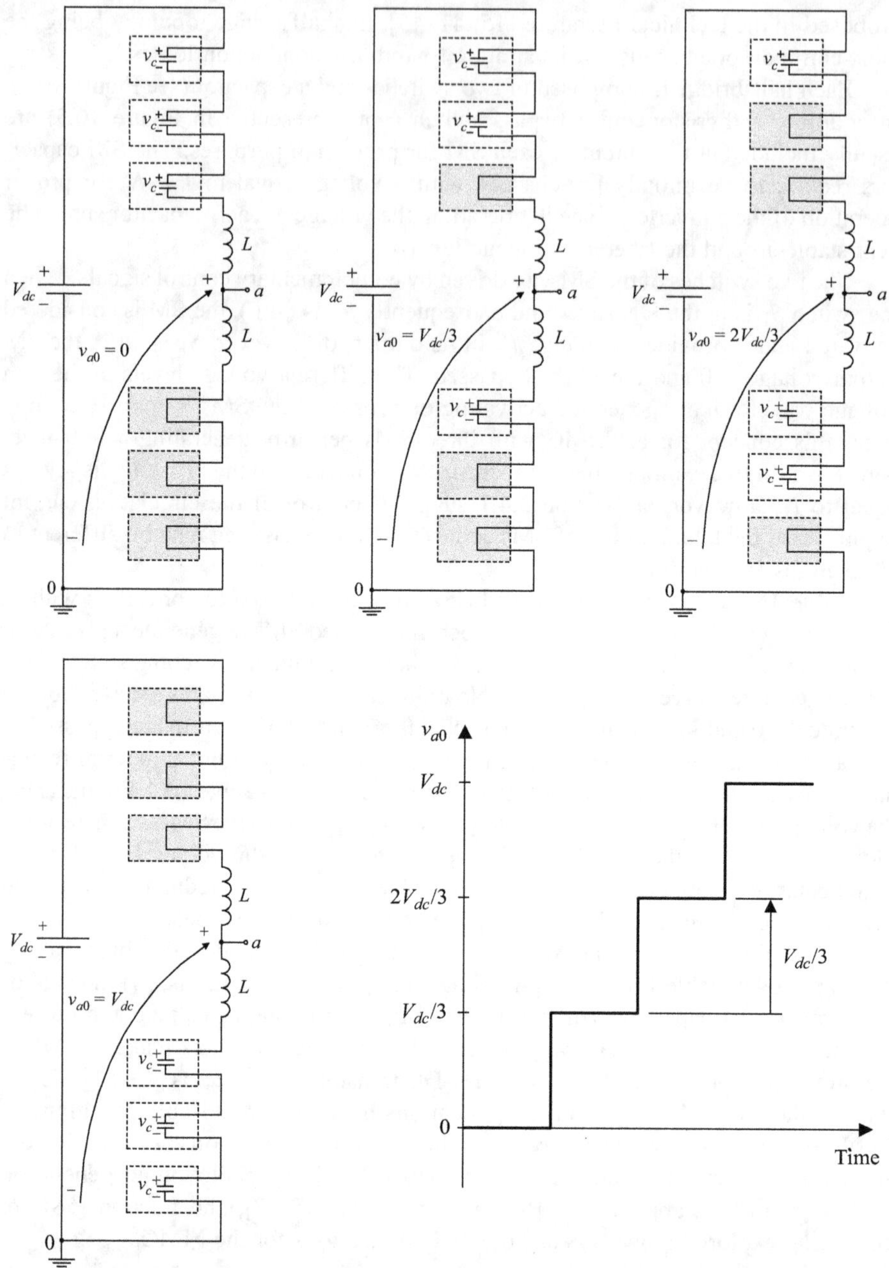

Figure 10.6 Derivation of the multilevel output voltage from a modular multilevel converter with three SMs per arm

Table 10.4 Number of inserted SMs, arm voltage levels, and switching states redundancies for an MMC converter with N SMs per arm

Number of SMs activated in a arm	Arm voltage v_{arm}	Number of redundant switching states
0	0	This switching state is unique. There is no possible redundancy
1	V_{dc}/N	N possible redundant switching states to generate this voltage level
2	$2V_{dc}/N$	$\frac{N(N-1)}{2}$ possible redundant switching states to generate this voltage level
...	...	...
α	$\alpha V_{dc}/N$	$\frac{N!}{\alpha!(N-\alpha)!}$ possible switching states to generate this voltage level
...	...	...
N	V_{dc}	This switching state is unique. There is no possible redundancy.

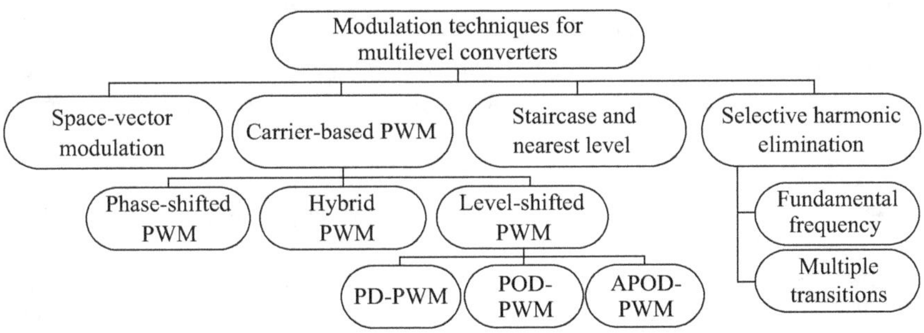

Figure 10.7 Classification of PWM for multilevel converters

each one with its own characteristics, advantages, and drawbacks. A classification of these techniques is made in Figure 10.7 [1, 21].

10.3.1 Low switching frequency modulation techniques

In low switching frequency modulation techniques the semiconductors only commutate a few times per cycle. Hence, the switching losses are low and the efficiency increases. However, the quality of the voltage waveform generated is relatively poor and can be improved. Staircase modulation [59, 60], nearest vector control [61, 62], and selective harmonic elimination (SHE) [63–67] are the best-known and well-established low switching frequency modulation strategies that are typically applied to multilevel converters. Generally, low frequency modulation techniques are used when the converter has a large number of levels.

Staircase modulation and nearest vector control approximate the normalised reference waveform ($v_{ref} \in [-1, 1]$) by rounding the output of the multilevel converter to the level or vector nearest to it with the output voltage levels given by

$$v_{a0} = \frac{V_{dc}}{N} \operatorname{round}\left(N \frac{1 + v_{ref}}{2} \right) \tag{10.1}$$

On the other hand, SHE-PWM is based on the decomposition of the multilevel voltage waveform into its Fourier coefficients and calculation of the required voltage level and switching instants in order to eliminate the low-order harmonics from the output voltage. The objective functions of the technique can be modified to target (i) minimisation of voltage THD, (ii) compliance of injected harmonics with grid codes [66] or (iii) reduction of current harmonic ripple. The complexities of SHE-PWM lie in finding solutions to the complex, non-linear system of equations that define the SHE-PWM problem over the complete range of operating points with sufficient solution continuity. Reference 67 provides a comprehensive review of SHE-PWM for multilevel converters and identifies various methods for mathematically defining the SHE-PWM problem.

10.3.2 High switching frequency modulation techniques

High switching frequency modulation techniques have also been developed for multilevel converters. In this case, the semiconductors commutate several times per cycle. Consequently, the generated switching losses are higher than in their low-frequency counterparts, but the quality of the output voltages improves. There are two complementary approaches to implement high switching frequency modulation techniques: space-vector modulation (SVM) and carried-based pulse-width modulation (CB-PWM).

The basics of SVM and CB-PWM are presented in detail in the following sections. The reader is referred to References 68–83 to further acquaint with the principles of SVM and CB-PWM strategies for multilevel converters, respectively. References 1 and 20–22 also offer interesting overviews on these modulation strategies.

10.3.2.1 SVM

This section introduces the principles of SVM for three-level converters. It also describes step-by-step a simple modulation algorithm to calculate the duty cycles that reduces the computational burden and facilitates the implementation of this technique in a digital controller. The reader is required to be acquainted with the space-vector representation and the Clarke transformation to be able to follow this section.

The space-vector diagram of a three-level converter in the $\alpha\beta$ plane is represented in Figure 10.8. This diagram is obtained applying the Clarke transformation to the voltages generated by all the possible switching states of a three-level converter. The vector *xyz* indicates the state of the three phases. For instance, the vector 210 indicates that the phases *a*, *b*, and *c* are connected to the positive, *NP*, and negative dc-bus rails, respectively (see Figure 10.2(a)). The vector diagram is

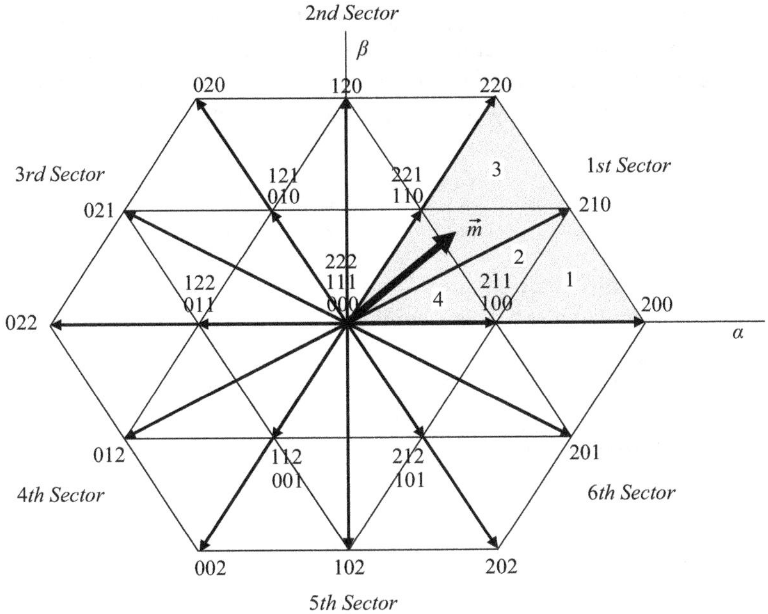

Figure 10.8 Space-vector diagram of a three-level converter

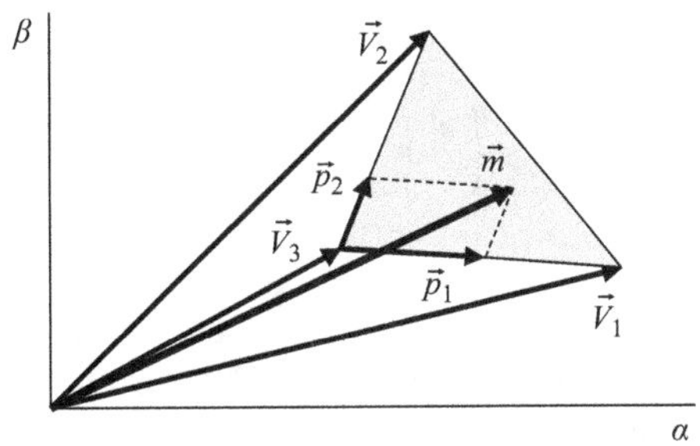

Figure 10.9 Limiting area to generate the reference vector $\vec{m}$ by using three vectors

divided into six sectors, each of which, in turn, is split into four regions. There are 27 vectors in the vector diagram, but some of them are redundant. Redundant vectors are those produced by different switching states, but that generate the same line-to-line voltages. For instance, vectors 221 and 110 shown in Figure 10.8 are redundant. These vectors introduce degrees of freedom that will be used to balance the NP voltage of an NPC converter.

The objective of SVM is to generate any reference vector $\vec{m}$ using a PWM-averaged combination of the switching vectors. The available switching vectors to generate a given reference vector $\vec{m}$ are those that contain it in the triangular area bounded connecting their tips. For instance, reference vector $\vec{m}$ in Figure 10.9 can be generated by the switching vectors $\vec{V}_1$, $\vec{V}_2$, and $\vec{V}_3$, since $\vec{m}$ lies on the triangular area bounded by these three vectors.

The average reference vector can be obtained by sequentially applying the switching vectors in a sampling period in accordance with

$$\frac{1}{T_s}\int_0^{T_s}\vec{m}dt = \frac{1}{T_s}\int_0^{T_1}\vec{V}_1 dt + \frac{1}{T_s}\int_{T_1}^{T_1+T_2}\vec{V}_2 dt + \frac{1}{T_s}\int_{T_1+T_2}^{T_s}\vec{V}_3 dt \tag{10.2}$$

where T_s is the sampling period, and T_1 and T_2 the time intervals that vectors $\vec{V}_1$ and $\vec{V}_2$ are applied within the sampling period. The following condition $T_1 + T_2 \leq T_s$ has to be fulfilled.

Assuming that $\vec{m}$ remains approximately constant during a sampling period, which is acceptable if T_s is much smaller than the fundamental period (T), then (10.2) can be approximated as

$$\vec{m} = d_1\vec{V}_1 + d_2\vec{V}_2 + d_3\vec{V}_3 \tag{10.3}$$

where $d_1 = \frac{T_1}{T_s}$, $d_2 = \frac{T_2}{T_s}$, and $d_3 = \frac{T-T_1-T_2}{T_s}$ are the duty cycles of the switching vectors, $\vec{V}_1$, $\vec{V}_2$, and $\vec{V}_3$, respectively. The duty cycles satisfy the following conditions:

$$d_1 + d_2 + d_3 = 1 \text{ and also } d_1 \geq 0,\ \ d_2 \geq 0 \text{ and } d_3 \geq 0. \tag{10.4}$$

Breaking down (10.3) into the $\alpha\beta$ components, the following expression is obtained:

$$\begin{bmatrix} d_1 \\ d_2 \\ d_3 \end{bmatrix} = \begin{bmatrix} V_{1\alpha} & V_{2\alpha} & V_{3\alpha} \\ V_{1\beta} & V_{2\beta} & V_{3\beta} \\ 1 & 1 & 1 \end{bmatrix}^{-1} \begin{bmatrix} m_\alpha \\ m_\beta \\ 1 \end{bmatrix} \tag{10.5}$$

Expression (10.5) provides the duty cycles of the switching vectors. It represents the traditional approach followed to solve the SVM problem. However, the calculation process requires inverting a matrix, which complicates the application of this method in a real-time processor system. Additionally, the equation system may need to be solved more than once per sampling period (T_s), since the region where the reference vector lies is previously unknown. A simplified way to calculate the duty cycles is introduced next.

Calculation of duty cycles by projections

The reference vector $\vec{m}$, shown in Figure 10.9, can be geometrically expressed as a sum of vectors:

$$\vec{m} = \vec{p}_1 + \vec{p}_2 + \vec{V}_3 \tag{10.6}$$

where the vectors $\vec{p}_1$ and $\vec{p}_2$ in Figure 10.9 are the projections of the reference vector $\vec{m}$ onto the segments that join the extreme of $\vec{V}_3$ to $\vec{V}_1$ and to $\vec{V}_2$, respectively.

Expression (10.6) can be manipulated and rewritten as

$$\vec{m} = \frac{|\vec{p}_1|}{|\vec{V}_1 - \vec{V}_3|}\vec{V}_1 + \frac{|\vec{p}_2|}{|\vec{V}_2 - \vec{V}_3|}\vec{V}_2 + \left(1 - \frac{|\vec{p}_1|}{|\vec{V}_1 - \vec{V}_3|} - \frac{|\vec{p}_2|}{|\vec{V}_2 - \vec{V}_3|}\right)\vec{V}_3 \quad (10.7)$$

Comparing (10.3) and (10.7), the duty cycles of the vectors can be directly deduced as follows:

$$d_1 = \frac{|\vec{p}_1|}{|\vec{V}_1 - \vec{V}_3|} \quad d_2 = \frac{|\vec{p}_2|}{|\vec{V}_2 - \vec{V}_3|} \quad d_3 = 1 - \frac{|\vec{p}_1|}{|\vec{V}_1 - \vec{V}_3|} - \frac{|\vec{p}_2|}{|\vec{V}_2 - \vec{V}_3|} \quad (10.8)$$

Expression (10.8) allows the calculation of the duty cycles fast and directly using the projections of the reference vector onto the segments that join the extreme of the switching vectors. If it were possible to calculate these projections with a low computational burden, the resulting method would be easily implementable in a digital processor. Reference 77 follows this approach and proposes a method to calculate the duty cycles efficiently. This method is described the next subsection.

Simplified calculation of duty cycles using gh components

Figure 10.8 shows the vector diagram of a three-level converter. In addition to the $\alpha\beta$ axes already shown in Figure 10.8, a new non-orthogonal reference frame is represented in Figure 10.10. This new reference frame is formed by the so-called *gh* axes, that have an angular difference of 60°. In accordance with (10.8), the projection of

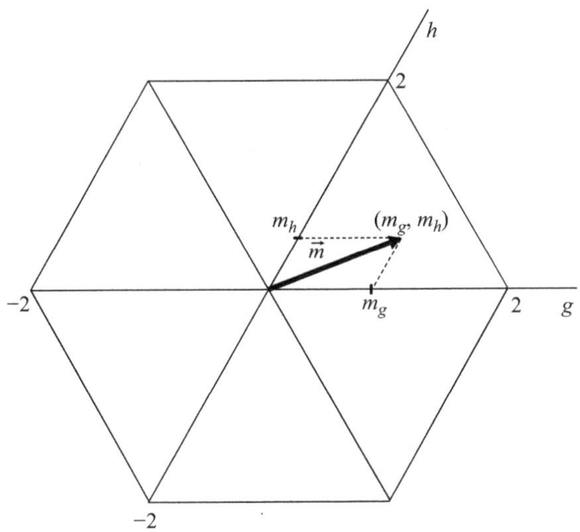

Figure 10.10 Space-vector diagram of a three-level converter. gh reference frame

the reference vector $\vec{m}$ over the gh axes (m_g and m_h) allows a simplified calculation of the duty cycles. The procedure to implement the SVM strategy in a digital controller using the gh axes has the following steps:

Step 1: Calculation of the m_g and m_h components of the reference vector
The projection of the reference vector $\vec{m}$ onto the gh reference frame can be calculated by means of

$$\begin{bmatrix} m_g \\ m_h \end{bmatrix} = \frac{2}{V_{dc}} \begin{bmatrix} v_a - v_b \\ v_b - v_c \end{bmatrix} \tag{10.9}$$

where v_a, v_b, and v_c are the line-to-neutral output voltages and V_{dc} is the dc-bus voltage. Expression (10.9) normalises the vector diagram so that the longitudes of segments that form the regions are equal to 1, i.e. the hexagon is two-unit sided (Figure 10.10). This is done to facilitate the calculation of the duty cycles as will be discussed later in Step 3.

Step 2: Determination of the sector where the reference vector lies
Once the gh components of the reference vector have been calculated it is necessary to determine what sector of the vector diagram contains the reference vector. This can be achieved following the rules of Table 10.5.

Step 3: Determination of the region and calculation of the vector duty cycles
Once the sector that contains the reference vector $\vec{m}$ has been determined, the particular vectors to be used in the modulation have to be found and their duty cycles calculated. To calculate the duty cycles, it is necessary to determine the region of the sector where the reference vector lies, and then apply the method based on projections using (10.8). Figure 10.11 shows the four regions that form the first sector. The figure also shows the projections of the reference vector over the segments that limit the regions. For instance, if the reference vector lies on Region 1 (Figure 10.11(a)), its projection over the segment that connects the redundant vectors 211-100 with the vector 210 will be m_h. Similarly, the projection of the reference vector over the segment that connects the redundant vectors 211-100 with the vector 200 is $m_g - 1$. In accordance with (10.8), these projections define the duty cycles of the switching vectors 210 and 200, respectively. Consequently, the duty cycle of the redundant vectors 211-100 is also

Table 10.5 Determination of the sector

gh components (m_g, m_h)	Sector
$m_g \geq 0$ $m_h \geq 0$	1st
$m_g < 0$ $m_h \geq 0$ $m_g + m_h \geq 0$	2nd
$m_g < 0$ $m_h \geq 0$ $m_g + m_h < 0$	3rd
$m_g < 0$ $m_h < 0$	4th
$m_g \geq 0$ $m_h < 0$ $m_g + m_h < 0$	5th
$m_g \geq 0$ $m_h < 0$ $m_g + m_h \geq 0$	6th

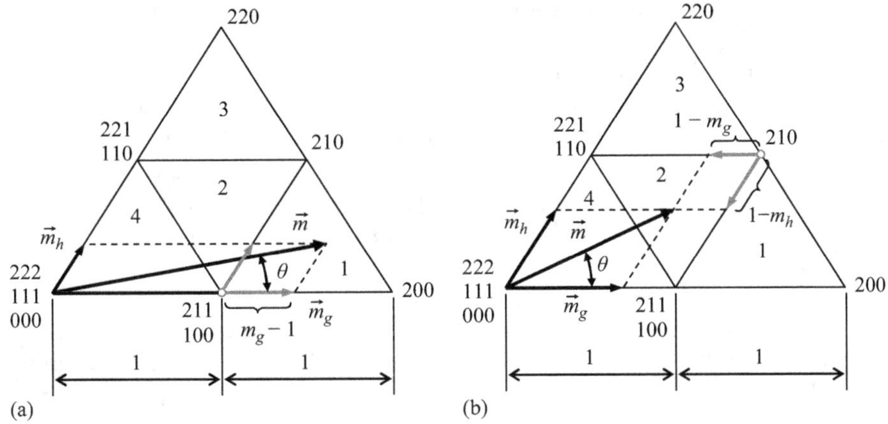

Figure 10.11 Projections for Regions (a) 1 and (b) 2

Table 10.6 Duty cycle calculation when the reference vector is located in the first sector

Case	Region	Duty cycles
$m_g > 1$	1	$\begin{cases} d_{200} = m_g - 1 \\ d_{210} = m_h \\ d_{100/211} = 2 - m_g - m_h \end{cases}$
$m_g \leq 1$ $m_h \leq 1$ $m_g + m_h > 1$	2	$\begin{cases} d_{100/211} = 1 - m_h \\ d_{110/211} = 1 - m_g \\ d_{210} = m_g + m_h - 1 \end{cases}$
$m_h > 1$	3	$\begin{cases} d_{210} = m_g \\ d_{220} = m_h - 1 \\ d_{110/221} = 2 - m_g - m_h \end{cases}$
$m_g \leq 1$ $m_h \leq 1$ $m_g + m_h \leq 1$	4	$\begin{cases} d_{100/211} = m_g \\ d_{110/221} = m_h \\ d_{111} = 1 - m_g - m_h \end{cases}$

univocally defined. Figure 10.11(b) shows the case where the reference vector is located in Region 2. A similar process can be applied to the other regions.

Table 10.6 summarises the information needed to ascertain the region where the reference vector lies and the expressions to calculate the duty cycles of the nearest switching vectors in the first sector. The expressions shown in Table 10.6 are based on the projections shown in Figure 10.11. A similar procedure can be followed for the remaining five sectors. The reader is referred to Reference 84 for a detailed description on how to proceed in the other sectors.

Step 4: Determination of the redundant vectors. NTV modulation for NPC converters

Table 10.6 allows the calculation of the duty cycles of the switching vectors. However, it does not provide a criterion on how to distribute the duty cycle between the redundant vectors. Different methodologies have been published in the technical literature to this end, each one trying to optimise different parameters such as for instance the switching losses [85] or the quality of the output waveforms [62, 70]. In this section we will introduce the switching known as nearest-three-vectors (NTV).

The NTV modulation technique uses only three of the closest vectors per modulation cycle. Thus, only a redundant vector will be selected from each pair. In the case of the NPC converter, the choice of the redundant vector is made according to the objective of maintaining balanced voltages in the dc-link capacitors; therefore, the voltage of the dc-link capacitors and the direction of the instantaneous output currents must be measured in each switching period.

To describe how the selection between the redundant vectors is made, let us suppose the following example where the particular vector between the redundant vectors 100-211 will be selected. Both vectors generate the same line-to-line output voltages, but each of them imposes an opposite current flowing through the NP (i_{NP}). This is illustrated in Figure 10.12. Vector 100 imposes an NP current equal to i_a ($i_{NP} = i_a$). Conversely, vector 211 imposes an NP current in the opposite direction, that is $i_{NP} = i_b + i_c = -i_a$ (supposed that $i_a + i_b + i_c = 0$). Consequently, the effect of each redundant vector over the dc-link capacitor voltages is the opposite. For example, assuming that $i_a > 0$ and $v_{C1} < v_{C2}$, the vector 211 should be selected. Conversely, if $v_{C1} > v_{C2}$, vector 100 should be used instead.

Table 10.7 generalises this example and shows the criterion that should be followed to select the redundant vectors in the first sector. In Table 10.7, 1 means true and 0 means false. Similar criterion can be found for the remaining redundant vectors of the other sectors.

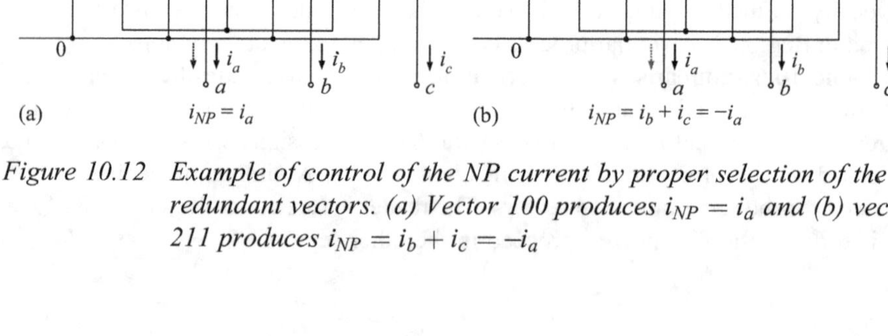

Figure 10.12 Example of control of the NP current by proper selection of the redundant vectors. (a) Vector 100 produces $i_{NP} = i_a$ and (b) vector 211 produces $i_{NP} = i_b + i_c = -i_a$

Table 10.7 Criteria for selecting between the vectors 100 and 211, and the vectors 110 and 221 ("1" and "0" mean true and false, respectively)

Selection between vectors 100 and 211		
$v_{C1} > v_{C2}$	$i_a > 0$	**Vector**
0	0	100
0	1	211
1	0	211
1	1	100
Selection between vectors 110 and 221		
$v_{C1} > v_{C2}$	$i_c > 0$	**Vector**
0	0	221
0	1	110
1	0	110
1	1	221

Step 5: Defining sequences of switching vectors
Once the duty cycles have been calculated and the redundant vectors selected, it only remains to sort out the switching vectors and apply them in a correct sequence to reduce the number of commutations.

Table 10.7 shows the sequences of vectors in the first sector that should be applied to minimise the switching frequencies of the power devices. These sequences depend on the redundant vectors that are selected according to voltage-balance requirements. The number of changes or switching steps between consecutive vectors associated with each sequence is also indicated in this table. The worst cases are Region 2 (vectors 100-221) and Region 4 (vectors 100-221), both of which require four switching steps, which is twice the number required by the other sequences. When a sequence is repeated in a subsequent switching period, the sequence is flipped in order to minimise the number of commutation steps from one cycle to the next one.

Following these steps, which consist of programming (10.9) and the Tables 10.5–10.8, the modulation strategy can be implemented easily in a digital controller. Because of the simplicity of the calculation process, the computational burden to execute the whole modulation algorithm is relatively low. It can be concluded that calculating the duty cycles with the *gh* components represents a powerful tool for the implementation of the SVM in a real-time controller.

10.3.2.2 Carrier-based PWM

Carrier-based PWM (CB-PWM) techniques are based on comparing a reference signal with one or more triangular or saw-tooth carriers. If the reference signal is sinusoidal it is referred to as sinusoidal PWM (SPWM). In such a modulation technique the width of the pulses generated at the output of the converter change according to the sinusoidal reference signal.

Table 10.8 Sequences of vectors in the first sector for the NTV modulation

Region	Redundant vectors	Vector sequence		Commutation steps
		Switching period 1	Switching period 2	
1	100	100-200-210	210-200-100	2
	211	200-210-211	211-210-200	2
2	100-110	100-110-210	210-110-100	2
	100-221	100-210-221	221-210-100	4
	211-110	110-210-211	211-210-110	2
	211-221	210-211-221	221-211-210	2
3	110	110-210-220	220-210-110	2
	221	210-220-221	221-220-210	2
4	100-110	100-110-111	111-110-100	2
	100-221	100-111-221	221-111-100	4
	211-110	110-111-211	211-111-110	2
	211-221	111-211-221	221-211-111	2

In a CB-PWM implementation, the switching frequency of the power devices is essentially dependent on the carrier frequency. If the carrier frequency increases, the switching frequency of the power devices increases too, and so do the switching power losses. On the other hand, the larger the carrier frequency, the better the quality of the output voltages. Therefore, there is a trade-off between output quality and power losses when choosing the carrier frequency. In an P-phase inverter, P evenly phase-shifted reference signals are compared to the carrier signal to obtain the control signals for the switches. In the case of a three-phase inverter, the harmonics multiple of three (triplen) that appear in the phase voltages disappear in the line-to-line voltages. This is because these harmonics are of zero-sequence, i.e. they are common to the three phases. A specific zero-sequence can be added on purpose to the modulation signals to achieve larger amplitudes of the output voltage fundamentals under linear operation modes, and to lower the actual switching frequency in the power devices [72, 86].

In order to avoid over-modulation, the amplitude of the reference signals should never exceed the amplitude of the carrier signal. Over-modulation is a non-linear operational mode of the inverter where the amplitudes of the fundamental output voltages are no longer proportional to the amplitude of the reference signal. It also produces low-frequency harmonic distortion.

In multilevel converters, more than one carrier signals are usually needed. In general terms, in the case of an n-level converter, $n-1$ carriers with the same amplitude and frequency are needed. Several dispositions of the carriers are possible and, depending on the type of disposition, multicarrier PWM can be categorised into phase-shifted PWM (PS-PWM) and level-shifted PWM (LS-PWM).

PS-PWM

In PS-PWM, the $n-1$ carriers of an n-level converter are phase-shifted among themselves with an angle of $360°/n$. Figure 10.13(a) shows the disposition of these carriers and the reference signal for a five-level converter. Each carrier is usually associated to a pair of switches, although they may rotate with the objective of sharing

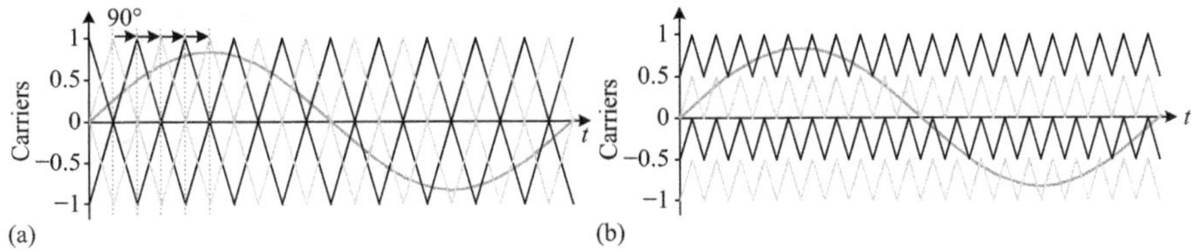

Figure 10.13 Carrier-based PWM for multilevel converters. (a) PS-PWM and (b) PD-PWM

losses in the power devices. PS-PWM is usually applied to modular topologies such as the H-bridge cascade converter and the MMC, but it is also applied to the FC converter.

LS-PWM

In LS-PWM, the carriers are arranged in level-shifted bands. Three main alternatives for LS-PWM have been developed in the technical literature [68, 87, 88]. Among them, the most widely used is phase-disposition PWM (PD-PWM), since the output voltages generated with this modulation have better spectra [89, 90]. In PD-PWM, all the carrier signals are in phase, as shown in Figure 10.13(b) for a five-level converter. A sinusoidal reference signal (v_{ref}) is compared with the four triangular carriers to define the voltage level that has to be generated at the output.

10.3.3 MMC: circulating current control and capacitor voltage balance

In Section 10.3.2.1, an example of SVM applied to the NPC converter was shown. In this section, PD-PWM is applied to the MMC. Also some details for the control of this converter topology are explained in this section.

A particular feature of the MMC is that it produces a circulating current within each phase-leg that does not appear in the output current. Although the circulating currents are naturally established, they can be reshaped by a proper circulating current control to improve the performance of the MMC.

In order to analyse the circulating currents of the MMC, each phase-leg of the converter is divided into two circuits; the common and the differential circuits [91, 92].

10.3.4 Common and differential circuits

Figure 10.14(a) shows a general circuit representation of an MMC phase-leg, while Figure 10.14(b) represents a phase-leg showing only the activated SMs in the arms. Since the output voltage of the non-activated SMs is zero, those SMs do not insert their capacitors in the arms, and this is why they are not included in Figure 10.14(b).

The common and differential voltages applied to the extremes of the inductors are

$$v_{\text{comm}} = \frac{v_u + v_l}{2} \tag{10.10}$$

and

$$v_{\text{diff}} = \frac{v_u - v_l}{2} \tag{10.11}$$

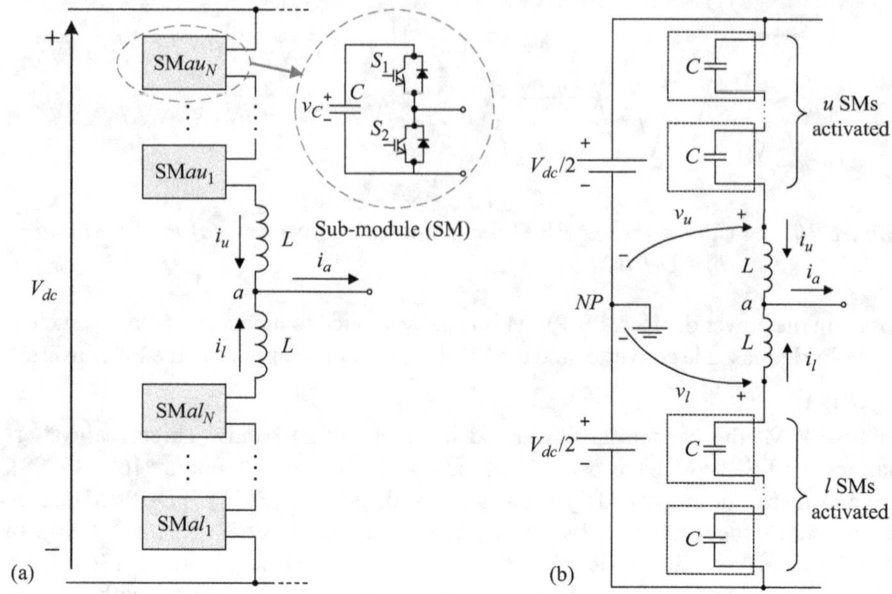

Figure 10.14 MMC phase-leg. (a) Circuit diagram and (b) circuit with activated SMs

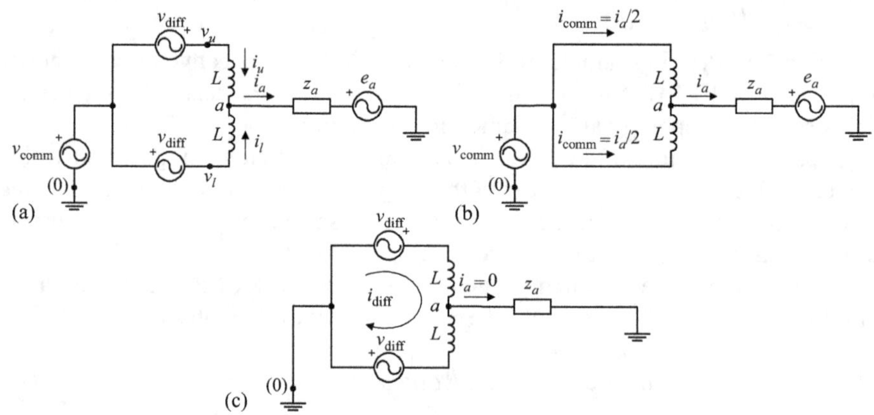

Figure 10.15 MMC phase-leg: (a) equivalent circuit, (b) common, and (c) differential mode circuits

Assuming that the phase-leg is connected to a grid voltage e_a through an impedance z_a and applying the principle of superposition, common and differential mode circuits can be obtained, as shown in Figure 10.15.

The common and differential arm currents are

$$i_{\text{comm}} = \frac{i_u + i_l}{2} = \frac{i_a}{2} \tag{10.12}$$

and

$$i_{\text{diff}} = \frac{i_u - i_l}{2} \tag{10.13}$$

From (10.12) and (10.13) the arm currents can be deduced as follows:

$$i_u = i_{\text{comm}} + i_{\text{diff}} = \frac{i_a}{2} + i_{\text{diff}} \tag{10.14}$$

$$i_l = i_{\text{comm}} - i_{\text{diff}} = \frac{i_a}{2} - i_{\text{diff}} \tag{10.15}$$

The circuits in Figure 10.15(b) and (c) can be analysed independently of each other. The differential voltage can be used to control the differential current within the phase-leg, without producing any distortion to the output current.

According to Figure 10.14(c), the differential current will be

$$i_{\text{diff}} = \frac{1}{L}\int_0^t v_{\text{diff}} dt + I_{\text{diff}_0} \tag{10.16}$$

where I_{diff_0} is the initial value of the differential current.

The differential current is indeed the circulating current. It contains a dc component (I_{dc}) that is essential to keep the arms energised, i.e. to maintain the capacitor voltages around their reference value. On the other hand, some ac current components ($i_{\text{diff}_{ac}}$) exist in the circulating current. When the circulating current is controlled, the ac components can be defined to meet certain objectives, such as minimizing the voltage ripples in the SM capacitors or the rms value of the arm currents to improve the MMC efficiency. Therefore, generally speaking the differential current will be composed by the following two terms:

$$i_{\text{diff}} = I_{dc} + i_{\text{diff}_{ac}} \tag{10.17}$$

10.3.4.1 Circulating current control

The circulating current of a phase-leg can be controlled through the differential voltage v_{diff} without affecting the output voltage and current of that phase-leg. Different references can be provided to the circulating current. If only the dc component is imposed ($i_{\text{diff}} = I_{dc}$), the rms value of the circulating current is minimised and therefore the power losses of the MMC are reduced. On the other hand, introducing some harmonics into the reference of the circulating current, especially a second harmonic, the SM capacitor voltage ripples at the fundamental frequency can be reduced. This enables for decreasing the capacitance values, which is an interesting target considering the large amount of capacitors included in the topology. A typical reference provided to the circulating current, which is based on instantaneous information [92], is

$$i_{\text{diff}} = \frac{v_{\text{ref}} i_a}{2} \tag{10.18}$$

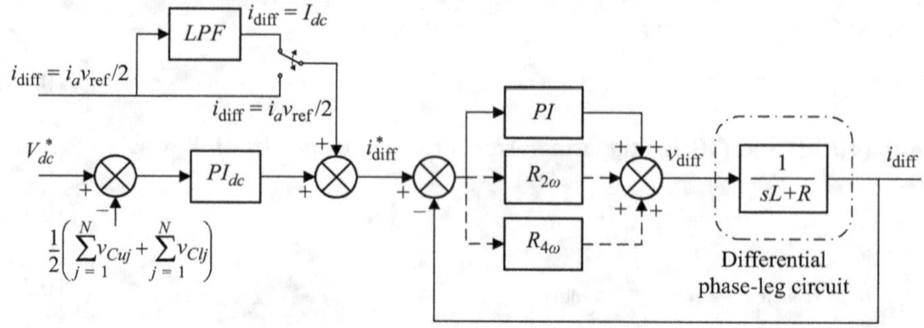

Figure 10.16 Circulating current control diagram

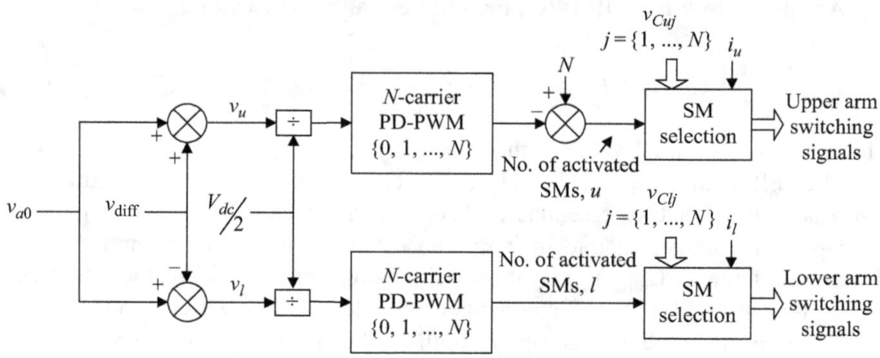

Figure 10.17 Implementation diagram to generate control signals for the SMs in a phase-leg

where v_{ref} is the normalised voltage reference signal that ranges in the interval $v_{ref} \in [-1, 1]$ under linear operation mode, and i_a the output current.

Figure 10.16 shows a possible implementation of circulating current control where the circulating current reference can be selected between the reference given by (10.18) and a dc component only (I_{dc}). The dc component is obtained by filtering the reference signal (10.18) using a low-pass filter (LPF), such as a moving average filter (MAF) tuned at twice the output frequency. This provides a draft estimation of the dc component of the circulating current.

The circulating current reference is complemented with a control loop that regulates the sum of the capacitor voltages of the phase-leg to twice the dc-link voltage. This control loop is based on a proportional-integral regulator (PI_{dc}). The circulating current reference is imposed by a proportional-integral controller (PI in Figure 10.16). Resonant controllers ($R_{2\omega}$ and $R_{4\omega}$) tuned at the frequencies of the main harmonics to be imposed/removed in the circulating current may also be included in the control loop.

Figure 10.17 shows an implementation diagram to inject the voltage v_{diff} for circulating current control. One can observe that this control signal is added to and subtracted from the general reference signal of the phase-leg (v_{a0}) for the upper and lower arms, respectively. The number of SMs that needs to be activated in the upper and lower arm is defined by the variables u and l, respectively. The particular SMs that need to be activated/deactivated in each arm are defined according to the objective of achieving capacitor voltage balance.

10.3.4.2 SM capacitor voltage balance

The circulating current provides the energy needed to maintain the SM capacitors charged. However, this energy should flow into the capacitors of each arm evenly in order to achieve capacitor voltage balance. This is achieved by selecting the SMs to be activated and deactivated according to their capacitor voltages and the direction of the arm current [16, 95]. The general rule is

If the arm current is in the charging direction:

- and the PWM method requires the addition of one SM in the arm, the SM with the lowest voltage that is not connected to the arm will be selected and added to the arm.
- and the PWM method requires the removal of one SM in the arm, the SM with the highest voltage that is connected to the arm will be selected and removed from the arm.

The way to proceed is the opposite if the arm current is in the discharging direction. Some variations of this strategy have been proposed in order to minimise the number of SMs that switch [96, 97] and therefore the power losses of the MMC.

An implementation algorithm to perform capacitor voltage balance in the upper arm is shown in Figure 10.18 [97]. The diagram will be similar for the lower arm. The algorithm uses two cascaded index sorting structures and a simple comparator for the selection of the SMs and generation of the arm switching states. As a first step, the capacitor voltages are multiplied with the sign of the arm current under the convention that a positive sign denotes a current within the arm that charges the SM capacitors. The modified SM capacitor voltages (v'_{Cuj}) are

$$v'_{Cuj} = -\text{sgn}(i_a)v_{Cuj} \tag{10.19}$$

The first sorting stage uses the modified SM capacitor voltage measurements (v'_{Cuj}) and sorts the SM indices (j) in a descending manner, providing a sorted list of

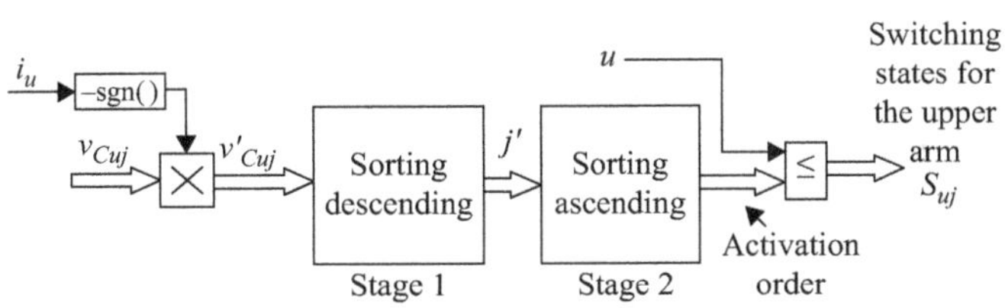

Figure 10.18 Implementation of a capacitor voltage balancing algorithm

indices based on SM voltages (j) from the highest to the lowest input values as derived from (10.19). The second sorting stage uses the output of the first (j) as input, based on ascending order. The output of the second stage indicates the activation order of each SM and the elements of the list directly relate to a particular SM with the first element indicating the activation order of SM_{au_1} and the Nth element indicating the activation order of SM_{au_N}. The SMs that need to be connected are those with an activation order below or equal to the number of required SMs (u) as defined by the modulation stage. A simple logical comparison of the activation order with u provides the SM switching states for the arm. A simple modification in this algorithm to achieve minimum switching transitions is also presented in Reference 97.

10.3.4.3 Capacitor voltage ripples

Figure 10.19 shows output voltage and arm currents (Figure 10.19(a)), and the capacitor voltages of the SMs of the upper arm (Figure 10.19(b)) when injecting only a dc circulating current. Figure 10.20(a) presents the ratio of capacitor voltage ripple amplitudes between applying (10.18) as a circulating current reference and only a dc component. One can observe the significant reduction achieved for large modulation indices when using (10.18). Figure 10.20(b) shows the ratio of rms arm currents under such conditions.

The frequency of the capacitor voltage ripples is the fundamental output one. This limits the application of the MMC to variable-speed motor drives because large capacitor voltage ripples appear at low speed/frequency. Nevertheless, the low-frequency capacitor voltage ripple can be eliminated by injecting a higher frequency component into the circulating currents in combination with a zero-sequence of the same frequency in the voltage reference signals [55, 93]. Discontinuous modulation is another way of reducing capacitor voltage ripples [94]. These techniques allow the MMC to be applied to variable-speed motor drives.

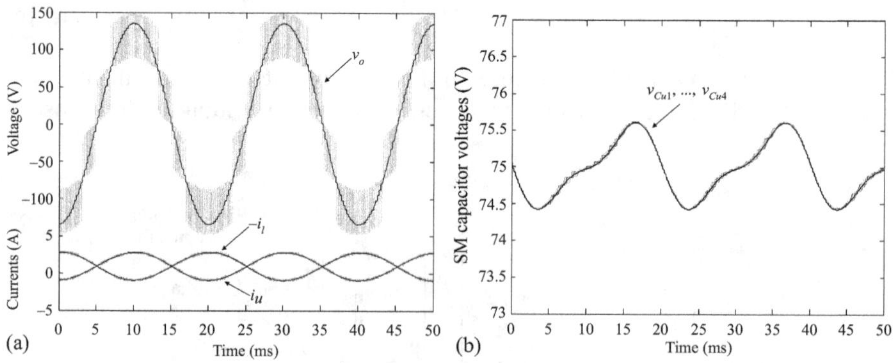

Figure 10.19 (a) ac-side voltage and arm currents and (b) SM capacitor voltages of the upper arm

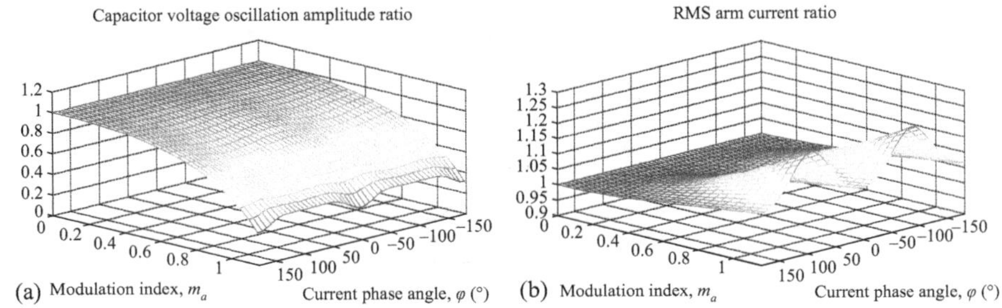

Figure 10.20 Ratios when controlling the circulating current with and without a second order harmonic: (a) capacitor voltage ripple amplitude and (b) rms arm currents

10.4 Digital controller implementations for multilevel converters

This section introduces a description of well-established digital hardware architectures to control multilevel converters. Depending on the converter complexity, different hardware implementations are used.

10.4.1 Centralised digital controllers for converters with a low number of levels

If the converter has a low number of levels, the digital controller can be centralised in a single unit. The main elements included in this unit are displayed in Figure 10.21. It usually includes a digital signal processor (DSP) and a field programmable gate array (FPGA). An additional microprocessor carrying out the tasks related with monitoring and communicating with external devices is occasionally used.

The control and the modulation algorithms are split between the DSP and the FPGA. The FPGA carries out those tasks that require high time accuracy, such as the generation of the PWM patterns to activate the semiconductors or the management of high priority error conditions, while the DSP executes analytical tasks such as processing the control and modulation algorithms. Figure 10.22 illustrates this concept. It displays the sequence of actions that takes place in each modulation period and shows the actions that are carried out by the DSP and the FPGA. Typically, in each sampling period, the FPGA enables the analog-to-digital converters (ADCs) and acquires the measurements needed to control the system. Once the measurements are properly sampled, the FPGA raises an interrupt condition that is attended by the DSP. The DSP takes the measurements and performs the control subroutines to regulate the dc-bus voltage and/or the active and reactive power if the converter is grid connected, or the speed and the electrical torque if the converter drives a motor. Other controllers for specific applications, such as maximum power point tracking (MPPT) algorithms in solar photovoltaic systems, motor position controllers, are also implemented in the DSP. The internal regulators used to control all these magnitudes generate a set of voltage references that are sent to the modulation block. The modulation block takes

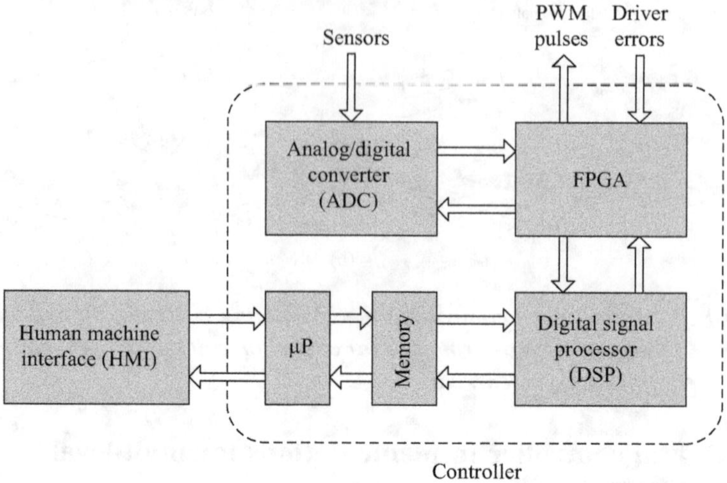

Figure 10.21 Centralised controller architecture

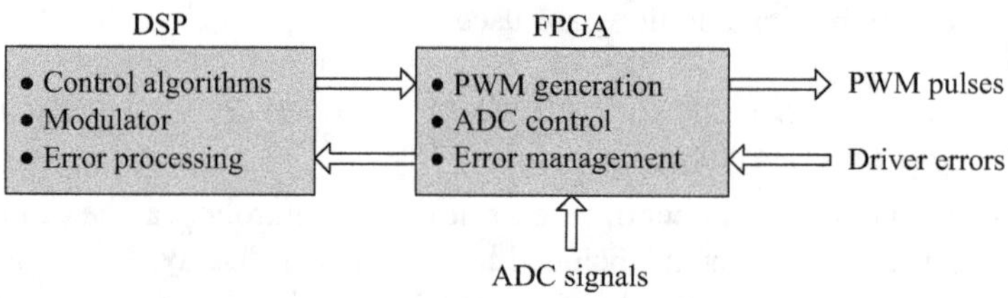

Figure 10.22 Tasks typically performed by the DSP and FPGA

these voltage references and processes them to finally come up with a set of ac reference voltages (if PWM modulation techniques are used) or duty cycles (if SVM techniques are used) that are allocated in the RAM memory blocks of the FPGA.

Based on this information, the FPGA generates the PWM pattern signals to activate the semiconductors, keeping the required switching sequences and dead times to avoid short-circuits in the commutation processes. These control signals are usually isolated and sent to the drivers. In addition, the FPGA also manages high priority error conditions to shut down the system in a fast and secure mode in case it was necessary.

This digital control architecture concept, with slight modifications depending on the specific application, is widely used in commercial products. However, the density of transistors that can be integrated nowadays in a single chip enables the use of more compact solutions based on system on chip (SoC) technologies [98–100]. SoCs based on FPGA include several complex blocks or cores into a single integrated circuit, thus allowing the integration of the complete digital controller into a single chip. Consequently, the resulting digital controller is simpler and more compact, thus allowing a reduction on the cost. In addition, the use of an

FPGA facilitates the parallel execution of tasks, hence increasing the execution speed. This technology, despite not being widely used yet, is suitable to control power converters and, due to the above-mentioned characteristics, is attracting increased interest to the power converter industry.

10.4.2 Distributed digital controllers for converters with large number of levels

When the converter has a modular structure with a large number of connected SMs, as in the case of the CHB converter and the MMC, the use of a digital controller with a centralised architecture might not be an optimum solution due to, among others, the following reasons:

- Centralised control systems have low expansion capability and no modularity. Consequently, it is not straightforward to update a controller already designed to be able to cope with higher numbers of SMs.
- The number of control wires/optic fibre lines running between the SMs and the centralised controller is large. In addition, since the central controller is usually located far from the SMs to avoid dealing with a high voltage environment, the length of the control lines tends to be large. These two factors decrease reliability and increase unnecessarily the cost of the control system.

Consequently, for converters with a high number of SMs, using a distributed digital control architecture with a master-slave configuration becomes a more interesting solution. Distributed control architectures are modular, flexible, and easily expandable. Therefore, they suit very well the requirements to control CHB converters and MMCs with a large number of SMs.

Several distributed control systems, relaying in different network communication topologies, have been proposed in the technical literature for CHB and MMCs [101–110]. Among them, control architectures based on ring or point-to-point star network communication topologies have concentrated most of the attention. These two configurations are described in the following paragraphs.

Figure 10.23(a) shows a representation of the distributed control architecture with a ring communication network proposed and analyzed in [101–104]. It consists of a master board as a central controller and as many slave boards as local controllers as SMs the converter has. The slaves are connected to the master board using a ring configuration. For the communication between the master and the slaves, open source real-time Ethernet protocols such as EtherCAT or PROFINET IRT are suggested. Optic fibre cables are preferred over copper wires to provide the galvanic isolation between the SMs. To increase the reliability of the communication architecture, avoiding a shut-down of the converter when one SM or an optic fibre fails, a double ring structure is usually implemented [101–103].

As shown in Figure 10.23(a), the master board includes a DSP and an FPGA, while the slave boards contain a ADC to monitor the voltage of the SM and a local FPGA (sometimes also a local DSP is included). In distributed control architectures, the master board performs the central control algorithms. The distribution

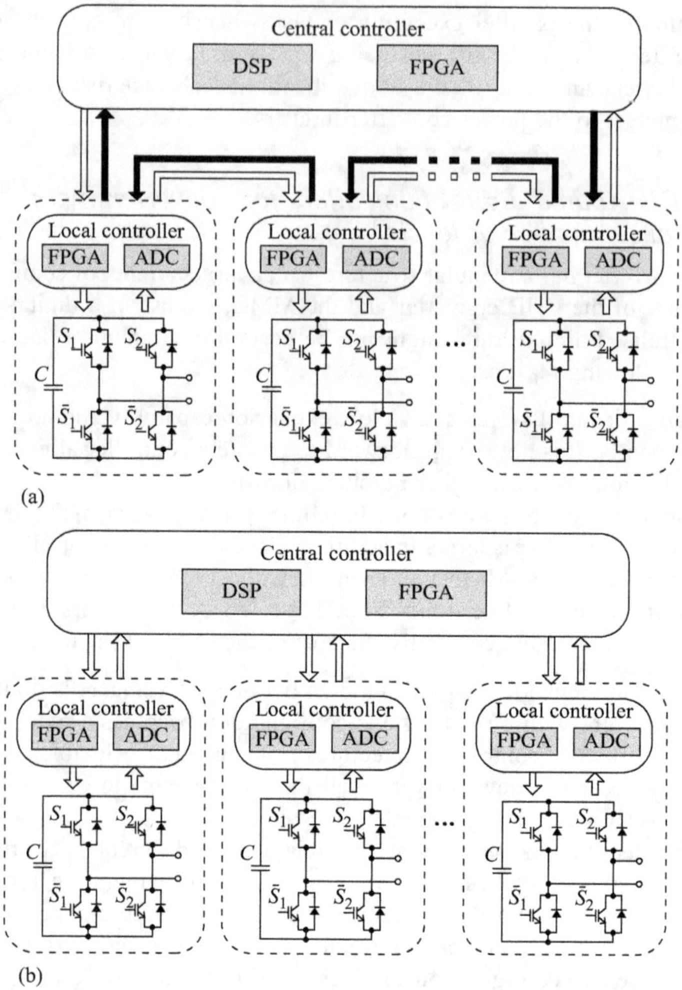

Figure 10.23 Distributed control architectures: (a) double ring topology and (b) star topology

of tasks between the FPGA and the DSP depends on the type of modulation strategy to be implemented. However, in the most general case, the DSP will carry out the control of the dc-bus voltage and output currents of the converter, while the FPGA will perform the SM voltage balancing tasks and will calculate the duty cycle of each SM. The duty cycles are then sent through the communication network to each specific SM. The FPGA of the SM will receive this information and generate the PWM signals to activate the semiconductors. Since the PWM generation is done locally in each SM, no high bandwidth is required for the communication network. The protection and failure management of each specific SM are also carried out in the local FPGA. The SM state will be reported to the master board and it will carry out the corresponding actions at a converter level if one or more SMs fail.

Figure 10.23(b) shows a representation of the distributed control architecture with a point-to-point star network communication topology introduced in References 105–108. As in the ring topology, it consists on a master board and as many slave boards as SMs the converter has. The task distribution between the master and slave boards to control the converter is the same that for the ring topology. In the star configuration, the slave boards associated to each SM are connected with a central hub or directly to the master board. This fact might increase the reliability of the star configuration with regard to the ring topology. However, it usually implies the use of longer cables. Nonetheless, this is not always the case, since it depends a lot on the specific implementation of the star configuration. Different implementations have been suggested in the technical literature. Among them, implementations based on controlled area network (CAN), serial communications, and optical passive networks have been proposed in References 105, 106, and 108, respectively.

Some other network topologies, such as daisy-chain networks using system packet interface (SPI) or EtherCAT, or hybrid solutions using a parallel combination of CAN and a high speed bus have also been proposed. The reader is referred to References 104, 109, and 110 to find information on these topologies.

10.5 Conclusions

This chapter has introduced the main multilevel converter topologies, i.e. the DCC, the FC, and the CHB converters, as well as the MMC, which is a recent topology that is expected to be widely applied in HVDC transmission systems. After classifying the modulation techniques, the basics of SVM and CB-PWM for multilevel converters have been presented. In the case of SVM, a modulation algorithm prone to be implemented in a digital processor has been applied to the three-level DCC (or NPC converter) and described in detail. A capacitor voltage balancing strategy based on selecting proper redundant vectors has also been embedded in the modulation. In the case of implementing a CB-PWM, several carrier signals are required for multilevel converters. Different carrier dispositions have been shown, including PS-PWM and LS-PWM. CB-PWM is especially interesting for modular converter topologies, such as the CHB converter and the MMC.

The circulating currents within the phase-legs is a particular feature of the MMC. These currents do not appear in the output current. Although the circulating currents are naturally established, a circulating current control has been presented, which improves the performance of the MMC regarding capacitor voltage ripples and power losses.

Two different digital control architectures for multilevel converters have been presented in this chapter: (i) centralised and (ii) distributed digital controls. The centralised control is interesting for converters with a low number of levels, while the distributed control should be applied to converters with a large number of levels, such as the modular topologies (CHB and MMC). In the centralised architecture, the control and modulation algorithms are implemented in a central unit that may include DSPs and FPGAs. On the other hand, the distributed control architectures are usually based on a master-slave configuration, where each SM board includes an FPGA to generate the PWM and an ADC to regulate the SM capacitor voltage. Several distributed control architectures have been proposed in the technical literature. Among them, the double ring topology and the star topology, are the most extended ones. Both configurations are modular, flexible, reliable, and easily expandable, suiting well the control requirements of converters with many SMs.

References

[1] J. Rodríguez, L. G. Franquelo, S. Kouro, *et al.*, "Multilevel converters: an enabling technology for high-power applications," *Proc. IEEE*, vol. 97, no. 11, pp. 1786–1817, Nov. 2009.

[2] R. Menz and F. Opprecht, "Replacement of a wound rotor motor with an adjustable speed drive for a 1400 kW kiln exhaust gas fan," in *Proc. 44th IEEE IAS Cement Ind. Tech. Conf.*, pp. 85–93, May 2002.

[3] W. C. Rossmann and R. G. Ellis, "Retrofit of 22 pipeline pumping stations with 3000-hp motors and variable-frequency drives," *IEEE Trans. Ind. Appl.*, vol. 34, pp. 178–186, Jan. 1998.

[4] P. Schmitt and R. Sommer, "Retrofit of fixed speed induction motors with medium voltage drive converters using NPC three-level inverter high voltage IGBT based topology," in *Proc. IEEE Int. Symp. Ind. Electron.*, vol. 2, pp. 746–751, Jun. 2001.

[5] J. Rodríguez, J. Pontt, G. Alzamora, N. Becker, O. Einenkel, and A. Weinstein, "Novel 20 MW downhill conveyor system using three-level converters," *IEEE Trans. Ind. Electron.*, vol. 49, no. 5, pp. 1093–1100, Oct. 2002.

[6] J. Rodríguez, J. Dixon, J. Espinoza, and P. Lezana, "PWM regenerative rectifiers: state of the art," *IEEE Trans. Ind. Electron.*, vol. 52, no. 1, pp. 5–22, Jan. 2005.

[7] F. Hernández, L. Morán, J. Espinoza, and J. W. Dixon, "A multilevel active front end rectifier with current harmonic compensation capability," in *Proc. IEEE Ind. Electron. Conf.*, vol. 2, pp. 1446–1451, Nov. 2004.

[8] J. M. Carrasco, L. G. Franquelo, J. T. Bialasiewicz, *et al.*, "Power-electronic systems for the grid integration of renewable energy sources: a survey," *IEEE Trans. Ind. Electron.*, vol. 53, no. 4, pp. 1002–1016, Jun. 2006.

[9] M. Calais, V. G. Agelidis, L. J. Borle, and M. S. Dymond, "Transformerless five level cascaded based single phase photovoltaic system," in *Proc. IEEE 31st Annu. Power Electron. Spec. Conf.*, vol. 3, pp. 1173–1178, Jun. 2000.

[10] S. Alepuz, S. Busquets-Monge, J. Bordonau, J. Gago, D. González, and J. Balcells, "Interfacing renewable energy sources to the utility grid using a three-level inverter," *IEEE Trans. Ind. Electron.*, vol. 53, no. 5, pp. 1504–1511, Oct. 2006.

[11] L. A. Tolbert, F. Z. Peng, T. Cunnyngham, and J. N. Chiasson, "Charge balance control schemes for cascade multilevel converter in hybrid electric vehicles," *IEEE Trans. Ind. Electron.*, vol. 49, no. 5, pp. 1058–1064, Oct. 2002.
[12] J. Dixon, M. Ortuzar, and L. Morán, "Drive system for traction applications using 81 level converter," in *Proc. IEEE Veh. Power Prop.*, Oct 2004.
[13] J. Dixon and L. Morán, "A clean four quadrant sinusoidal power rectifier using multistage converters for subway applications," *IEEE Trans. Ind. Electron.*, vol. 52, no. 3, pp. 653–661, Jun. 2005.
[14] H. J. Knaak, "Multilevel converters and HVDC/FACTS: a success story," in *Proc. EPE*, Aug.–Sep. 2011.
[15] H. Akagi, "Classification, terminology, and application of the modular multilevel cascade converter (MMCC)," *IEEE Trans. Power Electron.*, vol. 26, no. 11, pp. 3119–3130, Nov. 2011.
[16] M. Saeedifard and R. Iravani, "Dynamic performance of a modular multilevel back-to-back HVDC system," *IEEE Trans. Power Del.*, vol. 25, no. 4, pp. 2903–2912, Oct. 2010.
[17] J. Dixon, L. Morán, J. Rodríguez, and R. Domke, "Reactive power compensation technologies: state of the art review," *Proc. IEEE*, vol. 93, no. 12, pp. 2144–2164, Dec. 2005.
[18] P. Flores, J. Dixon, M. Ortuzar, R. Carmi, P. Barriuso, and L. Morán, "Static var compensator and active power filter with power injection capability, using 27-level inverters and photovoltaic cells," *IEEE Trans. Ind. Electron.*, vol. 56, no. 1, pp. 130–138, Jan. 2009.
[19] M. Ortuzar, R. Carmi, J. Dixon, and L. Morán, "Voltage source active power filter, based on multilevel converter and ultracapacitor DC link," *IEEE Trans. Ind. Electron.*, vol. 53, no. 2, pp. 477–485, Apr. 2006.
[20] L. G. Franquelo, J. Rodríguez, J. I. León, S. Kouro, R. Portillo, and M. A. Prats, "The age of multilevel converters arrives," *IEEE Ind. Electron. Magazine*, vol. 2, no. 2, pp. 28–39, Jun. 2008.
[21] J. Rodríguez, J.-S. Lai, and F. Z. Peng, "Multilevel inverters: a survey of topologies, controls and applications," *IEEE Trans. Ind. Electron.*, vol. 49, no. 4, pp. 724–738, Aug. 2002.
[22] S. Kouro, M. Malinowski, K. Gopakumar, *et al.*, "Recent advances and industrial applications of multilevel converters," *IEEE Trans. Ind. Electron.*, vol. 57, no. 8, pp. 2553–2580, Aug. 2010.
[23] A. Nabae, I. Takahashi, and H. Akagi, "A new Neutral-Point-Clamped PWM inverter," *IEEE Trans. Ind. Appl.*, vol. IA-17, no. 5, pp. 518–523, Sep.–Oct. 1981.
[24] X. Yuan and I. Barbi, "Fundamentals of a new diode clamping multilevel inverter," *IEEE Trans. Power Electron.*, vol. 15, no. 4, pp. 711–718, Jul. 2000.
[25] B.-S. Suh and D.-S. Hyun, "A new N-level high voltage inversion system," *IEEE Trans. Ind. Electron.*, vol. 44, no. 1, pp. 107–115, Feb. 1997.
[26] T. Brckner, S. Bernet, and H. Gldner, "The active NPC converter and its loss-balancing control," *IEEE Trans. Ind. Electron.*, vol. 52, no. 3, pp. 855–867, Jun. 2005.

[27] M. Marchesoni and P. Tenca, "Theoretical and practical limits in multilevel MPC inverters with passive front ends," in *Proc. EPE01*, Aug. 2001.

[28] S. Busquets-Monge, S. Alepuz, J. Bordonau, and J. Peracaula, "Voltage balancing control of diode-clamped multilevel converters with passive front-ends," *IEEE Trans. Power Electron.*, vol. 23, no. 4, pp. 1751–1758, Jul. 2008.

[29] S. Busquets-Monge, J. Bordonau, and J. Rocabert, "A virtual-vector pulse-width modulation for the four-level diode-clamped DC-AC converter," *IEEE Trans. Power. Electron.* vol. 23, no. 4, pp. 1964–1972, Jul. 2008.

[30] S. Busquets-Monge, R. Maheshwari, J. Nicolas-Apruzzese, E. Lupon, S. Munk-Nielsen, and J. Bordonau, "Enhanced DC-link capacitor voltage balancing control of DC-AC multilevel multileg converters," *IEEE Trans. Ind. Electron.*, vol. 62, no. 5, pp. 2663–2672, May 2015.

[31] Z. Pan, F. Z. Peng, K. A. Corzine, V. R. Stefanovic, J. M. Leuthen, and S. Gataric, "Voltage balancing control of diode-clamped multilevel rectifier/inverter systems," *IEEE Trans. Ind. Appl.*, vol. 41, no. 6, pp. 1698–1706, Nov.–Dec. 2005.

[32] T. A. Meynard and H. Foch, "Multi-level choppers for high voltage application," *EPE Journal*, vol. 2, pp. 45–50, Mar. 1992.

[33] T. A. Meynard and H. Foch, "Multi-level conversion: high voltage choppers and voltage-source inverters," *Proc. IEEE-PESC92*, vol. 1, pp. 397–403, Jun.–Jul. 1992.

[34] P. Barbosa, P. Steimer, J. Steinke, L. Meysenc, M. Winkelnkemper, and N. Celanovic, "Active neutral-point-clamped multilevel converters," *Proc. Power Electron. Specialist Conf. PESC05*, pp. 2296–2301, Jun. 2005.

[35] M. Marchesoni, M. Mazzucchelli, and S. Tenconi, "A nonconventional power converter for plasma stabilization," *IEEE Trans. Power Electron.*, vol. 5, no. 2, pp. 212–219, Apr. 1990.

[36] L. Tolbert and F. Z. Peng, "Multilevel converters as a utility interface for renewable energy systems," *IEEE Power Eng. Soc. Summer Meeting*, vol. 2, pp. 1271–1274, Jul. 2000.

[37] N. P. Schibli, T. Nguyen, and A. C. Rufer, "A three-phase multilevel converter for high-power induction motors," *IEEE Trans. Power Electron.*, vol. 13, no. 5, pp. 978–986, Sep. 1998.

[38] M. D. Manjrekar, P. K. Steimer, and T. A. Lipo, "Hybrid multilevel power conversion system: a competitive solution for high-power applications," *IEEE Trans. Indus. Appl.*, vol. 36, no. 3, pp. 834–841, May/Jun. 2000.

[39] Y.-F. Liu, Z.-M. Zhao, C. Qian, and X. Zhou, "A novel three-phase multilevel voltage source converter," in *Proc. Power Electronics and Motion Control Conf. IPEMC00*, vol. 3, pp. 1172–1175, Aug. 2000.

[40] A. Leniscar and R. Marquardt, "An innovative modular multilevel converter topology suitable for a wide power range," in *Proc. Power Tech Conf.*, Jun. 2013.

[41] G. Konstantinou, J. Zhang, S. Ceballos, J. Pou, and V. G. Agelidis, "Comparison and evaluation of sub-module configurations in modular multilevel converters," in *Proc. of IEEE PEDS,* 2015.

[42] A. Nami, J. Liang, F. Dijkhuizen, and G. D. Demetriades, "Modular multilevel converters for HVDC applications: review on converter cells and functionalities," *IEEE Trans. Power Electron.*, vol. 31, no. 1, pp. 1–16, Jan. 2015.

[43] E. Solas, G. Abad, J. Barrena, S. Aurtenetxea, A. Carcar, and L. Zajac, "Modular multilevel converter with different submodule concepts – PII: experimental validation & comparison for HVDC applications," *IEEE Trans. Ind. Electron.*, vol. 60, no. 10, pp. 4536–4545, Oct. 2013.

[44] A. K. Sahoo, R. Otero-De-Leon, V. Chandrasekaran, and N. Mohan, "New 3-level submodules for a modular multilevel converter based HVDC system with advanced features," in *Proc. IECON 2013*, pp. 6269–6274, Nov. 2013.

[45] X. Li, W. Liu, Q. Song, H. Rao, and S. Xu, "An enhanced MMC topology with DC fault ride-through capability," in *Proc. IECON 2013*, pp. 6182–6188, Nov. 2013.

[46] B. Gemmell, J. Dorn, D. Retzmann, and D. Soerangr, "Prospects of multi-level VSC technologies for power transmission," in *Proc. IEEE/PES T&D Conf. Expo.*, pp 1–16, Apr. 2008.

[47] SIEMENS, HVDC PLUS Basics and principle of operation. [Online]. Available: http://www.energy.siemens.com/br/pool/br/transmissao-de-energia/transformadores/hvdc-plus-basics-and-principle-of-operation.pdf

[48] K. Friedrich, "Modern HVDC PLUS application of VSC in modular multi-level converter topology," in *Proc. IEEE Int. Symp. Ind. Electron (ISIE 2010)*, pp. 3807–3810, Jul. 2010.

[49] V. Hussennether, J. Rittiger, A. Barth, *et al.*, "Projects BorWin2 and HelWin1 large scale multilevel voltage-sourced converter technology for bundling of offshore wind power," in *Proc. CIGRE,* 2012.

[50] J. Dorn, H. Gambach, J. Strauss, T. WesterWeller, and J. Alligan, " 'Trans bay cable' a breakthrough of VSC multilevel converters in HVDC Transmission," in *Proc. CIGRE*, 2012.

[51] T. WesterWeller, K. Friedrich, U. Armonies, A. Orini, D. Parquet, and S. When, " 'Trans bay cable' world's first HVDC system using multilevel voltage-sourced converter," in *Proc. CIGRE,* 2010.

[52] H. Mohammadi and M. Tavakoli Bina, "A transformerless medium voltage STATCOM topology based on extended modular multilevel converters," *IEEE Trans. Power Electron.*, vol. 26, no. 5, pp. 1534–1545, May 2011.

[53] J. Mei, B. Xiao, K. Shen, L. M. Tolbert, and J. Y. Zheng, "Modular multi-level inverter with new modulation method and its application to photo-voltaic grid-connected generator," *IEEE Trans. Power Electron.*, vol. 28, no. 11, pp. 5063–5073, Nov. 2013.

[54] S. Debnath and M. Saeedifard, "A new hybrid modular multilevel converter for grid connection of large wind turbines," *IEEE Trans. Sustain. Energy*, vol. 4, no. 4, pp. 1051–1064, Oct. 2013.

[55] A. J. Korn, M. Winkelnkemper, and P. Steimer, "Low output frequency operation of the modular multi-level converter," in *Proc. IEEE Energy

Conversion Congress and Exposition (ECCE), Atlanta, GA, USA, pp. 3993–3997, Sep. 2010.

[56] J. Kolb, F. Kammerer, and M. Braun, "Straight forward vector control of the modular multilevel converter for feeding three-phase machines over their complete frequency range," in *Proc. IEEE Ind. Electron. Conf. (IECON), Melbourne, Australia*, pp. 1596–1601, Nov. 2011.

[57] K. Wang, Y. Li, Z. Zheng, and L. Xu, "Voltage balancing and fluctuation suppression methods of floating capacitors in a new modular multilevel converter," *IEEE Trans. Ind. Electron.*, vol. 60, pp. 1943–1954, May 2013.

[58] M. Glinka and R. Marquardt, "A new AC/AC multilevel converter family," *IEEE Trans. Ind. Electron.*, vol. 52, no. 3, pp. 662–669, Jun. 2005.

[59] S. Kouro, R. Bernall, H. Miranda, C. Silva, and J. Rodríguez, "High-performance torque and flux control for multilevel inverter fed induction motors," *IEEE Trans. Power Electron.*, vol. 22, no. 6, pp. 2116–2123, Nov. 2007.

[60] M. Pérez, J. Rodríguez, J. Pontt, and S. Kouro, "Power distribution in hybrid multicell converter with nearest level modulation," in *Proc. IEEE Int. Symp. Ind. Electron (ISIE 2007)*, pp. 736–741, Jun. 2007.

[61] J. Rodríguez, L. Morán, J. Pontt, P. Correa, and C. Silva, "A high-performance vector control of an 11-level inverter," *IEEE Trans. Ind. Electron.*, vol. 50, no. 1, pp. 80–85, Feb. 2003.

[62] J. Rodríguez, L. Morán, P. Correa, and C. Silva, "A vector control technique for medium voltage multilevel inverters," *IEEE Trans. Ind. Electron.*, vol. 49, no. 4, pp. 882–888, Aug. 2002.

[63] S. Sirisukprasert, J.-S. Lai, and T.-H. Liu, "Optimum harmonic reduction with a wide range of modulation indexes for multilevel converters," *IEEE Trans. Ind. Electron.*, vol. 49, no. 4, pp. 875–881, Aug. 2002.

[64] L. Li, D. Czarkowski, Y. Liu, and P. Pillay, "Multilevel selective harmonic elimination PWM technique in series-connected voltage inverters," *IEEE Trans. Ind. Appl.*, vol. 36, no. 1, pp. 160–170, Jan.–Feb. 2000.

[65] Z. Du, L. M. Tolbert, and J. N. Chiasson, "Active harmonic elimination for multilevel converters," *IEEE Trans. Power Electron.*, vol. 21, no. 2, pp. 459–460, Mar. 2006.

[66] L. G. Franquelo, J. Nápoles, R. Portillo, J. I. León, and M. Aguirre, "A flexible selective harmonic mitiguation technique to meet grid codes in three-level PWM converters," *IEEE Trans. Ind. Electron.*, vol. 54, no. 6, pp. 3022–3029, Dec. 2007.

[67] M. S. A. Dahidah, G. Konstantinou, and V. G. Agelidis, "A review of multilevel selective harmonic elimination PWM: formulations, solving algorithms, implementation and applications," *IEEE Trans. Power Electron.*, vol. 30, no. 8, pp. 4091–4106, Aug. 2015.

[68] G. Carrara, S. Gardella, M. Marchesoni, R. Salutari, and G. Sciutto, "A new multilevel PWM method: a theoretical analysis," *IEEE Trans. Power Electron.*, vol. 7, no. 3, pp. 497–505, Jul. 1992.

[69] B. P. McGrath, D. G. Holmes, and T. A. Lipo, "Optimised space vector switching for multilevel inverters," *IEEE Trans. Power Electron.*, vol. 18, no. 6, pp. 1293–1301, Nov. 2003.

[70] B. P. McGrath, D. G. Holmes, and T. Meynard, "Reduced PWM harmonic distortion for multilevel inverters operating over a wide modulation range," *IEEE Trans. Power Electron.*, vol. 21, no. 4, pp. 941–949, Jul. 2006.

[71] N. Celanovic and D. Boroyevich, "The feedforward method of controlling three-level diode clamped converters with small DC-Link capacitors," in *Proc. Power Electronics Specialist Conference PESCÂ'01*, vol. 3, pp. 1357–1362, Jun. 2001.

[72] J. Pou, J. Zaragoza, S. Ceballos, M. Saeedifard, and D. Boroyevich, "A carrier-based PWM strategy with zero-sequence voltage injection for a three-level neutral-point-clamped converter," *IEEE Trans. Power Electron.*, vol. 27, no. 2, Feb. 2012.

[73] I. Lopez, S. Ceballos, J. Andreu, I. Kortabarria, and J. Pou, "PWM algorithm with adaptive offset for three-level multi-phase neutral-point-clamped converters," in *Proc. IEEE Ind. Electron. Conf. (IECON)*, pp. 1183–1188, Nov. 2013.

[74] J. Pou, J. Zaragoza, P. Rodriguez, *et al.*, "Fast-processing modulation strategy for the neutral-point-clamped converter with total elimination of the low-frequency voltage oscillations in the neutral point," *IEEE Trans. Ind. Electron.*, vol. 54, no. 4, pp. 2288–2294, Aug. 2007.

[75] J. Zaragoza, J. Pou, S. Ceballos, E. Robles, C. Jaén, and M. Corbalan, "Voltage-balance compensator for a carrier-based modulation in the neutral-point-clamped converter," *IEEE Trans. Ind. Electron.*, vol. 56, no. 2, pp. 305–309, Feb. 2009.

[76] J. Zaragoza, J. Pou, S. Ceballos, E. Robles, P. Ibañez, and J. L. Villate, "A comprehensive study of a hybrid modulation technique for the neutral-point-clamped converter," *IEEE Trans. Ind. Electron.*, vol. 56, no. 2, pp. 294–304, Feb. 2009.

[77] N. Celanovic and D. Boroyevich, "A fast space-vector modulation algorithm for multilevel three-phase converters," *IEEE Trans. Ind. Appl.*, vol. 37, no. 2, pp. 637–641, Mar.–Apr. 2004.

[78] N. Celanovic and D. Boroyevich, "A comprehensive study of neutral-point voltage balancing problem in three-level neutral-point-clamped voltage source PWM inverters," *IEEE Trans. Power Electron.*, vol. 15, no. 2, pp. 242–249, Mar. 2000.

[79] J. Pou, R. Pindado, D. Boroyevich, and P. Rodríguez, "Evaluation of the low-frequency neutral point voltage oscillations in the three-level inverter," *IEEE Trans. Ind. Electron.*, vol. 52, no. 6, pp. 1582–1588, Dec. 2005.

[80] J. Pou, D. Boroyevich, and R. Pindado, "New feedforward space-vector PWM method to obtain balanced output voltages in a three-level neutral-point-clamped converter," *IEEE Trans. Ind. Electron.*, vol. 49, no. 5, pp. 1024–1034, Oct. 2002.

[81] J. Pou, D. Boroyevich, and R. Pindado, "Effects of imbalances and nonlinear loads on the voltage balance of a neutral-point-clamped inverter," *IEEE Trans. Power Electron.*, vol. 20, no. 1, pp. 123–131, Jan. 2005.

[82] J. Pou, P. Rodríguez, V. Sala, S. Busquets-Monge, and D. Boroyevich, "Algorithm for the virtual vectors modulation in three-level inverters with voltage-balance control loop," in *Proc. EPE05*, Sep. 2005.

[83] J. Pou, R. Pindado, and D. Boroyevich, "Voltage-balance limits in four-level diode-clamped converters with passive front ends," *IEEE Trans. Ind. Electron.*, vol. 52, no. 1, pp. 190–196, Feb. 2005.

[84] J. Pou, "Modulation and control of three-phase PWM multilevel converter," Ph.D. dissertation, Technical University of Catalonia, Electronic Engineering Department, Nov. 2002.

[85] B. Kaku, I. Miyashita, and S. Sone, "Switching loss minimized space vector PWM method for IGBT three-level inverter," in *Proc. IEEE Electric Power Appl.*, vol. 144, no. 3, pp. 182–190, May 1997.

[86] V. G. Agelidis and H. C. Goh, "A low-distortion variable-level PWM technique," *IEEE Proc. Int. Conf. on Ind. Technology*, pp. 247–251, Dec. 1996.

[87] B. P. McGrath and D. G. Holmes, "Multicarrier PWM strategies for multilevel inverters," *IEEE Trans. Ind. Electron.*, vol. 49, no. 4, pp. 858–867, Aug. 2002.

[88] R. Naderi and A. Rahmati, "Phase-shifted carrier PWM technique for general cascaded inverters," *IEEE Trans. Power Electron.*, vol. 23, no. 3, pp. 1257-1269, May 2008.

[89] B. Cougo, G. Gateau, T. Meynard, M. Bobrowska-Rafal, and M. Cousineau, "PD modulation scheme for three-phase parallel multilevel inverters," *IEEE Trans. Ind. Electron.*, vol. 59, no. 2, pp. 690–700, Feb. 2012.

[90] B. M. H. Jassim, B. Zahawi, and D. Atkinson, "Simple control method for parallel connected three-phase PWM converters," in *Proc. IET Int. Conf. PEMD, Bristol, U.K.*, pp. 1–5, Mar. 2012.

[91] M. Hagiwara and H. Akagi, "Control and experiment of pulsewidth-modulated modular multilevel converters," *IEEE Trans. Power Electron.*, vol. 24, no. 7, pp. 1737–1746, Jul. 2009.

[92] J. Pou, S. Ceballos, G. Konstantinou, V. G. Agelidis, R. Picas, and J. Zaragoza, "Circulating current injection methods based on instantaneous information for the modular multilevel converter," *IEEE Trans. Ind. Electron.*, vol. 62, no. 2, pp. 777–788, Feb. 2015.

[93] A. Antonopoulos, L. Ångquist, S. Norrga, K. Ilves, L. Harnefors, and H.-P. Nee, "Multilevel converter AC motor drives with constant torque from zero to nominal speed," *IEEE Trans. Ind. Appl.*, vol. 50, no. 3, pp. 1982–1993, May/Jun. 2014.

[94] R. Picas, S. Ceballos, J. Pou, J. Zaragoza, G. Konstantinou, and V. G. Agelidis, "Closed loop discontinuous modulation technique for capacitor voltage ripples and switching losses reduction in modular multilevel converters," *IEEE Trans. Power Electron.*, vol. 30, no. 9, pp. 4714–4725, Sep. 2015.

[95] M. Glinka and R. Marquardt, "A new AC/AC-multilevel converter family applied to a single-phase converter," in *Proc. IEEE Power Electronics and Drive Systems (PEDS)*, pp. 16–23, Nov. 2003.

[96] Q. Tu, Z. Xu, and L. Xu, "Reduced switching-frequency modulation and circulating current suppression for modular multilevel converters," *IEEE Trans. Power Delivery*, vol. 26, no. 3, pp. 2009–2017, Jul. 2011.

[97] R. Darus, J. Pou, G. Konstantinou, S. Ceballos, R. Picas, and V. G. Agelidis, "A modified voltage balancing algorithm for the modular multilevel converter: Evaluation of staircase and phase-disposition PWM," *IEEE Trans. Power Electron.*, vol. 30, no. 8, pp. 4119–4127, Aug. 2015.

[98] R. Joost and R. Salomon, "Advantages of FPGA-based multiprocessor systems in industrial applications," in *Proc. of Industrial Electronics Society Conference (IECON)*, Nov. 2005.

[99] Z. Zalewski, "Winning with SoC designs in FPGAs," *Embedded System Engineering*, vol. 14, no. 8, pp. 30–31, 2006.

[100] E. Ormaetxea, J. Andreu, I. Kortabarria, I. Martínez de Alegra, E. Ibarra, and E. Olaguenaga, "System on chip cores for protection and commutation of a matrix converter," in proc. power electronics and motion control conference (epe/pemc), pp. 15–21, Sep. 2010.

[101] R. Teodorescu, E.-P. Eni, L. Mathe, and P. Rodríguez, "modular multilevel converter control strategy with fault tolerance," in *Proc. International Conference on Renewable Energies and Power Quality (ICREPQ13)*, Mar. 2013.

[102] E.-P. Eni, "Fault tolerant distributed control strategy for modular multilevel converter in HVDC applications," Master thesis, Department of Energy Technology, Aalborg University, Jun. 2013.

[103] C. L. Toh and L. E. Norum, "Implementation of high speed control network with fail-safe control and communication cable redundancy in modular multilevel converter," in *Proc. EPE13*, Sep. 2013.

[104] S. Huang, "Distributed modulation and control of modular multilevel converter for HVDC application," Master thesis, Department of Energy Technology, Aalborg University, May 2013.

[105] Y.-M. Parky, H.-S. Ryu, H.-W. Lee, M.-G. Jung, and S.-H. Lee, "Design of a cascaded H-bridge multilevel inverter based on power electronics building blocks and control for high performance," *Journal of Power Electronics*, vol. 10, no. 3, pp. 262–269, May 2010.

[106] S.-Y. Lee, M.-H. Woo, J.-K. Kim, and S.-P. Ryu, "A novel method to improve output voltage quality of grid-connected cascaded H-bridge multilevel converter with phase-shifted PWM and serial bus communication," in *Proc. Energy Conversion Congress and Exposition (ECCE)*, pp. 4513–4518, Sep. 2012.

[107] W. Liu, "Distributed modular controller architecture for high power converter applications," Master thesis, Graduate Faculty of North Carolina State University, 2005.

[108] C. C. Davidson, A. C. Lancaster, A. J. Totterdell, and C. D. M. Oates, "A 24 MW level voltage source converter demonstrator to evaluate different converter topologies," in *Proc. CIGRE 2012*, Aug. 2012.

[109] S. Huang, R. Teodorescu, and L. Mathe, "Analysis of communication based distributed modulation and control of MMC for HVDC," in *Proc. EPE13*, Sep. 2013.

[110] Y. Zhou, D. Jiang, P. Hu, J. Guo, Y. Liang, and Z. Lin, "A prototype of modular multilevel converters," *IEEE Trans. Power Electron.*, vol. 29, no. 7, pp. 3267–3278, Jul. 2014.

Part III

New trends in control circuits for power electronics

Chapter 11

State-of-the-art intelligent gate drivers for IGBT power modules – monitoring, control and management at the heart of power converters

Bing Ji[1], *Wenping Cao*[2,3] *and Volker Pickert*[4]

11.1 Introduction to gate drivers

Power semiconductors are the most important and enabling components in power converters, similar to the "muscle" in a living body. However, they are controlled by the central control unit, which is similar to the "brain" in a living body. Power semiconductors in power electronic systems process and control power flow. Largely different from their electronic counterpart used in signal processing, power semiconductor devices handle much higher voltages and currents for power flow control. In order to improve the system efficiency, power semiconductor devices are operated in a switching mode (On/Off).

As the power semiconductor devices are continuously developed for electrical power applications, their performance is gradually approaching ideal switches manifesting the following aspects:

- high current (DC, pulse, rms, average, peak, surge)
- high blocking voltage
- fast switching (short on/off delays, short rise/fall times, short turn-on/turn-off times)
- high efficiency (low on/off conduction losses, low on/off switching losses, low control losses)
- ease of use (low cooling effort, low control power, simple gate driver control or protection)
- high reliability (low random failure, long lifespan and robustness).

[1]Department of Engineering, University of Leicester, UK
[2]School of Engineering and Applied Science, Aston University, Birmingham, UK
[3]Department of Electrical Engineering and Computer Science, Massachusetts Institute of Technology, USA
[4]School of Electrical and Electronic Engineering, Newcastle University, UK

Since the characteristics of power semiconductors are getting closer to their physical limits, further improvements in performance and reliability pose a particular challenge whilst gate driver unit (GDU) technologies have been well developed to fit the bill. The GDU is an integral part of a power electronics system, as an extension of the "brain." It provides essential control and many functions to deliver the required specifications for power semiconductor devices/modules and power electronics systems.

Various types of power semiconductors have been developed and used for a wide range of applications. Because of their different gate characteristics and control requirements, tailored designs dedicated to specific types of devices and operational conditions are gaining in popularity in industry. Among these, insulated gate bipolar transistors (IGBTs) have shown their superior performance and are thus widely used in applications such as energy conversion, transportation, industrial and home appliances. Their gate drivers are the focus of this chapter.

11.1.1 Power electronic systems, IGBTs and gate driver units

Power electronic systems consist of a variety of elements to deal with electrical energy conversion. A block diagram of a typical power converter is shown in Figure 11.1. It includes three parts: the power circuit formed by the power semiconductor devices and passive components, a centralized control unit providing functions such as communication (input/output), data storage, signal/data processing, control algorithm and command, and an interface formed by gate drivers and sensing units. The control unit operates at much lower voltage and power levels. An interface is therefore required for the control unit to receive sensor signals and issue control commands to gate drivers.

IGBTs are packaged for various purposes (power rating, cost, footprint, etc.). Some commonly used packaging types include "Transistor Outline" (TO) packages, "Single In-Line" (SIL) packages, "Dual In-Line" packages (DIL/DIP), standard power modules and press-pack modules (also called capsules). Figure 11.2a shows various existing technologies for different applications and power ratings. Discrete packages are dominant in low power applications. Standard power modules are adopted in vast majority of medium and high power markets due to their easiness and flexibility in manufacturing, assembly, function integration and power scaling. The press-pack IGBTs (PPIs) have also become a mature technology to manipulate very high power ratings (several MW) and are frequently found in use as series connection for power transmission and distribution applications. IGBT module packages normally contain IGBTs and freewheeling diodes (FWDs) with a defined IGBT/FWD ratio for an inductive load and their circuit symbols are shown in Figure 11.2b.

The GDU is an integral interface between the high voltage (HV) power stage (IGBT devices/modules) and the low voltage control unit of the power converter which provides essential control and protection functions. A typical IGBT GDU comprises a low-voltage side (interfacing the control unit) and a HV side (interfacing the IGBT). The former consists of a low voltage logic device, or a logic-to-driver interface (LDI) and auxiliary components such op-amps and comparators.

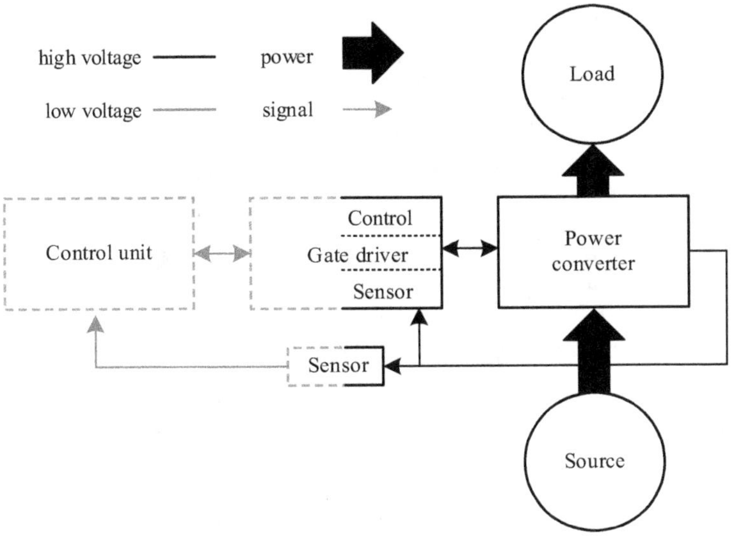

Figure 11.1 Block diagram of a power converter

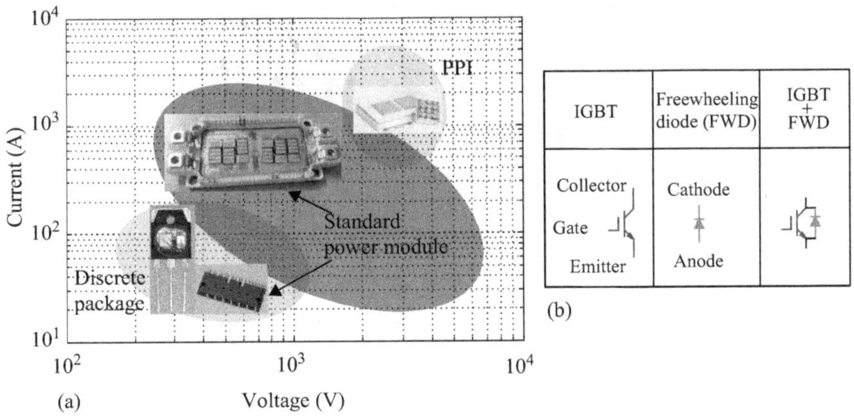

Figure 11.2 Commercially available IGBTs. (a) Different packaging techniques. (b) Circuit symbols

It generates the control signals (On/Off), processes the feedback signals and communicates with the master controller. The latter is the output stage that controls and interfaces the HV power semiconductor devices. The output stage power circuit is tailored to match to the individual IGBT device/module for specific applications. Figure 11.3 shows a commercial GDU and an in-house made GDU.

In general, the HV side of conventional GDUs provides five basic functions (rectangular boxes in Figure 11.4) and other features have evolved for that of the innovative GDUs (hexagon boxes in Figure 11.4). First, the HV side is an amplifier

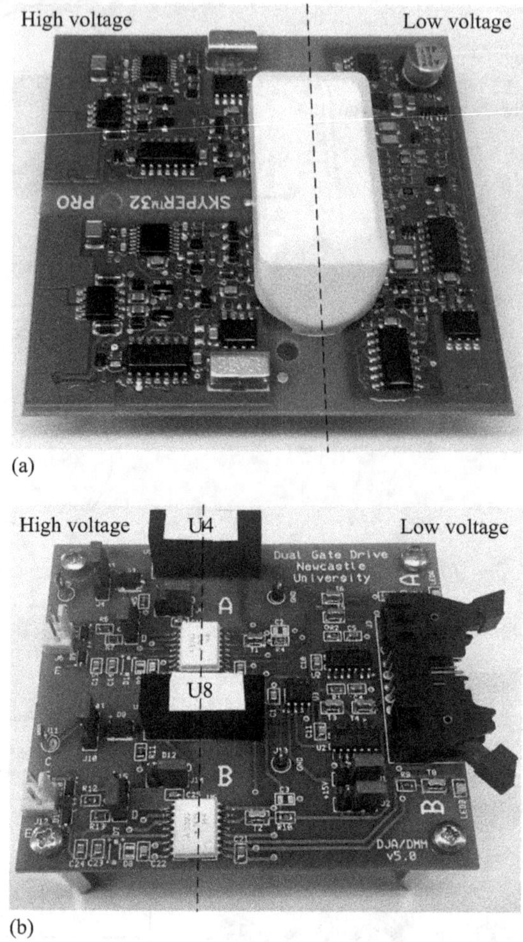

Figure 11.3 Standard gate driver for IGBT modules. (a) A commercial GDU. (b) An in-house made GDU

or a booster. Since a high driving power is required for IGBT gate capacitance charging/discharging during On/Off switching transients, the instantaneous gate current can reach up to tens of amps and volts. In addition, isolation and/or level shifting is required for both signal and power transmissions. The control signal is transmitted through communication channels so that the switching patterns is mapped from the low-voltage side to the HV power converters while the HV side feedback signal (status or sensor information) is also sent back. An isolated power supply is also normally required for the output stage in most converter topologies. For instance, the collector of the top (high-side) IGBT in a half bridge is connected to DC+ bus. The emitter of that IGBT floats with respect to DC− to follow IGBT's output characteristics. This in turn requires the use of an isolated gate driver in

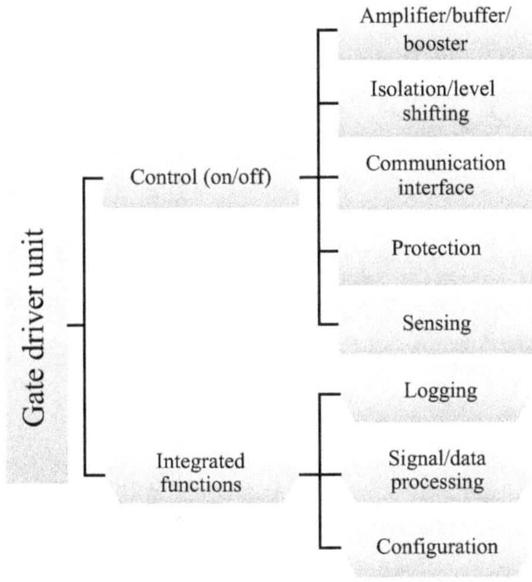

Figure 11.4 Examples of basic (rectangular) and extensive (hexagon) functions in a gate driver unit

order to isolate the low-voltage inputs/outputs of the control unit from the HVs of the IGBTs. The isolated gate-drivers are also used for the bottom (low-side) IGBTs' control. Common signal transmission methods use transformers, coreless transformers, optocouplers, optic fibres, capacitive barriers and giant magnetoresistive (GMR) isolators. Some examples of the voltage supply methods can be found in Table 11.1. The HV side electronics employ the sensing and monitoring mechanisms (e.g. voltage, current or temperature monitoring) to protect IGBTs from hazards during critical destructive operations (e.g. shoot-through faults). A post-fault protection scheme is activated under fault detection and the alarm is generated to prevent failure propagation.

11.1.2 Sensing and control systems

Typical power electronic converters are equipped with a number of sensors and instrumentation electronics in order to deliver the expected system performance. Figure 11.5 shows a typical block diagram of a full-bridge power inverter in an AC motor drive system.

It consists of a rectifier, an inverter, DC bus, sensor network and CPU (e.g. microcontroller, microprocessor). Typical sensors and their measurement points in a power inverter system are presented in Table 11.2. Voltage, current and temperature readings are collected and transmitted through signal conditioning electronics to the control unit for signal processing and control algorithm execution. It then outputs appropriate switching signals to provide control and fault management at the system level. A combination of different sensors can be selected to achieve the control targets associated with the specific system profile.

Table 11.1 GDU voltage supply

Supply and functional diagram	Key features
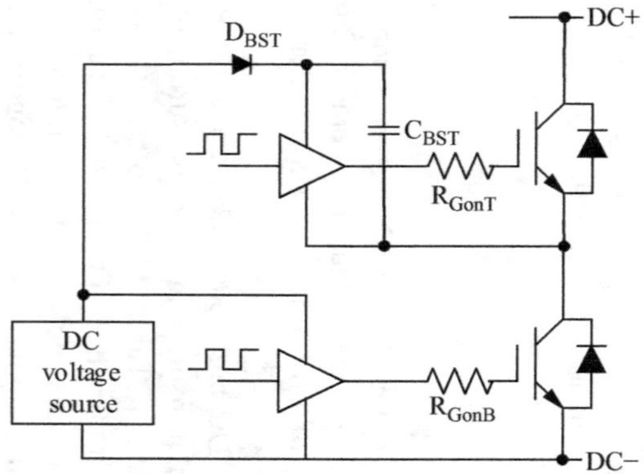 Bootstrap	Cost effectiveness. No isolation. Duty cycle and on time are both constrained to recharge and discharge the bootstrap capacitor. The bootstrap capacitor voltage decreases until a new charging state due to both the gate leakage current and charging energy. Only unipolar DC source is available. Start-up scheme required.
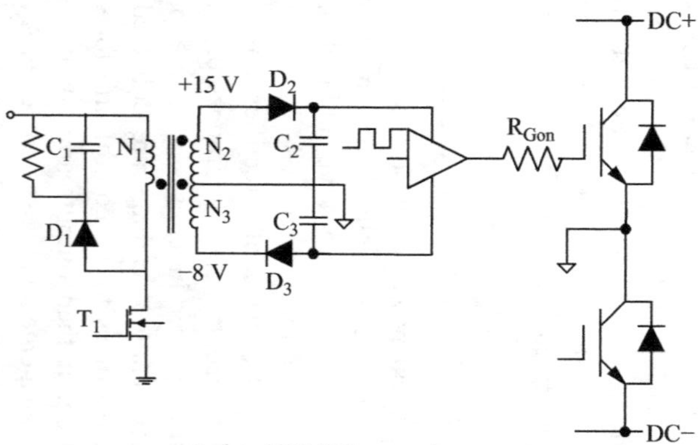 Isolated DC-DC converter	Isolated supply voltage. Constant driving voltage is available for very low frequency applications. The unwanted displacement current is induced by the high *dv*/*dt* acting on the coupling capacitance of the converter isolation barrier.

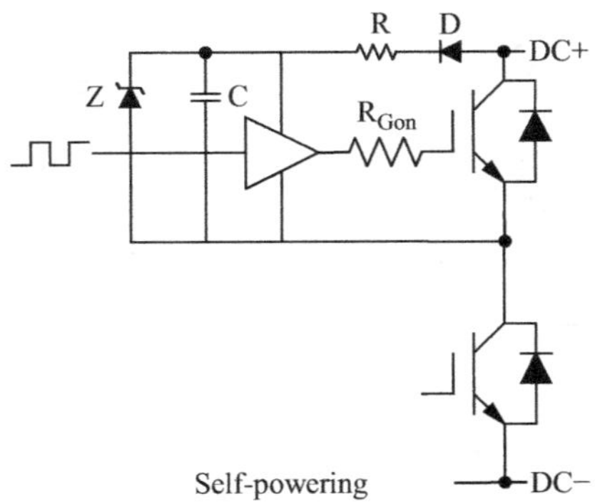

Self-powering

Self-power supply based on linear regulation. The efficiency of the linear regulator should be noted. Good compatibility with the main power semiconductors. Storage capacitor voltage decreases until a new charging state due to both the gate leakage current and charging energy. Only unipolar DC source is available.

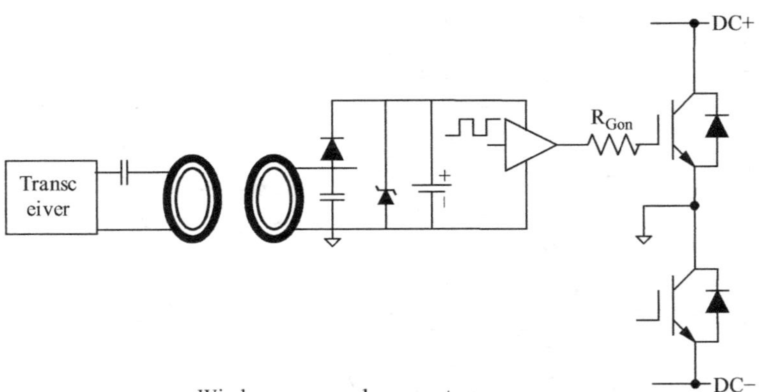

Wireless power and energy storage

Coreless inductive power transmission for harsh environments. Safe solution for isolated measurements with respect to high and/or transient reference voltages. Supply efficiency is largely reduced with increased power transmit distance and reduced alignment.

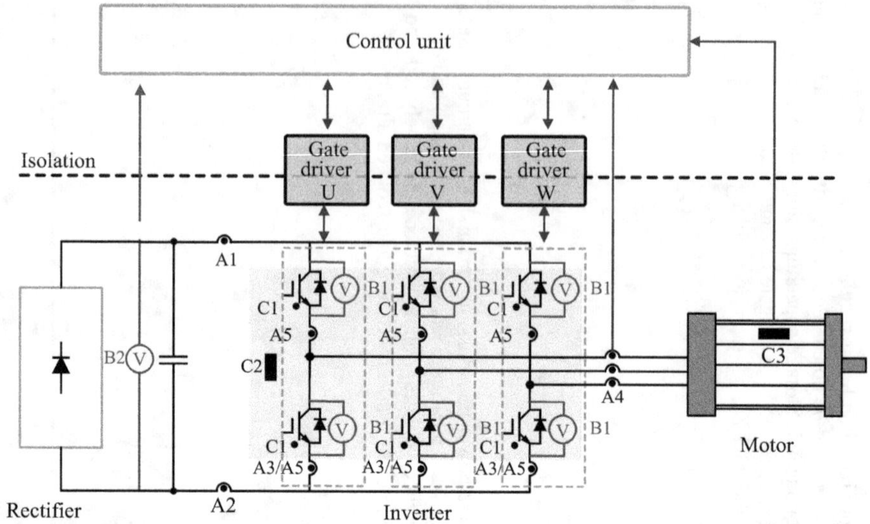

Figure 11.5 Block diagram of a typical full-bridge power inverter in an AC motor drive

11.2 Innovative gate driver and system architecture

Recently, the power electronics market has moved towards innovative gate driver features and optimized solutions. Extension of the GDU performance and functions is necessary and this is largely associated with product-driven performance requirements. It is already known for the low power consumer market, the cost and robustness of GDUs are critically important while in the high-end market (mainly for medium/high power applications), their performance and controllability are of major concerns. High-temperature performance is sometimes intended for harsh environments and extended lifetime. With the advances in materials (SiC, GaN, etc.) and design/fabrication/packaging techniques, some high-temperature power semiconductors are now commercially available, pushing the operational temperature over 200 °C.

11.2.1 System integration

Integration of elements and functions is a current trend in power electronic systems. It helps achieve system miniaturization (footprint, volume and weight), reduced interconnection, reduced assembly effort and increased power density. Intelligent power modules (IPMs) have been well developed mainly for low- and medium-power applications. They are normally comprised of power semiconductors in a certain circuit topology, associated gate drivers integrating various functions, interconnections (wires, copper tracks, etc.) and terminal leads interfacing the power circuit and the control unit.

Table 11.2 Typical sensors and their measurement points in power converters

Measurement	Symbol	Measurement point	Features
Current	A1	DC+ bus	Short circuit faults (line-to-line, ground and shoot through) can be monitored. Isolation is required.
	A2	Ground reference	Measurement is taken at ground reference and isolation is not required. Ground fault cannot be monitored.
	A3	Emitter side cable of bottom IGBTs	Measurement is taken at ground reference and isolation is not required. Ground fault cannot be monitored.
	A4	Inverter output phases	Typical measurement for high-end motor drive. The third phase current sensor can be omitted in normal operation.
	A5	Emitter/auxiliary emitter of six IGBTs	Current sense IGBTs or emitter inductance integrator is required. Integrated in GDUs.
Voltage	B1	Collector-emitter of six IGBTs	Desaturation monitoring feeding the integrated comparator to prevent short circuit faults. Integrated in GDUs.
	B2	DC+ bus	Measured at both capacitor pre-charge and normal operation phases. Resistive divider with or without isolation is used to match the DC bus with the read-out electronics.
Temperature	C1	Six IGBT dies	IGBT temperature is measured by the integrated temperature sense pn-diode but not for freewheeling diodes. Integrated in GDUs.
	C2	Heat sink or baseplate	Temperature reference measured by contact sensors such as thermistor (e.g. NTC/PTC) and resistance temperature detector (RTD, e.g. Pt100)
	C3	Motor	Temperature reference for motors

The existing gate driver integration technologies are developed at different levels. Due to the well-developed silicon-on-insulator (SOI) technology with HV blocking capability, the monolithic gate driver can be directly integrated on the direct copper bonded (DCB) substrate of power modules [1], as shown in Figure 11.6a. A key advantage of this technology is the good thermal conductivity to assist cooling for both the power semiconductors and the gate drivers. However, the asymmetric propagation delay of driving signals arising from the restricted copper track layouts should be considered. Alternatively, the monolithic gate driver can be integrated in the printed circuit board (PCB) inside the IPM as in Figure 11.6b. This is compatible with the current gate driver integrated circuit (IC) design. In Figure 11.6c, a combination of power semiconductors and a discrete gate driver IC with peripheral components (capacitors and resistors) are integrated into a

Figure 11.6 Different IPMs with integrated gate drivers. Courtesy of (a) SEMIKRON International GmbH, (b) Fuji Electric and (c) Vincotech GmbH

module package. A "thick film" technology is used for the integration, where different layers of conducting and isolating materials are printed on a ceramic sheet, similar to PCBs for tracks, pads, ICs, capacitors and resistors.

Since the embedded passive integrated circuit (emPIC) techniques are still under development, realization of isolation inside the module package results in complexity with increased manufacturing effort. The integrated voltage supply for IPM gate drivers are mainly powered by the bootstrap circuits, resulting in a limited blocking voltage (up to 1200 V). To properly operate the IPM under all specified conditions, an external power supply is needed. Therefore, gate drivers integrated with galvanically separated power supplies sufficient for higher voltages are to be developed in the future.

11.2.2 High temperature operation

Silicon (Si) is by far the most commonly used semiconductor material and the operating temperature of traditional Si-based devices range from −40 °C to 150 °C. Higher temperature may be desired for specific applications. Currently, high-temperature IGBTs show good robustness at a junction temperature of 200 °C. Wide band-gap (WBG) devices such as SiC and GaN have intrinsic temperature capabilities and push the operating temperature to the range of 200–300 °C. The advances in packaging materials and WBG devices (e.g. diamond) promise further improvement over operating temperature of 300–500 °C. As a potential disruptive technology, high-temperature power semiconductors have received much attention in harsh-environment and power-dense applications, including automotive, aerospace, down-hole drills and nuclear. For instance, there is a trend for more electric aircraft (MEA) to replace conventional hydraulic systems with electric actuators, which promises reduced weight and improved reliability. However, other associated high temperature components and control electronics are still challenging. The GDU is within the most closed vicinity of power devices (heat source) and is therefore required to sustain high temperature operation.

The evolution of high-temperature IGBTs has led to the development of high temperature gate driver ICs and GDUs. A SOI-based gate driver is proposed in Reference 2 and its driving capability is demonstrated for SiC MOSFET (1200 V/100 A) at the ambient temperature of 200 °C and the junction temperature of 250 °C. The high temperature integrated SOI gate driver chip provides a multitude of functions including logic control, output stage, on-chip power supply and protection, as shown in Figure 11.7a. An on-chip temperature supervisory circuit can also be integrated for over temperature protection. Another high temperature gate driver solution using hermetic chipset (see Figure 11.7b) is commercially available for a chip junction temperature of 225 °C. These innovations offer generous freedom to lay gate drivers next to the power transistors to minimize the parasitic components and maximize the switching speed. They can also be integrated into IPMs based on high-temperature semiconductors to further exploit their high-temperature and high-speed capability.

11.3 Integrated data acquisition methods

The centralized control architecture with the sensing elements and GDUs is widely used for conventional power electronics converter systems to provide system level control and protection. For the interest of high performance IGBT modules and their intelligent gate drivers, innovative data acquisition functions are incorporated to enhance condition monitoring and control optimization. In addition to treating the whole converter system as a control target, the performance of individual power semiconductor switches and their controllability also attract much attention. Conventional sensing and instrumentation technologies for power converter control might not be capable of meeting the new data acquisition requirement and require further considerations. Before identifying the appropriate sensing techniques for various parametric measurements such as current, voltage, temperature, power semiconductor characteristics in an operating converter are needed to understand. As shown in Figure 11.8, IGBT devices in power converters are characterized with dramatic change in magnitude and frequency, especially within their switching transients.

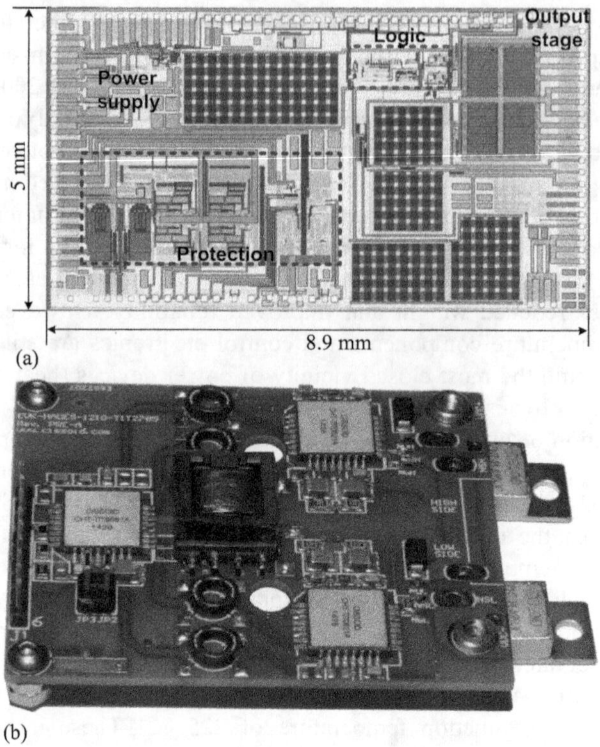

Figure 11.7 High-temperature gate driver unit. (a) SOI gate driver chipsets and (b) HADES® chipsets (Courtesy of CISSOID S.A.). (a) ©2015 IEEE. Reproduced with permission from Reference 2

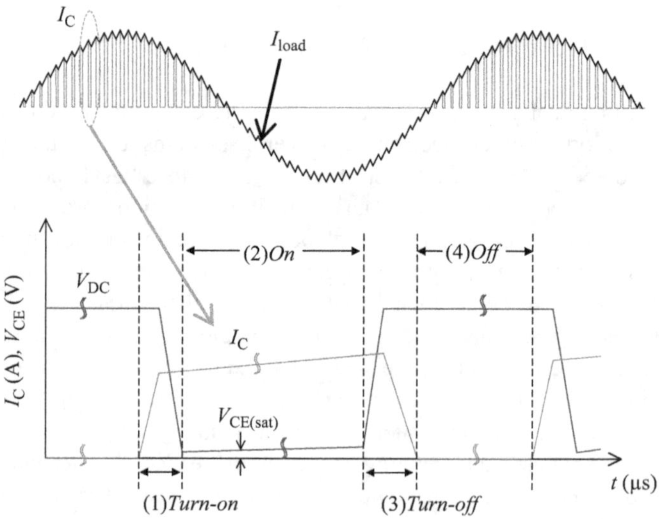

Figure 11.8 A typical PWM chopped sinusoidal waveform for IGBT switches

Typical voltage and current waveforms of an IGBT switch from a pulse width modulated (PWM) chopped sinusoidal waveform are shown in Figure 11.8. For a high fidelity measurement, both DC and high bandwidth parametric measurement are required. In the static states of (2) and (4), current varies between nil and its peak ratings while voltage magnitude varies to the order of 2 or 3. During the dynamic transients of (1) and (3), current and voltage are traversing a large magnitude within a few nanoseconds to us resulting high di/dt and dv/dt up to tens or hundreds of A/ns and V/ns, respectively. A dedicated measurement technique is therefore required to cope with the challenging conditions including EMI, distortion and noise. Moreover, it should cause neither damage nor intrusion (performance, control strategy, etc.) to operating power converters. For example, the measurement circuit design should take into account of design challenges such as insulation, short fault in the switching operation and should avoid induced anomalies to the gate drives and converter systems.

11.3.1 Voltage measurement

The IGBT collector-emitter voltage is a commonly monitored performance parameter. An IGBT is used as a power switch and it traverses from several volts to hundreds or thousands of volts for on-state and off-state with a dynamic dv/dt up to a few hundred *V*/ns. To accurately quantify its performance characteristic, high-fidelity voltage measurement has to be made on both the switching transients and the on-state conditions.

In order to acquire the transient behaviour, oscilloscopes with high bandwidth (in MHz) voltage probes are normally used. To precisely obtain the IGBT transient voltage characteristics is challenging. Recently, some field programmable devices (e.g. field programmable gate arrays (FPGAs), complex programmable logic devices (CPLDs)) with front-end high-speed electronics have been integrated into gate drivers to provide in-situ measurement [3, 4]. The data-acquisition subsystem can be used in a burst mode to adapt the periodic transient characterization, where higher sample rates only apply for shorter periods to reduce the throughput demands. However, it is still extravagant to achieve a high-fidelity switching transient measurement. An alternative method is to prolong the switching transient period (e.g. by increasing the gate resistance) but this seriously impact the IGBT performance and generates more switching losses as well [5]. Yet others obtain transient characteristics with cost-effective solutions by sacrificing the signal integrity. Commonly used measurement techniques include reduced sampling rate and event-triggered measurement. However, the critical information in the high frequency contents of the analogue signal may be lost. For example, peak detectors screen and pinpoint the specific features in IGBT switching waveforms [6].

For IGBT on-state voltage measurement with their intermittent exposure to high blocking voltage at off-states, read-out electronics can be provided with voltage dividers or scaling functions to avoid overdrive. To detect the comparatively small on-state voltage changes in a very large overall voltage (voltage surges, off-state voltage, turn-off voltage transients, etc.), the sensitivity might be insufficiently low.

$$\text{Sensitivity} = \frac{\text{Full Scale}}{2^{\text{ADC's resolution in bits}}} \tag{11.1}$$

For instance, a 400-V DC-link voltage is assumed as the full scale measurement voltage and a 10-bit ADC provides a sensitivity of $400/2^{10} \approx 0.39$ V.

In order to achieve high accuracy measurement (in the level of a few millivolts) in a kilovolt range, a practical way is to block or clamp the HV in order to maximize the effective full scale. As for the previous example when the maximum scale set to 5 V the new sensitivity is 5 V/$2^{10} \approx 4.9$ mV.

There have been some practical solutions proposed in the literature, as demonstrated in Figure 11.9. The first possible approach is to apply a decoupling diode to block the HV as shown in Figure 11.9a. It is a simple and cheap solution with high bandwidth, but the voltage across the diode is included in the output and an external voltage reference (V_r) is required. The second category introduces a series connection of a diode and a Zener diode after a limiting resistor as displayed in Figure 11.9b. The output voltage is clamped by the Zener diode when the power module is turned off [7]. This is still a simple and economic solution but the bandwidth is limited by the current limiting resistor and performance of the diodes. In Reference 8, the JFET (N-channel) current generators cells are used instead of the current limiting resistors, which could make the voltage clamping up to 100 kHz. The third category is employing a relay or an active switch to connect and disconnect the measuring circuit, as shown in Figure 11.9c. Potential problems with this circuit are the limited bandwidth due to the dead time and the need for additional gate signals for the control switch.

11.3.2 Current measurement

Current measurements in conventional power electronics converters are taken for two purposes, achieving the converter control target and protecting the power semiconductor devices from fault (e.g. short fault, overcurrent). A number of current measurement methods with different performance index can be determined depending on the measurement goals. For instance, system level dynamic control requires a sampling rate of PWM frequency (normally varies from DC to tens of kHz) and accuracy of approximately 2~5%. However, for the purpose of IGBT characterization especially during switching transients, most specific requirement in current measurement should be noted.

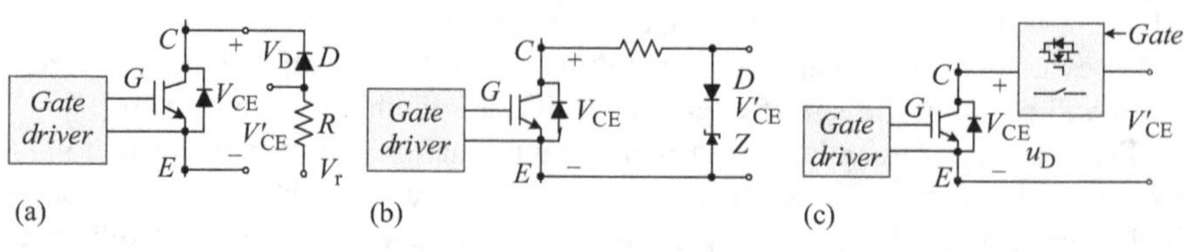

Figure 11.9 V_{CE}(on) measurement circuit. (a) Decoupling diode. (b) Zener clamp. (c) Control switch

(i) The bandwidth of the sensor should be high enough to accurately track the rising and falling transients as rapid as tens of nanoseconds.
(ii) The sensor dynamic range should match the peak amplitude of the measured signal while maintaining sufficient sensitivity at small current and out of saturation at peak current.
(iii) For the invasive approaches, the interference due to the added impedance should be minimized.

Current sensing techniques can be divided into two main categories: the non-isolated and the isolated. The former embraces the two-wire shunt, four-wire shunt (including coaxial shunts or pseudo four-wire), and current sense IGBT using split cells. In particular, subsequent read-out electronics is also integrated to provide analogue-digital converter (ADC) and isolation. They are illustrated in Figure 11.10. The current shunt based sensing technology is known to achieve high bandwidth. However, the parasitic effects due to internal inductance need to be minimized or compensated for accurate measurement. Moreover, their power losses and thermal issues in high current applications should be addressed. In particular, the split-cells in IGBTs are fabricated by some manufactures to reduce the shunt power losses. In order to make shunt sensing signal available for control units while minimizing distortions, digitalized isolation techniques such as optocouplers or pulse transformers are used. For instance, some sense IGBTs employ the Sigma/Delta-ADC technique with the built-in integrator and comparator for shunt front-end measurement, which allows for significant compromise to be introduced between the effective sampling rate and signal resolution. The V_{CEsat} monitoring technique is commonly found in gate drivers and makes use of the IGBT itself as a shunt. The collector-emitter voltage is monitored based on the IGBT output characteristics for short protections and a fault warning is issued when the IGBT runs into desaturation, i.e. a surge of effective IGBT resistance occurs. For short protection typical shunts feature cost effectiveness and design simplicity, they may suffer from a demanding surge for accurate measurement

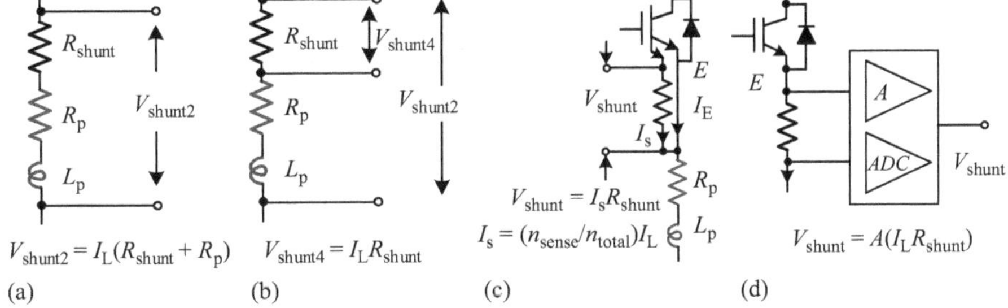

Figure 11.10 Typical non-isolated current measurement circuits. (a) Two-wire shunt. (b) Four-wire shunt. (c) Current split cell. (d) Integrated Sigma/Delta-ADC

during rapid switching transients where ultra-low inductance shunts and high performance read-out electronics (e.g. compensation and isolation features) are required.

The latter isolated technologies are more expensive than current shunts. Hall Effect based current transducers categorized as both open-loop and closed-loop types are widely used for converter control but their comparatively low bandwidth may hinder their applications in dynamic current transients. In case of the short-current fast protection (e.g. normally within 10 us), gate drivers with shunt or V_{CE} measurement are required. Current transformer (CT) is used for pulse and frequency alternating current applications. Present products can measure the current up to 250 MHz with hundreds of amps by applying advanced material and techniques. However, its comparatively large footprint makes it hard to be implemented in practical applications. Moreover, the DC and low frequency components of the measured current may cause core magnetic saturation [9]. A Rogowski coil also measures pulse and frequency alternating current by using an air toroid with an additional integrator. Its small footprint leads to an easy circuit integration. The iron coreless design results in a wider current measuring range compared to the CT, however, signal bandwidth is largely limited by their self-inductance and the capacitive effect in windings and integration electronics. In Reference 10, the experimental results of comparison between Ultra Mini Rogowski and Pearson current monitor (a commercial CT) show that the latter has better performance for high dynamic switching current measurement (e.g. fast transient response, less harmonic). Since both methods stated earlier can only be employed in the AC measurement, authors in Reference 11 present a highly dynamic current measuring method by applying the inductive current sensor for high-frequency component measurement and low dynamic DC-current probe for calibration. In order to get the accurate results for the current flow of extremely fast switching semiconductors, the Q-algorithm method is utilized to calculate the final current waveform after offset correction and calibration. Direct-Current Current Transformers (DCCTs) are capable of measuring DC and AC currents over the magnetic coupled device by incorporating novel principles of operation and technology [12]. However, due to the magnetic modulator principle applied in DCCTs, an accurate measurement of current is ranging from DC to a few hundred kHz, which becomes one of its limitations. Magnetoresistance (MR) current sensors, based on magneto-resistance effect or giant MR where constructive material resistance changes with applied magnetic field, are developed for power converter applications to meet high bandwidth, accuracy and insulation performances but have limited bandwidth for switching transients. It is exempt from iron core and achieves high linearity and accuracy [13]. Magnetoimpedance (MI) current sensors also emerge, showing more potential advantages than MR sensors in terms of cost, flexibility, size, sensitivity and bandwidth [14]. Fibre-optic current sensors, based on Faraday magneto-optic effect, perfectly integrate the magnetic field along the sensing fibre and directly measure the current [15]. Its measurement range is up to several hundred kA but bandwidth is still limited as shown in Figure 11.11. An overview of current sensing technologies that are suitable for packaging into integrated power electronics modules is presented in Reference 16. The integrated planar shunts and planar

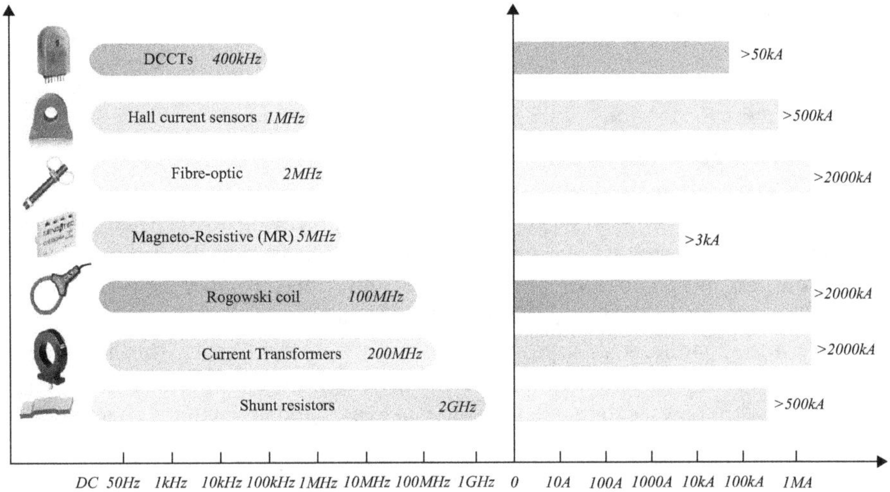

Figure 11.11 Bandwidth and measurement range of typical current sensors

embedded Rogowski coils technologies [17] are proposed as two candidates for fully integrated current sensors in power modules.

The popular WBG power semiconductors and their higher switching speed are leading the way to high bandwidth current sensors especially for the switching current measurement. The appropriate current sensors (together with their read-out electronics) must be selected to avoid the tremendous interference to the semiconductor characteristics. Most conventional current sensors may have their own limit for this specific application. Different characteristics of these sensors should be taken into account in terms of sensing elements, operating frequency range, accuracy, high current measuring capacity, temperature sensibility, power consumption, relative cost, integratability and current type.

11.3.3 Temperature measurement

The semiconductor chip junction temperature (T_j) is an important parameter for the thermal and electrical characterization and it is generally used for advanced condition monitoring and control. Traditionally, three different methods can be used for in-situ measurement, namely optical instrumentation, physical contacts and temperature sensitive electrical parameters (TSEPs). TSEP provides a global value for the entire chip, so-called "virtual" junction temperature, without a significant intrusion to the conventional power modules. It can generally be divided into two groups in terms of device operational conditions: static and dynamic parameters. Although the dynamic TSEPs have potential ability to be utilized for normal operational drives, they have a demanding requirement to read-out electronics and signal conditioning circuits. Their accuracy and resolution can hardly meet the health monitoring purpose without rigorous examinations. A number of static TSEPs are considered and their pros and cons are briefly discussed in Reference 18.

It should also be noted that "virtual" junction temperature only provides a medium temperature while the temperature distribution occurring in the monolithic chip and multiple parallel connected chips can hardly be derived.

11.4 Intelligent control

Power electronic systems advance with the development of power semiconductor components, signal processing units and control algorithms. With the trends of integration and intelligence of power control, high performance of gate drivers, which takes a central place in an inverter to interface between the control unit and the power semiconductor devices, will inevitably demand. Apart from the typical control, protection and communication functions, novel GDUs incorporated digital technology with new features developed such as data acquisition, data logger, signal processing and in-system configuration. The digitalized gate driver provides more functionality, flexibility and controllability. Figure 11.12 shows the schematic diagram of innovative GDUs integrated with main intelligent control features.

11.4.1 Condition monitoring

With the extensive development of microelectronics (analogue and digital electronics), signal conditioning and processing techniques, protection and condition monitoring functions have been achieved and improved by integrating a multitude of advanced sensing, data processing and control functionality into GUDs. The localized sensing, computing and control are completed within the vicinity of IGBTs in the gate drivers, which can relieve the monolithic core of the master

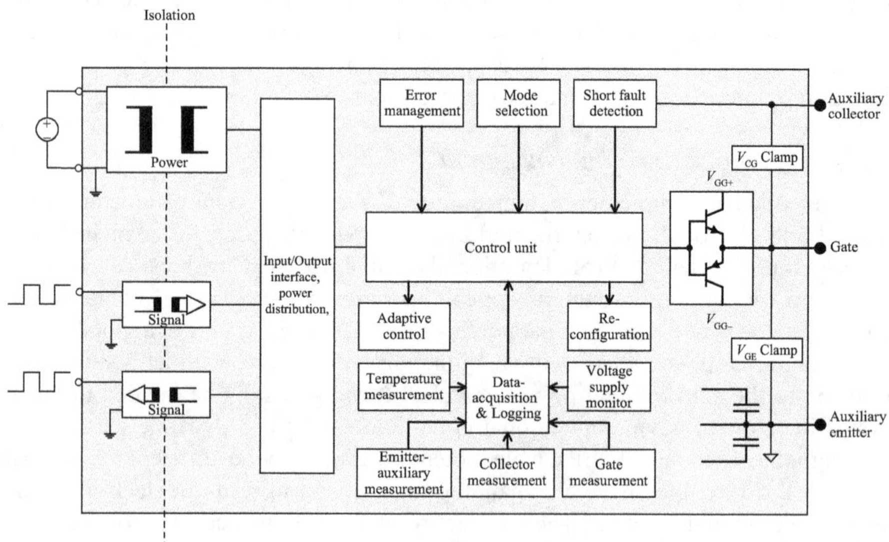

Figure 11.12 Schematic diagram of the innovative digitalized GDUs

control unit, reduce signal propagation delay and simplify communication demands over long distance. Advanced GDUs allow the IGBT and power converter parameters in-situ monitored and logged, which can then be post-processed for diagnostic and prognostic purposes.

Since IGBTs are subject to significant stresses in line with their load and environmental conditions, both abrupt and degraded failure modes can occur. To improve reliability, in-situ condition monitoring techniques have been integrated into gate drivers to provide insights into performance and health conditions. The integrated condition monitoring and health management techniques enable in-service root failure reasoning, early warnings of incipient failures, preventative maintenance scheduling and adaptive health management. Some common precursor parameters in respect of degradation failure modes include forward voltage, $V_{CE(sat)}$, threshold voltage, V_{TH}, thermal resistance, R_{th}, etc.

11.4.2 Control of switching characteristics

With the advanced in-system monitoring electronics and complex digital control unit, the output stage of the intelligent GDUs can be adaptively configured or controlled. This helps to achieve optimal IGBT performance under both normal and fault operating conditions and the operational trade-off (e.g. delay time, Miller plateau duration, electrical stresses, losses, temperature). Since the switching transients (turn-on and turn-off) are the key states for operating IGBTs and their characteristics are interrelated with their switching losses, safe operating area (SOA), electromagnetic interference (EMI), commutation behaviour, etc., they are among the most attractive control targets for intelligent GDUs.

The output stage power source can be actively controlled during IGBT switching transients to modify the switching waveforms. In particular, the derivatives of the IGBT collector-emitter voltage and collector current can be adjusted with control parameters and some common ways to achieve this are shown in Figure 11.13. Figure 11.13a shows an example of output stage with different gate resistors. The source follower receives the logic signal to activate/deactivate the gate charging. The additional current injection can be switched in or out at certain stages to minimize the turn-on delay time and switching losses. Gate discharge during turn-off transients can be regulated by current removal electronics in the same manner and omitted for clarity. Figure 11.13b shows an output stage with a gate current booster. It comprises an extra voltage source with higher amplitude and a digitally controlled switch. Figure 11.13c shows a voltage-controlled current source (VCCS) to adjust the gate current based on a digital to analogue converter (DAC) output signal. The DAC outputs a gate current command within ± 10 V, which is determined by control unit at individual subdivided stages of a switching period. Figure 11.13d shows the use of an external gate-emitter capacitor, which is normally used to reduce the IGBT parasitic turn-on and optimize the switching characteristics. A Zener diode with its internal resistance is used for damping the possible resonance and counteracts the slow turn-off. All the output stage topologies can follow a predefined control profile or be based on an event feedback.

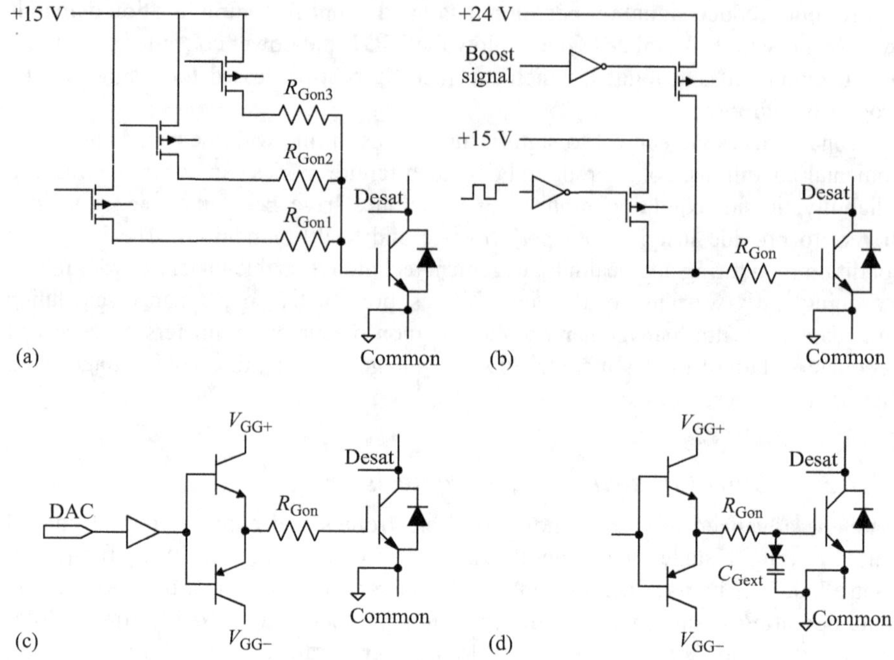

Figure 11.13 Circuit diagrams of the IGBT gate driver active control. (a) Selectable gate resistor. (b) Gate current booster. (c) Voltage-controlled current source. (d) External gate-emitter capacitor

By subdividing the switching transients into different intervals, the switching performance at each stage can be to a large extent individually tuned, which includes delay, voltage and current derivatives, voltage and current overshoots, etc.

An active gate driver that influences the switching behaviour of an IGBT by adjusting the gate current profiles is presented in Reference 3. The desired gate current is provided by a VCCS that follows the voltage command ($v_{\text{actuating}}$) at different switching stages. The current slope dI_C/dt during turn-on and voltage slope dV_{CE}/dt during turn-off in respect of various gate currents are influenced as shown in Figure 11.14. The current slope dI_C/dt and the voltage slope dV_{CE}/dt increase in line with the amplitude of the gate current. The turn-on and turn-off times are reduced and same for the switching losses.

Since the IGBT characteristics and performance are largely determined by the device itself as well as being subject to the operating point (V, I, T) and conditions (load, filter, snubber, parasitic aspects, etc.), the GDU control parameters should be adaptively determined to optimize the IGBT performance. Several adaptive strategies have been compared and discussed in Reference 3 in order to determine the optimal switching transients at individual operating point (i.e. load conditions, temperature, variations of circuit topology). The closed-loop concept has applied to achieve optimized control based on reference and feedback signals. However, the

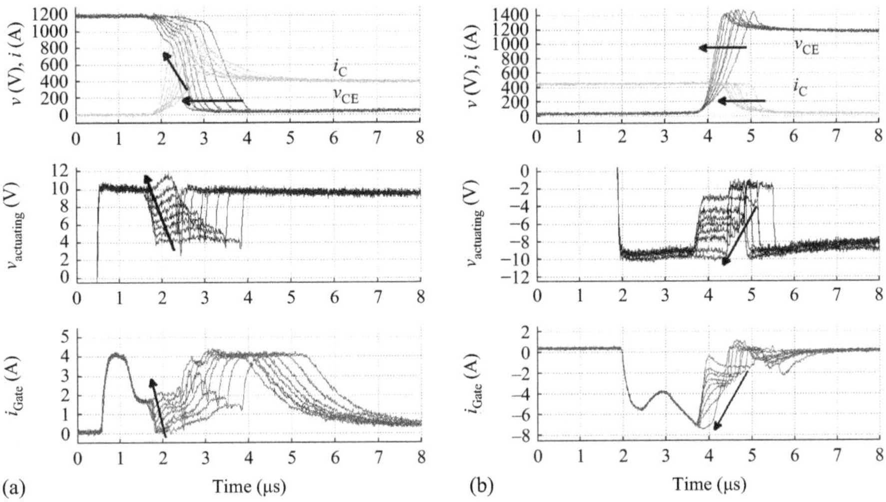

Figure 11.14 Switching waveforms in respect of various gate currents. (a) Turn-on. (b) Turn-off. ©2013 IEEE. Reprinted with permission from Reference 3

GDU complexity and cost are increased accordingly. One should note that the system stability should be justified before in-field implementation.

11.4.3 Series connection

HV IGBT modules with voltage blocking capability of up to 6.5 kV are used in many high power applications. For higher blocking voltages, a common way is to employ circuits with cascaded or multilevel structures where the voltage rating of individual semiconductors is not critical. Alternatively, power switches may be constituted by IGBT power modules in series connections. For instance, to form a 13 kV switch for the rail traction drive, a number of HV-IGBT modules can be connected in series (e.g. 3 × 4.5 kV or 2 × 6.5 kV HV-IGBTs are connected in series). Voltage unbalance between individual IGBT modules might occur during both static and dynamic states, which necessitates the de-rating and even results in power switch breakdown. This is because of the production tolerances of the IGBT chips and packages that lead to the dispersions between the characteristics of the different IGBT devices/modules. This can also be attributed to the uneven module temperature as well as characteristic differences of the associated gate drivers. Mismatches of main circuit (e.g. layout) and parasitic components (e.g. inductance, capacitance) can also influence.

For static state voltage balancing when the switch is turned off, balancing resistors in parallel with the IGBT devices are used (see Figure 11.15a). The leakage (or cut-off) current I_{CES} in the blocking IGBT module, i.e. the sum of collector-emitter blocking current of the IGBT and the reverse blocking current of the diode, is compensated by the parallel connected balancing resistors.

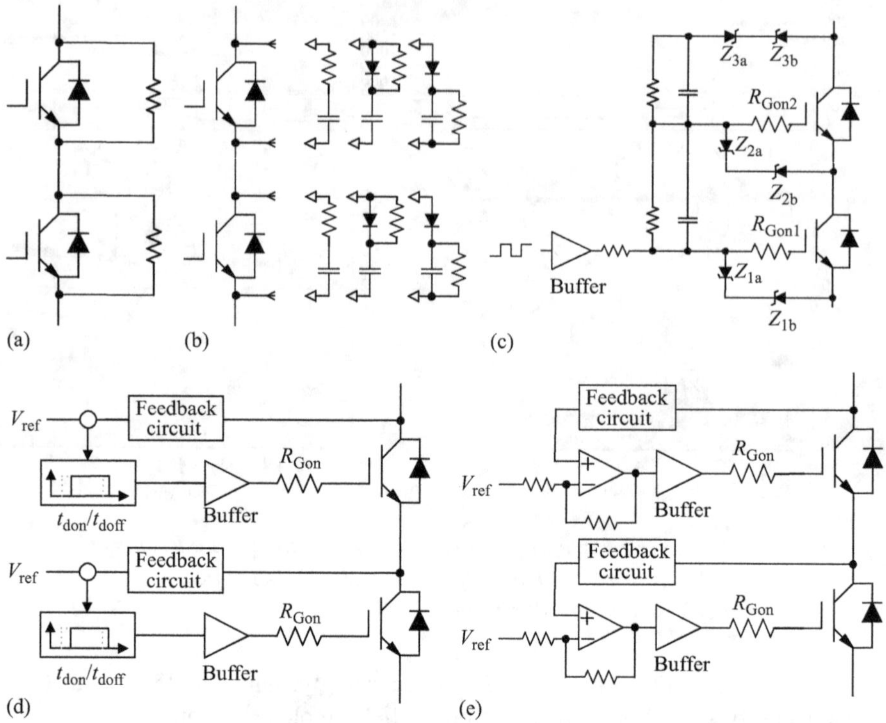

Figure 11.15 Diagram of the voltage balancing methods. (a) Balancing resistors. (b) RC and RCD Snubbers. (c) Master–slave quasi-active control. (d) Delay time control. (e) dV_{CE}/dt control. (c) © 2014 IEEE. Reproduced with permission from Reference 20

By rating the resistor current 3~5 times of I_{CES}, the symmetry of effective blocking resistance and thus the static voltage balancing can be achieved. The voltage balancing during dynamic states is achieved with power switch layout/structure optimization, snubber circuits and advanced gate control. In order to minimize the negative effects of parasitic capacitances within gate drivers and the main power circuit, a self-powered gate driver with a vertical structure for IGBT strings is reported in Reference 19, as compared to the ground-referenced isolated power supply with a horizontal structure.

Passive snubber networks (either RC or RCD) in parallel with power devices to slow down the switching are used (see Figure 11.15b). However, their performance is very much related to the operating point and it also causes losses and reliability concerns. A quasi-active gate control circuit is implemented in Figure 11.15c that is consisting of only passive components with a master–slave structure. It provides dynamic and static voltage sharing by using a simple RC balancing network. Only the bottom switch (master) receives pulse control and the RC balancing network induces the switching operation of the top switch (slave). The Zener diodes Z_1 and Z_2 are used to protect the gate from over-voltage while Z_3 prevents the gate circuit to flow the collector circuit during static conduction [20]. Some active control techniques are also carried out for series-connected IGBTs to eliminate the snubber networks. The delay times t_{don} and t_{doff} for turn-on and turn-off are set relative to one another by a dynamic balancing controller as shown in Figure 11.15d. A compensation time can be generated according to the status feedback, which interprets the voltage unbalance between individual IGBTs [21]. Figure 11.15e shows the dV_{CE}/dt control during dynamic voltage transition. Both turn-on and turn-off transitions are controlled to follow a predetermined dV_{CE}/dt. The IGBT switch trajectory can be regulated by a tailored voltage reference, V_{ref}, via the feedback loop [22]. A number of compensated control signals are generated by the feedback circuit integrating scaling and limiting circuits and these switching parameters generally includes V_{CE}, i_C, V_{GE}, dV_{CE}/dt, di_C/dt or their combinations.

11.4.4 Parallel connection

The maximum current density and chip size of IGBTs are limited by thermal and manufacturing capacity. The current density of commercially available IGBT chips ranges from 100 $A{\cdot}cm^{-2}$ to 200 $A{\cdot}cm^{-2}$. The chip size extends to 3 cm^2 since concerns associated with both the yield rate (resulted by single cell defect for high cell densities) and isothermal chip conditions manifest. To achieve the high power and current ratings, parallel arrangement is consequently implemented at various levels. Power modules (up to a few MW) possessing multiple semiconductor chips in parallel are manufactured in large quantities for emerging market sectors such as renewable energy and electric vehicles. The switching cells making use of standard power modules in parallel connections are used at the highest power levels up to several gigawatts.

In order to maximize the overall power capability and ensure the safe operation of the shunt-connected switches, the current balancing between IGBT modules should be taken care of under both static (i.e. conduction state) and dynamic conditions. The reasons for unbalancing current distribution are due to the differences either in the control input or in the power circuit loop. More specifically, it is because of tolerances in the IGBT modules (transfer characteristics, gate threshold voltage, etc.), switching cells (asymmetry of current paths and parasitic elements) or the commutation circuits (including bus-bar, FWDs, conductors and cables). This can also be attributed to the gate driver circuitry differences (signal propagation times, voltage supply tolerances). The secondary effects of the inhomogeneous temperature also cause the uneven drift of IGBT/FWD electrical properties.

Although with optimal selection and system layout, tolerances of IGBT modules as well as their gate drivers and cooling circuit performance, a minimum derating of 10% is normally required for power modules of parallel connections and it can be even higher in asymmetric layouts/arrangements.

For static current balancing in parallel connections, IGBT modules with low tolerances of I–V characteristics are generally selected. The current asymmetry can also be mitigated with the positive temperature coefficient property of the on-state voltage over majority current range.

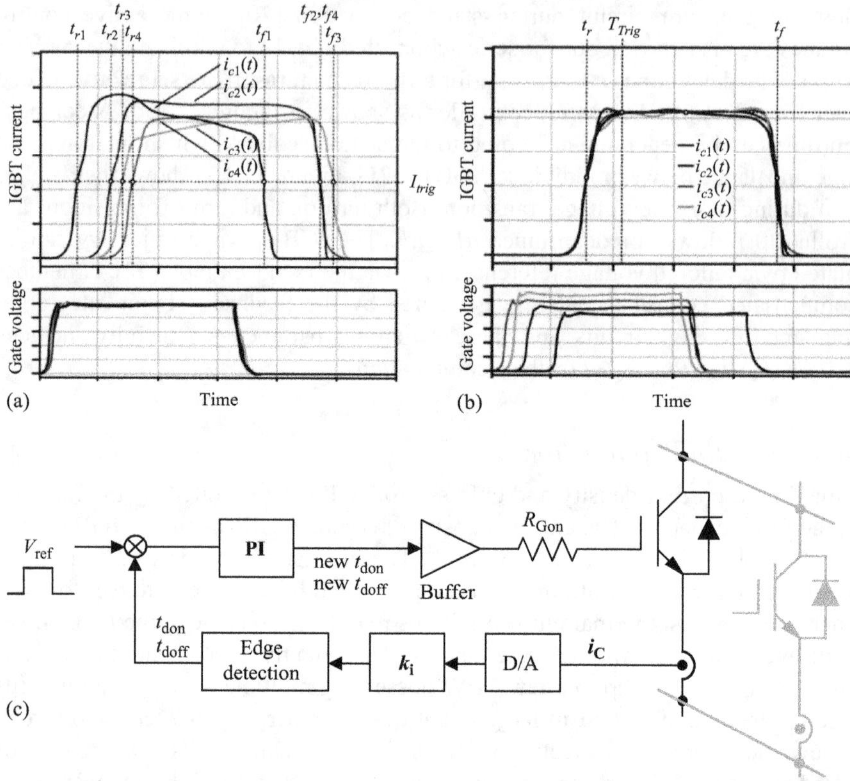

Figure 11.16 Example illustration of current balancing of parallel-connected IGBTs. (a) Without active gate control; (b) with amplitude and the firing edge control and (c) simplified control loop block diagram. (a) and (b) © 2008 IEEE. Reprinted with permission from Reference 23

For dynamic current balancing, some active gate control techniques are implemented. Based on the parameters monitored from the gate driver, IGBTs and the system, the driver output stage can be adaptively reconfigured and controlled for optimal balancing. The collector currents i_C as well as its rising and falling edges between the directly parallel connected IGBTs were measured through a PCB Rogowski coil and the delay time is determined on the detection of a predefined current trigger, as shown in Figure 11.16a [23]. The resulting delay time and current signals feed a control unit (FPGA/microcontroller), which adjusts both the amplitude and the firing edges of different gate drivers to in order to achieve balanced collector currents i_C, as shown in Figure 11.16b. Based on the detected current edge from the feedback loop of Figure 11.16c, the on/off pulse edges are shifted accordingly for the consequent pulses. In another example, a delay time compensator is implemented for the current balancing of the parallel connected IGBTs. The measurement of emitter-to-auxiliary emitter voltage enables feature extraction and the consequent delay time compensation. It also allows removing the high bandwidth dynamic current measurement and is less demanding for the control unit [24].

11.5 Summary

In this chapter, the current and future developments of gate driver techniques have been presented and extensively discussed. The focus is placed on the technical advances in data-acquisition and control functions adapted for gate drivers. The direction of technology development of power electronics is towards higher power density, higher efficiency, higher reliability, more informatics at lower cost. For the low power applications, the integration of both electronics and power electronics is the trend. The optimization and system integration of different elements (e.g. gate resistance, temperature measurement, driver, monitoring and protective functions) are under development.

In the realm of medium to high power applications, the performance of power modules and their GDUs play an important role and their reliability is a key issue. The application of embedded digital control to the distributed gate driver allows in-situ measurement of the gate driver and IGBT parameters. These parameters provide insights into the performance of the gate drivers, IGBT power modules and the power electronics system. Extended solutions dedicated to specific issues are available with innovative intelligent monitoring and adaptive control capability of the intelligent gate drivers.

Acknowledgements

The authors would like to thank the researchers who supported this work with helpful comments and discussions. Special thanks go to Haimeng Wu, Dave Atkinson and Chen Wang from Newcastle University.

References

[1] B. Vogler, M. Rossberg, R. Herzer, L. Reusser, "Integration of 1200V SOI gate driver ICs into a medium power IGBT module package," *22nd International Symposium on Power Semiconductor Devices & IC's (ISPSD)*, pp. 97–100, 6–10 June 2010.

[2] Z. Wang, X. Shi, L. M. Tolbert, *et al.*, "A high temperature silicon carbide mosfet power module with integrated silicon-on-insulator-based gate drive," *IEEE Transactions on Power Electronics*, Vol. 30, No. 3, pp. 1432–1445, March 2015.

[3] L. Dang, H. Kuhn, A. Mertens, "Digital adaptive driving strategies for high-voltage IGBTs," *IEEE Transactions on Industry Applications*, Vol. 49, No. 4, pp. 1628–1636, July–August 2013.

[4] M. Shahbazi, P. Poure, S Saadate, M. R. Zolghadri, "FPGA based fast detection with reduced sensor count for a fault-tolerant three-phase converter," *IEEE Transactions on Industrial Informatics*, Vol. 9, No. 3, pp. 1343–1350, August 2013.

[5] H. Chen, B. Ji, V. Pickert, W. Cao, "Real-time temperature estimation for power MOSFET considering thermal ageing effects," *IEEE Transactions on Device and Materials Reliability*, Vol. 14, No. 1, pp. 220–228, March 2014.

[6] V. K. Sundaramoorthy, E. Bianda, R. Bloch, F. Zurfluh, "Simultaneous online estimation of junction temperature and current of IGBTs using emitter-auxiliary emitter parasitic inductance," *International Exhibition and Conference for Power Electronics, Intelligent Motion (PCIM Europe'14), Renewable Energy and Energy Management*, 20–22 May 2014.

[7] B. Ji, V. Pickert, W. Cao, B. Zahawi, "In situ diagnostics and prognostics of wire bonding faults in IGBT modules for electric vehicle drives," *IEEE Transactions on Power Electronics*, Vol. 28, No. 12, pp. 5568–5577, December 2013.

[8] A. Calmels, "VDS(on), VCE(sat) measurement in a high voltage, high frequency system," *Application Note APT0407, Advanced Power Technology*, 2004.

[9] M. Hartmann, J. Biela, H. Ertl, J. W. Kolar, "Wideband current transducer for measuring AC signals with limited DC offset," *IEEE Transactions on Power Electronics*, Vol. 24, No. 7, pp. 1776–1787, 2009.

[10] H. Li, M. N. Stig, "Challenges in switching SiC MOSFET without ringing," *The International Exhibition and Conference for Power Electronics, Intelligent Motion (PCIM Europe'14), Renewable Energy and Energy Management*, pp. 1–6, 2014.

[11] S. Hain, M. M. Bakran, "Highly dynamic current measurements with inductive current sensors – a numerical recipe," *The International Exhibition and Conference for Power Electronics (PCIM Europe'14), Intelligent Motion, Renewable Energy and Energy Management*, pp. 1–8, 2014.

[12] G. Hudson, K. Bouwknegt, "4–13 kA DC current transducers enabling accurate in-situ calibration for a new particle accelerator project, LHC," *2005 European Conference on Power Electronics and Applications,* pp. 1–8, 2005.

[13] S. Scherner, R. Slatter, "New applications in power electronics for highly integrated high-speed magnetoresistive current sensors," *The 8th International Conference on Integrated Power Systems (CIPS),* pp. 1–7, 2014.

[14] A. Asfour, M. Zidi, J. Yonnet, "High frequency amplitude detector for GMI magnetic sensors," *Sensors*, Vol. 14, No. 12, pp. 24502–24522, 2014.

[15] P. Dreler, P. Fiala, "Utilization of Faraday mirror in fiber optic current sensors," *Radioengineering*, Vol. 17, No. 4, pp. 101–107, 2008.

[16] C. Xiao, L. Zhao, T. Asada, W. G. Odendaal, J. D. van Wyk, "An overview of integratable current sensor technologies," *The 38th IAS Annual Meeting. Conference Record of Industry Applications Conference*, Vol. 2, pp. 1251–1258, 2003.

[17] A. Ahmed, L. Coulbeck, A. Castellazzi, C. M. Johnson, "Design and test of a PCB Rogowski coil for very high dI/dt detection," *15th International Power Electronics and Motion Control Conference (EPE/PEMC),* pp. DS1a.2-1–DS1a.2-4, 2012.

[18] B. Ji, X. Song, H. Wu, V. Pickert, W. Cao, "Integrated condition monitoring for vehicle-ready power modules," *The IEEE Transportation Electrification Conference and Expo (ITEC) Asia-Pacific,* Beijing, August 2014.

[19] N. Van, P. Jeannin, E. Vagnon, D. Frey, J. C. Crebier, "Series connection of IGBTs with self-powering technique and 3-D topology," *IEEE Transactions on Industry Applications*, Vol. 47, No. 4, pp. 1844–1852, July–August 2011.

[20] N. Teerakawanich, C. M. Johnson, "Design optimization of quasi-active gate control for series-connected power devices," *IEEE Transactions on Power Electronics*, Vol. 29, No. 6, pp. 2705–2714, June 2014.

[21] S. Ji, T. Lu, Z. Zhao, H. Yu, L. Yuan, "Series-connected HV-IGBTs using active voltage balancing control with status feedback circuit," *IEEE Transactions on Power Electronics*, Vol. 30, No. 8, pp. 4165–4174, August 2015.

[22] T. C. Lim, B. W. Williams, S. J. Finney, P. R. Palmer, "Series-connected IGBTs using active voltage control technique," *IEEE Transactions on Power Electronics*, Vol. 28, No. 8, pp. 4083–4103, August 2013.

[23] D. Bortis, J. Biela, J. Kolar, "Active gate control for current balancing of parallel-connected IGBT modules in solid-state modulators," *IEEE Transactions on Plasma Sciences*, Vol. 36, pt. 3, No. 5, pp. 2632–2637, October 2008.

[24] N. Idir, R. Bausiere, J. J. Franchaud, "Active gate voltage control of turn-on di/dt and turn-off dv/dt in insulated gate transistors," *IEEE Transactions on Power Electronics*, Vol. 21, No. 4, pp. 849–855, July 2006.

Chapter 12

Control of integrated switched capacitor power converters

D. Brian Ma[1] *and Lei Chen*[1]

12.1 Introduction

Recent research efforts on system miniaturization and energy efficiency have been frequently highlighted in the growing areas including wireless sensing, ubiquitous computing, implantable bioengineering, homeland security, and so on. In many of these areas, renewable energy, which is harvested by a variety of mechanisms and modalities such as solar cells, piezoelectric devices, and wireless power transmission, has been successfully utilized, supplementing traditional batteries to accomplish system self-powering. However, compared to conventional battery-powered devices, these self-powered ones demand much more adaptable and robust operations on the power supply circuits. With continuously varying environment, the operation conditions (such as source voltage and current) of the power source may experience significant changes. A power supply circuit has to robustly accommodate the changes and deliver low noise, steady power with fast transient response. Moreover, in order to improve the energy efficiency, it is highly desirable if the power supply circuit would interface with both high-level power management schemes and device level transducers and output load applications to deliver constantly optimized operation. On the other hand, with stringent requirements on system volume by emerging applications, the power circuit is highly preferred to be fully integrated on-chip, despite the employment of bulky power devices.

Existing on-chip power supply circuits can be categorized into three major types: linear regulators, switch mode power converters, and switched capacitor (SC) power converters. Historically, linear regulators are not favorable choices for such self-powered applications due to restricted operating voltage range and dropout voltages. With respect to switch mode power converters, they require inductor(s) or transformer(s) to store and transfer power from one voltage level to another. In many cases, the voltage conversion gain (CG) of such a converter is only related to duty ratio and load condition, but independent of power stage topology. Therefore, by varying the duty ratio through a feedback controller, a switch mode

[1]Department of Electrical Engineering, University of Texas at Dallas, Richardson, TX, USA

power converter is capable of delivering efficient voltage conversion over wide and continuous input and output voltage ranges with a fixed circuit topology.

SC power converters, on the other hand, use capacitors as temporary energy storage elements. As an obvious disadvantage, compared to a switch mode power converter, a SC power converter can only provide efficient power conversion at limited discrete CGs, which is highly topology dependent. Similar to a linear regulator, if the output voltage deviates from the desired level defined by existing CGs, the efficiency of an SC power converter drops significantly. Nevertheless, an SC power converter demonstrates some attractive features, which its inductor-based counterparts are lack of. Firstly, as it does not employ any magnetic components which could introduce considerable electromagnetic interference (EMI) noise, it is highly preferred in noise-sensitive applications. Secondly, the quiescent current in SC power converters is significantly low, due to its simple controller implementation. A wireless sensor or a portable device works in either sleep or standby modes for most of time. In such periods, load power is usually low and the power dissipation of the controller circuit becomes comparable to the output power. Since this power consumption directly relates to the battery lifetime, it is very important to employ controller circuits with low quiescent current. Thus, for self-powered microsystems that have low power budget, SC power converters have become more preferable. Thirdly, from the perspective of implementation, with Moore's Law, semiconductor power devices have significantly reduced the turn-on resistance, with smaller channel lengths. This allows more switches to be implemented within a given silicon real estate. The increase in switch count allows more efficient CGs to be realized, leading to more supply power that can be delivered to the output efficiently. Lastly, with today's semiconductor technology, it has become practical to fully integrate an SC power converter on-chip. This is highly desirable for a wireless sensor or an implantable device, placing considerable impact on low form factor requirements.

12.2 Charge pump design considerations

As mentioned earlier, an SC power converter is a DC–DC power converter that converts one voltage level to another with the use of switches and capacitors. It usually consists of a power stage (known as charge pump) and a feedback/feed-forward controller that regulates the output to a desired voltage value. In general, a charge pump employs pumping capacitors as temporary energy storage medium. To achieve the desired power conversion, the capacitors are energized from the input power source during charge periods and then discharge the stored energy to the output during discharge periods. This is implemented with the aid of the controller, which appropriately turns on/off the power switches in each operation phase, so that proper amount of power can be delivered to the load at desired voltage level.

Since charge pumps handle the power transfer operations between the input power source and the output power load, they are critical for efficiency,

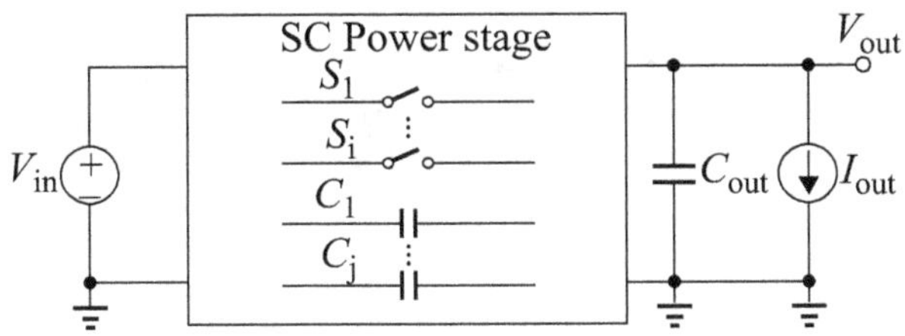

Figure 12.1 A generic charge pump circuit

robustness, and other operation performance of SC power converters. Thus, a fundamental understanding of charge pumps is necessary before advanced circuit topologies and control schemes are introduced. To aid this discussion, consider a generic charge pump, illustrated in Figure 12.1. V_{in} is the input voltage of the charge pump, while the output voltage is V_{out} and supplies a load current I_{out}. The charge pump employs pumping capacitors C_1 to C_j along with switches S_1 to S_i to achieve the desired voltage conversion. One most critical parameter in defining charge pumps is the CG. It is defined as the ratio of the output voltage to the input supply voltage,

$$\mathrm{CG} = \frac{V_{out}}{V_{in}}. \tag{12.1}$$

The magnitude of a CG in an SC power converter can be either lower than one for a step-down voltage conversion (Figure 12.2(a)) or higher than one for a step-up voltage conversion (Figure 12.2(b)). Meanwhile, its output voltage polarity can be opposite to the input voltage, as shown in Figure 12.2(c). These provide a large variety and flexibility of voltage conversion options among SC power converters.

As power converter is a voltage regulation device, the quality of its output voltage is usually measured by the output voltage ripple, which is defined as the peak-to-peak output voltage variation in steady state and with a constant load current. The output voltage ripple is affected by the output filtering capacitor C_{out}, its equivalent series resistance and inductance, and the switching frequency f_s. Typically, the ripple is expected to be less than a few tens of millivolts. A large voltage ripple impacts the operation performance and robustness of load applications and is of critical concern in noise-sensitive modules.

In addition to CG and ripple voltage, another critical but very different parameter is the efficiency, which is defined as

$$\eta = \frac{P_{out}}{P_{in}} \times 100\% = \frac{P_{out}}{P_{loss} + P_{out}} \times 100\%. \tag{12.2}$$

Here, P_{out} and P_{in} are the output and the input power, respectively. P_{loss} is the power dissipation of the power converter itself. With the proliferation of ubiquitous

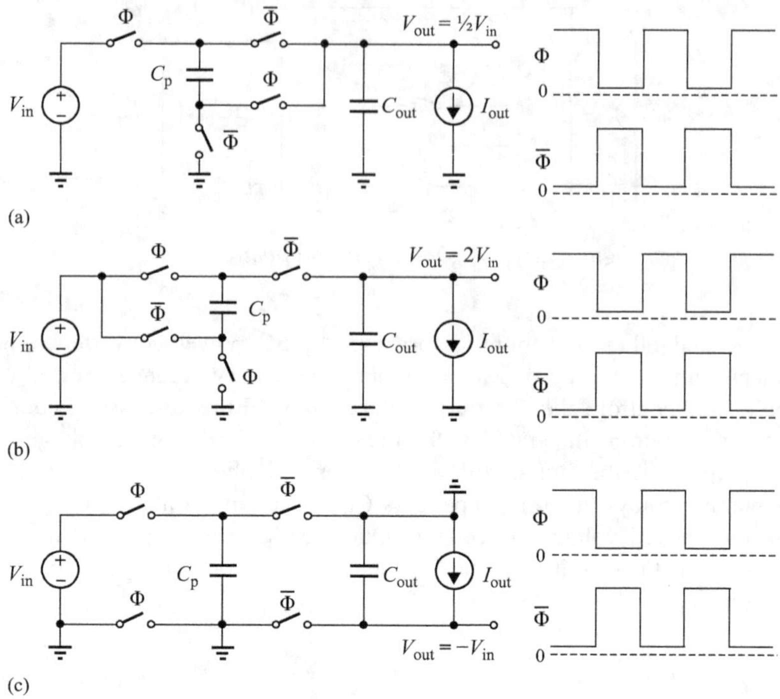

Figure 12.2 (a) A step-down charge pump, (b) a step-up charge pump, and (c) a voltage inverter

sensing and computing, power level of many devices have dropped to only a few milliwatts or below, certain tiny power losses, which used to be ignorable in traditional designs, become critical. Take a highly popular cross-coupled voltage doubler [1, 2] as an example. As depicted in (12.2), the total internal power loss P_{loss} that determines the efficiency of a converter consists of the power losses of the power stage, the gate drivers, and the feedback controller. With modern IC technology, an integrated controller could be designed and implemented with very low static power. Meanwhile, transistor sizes in gate drivers and the controller are much smaller than those in the power stage. Hence, the power loss of the power stage normally dominates P_{loss}.

Based on power loss generation mechanisms, the power loss in a power stage mainly consists of redistribution loss, conduction loss, switching loss and reversion loss [2–5]. The redistribution loss is caused by charge redistribution between the capacitors in a charge pump. For example, in a voltage doubler of Figure 12.3(a), when $\mathbf{\Phi}_A$ is low and $\mathbf{\Phi}_B$ is high, the PMOS power switch M_{pB} is turned on. V_B is charged to $2V_{in}$. However, due to the discharging I_{out} on C_{out}, V_{out} is slightly lower than $2V_{in}$. The voltage difference between V_B and V_{out} forces the charge flowing from C_B to C_{out}, leading to the redistribution loss, which is computed as

$$\frac{1}{2}f_s\left(C_B(V_B - V_{DD})^2 + C_{out}V_{out}^2 - \frac{(C_B V_B + C_{out}V_{out})^2}{C_B + C_{out}}\right).$$

When a MOSFET switch is conducting, the ON resistance causes the conduction loss, which is determined by $\frac{L}{\mu C_{OX}(V_{GS} - V_{TH})W} f_s \int_0^{DT} i(t)^2 dt$. The conduction loss can be reduced by increasing the aspect ratio W/L of the transistor. On the other hand, a MOSFET transistor is a voltage-controlled device. Its ON/OFF states as a switch are accomplished by charging or discharging the gate capacitor C_{GS}. As C_{GS} is proportional to the equivalent conduction channel area, the switching loss is usually proportional to the product of W and L. Hence, with a given channel length L, there is an optimal width which gives the minimum sum of the conduction loss and the switching loss in a MOSFET power switch. At last, a reversion loss occurs when shorting a higher voltage node to a lower one, forming a reversion current against the original power flow. In this case, charge is transferred from a higher voltage circuit node to a lower one. For example, in Figure 12.3(b), when both Φ_A and Φ_B are low during the transition period of t_{Low}, power switches M_{pA} and M_{pB} are both turned on. V_A and V_B are both equal to V_{in}. A current flows from V_{out} to C_A and C_B reversely, degrading the efficiency. The reversion loss is then equal to $\frac{1}{2}f_s\frac{C_B C_{out}(V_{out} - V_B)^2}{C_B + C_{out}}$. In summary, it is crucial to understand the power losses and their causing mechanisms, which plays a vital role in appreciating the topologies and control schemes of SC power converters.

12.3 Control schemes

Unlike the control schemes addressed for conventional SC power converters, the ones discussed here have strong focus on the emerging technologies, driven by fast-growing self-powered devices and implantable biomedical microsystems. Such an SC power converter is usually expected to operate robustly at ultra-low power levels, improve operation lifetime, and reduce size, cost, and PCB footprint with highly integrated implementation, by overcoming new design challenges coming along with new operation environments and design specifications. For example, when a self-powered device scavenges energy from the ambient environment, the harvested power level is usually low and the voltage level can fluctuate largely due to varying intensity and availability of the energy sources. Hence, even for a fixed V_{out}, the efficiency can change considerably due to the fluctuating input voltage level. In the meantime, because of inherent switching actions, power supplies such as SC power converters generate voltage and current ripples at both inputs and outputs, causing substantial switching noise. This noise can be coupled into the entire IC through the supply metal lines or through the

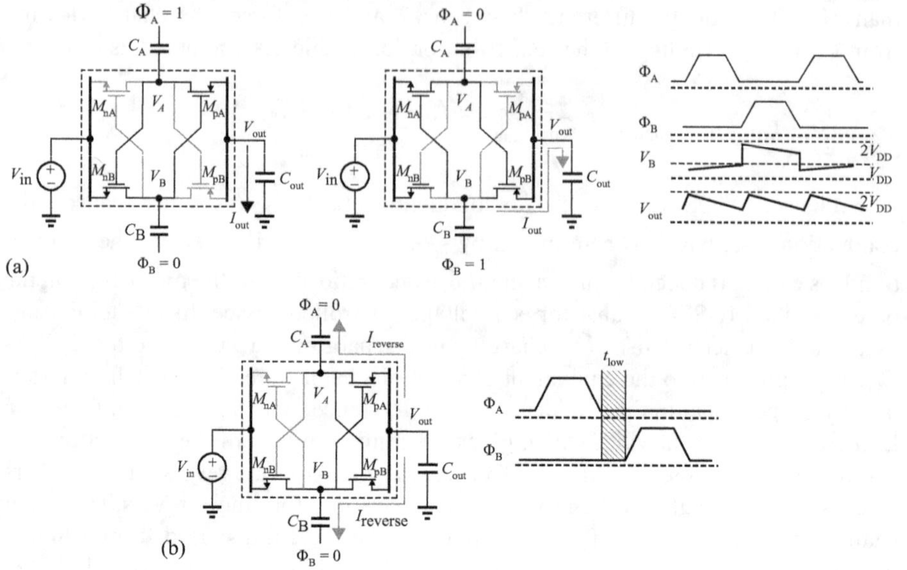

Figure 12.3 Illustration of (a) redistribution loss generation, and (b) reversion loss generation.

substrates of the power transistors. As a result, the reliability and performance of load applications is severely impacted. Furthermore, to effectively utilize the limited harvested energy, self-powered devices are operated with smart power management techniques, such as dynamic voltage frequency scaling (DVFS). This requires the adaptive adjustment of the output voltage with respect to the instantaneous operating conditions, such as workload and power demand. However, one of the major drawbacks with traditional SC power converters is their ability to provide only a fixed CG. If the output voltage deviates from this desired level, the efficiency of the converter drops. Hence, to accommodate large input and output voltage range applications, a conventional SC power converter with a fixed CG is insufficient and impractical.

This section introduces the control schemes that contribute to robust, efficient, low noise, and fast response operations in SC power converters. The section starts with two-stage regulation strategies involving linear analog devices and circuits, with focus on switching noise and in-rush current suppression. To adapt to largely varying input and output voltage applications, schemes of reconfiguring the power stages are then addressed. Based on the instantaneous operating conditions, SC converters are capable of undergoing dynamic reconfiguration the optimal power stage topology, which minimizes power losses and provides enhanced power regulation. Finally, in order to improve the transient response, ripple voltage and system robustness, frequency and/or phase modulation schemes, as well as multiphase interleaving schemes are discussed. While each of the control schemes can regulate an SC power converter independently, it is feasible and encouraging to apply multiple schemes in one SC power converter to achieve jointly optimized performances. As the schemes as such are hundreds, the chapter only provides a sneak peek of their fundamental operation principles.

12.3.1 Two-stage regulation strategies

In the conventional applications, since the input and output voltages variations are not significant, many SC power converters only achieve single-fixed CG. For simplicity, a feedback controller is avoided. Following a charge pump, an area-efficient low dropout (LDO) linear regulator is usually utilized for post-regulation, as illustrated in Figure 12.4. Although a charge pump cannot ensure accurate and low-noise voltage regulation, it can be accomplished by the LDO regulator. Clearly, such a strategy is straightforward and easy to implement. Hence, it still remains popular today, especially when variations at both input and output voltages are relatively low.

As CMOS technology enters nanometre regime, such a two-stage topology demonstrates new advantages over a single-stage SC topology. With nanoscale CMOS technology, although circuit speed and chip density are highly improved, power transistors breakdown voltages are scaled down, causing potential breakdown failures such as oxide breakdown and source/drain-substrate junction diode breakdown. This jeopardizes system reliability and seriously narrows the operation voltage range of an SC power converter. For example, consider a single-stage charge pump, as shown in Figure 12.5. When it operates in the charge phase Φ, circuit node A is connected to the input V_{in} and the voltages across the thin gate oxide (V_{DG2}) and drain-source (V_{DS2}) of power switch M_2 are equal to $V_{in} - V_{out}$, which are 2 V in this example. In the discharge phase $\overline{\phi}$, the node A is connected to output V_{out} (1V), thus V_{DG} and V_{SD} stresses on switch M_1 are also 2 V. As the rated breakdown voltage is 1.8 V in this fabrication process, M_1 and M_2 both face multiple reliability risks, since their oxides and channels undergo considerable overvoltage stresses.

In order to mitigate breakdown risks and achieve robust operations, both device- and circuit-level measures can be taken. On the device level, applying thickened oxide or extended drains helps increase breakdown voltages. However, they require extra process steps, larger die area, and thus significantly increased

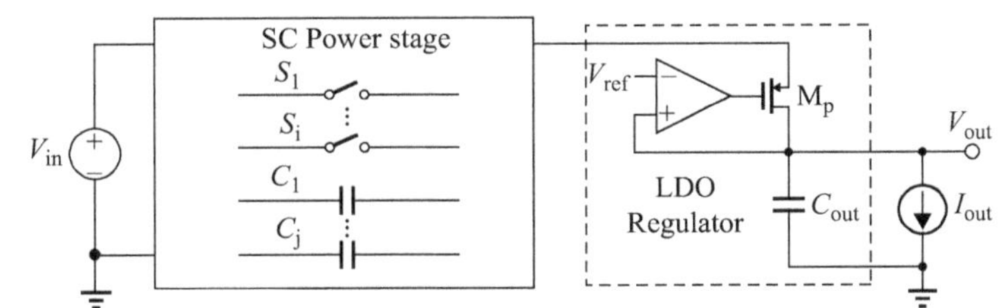

Figure 12.4 A cascaded SC-LDO topology using LDO for post-regulation

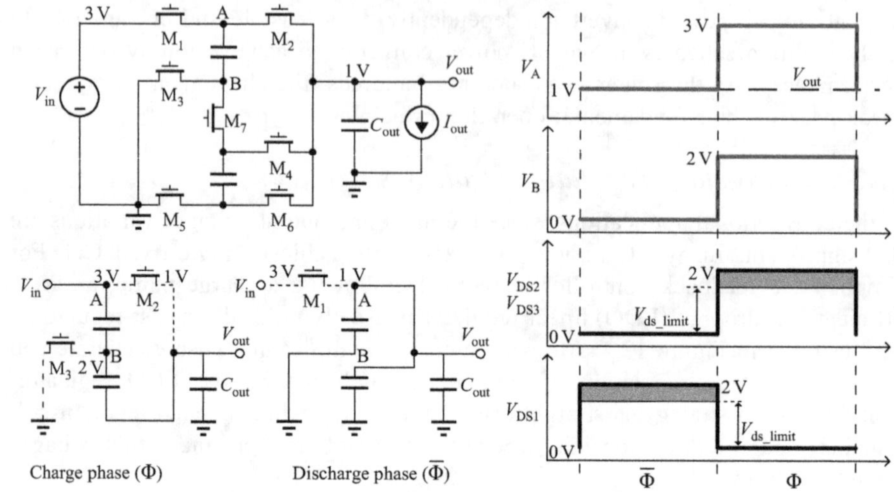

Figure 12.5 Illustration of breakdown risk of core power transistors in a charge pump

cost. On the circuit level, stacked transistor structures can be used. However, it increases power losses due to additional gate drivers and the stacked transistors.

To overcome the above challenges, a two-stage power converter [6], consisting of an SC power stage and an LDO regulator, can be employed, as illustrated in Figure 12.6. The main objective is to mitigate the breakdown risks imposed by high input voltage, while still receiving benefits from high performances by the thin-oxide core transistors. In contrast to the conventional converter in Figure 12.5, the two-stage cascaded topology allows the SC power stage to raise its output voltage level to V_{out1} (in the example, it is 1.2 V), whereas the second-stage LDO still regulates the output voltage V_{out} at the targeted voltage (1 V). A fundamental difference, however, is that all the power transistors in the SC power stage no longer suffer from breakdown risks in both charge and discharge phases, as illustrated in Figure 12.7. Meanwhile, the new topology is beneficial to achieve monolithic implementation. In an SC power converter as in Figure 12.5, the output filter capacitor is usually large to reduce the ripple voltage, which is difficult for monolithic integration. With the two-stage topology in Figure 12.6, the value of the capacitor can be significantly reduced due to high power supply rejection ratio of the LDO regulator.

Another important application of two-stage regulation strategies is to suppress undesirable input current surge. Such input current surge is often observed during start-up transients, causing potentially fatal damages to the power devices. In switch mode power converters, because the inrush current has to flow through the inductors before getting to the load, the current slew rate limit of an inductor confines the drastic changes of such an inrush current, especially with the aid of soft start-up mechanisms. However, for SC power converters, the problem appears

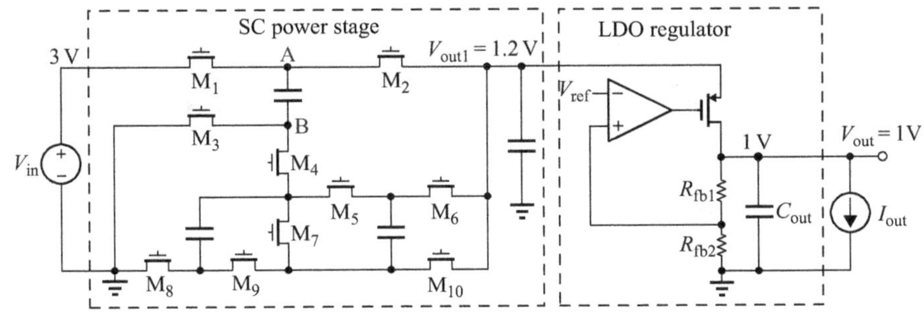

Figure 12.6 Breakdown-resilient two-stage SC-LDO cascaded topology in Reference 6

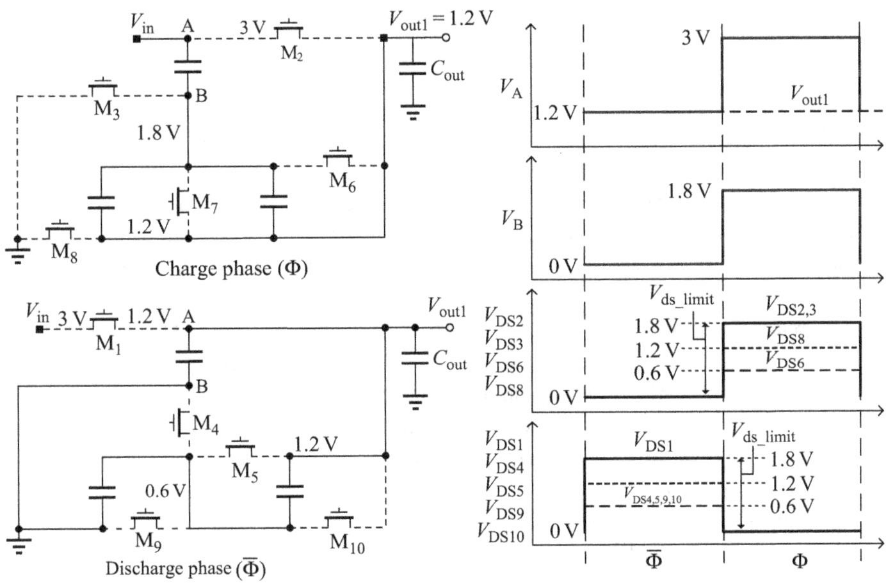

Figure 12.7 Illustration of reducing breakdown risk in the two-stage SC-LDO topology in Reference 6

to be more severe due to the absence of the inductors. Meanwhile, since the capacitors in charge pumps do not conduct continuous currents, they generate more current ripple and switching noise inherently.

To suppress the input inrush current in an SC power converter, a current-limiting device is essentially needed, which plays the similar role as an inductor in a switch mode power converter. Such a current-limiting circuit can be implemented in numerous ways. For example, it can be achieved by the voltage controlled current source (VCCS) shown in the two stages in Figure 12.8. In circuit level, the

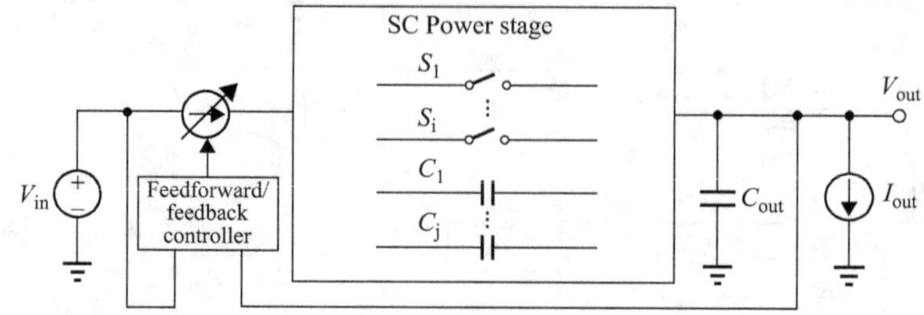

Figure 12.8 Two-stage LDO-SC power converter for in-rush current suppression

VCCS can be implemented by a MOSFET power transistor operating in saturation region or a BJT power transistor operating in linear region. In addition to current limiting, such a device also improves the power supply noise rejection in a power converter.

12.3.2 Reconfiguration schemes

In self-powered portable devices, the power source could be very unstable. Therefore, efficiency in an SC converter is difficult to maintain high at a single-fixed CG. It is expected that a converter should achieve superb line regulation for robust operation. Moreover, it should have the capability of reconfiguring its power stage and therefore accomplishing multiple CGs for high efficiency. Under these circumstances, reconfiguration schemes in SC power converter become very attractive.

A generic reconfigurable charge pump is illustrated in Figure 12.9. In general, the circuit consists of a switch-capacitor array. Each capacitor in the array is associated with six switches, which can flexibly connect either plate of the capacitor to V_{in}, V_{out}, ground, or another capacitor. For example, the top plate of C_{P1} can be connected to V_{in} by S_{11}, to V_{out} by S_{12}, or to the bottom plate of C_{PN} by S_{N6}. Meanwhile, the bottom plate of C_{P1} can be connected to V_{in} by S_{13}, to V_{out} by S_{14}, to the top plate of C_{P2} by S_{16}, or to ground by S_{15}. In general, with N pumping capacitors and $6N$ switches, an SC power converter can achieve $4N-5$ different CGs, with the options of 1 to N interleaving phases [7]. For step-up conversions, the CG can be represented as i/j, where $j = 1, 2, \ldots, N$, and $i = j, j+1, \ldots, N$. For step-down voltage conversions, the CG can be represented as i/j, where $j = 1, 2, \ldots, N$, and $i = 1, 2, \ldots, j$. In practical implementations, the generic architecture can be simplified in order to reduce the switch count and thus the cost. For example, if only step-down conversions are needed, the switches S_{i3} can be eliminated. The charge pump then offers $2N-2$ step-down CGs with N capacitors and $5N$ switches. Similarly, the switches S_{i4} can be eliminated if only step-up conversions are required. In this case, the charge pump can provide $2N-3$ CGs with N capacitors and $5N$ switches.

To illustrate the operation of an SC reconfigurable power converter, a step-up and step-down SC power converter with five CGs is illustrated in Figure 12.10.

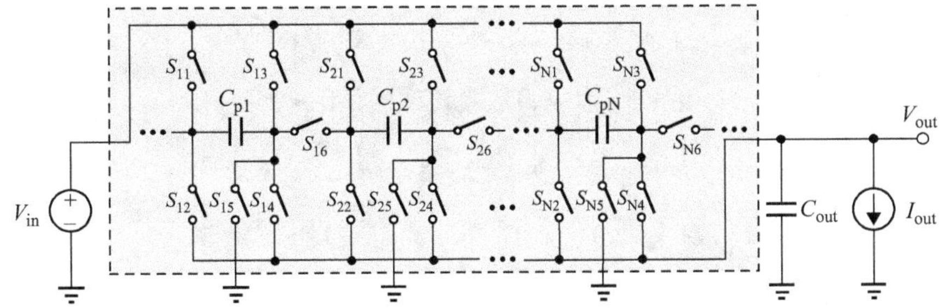

Figure 12.9 Generic architecture of a reconfigurable SC power converter with multiple CGs

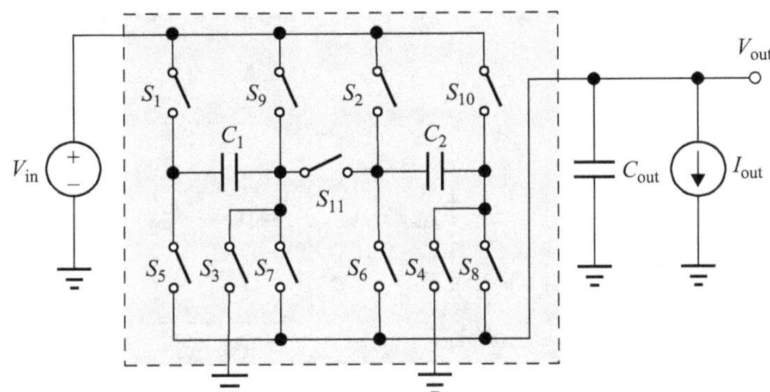

Figure 12.10 A reconfigurable step-up and step-down SC power converter achieving five CGs

The converter employs two pumping capacitors C_1 and C_2, and 11 switches that are controlled by a pair of complementary clock signals ϕ and $\overline{\phi}$. It can be reconfigured with five different CGs: 1/2, 2/3, 1, 3/2, 2. For example, as depicted in Figure 12.11(a), in order to achieve a CG of 3/2, during the charge phase ϕ, the power switches S_1, S_4, S_{11} are turned on. This creates a closed loop to charge the capacitors C_1 and C_2. During the discharge phase $\overline{\phi}$, the switches S_5, S_6, S_9, S_{10} are turned on. Accordingly, two discharge paths are established to deliver the power from C_1 and C_2 to V_{out}. Eventually, in steady state, a step-up voltage conversion of 3/2 is implemented between the input and the output. Similarly, Figure 12.11(b–e) demonstrates the reconfiguration of the converter in achieving other CGs 1/2, 2/3, 1, and 2, respectively.

In summary, reconfigurable SC power converters are capable of operating with multiple CG topologies and reliably accommodating the fluctuating input voltages. In addition, by supplying multiple output voltage levels, it makes DVFS operation

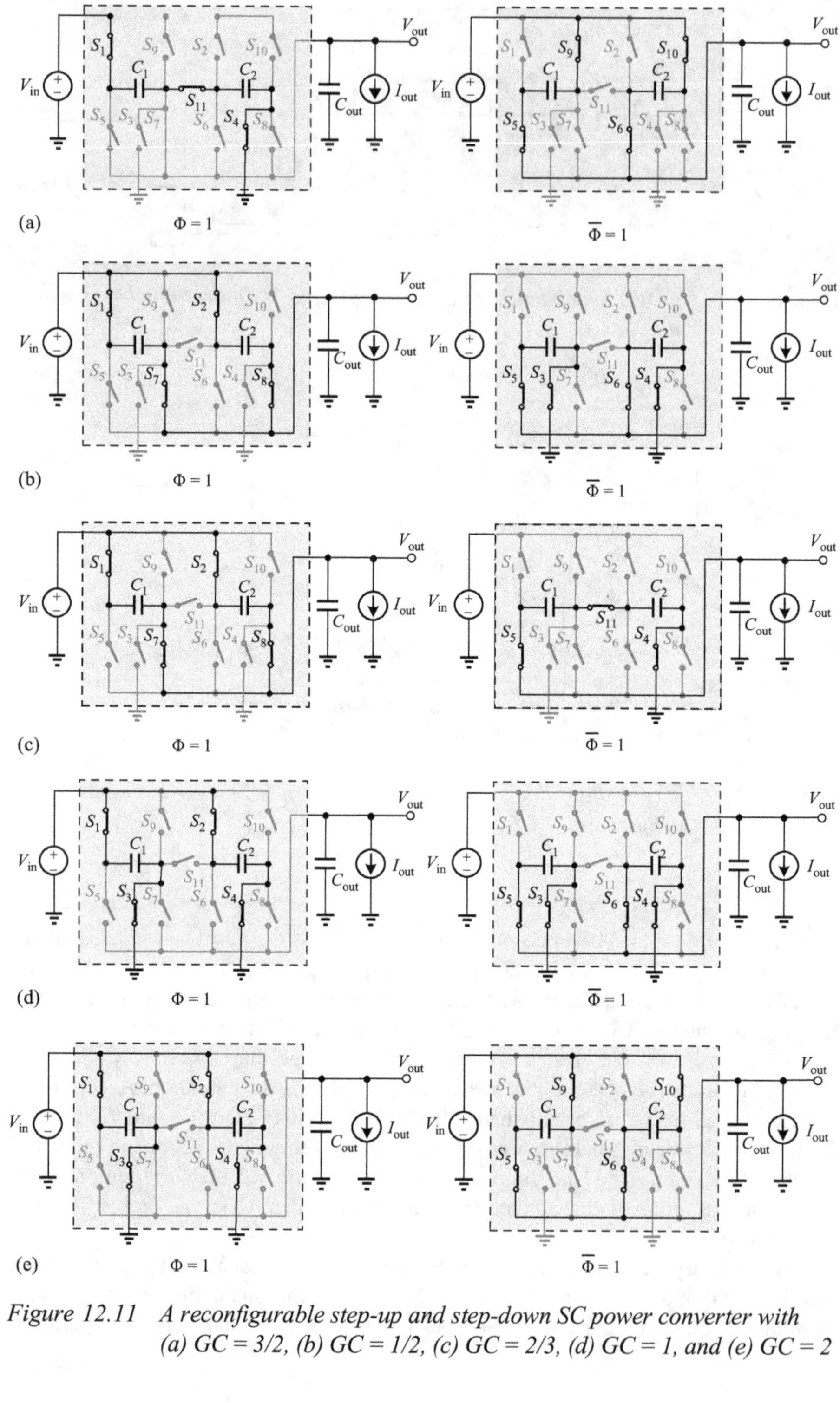

Figure 12.11 A reconfigurable step-up and step-down SC power converter with (a) GC = 3/2, (b) GC = 1/2, (c) GC = 2/3, (d) GC = 1, and (e) GC = 2

possible under an SC power supply. However, the use of a highly reconfigurable power stage significantly increases the number of power components, such as the pumping capacitors and power switches, along with the complexity of the controller. This introduces considerable challenges during the topology and efficiency optimization.

12.3.3 Pulse frequency modulation and pulse control schemes

While the CG of a switch mode power converter is set by the duty ratio, an SC power converter exhibits a constant CG confined by its own topology. In this sense, the control mechanism of an SC power converter has similarity to that of an LDO regulator, where the output resistance (R_{out}) is modulated to adaptively adjust the output voltage according to the load. In an SC converter, R_{out} is a function of switching frequency, duty ratio, and ON resistance of power switches.

From the perspective of circuit implementation, switching frequency is considered the most straightforward variable for voltage regulation among the three, resulting in the so-called pulse frequency modulation (PFM) schemes. Because R_{out} is reversely proportional to the switching frequency, regulation on the output voltage is achieved by scaling down the switching frequency of an SC converter (equivalently increasing R_{out}) at light load, and up at heavy load. However, as R_{out} becomes a weak function of switching frequency in high frequency domain, the PFM regulation become less effective to heavy load operation. Other control mechanisms, such as power transistor sizing and CG control, can then be added to compensate the regulation performance in PFM.

In terms of system implementation, hysteretic feedback control is a common way to operate an SC power converter in PFM mode. Different from a conventional pulse width modulation control, hysteretic control overcomes the problem of instability and achieves fast transient response. Because a hysteretic controller usually has very few or none analog component, the quiescent power can be very low. Power consumption in the hysteretic controller thus is much reduced.

Here, we introduce two hysteretic feedback controls for PFM operation. If the output capacitor has the same magnitude of the flying capacitors, the output voltage ripple would be significant, which can be directly compared with a voltage reference to generate the gate control signals. This is called the single bound hysteretic control. Figure 12.12 illustrates a simplified single bound hysteretic control SC power converter [8]. On the other hand, if an SC converter uses much larger output capacitor or multiphase topology, and thus has negligible output voltage ripple, a dual-bound hysteretic control can be used [9] as shown in Figure 12.13. When the output voltage surpasses the upper bound, the control clock CLK is disabled. No more charge action takes place. The output voltage drops due to load current discharging. When it drops below the lower bound, the clock CLK will be activated. The switching actions for charge delivery resume. As results, the output voltage is limited within the hysteresis window. With the employment of the clock-triggered comparators, the quiescent power consumption can be improved. Clearly, the hysteresis window size impacts the output voltage ripple. For small ripple

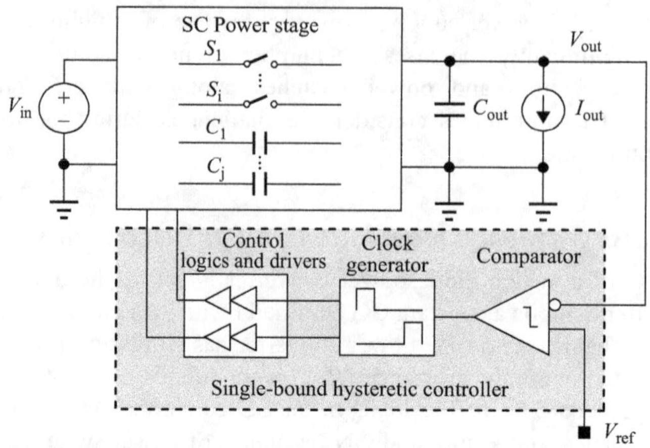

Figure 12.12 Block diagram of a single bound hysteretic controller SC power converter for PFM operation

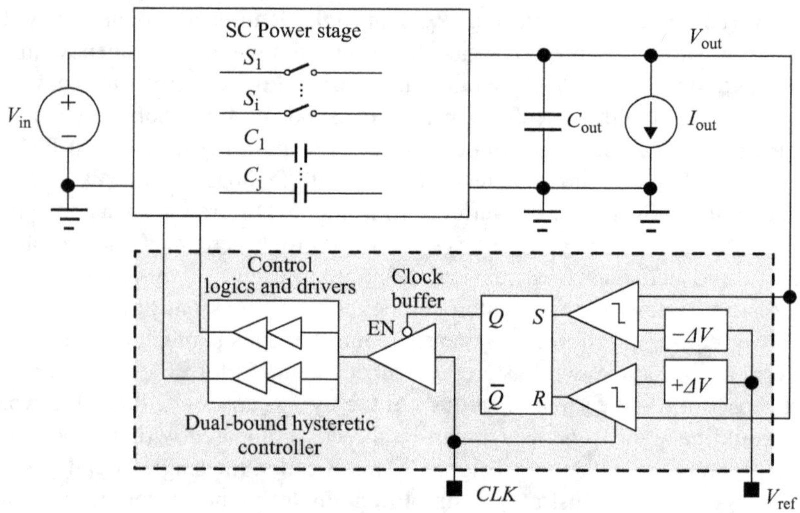

Figure 12.13 Block diagram of a dual-bound hysteretic controller SC power converter for PFM operation

voltage performance, a small hysteresis window and a high sampling rate for the comparator is preferred, which is at the cost of high switching power loss.

It should be noted that the reconfigurable schemes in Section 12.3.2 can be incorporated with PFM hysteretic control, which leads to a combination of pulse frequency control and CG control. Accordingly, an SC power converter with an adaptive gain–pulse (AG–AP) control is introduced [10]. The converter adaptively employs a reconfigurable SC power stage with adjustable CG and variable power pulses for efficient operation under a wide input range.

The method of topology reconfiguration is based on Figure 12.11 in Section 12.3.2. Unlike a switch mode power converter, the efficiency of the SC power converter is highly related to its CG. With a fixed CG, when the input power source is highly variable, the efficiency could drop dramatically. Hence, a reconfigurable SC converter with multiple CGs is highly desirable. This can be achieved through the circuit in Figure 12.14. The control of the converter is in fact a combination of AG control and AP control. Different CGs in the converter lead to different charge and energy transference capabilities.

As shown in Figure 12.14(a), the controller employs dual control paths. In the feed-forward path, the input ADC senses V_{in}, so that the controller determines the instant CG for optimized efficiency. As shown in Figure 12.11, the SC converter can achieve five CGs. However, if only the AG control is applied, the durations for

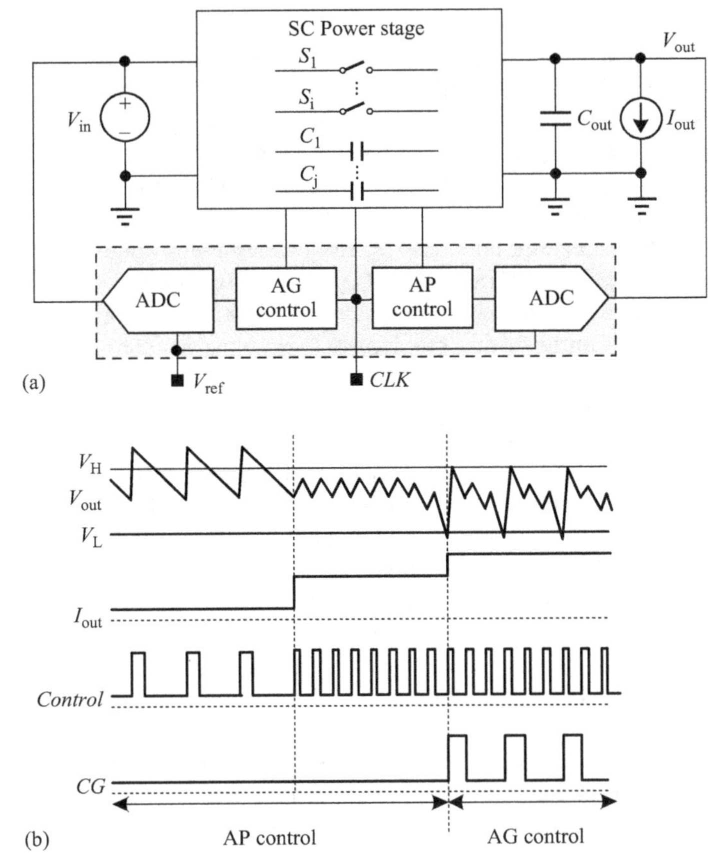

Figure 12.14 (a) Block diagram and (b) control timing diagram of an SC power converter with AG and AP control in Reference 10

charge and discharge phases are fixed. In steady state, if the energy delivered in charge phase is much higher than the actual load demand, the converter does not have a mechanism to reduce the charge current. As a result, ripple voltages increase. In addition, at light load, this fixed high frequency switching causes substantial switching power loss, degrading the efficiency. The AP control (PFM operation indeed) thus takes into effect in this scenario (Figure 12.14(b)). At light load, the load has no urgent energy demand. The controller adaptively reduces the switching frequency. Switching loss of the converter is then reduced, and the efficiency thus remains at a relatively high level. When the load has a sudden increase and AP control could not supply sufficient energy, AG control takes the charge. It reconfigures the converter with a higher CG for more charge and energy delivery.

12.3.4 Interleaving multiphase regulation

Inherent switching actions in both switch mode and SC power converters cause output voltage ripple and current ripple, which lead to substantial switching noise. To further understand the impact of voltage and current ripple generated by an SC power converter, consider a traditional voltage doubler in Figure 12.15(a). When charge phase Φ is active, pumping capacitor C_P is charged by the current from V_{in}. In this period, the output V_{out} is isolated from both V_{in} and C_P. Because load current I_{out} discharges C_{out}, V_{out} drops below the average level. In discharge phase $\overline{\Phi}$, the output V_{out} is connected to the top plate of C_P. A charge path is established from V_{in} to V_{out}, delivering current to charge up C_{out}. V_{out} rises above the average level at the end of the discharge phase. The ripples at V_{out} are clearly caused by the charge and discharge actions in the charge pump, contributing switching noise in an SC converter. One way to mitigate this problem is to employ an interleaving regulation technique.

With such a technique, a power converter is decomposed into multiple subconverters, which share the same input and output circuit nodes. Each subconverter processes a fraction of the total output power with a phase shift in between, leading to several benefits such as input and output ripple cancellation and improved transient response. This can be verified through the example in Figure 12.15(b) [11]. With the same switching frequency, each pumping capacitor and power transistor size are reduced by half, thus avoiding the penalty on extra silicon. Other advantages of this work lie in that the output ripples are reduced by nearly 50%, while the load transient response is largely improved. To further improve the ripple and transient response performance, more interleaving phases can be added with proper circuit adjustment such as in Figure 12.16.

As mentioned in section 12.2, power loss minimization is critical for the efficiency of an SC power converter. In order to minimize the power losses, the control clock signals have to be precisely designed. With more operation phases in a interleaving SC converter, the issue could become very complicated, which is considered the most challenging issue in interleaving multiphase SC power converter designs. For example, phase-delayed control is applied in the interleaving SC converters in

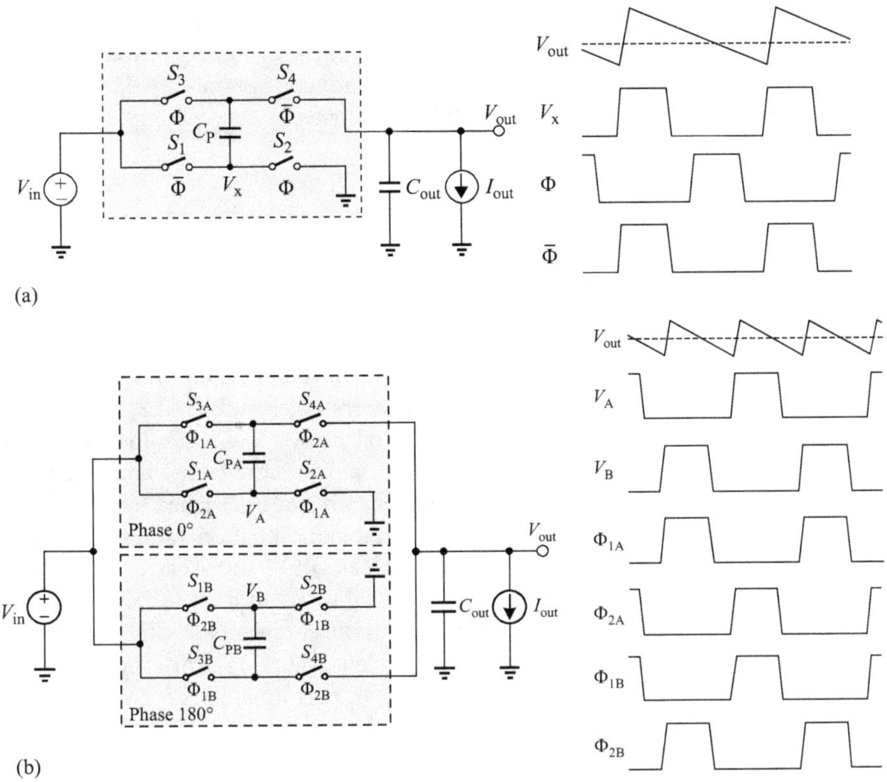

Figure 12.15 (a) A conventional CMOS voltage doubler with key waveforms, (b) a two-phase interleaving voltage doubler with key waveforms in Reference 11

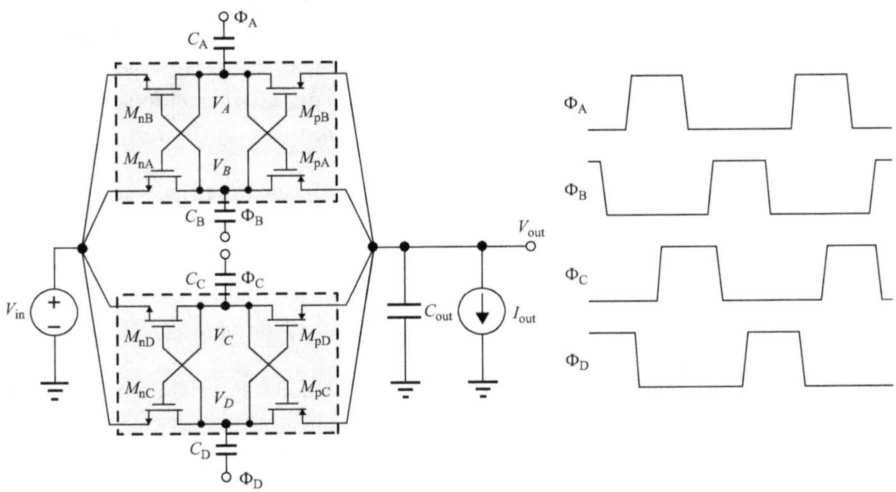

Figure 12.16 A CMOS cross-coupled interleaving charge pump in Reference 12

References 11 and 12. As depicted in Figure 12.16, by carefully managing the control clocks and the delays between them, an interleaving SC power converter effectively reduces the switching noise with a continuous output regulation. In the meantime, reverse power loss is successfully eliminated.

12.4 Conclusions

Driven by the emerging applications such as wireless sensors and self-powered biomedical implants, monolithic SC power converters have become a very attractive solution for on-chip power supplies. In this chapter, Section 12.1 first addresses the design challenges. To prepare the readers to fully understand the control and operation methods of these SC power converters, in Section 12.2, key design parameters and causing mechanisms of potential power losses are introduced. In Section 12.3, major control schemes for SC power converters are discussed. Focusing on reliability and inrush current issues, two-stage cascaded topologies are first explained. In order to adapt to variable input and output operation conditions, reconfigurable SC power converters are introduced. PFM and AG–AP control are then discussed to provide different regulation solutions at a wide range of load. To further suppress the ripples, improve the transient response and robustness, examples of interleaving regulations are provided. Many control schemes in the chapter can jointly operate to achieve optimal performance of an SC power converter.

References

[1] Favrat P., Deval P., Declercq M. 'A High-Efficiency CMOS Voltage Doubler'. *Journal of Solid State Circuits*. 1998; **33**(3): 410–416.

[2] Su F., Ki Y. H., Tsui C. Y. 'High Efficiency Cross-Coupled Doubler with No Reversion Loss'. *IEEE International Symposium on Circuits & Systems*; Kos, Greece, 2008, pp. 2761–2764.

[3] Ki Y. H., Tsui C. Y. 'Charge Redistribution Loss Consideration in Optimal Charge Pump Design'. *IEEE International Symposium on Circuits & Systems*; Kobe, Japan, 2005, pp. 1895–1898.

[4] Su L., Ma D. 'Design and Optimization of Integrated Low-Voltage Low-Power Monolithic CMOS Charge Pumps'. *International Symposium on Power Electronics, Electrical Drives, Automation and Motion*; Ischia, Italy, 2008, pp. 43–48.

[5] Ma D., Bondade R., *Reconfigurable Switched-Capacitor Power Converters*. New York: Springer; 2013.

[6] Geng Y., Ma D. A 100-MHz Breakdown-Resilient Power Converter with Fully Monolithic Implementation on Nanoscale CMOS Process. *Transaction on Industrial Electronics*, 2015; **62**(9): 5529–5538.

[7] Chowdhury I., Ma D., Design of Reconfigurable and Robust Integrated SC Power Converter for Self-Powered Energy-Efficient Devices. *Transaction on Industrial Electronics.* 2009; **56**(10): 4018–4028.
[8] Seeman M. D., *A Design Methodology for Switched-Capacitor DC-DC Converters*. University of California, Berkeley; 2009.
[9] Seeman M. D., Sanders S. R., Rabaey J. M. 'An Ultra-Low-Power Power Management IC for Wireless Sensor Nodes'. *IEEE Custom Integrated Circuits Conference*; San Jose, USA, 2007, pp. 567–570.
[10] Chowdhury I., Ma D. An Integrated Reconfigurable Switched-Capacitor DC-DC Converter with Dual-Loop Adaptive Gain-Pulse Control. *IEEE International Symposium on Circuits & Systems*; Seattle, USA, 2008, pp. 2610–2613.
[11] Su F., Ki Y. H., Tsui C. Y. 'Regulated Switched-Capacitor Doubler with Interleaving Control for Continuous Output Regulation'. *Journal of Solid State Circuits*. 2009; **44**(4): 1112–1220.
[12] Somasundaram M. S., Ma D. Low-Ripple CMOS Switched-Capacitor Power Converter with Closed-Loop Interleaving Regulation. *IEEE Custom Integrated Circuits Conference*; San Jose, USA, 2006, pp. 245–248.

Chapter 13

DSP-based natural frame control schemes for three-phase unity power factor rectifiers

Ramon Guzman[1], Luis Garcia de Vicuña[2] and Rafael Peña Alzola[3]

13.1 Introduction

Traditionally, digital signal processor (DSP) control algorithms for three-phase power converters are designed in rotating or stationary reference frames. These approaches require the use of rotation matrices and employ linear controllers such as proportional integral (PI) and proportional resonant. This chapter presents an alternative control solution developed in the natural reference frame and applied to a three-phase unity power factor rectifier (UPFR). This solution requires no transformation matrices, and by harnessing all computation capabilities of the modern DSPs, nonlinear techniques, such as sliding-mode control (SMC) [1, 2] and Kalman filter (KF) [3], can be employed. The main features, along with the advantages and the limitations, of this approach will be discussed in detail throughout the chapter.

UPFR's are characterized by a power factor very close to unity and fast output voltage dynamic response can be achieved if feedforward output-current control is implemented. Many solutions have been reported to improve the power factor in single-phase [4, 5] and three-phase rectifiers [6, 7]. A review of the state of the art about UPFRs can be found in Referenes 8–10.

As it is well known, sliding mode control inherently presents characteristics such as robustness, insensitivity to system parameter variation, and fast dynamic response. However, the drawbacks of this technique are a variable switching frequency and a heavy interference between phases in three-phase three-wire systems.

Fixed switching frequency in three-phase converters can be obtained using hysteretic modulators with variable band, digital hysteretic modulators being the most common solution [11–13]. However, a high sampling frequency is needed for the correct implementation of digital hysteretic modulators in order to make the

[1]Department of Signal Theory and Communication, Technical University of Catalonia, Vilanova i la Geltrú, Spain
[2]Department of Electronic Engineering, Technical University of Catalonia, Vilanova i la Geltrú, Spain
[3]Electrical and Computer Engineering, The University of British Columbia, Vancouver, BC, Canada

control algorithm effective. This problem can be solved if analog hysteretic comparators are used but at the expense of additional hardware [14]. Another solution can be adopted by using predictive control and the DSP timers to decide the switching time [15]. In any case, in order to achieve fixed switching frequency in the natural reference frame, the three controllers must be decoupled, so that each controller only can depend on its own control variable [14, 16].

In this chapter, an effective model-based solution, using a KF algorithm, is proposed in order to solve the phase-coupling problem of the UPFR and to estimate the required state variables and the grid-voltages required by the control algorithm. With this aim, a decoupled converter linear model is derived that allows to deal with a three-phase system as three independent single-phase systems from the control point of view. Moreover, this approach also allows using only current sensors, without needing grid voltage sensors, which reduces the cost and increases the reliability. Once the controller variables have been decoupled, hysteretic modulators with variable band and switching decision algorithm (SDA) can be used to achieve fixed switching frequency. This methodology allows to design sliding mode controllers, which provide linear dynamics in the output voltage and robustness against load changes.

This chapter is organized as follows: in the Section 13.2, a physical nonlinear model for a three-phase rectifier is presented. In Section 13.3, a conventional sliding mode control based on this model is implemented, which exhibits severe interferences between phases. A decoupled linear converter model is introduced in Section 13.4 in order to obtain a discretized model to be used in the KF algorithm. A sliding mode control scheme based on a KF algorithm is presented in Section 13.5. The sliding mode control strategy and the output voltage control are introduced in Section 13.6. Practical considerations to operate at fixed switching frequency using variable hysteresis band and SDA are shown in Section 13.7. Finally, experimental results are presented in Section 13.8.

13.2 Physical model of the power converter

A three-phase UPFR is represented in Figure 13.1.

From this circuit, the three-phase system differential equations for each phase-leg i, with $i \in \{a,b,c\}$, can be written as follows

$$L\frac{d\mathbf{i}}{dt} = \mathbf{v} - \frac{v_o}{2}\mathbf{u} + v_n\mathbf{1}^T \tag{13.1}$$

$$C\frac{dv_o}{dt} = i_a\frac{u_a}{2} + i_b\frac{u_b}{2} + i_c\frac{u_c}{2} - i_o \tag{13.2}$$

$$v_n = \frac{v_o}{2}\left(\frac{u_a + u_b + u_c}{3}\right) \tag{13.3}$$

where v_n is the neutral point voltage, $\mathbf{i} = (i_a\ i_b\ i_c)^T$ is the inductor current vector, $\mathbf{v} = (v_a\ v_b\ v_c)^T$ is the grid voltage vector, with $v_a + v_b + v_c = 0$, $\mathbf{u} = (u_a\ u_b\ u_c)^T$ is the control variables vector with $u_{a,b,c} \in \{\pm 1\}$, v_o is the output voltage, and $\mathbf{1}^T$ is a column vector defined as $[1\ 1\ 1]^T$.

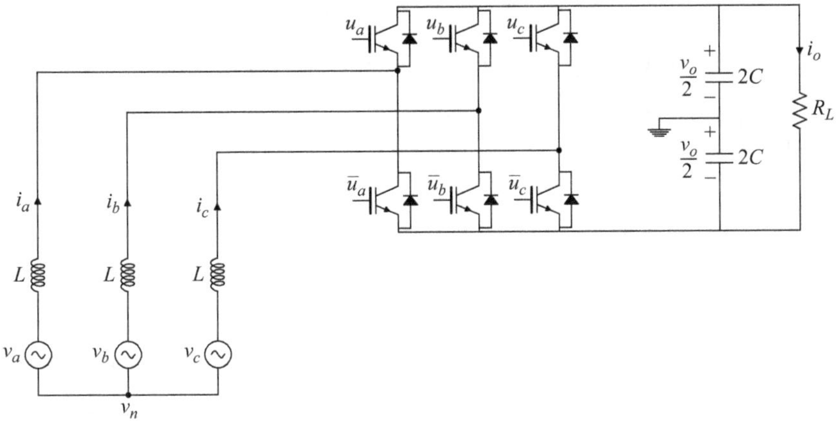

Figure 13.1 Circuit diagram of three-phase unity power factor rectifier

13.3 Conventional sliding mode control in three-phase converters

In this section, a sliding surface vector $\mathbf{S} = [S_a\ S_b\ S_c]^T$ is proposed to control the UPFR. In the conventional SMC, the sliding surface vector is expressed with the use of the following equation:

$$\mathbf{S} = \mathbf{i}^* - \mathbf{i} \tag{13.4}$$

where in an UPFR the reference current vector is proportional to the grid voltage vector $\mathbf{i}^* = k\mathbf{v}$.

The controllers' dynamics $\dot{\mathbf{S}}$ can be obtained by taking the time derivative of (13.4). Then using (13.1) and (13.3) the system dynamics for the sliding surfaces can be found. Note that a coupling term due to the neutral point voltage v_n is presented in each of the following expressions:

$$\frac{dS_a}{dt} = \frac{di_a^*}{dt} - \frac{1}{L}\left(v_a - \frac{v_o}{2}u_a + \frac{v_o}{2}\frac{u_a + u_b + u_c}{3}\right) \tag{13.5}$$

$$\frac{dS_b}{dt} = \frac{di_b^*}{dt} - \frac{1}{L}\left(v_b - \frac{v_o}{2}u_b + \frac{v_o}{2}\frac{u_a + u_b + u_c}{3}\right) \tag{13.6}$$

$$\frac{dS_c}{dt} = \frac{di_c^*}{dt} - \frac{1}{L}\left(v_c - \frac{v_o}{2}u_c + \frac{v_o}{2}\frac{u_a + u_b + u_c}{3}\right). \tag{13.7}$$

The aforementioned equations can be expressed in a matrix form as

$$\dot{\mathbf{S}} = \frac{d\mathbf{i}^*}{dt} - \frac{\mathbf{v}}{L} + \frac{v_o}{6L}\mathbf{C}\mathbf{u} \tag{13.8}$$

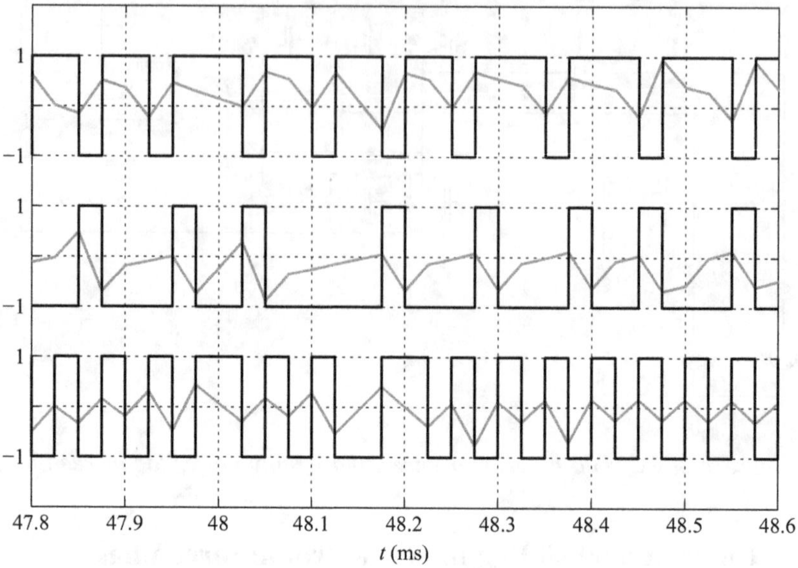

Figure 13.2 Three-phase control signals with corresponding sliding surfaces in a cross-coupled situation: top u_a and S_a, middle u_b and S_b, and bottom u_c and S_c

where

$$\mathbf{C} = \begin{pmatrix} -2 & 1 & 1 \\ 1 & -2 & 1 \\ 1 & 1 & -2 \end{pmatrix} \tag{13.9}$$

Due to the linear dependence between inductor currents, matrix $\mathbf{C}$ is singular (i.e., $det\{\mathbf{C}\} = 0$) and the system is cross-coupled provoking that each switching surface dynamics is affected by the commutations of the other phase-legs due to the neutral point voltage. This interference causes severe irregularities in the ordinary hysteresis operation. This phenomenon is shown in Figure 13.2. Each switching surface changes its slope several times while its control action u_i remains unchanged. These slope variations are due to the other control actions u_j, and sometimes they can produce losses in the sliding motion and, as a consequence, a reduction of the switching frequency.

13.4 Decoupled model of the power converter

In this section, a solution to the cross-coupling problem based on a decoupled model is introduced. Here, a control alternative is used to remove the neutral point influence. Indeed, if matrix $\mathbf{C}$ is diagonal the problem is solved. Thus, the design is reduced to finding a decoupled model for the UPFR and to using it in a state space observer in order to estimate the necessary variables. The proper model is obtained by removing the neutral point voltage from the inductor current equation in (13.1), yielding

$$\frac{dS_a}{dt} = \frac{di_a^*}{dt} - \frac{1}{L}\left(v_a - \frac{v_o}{2}u_a\right) \tag{13.10}$$

$$\frac{dS_b}{dt} = \frac{di_b^*}{dt} - \frac{1}{L}\left(v_b - \frac{v_o}{2}u_b\right) \tag{13.11}$$

$$\frac{dS_c}{dt} = \frac{di_c^*}{dt} - \frac{1}{L}\left(v_c - \frac{v_o}{2}u_c\right). \tag{13.12}$$

The aforementioned equations can be expressed in a matrix form in a similar way that in (13.5)–(13.7)

$$\dot{\mathbf{S}} = \frac{d\mathbf{i}^*}{dt} - \frac{\mathbf{v}}{L} + \frac{v_o}{2L}\mathbf{C}\mathbf{u} \tag{13.13}$$

where

$$\mathbf{C} = \begin{pmatrix} 1 & 0 & 0 \\ 0 & 1 & 0 \\ 0 & 0 & 1 \end{pmatrix} \tag{13.14}$$

As it can be seen, matrix **C**, which multiplies the control signal **u**, is diagonal and the cross-coupling is removed, making possible to implement three independent controllers, one for each phase-leg i. Figure 13.3 shows this decoupling effect, where each sliding surface only changes its slope when its control signal changes.

13.4.1 Decoupled model derivation

The decoupled model will be combined with a KF allowing us to estimate the control variables without the cross-coupling terms. The use of a KF presents also an opportunity to reduce the number of sensors in the control algorithm. This is the reason why the state space model is augmented with two additional states, v_i and its quadrature v_{iq}, in order to estimate the grid voltages by means a KF algorithm. Thus, two objectives are presented for the model derivation. The first one is to achieve three sliding surfaces S_a, S_b, and S_c, each one only depending on its control variable u_a, u_b, and u_c, respectively, and which dynamics are given by (13.10)–(13.12). The second objective is to reduce the number of sensors, which allows to reduce the cost and to increase the reliability. In order to derive this model, some important considerations must be taken into account. First, it must be ensured that the model is a decoupled model (i.e., each phase only depends on its own control variable), and second, the model must be linear, accurate, and simple

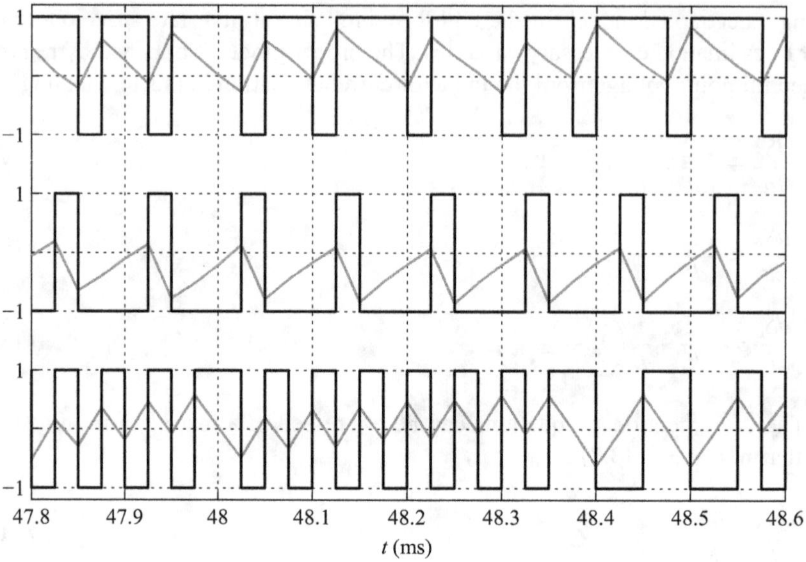

Figure 13.3 Three-phase control signals with corresponding sliding surfaces when the cross-coupled is removed: top u_a and S_a, middle u_b and S_b, and bottom u_c and S_c

in order to reduce the computational time. For these reasons, the assumptions listed below are considered:

- A decoupled model is obtained by removing the neutral point voltage from the converter model in (13.1). This fact has a negligible effect in the low-frequency dynamics of the closed-loop system. The reason is that v_n is a high-frequency signal with nearly zero average value; see (13.3).
- As the output capacitor is usually large, the output voltage has a slow dynamics. Hence, the state variable v_o can be assumed constant between different sampling instants of the same switching period. Besides, since this variable is measured, it can be considered as a parameter and its dynamics can be neglected. With this consideration, the new state space model for each phase-leg i can be considered linear in a switching period [17].

These considerations lead to the following differential equations for each phase-leg i:

$$L\frac{di_i}{dt} = v_i - \frac{v_o}{2}u_i \tag{13.15}$$

$$\frac{dv_i}{dt} = \omega_o v_{iq} \tag{13.16}$$

$$\frac{dv_{iq}}{dt} = -\omega_o v_i \tag{13.17}$$

The aforementioned equations can be expressed as a state space model including the measurement equation yielding:

$$\frac{d\mathbf{x}_i}{dt} = \mathbf{A}\mathbf{x}_i + \mathbf{B}v_o u_i \tag{13.18}$$

$$y_i = \mathbf{C}\mathbf{x}_i \tag{13.19}$$

where

$$\mathbf{x}_i = \begin{bmatrix} i_i & v_i & v_{iq} \end{bmatrix}^T \tag{13.20}$$

$$\mathbf{A} = \begin{pmatrix} 0 & \frac{1}{L} & 0 \\ 0 & 0 & \omega_o \\ 0 & -\omega_o & 0 \end{pmatrix} \tag{13.21}$$

$$\mathbf{B} = \begin{pmatrix} -\frac{1}{2L} & 0 & 0 \end{pmatrix}^T \tag{13.22}$$

The control algorithm does not use grid voltage sensors, and the only measured magnitudes are the grid currents. Hence, the value of the output matrix **C** can be defined as:

$$\mathbf{C} = \begin{pmatrix} 1 & 0 & 0 \end{pmatrix} \tag{13.23}$$

13.4.2 Controllability and observability of the proposed model

Once the model is derived, the controllability and observability should be proved. The measurement vector **C** yields to the following observability matrix:

$$\mathbf{O} = \begin{bmatrix} \mathbf{C} & \mathbf{CA} & \mathbf{CA}^2 \end{bmatrix}^T. \tag{13.24}$$

Moreover the controllability matrix γ is given by:

$$\gamma = \begin{bmatrix} \mathbf{B} & \mathbf{AB} & \mathbf{A}^2\mathbf{B} \end{bmatrix} \tag{13.25}$$

Matrices **O** and γ are both of full-rank, (i.e., $rank\{\mathbf{O}\} = rank\{\gamma\} = 3$), so that the system is controllable and can be observed using only the measured current i_i.

13.5 Sliding mode control scheme based on estimators

In this section, the model deduced previously will be combined with a KF in order to obtain the estimated variables for the implementation of the SMC. If the decoupled model is used in a KF, the estimated inductor currents are represented by (13.15). Thus, the sliding surface vector is defined as

$$\mathbf{S} = k\hat{\mathbf{v}} - \hat{\mathbf{i}} \tag{13.26}$$

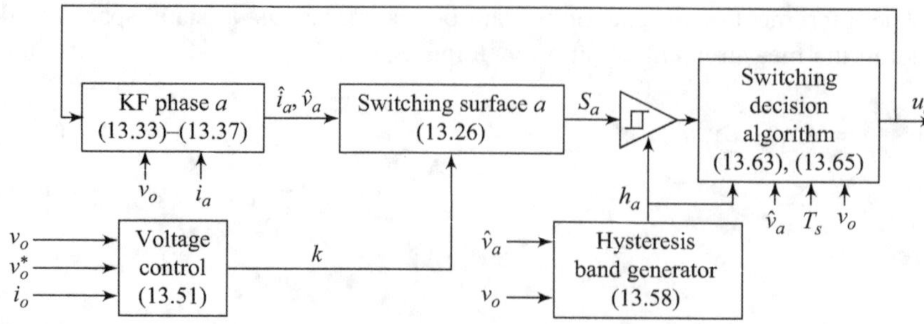

Figure 13.4 Proposed control system for phase-leg a of the UPFR

where $\hat{\mathbf{v}}$ and $\hat{\mathbf{i}}$ are the estimated grid voltage and inductor current vectors, respectively. The reference current vector is proportional to the estimated grid voltage (13.26). This proportional gain k is computed in an external control loop to regulate the output voltage, as shown in Section 13.6.2.

The dynamics of the three controllers are given by (13.10)–(13.12), and the three-phase system is considered decoupled from the control point of view. Once the system is decoupled, the SMC can operate with fixed switching frequency if a proper hysteretic control is designed. The hysteretic control will be explained in further sections.

The control diagram for each single phase-leg, in this case phase-leg a, is represented in Figure 13.4. The main features provided by the KF in this application can be summarized as follows:

- The three controllers are decoupled, since the model defined by (13.18)–(13.19) is used in the KF algorithm.
- All the variables used in the control algorithm are estimated and free of high-frequency noise. The Kalman estimators produce a filtered version of the inductor currents, which results in an improved sliding motion.
- The grid voltages are estimated, so the number of measurement sensors is reduced, which diminishes the cost and increases the reliability.

13.5.1 Discrete decoupled model

For the digital KF implementation, the state space model (13.18) must be discretized. The discrete model and measurement equations are given by the following expressions:

$$\mathbf{x}_{i_{k+1}} = \mathbf{A}_d \mathbf{x}_{i_k} + \mathbf{B}_d v_{o_k} u_{i_k} + \eta_{i_k} \tag{13.27}$$

$$y_{i_k} = \mathbf{C}\mathbf{x}_{i_k} + \mathbf{w}_{i_k} \tag{13.28}$$

where

$$\mathbf{A}_d = e^{AT_s} \cong \mathbf{I} + \mathbf{A}T_s = \begin{pmatrix} 1 & \dfrac{T_s}{L} & 0 \\ 0 & 1 & \omega_o T_s \\ 0 & -\omega_o T_s & 1 \end{pmatrix} \tag{13.29}$$

$$\mathbf{B}_d = \int_0^{T_s} Be^{A\lambda} d\lambda \cong \mathbf{B}T_s = \begin{bmatrix} -\frac{T_s}{2L} & 0 & 0 \end{bmatrix}^T \tag{13.30}$$

where $\mathbf{I}$ being the identity matrix and T_s being the sampling time. Here, it is assumed that v_{o_k} is constant between different samples within the same switching period, so that $v_{o_{k+1}} \cong v_{o_k}$, and the discrete model can be considered linear over the same switching period.

The process and the measurement noise vectors are η_k and $\mathbf{w}_k$, respectively, with the covariance matrices given by:

$$\mathbf{Q}_{i_k} = E\left\{\eta_{i_k}\eta_{i_k}^T\right\} \tag{13.31}$$

$$\mathbf{R}_{i_k} = E\left\{\mathbf{w}_{i_k}\mathbf{w}_{i_k}^T\right\}. \tag{13.32}$$

13.5.2 KF algorithm

The recursive computation of the Kalman algorithm is divided into two parts: (1) time update and (2) measurement update. The following equations show the recursive steps:

- The time update step predicts the state ahead and the error covariance ahead by using (13.33) and (13.34), respectively

$$\hat{\mathbf{x}}_{i_k}^- = \mathbf{A}_d\hat{\mathbf{x}}_{i_{k-1}}^- + \mathbf{B}_d v_{o_k}\mathbf{u}_k \tag{13.33}$$

$$\mathbf{P}_{i_k}^- = \mathbf{A}_d\mathbf{P}_{i_{k-1}}\mathbf{A}_d^T + \mathbf{Q}_{i_k} \tag{13.34}$$

- In the measurement update, an update for the predicted values is made by using the Kalman gain, which is computed as follows (13.35)

$$\mathbf{L}_{i_k} = \mathbf{P}_{i_k}^-\mathbf{C}^T(\mathbf{C}\mathbf{P}_{i_k}^-\mathbf{C}^T + \mathbf{R}_{i_k})^{-1} \tag{13.35}$$

This gain is calculated recursively in order to minimize the mean square error between the measured values and the predicted ones for the system states. When the gain is computed, the state estimation and the error covariance are updated using (13.36) and (13.37), respectively

$$\hat{\mathbf{x}}_{i_k} = \hat{\mathbf{x}}_{i_k}^- + \mathbf{L}_{ik}(i_{i_k} - \hat{i}_{i_k}^-) \tag{13.36}$$

$$\mathbf{P}_{i_k} = (\mathbf{I} - \mathbf{L}_{i_k}\mathbf{C})\mathbf{P}_{i_k}^-. \tag{13.37}$$

The estimated values $\hat{\mathbf{x}}_{i_k} = [\hat{i}_i \ \hat{v}_i \ \hat{v}_{iq}]$ are obtained from (13.36).

13.5.3 Practical considerations: selection of Q and R matrices

In the implementation of the Kalman algorithm, the covariance matrices $\mathbf{R}_{i_k}$ and $\mathbf{Q}_{i_k}$ must be computed. R and Q are usually selected as diagonal matrices because there is no information on the cross-correlation. The measurement noise covariance matrix $\mathbf{R}_{i_k}$, whose dimension is equal to the dimension of the product $\mathbf{CP}_{i_k}^{-}\mathbf{C}^T$, is usually estimated from measurements obtained from sensors. According to (13.23) and with the consideration that $\mathbf{P}_{i_k}^{-}$ is a square matrix, $\mathbf{R}_{i_k}$ is reduced to an scalar. Hence, R_{i_k} can be estimated using the unbiased mean rms power estimator

$$R_{i_k} = \frac{1}{NT_s}\sum_{k=0}^{N-1}|w_{i_k}|^2 \tag{13.38}$$

where w_{i_k} is the noise sample at time k, in phase-leg i, and N is the number of samples.

The noise vector w_{i_k} can be obtained from the reference DC-voltage used in the sensing system V_{ref_k}. This voltage is connected at one analog input of the DSP and is contaminated with the system noise when the rectifier is switching. The noise samples can be obtained by subtracting from V_{ref} measures, the exact value, V_{ref}^* (i.e., the value when the rectifier is not switching), yielding

$$R_{i_k} = \frac{1}{NT_s}\sum_{k=0}^{N-1}|V_{ref_k} - V_{ref}^*|^2 \tag{13.39}$$

Matrix $\mathbf{Q}_{i_k}$ is more difficult to estimate, and a tuning method is usually used [18]. For this application, $\mathbf{Q}_{i_k}$ is a diagonal matrix of dimension 3, and its value is obtained by means of simulations. For the present case, the simulation results provided an approximate value of $\mathbf{Q} = 0.005\mathbf{I}_3$, where $\mathbf{I}_3$ is a 3-dimensional identity matrix. All the conditions in the Kalman algorithm are initialized to zero except for the covariance matrix, which is initialized to the identity matrix.

13.5.4 Practical considerations: computational load reduction

A high computational load is usually a drawback in the KF algorithm implementation. Hence, to reduce the time used in the computation will be an important improvement in the controller design. Specially, (13.35) contains a matrix inversion that needs an important computational load. When the decoupled model (13.21) is used, only one measurement for each phase is needed, and the matrix inversion of (13.35) is reduced to an scalar inversion. Besides, the noise R_{i_k} is normally similar for all the phases and the Kalman gains $\mathbf{L}_{a_k}$, $\mathbf{L}_{b_k}$, and $\mathbf{L}_{c_k}$ have approximately the same value. Therefore, the Kalman gain (13.35) is computed only for one phase.

13.6 Sliding mode control of a UPFR

The control scheme for the UPFR consists of two different loops. The first one is a fast inner input current loop to ensure sinusoidal input currents in phase with the line voltages, $\mathbf{i} = k\mathbf{v}$. The second one is a slow outer loop, which is usually a PI controller, used to regulate the output voltage by modifying the input current amplitudes.

The control design methodology is dedicated to guarantee unity power factor and also provides a regulated output DC voltage with fast dynamic response against sudden changes in the load. Moreover, the amplitude control gain k is conceived to linearize the output voltage dynamics.

13.6.1 Inner control loop

The current controllers are derived by using the following switching surfaces

$$S_a = k\hat{v}_a - \hat{i}_a \tag{13.40}$$

$$S_b = k\hat{v}_b - \hat{i}_b \tag{13.41}$$

$$S_c = k\hat{v}_c - \hat{i}_c \tag{13.42}$$

where $\hat{\mathbf{v}} = [\hat{v}_a \; \hat{v}_b \; \hat{v}_c]$ and $\hat{\mathbf{i}} = [\hat{i}_a \; \hat{i}_b \; \hat{i}_c]$ are the estimated grid voltages and inductor currents.

In the ideal case of sliding motion, that is, infinite switching frequency, the average value of $\mathbf{u}$ is known as equivalent control, $\mathbf{u}_{eq}$. The equivalent control is deduced by imposing the sliding regime condition $\dot{\mathbf{S}} = 0$. Using (13.10)–(13.12), we can obtain:

$$\mathbf{u}_{eq} = \frac{2}{v_o}\left(\hat{\mathbf{v}} - kL\frac{d\hat{\mathbf{v}}}{dt}\right). \tag{13.43}$$

where $\mathbf{u}_{eq} = [u_{eqa} \; u_{eqb} \; u_{eqc}]$ is the equivalent control vector.

The main requirement in the design of SMC is to satisfy the reaching conditions and also to guarantee the existence of a sliding regime in the switching surfaces $\mathbf{S} = 0$. The most often used expression for the reaching condition in each phase-leg i is given by

$$S_i\dot{S}_i < 0 \tag{13.44}$$

which allows us to determine the control laws

$$u_i = \begin{cases} 1 & \textit{if} \quad S_i < 0 \\ -1 & \textit{if} \quad S_i > 0 \end{cases} \tag{13.45}$$

13.6.2 Outer control loop

In this section, the main steps to design the outer control loop will be explained. First, in order to avoid input current distortion, the outer control loop is designed with slow dynamic behavior. Second, an output-current feedforward signal can be introduced in the control loop to ensure a fast response when a sudden change in the load is produced. With this aim, an appropriate design of the gain k is proposed. This control gain k is also conceived to linearize, in a large signal sense, the output voltage dynamics as it will be seen later.

To design the outer loop, the zero dynamics concept [19] is used. The zero dynamics analyzes the dynamics of v_o, assuming that the control objective $\mathbf{i} = k\mathbf{v}$ is achieved. The output voltage differential equation in sliding regime can be found by using (13.43) in (13.2)

$$C\frac{dv_o}{dt} = \frac{k}{v_o}(\hat{v}_a^2 + \hat{v}_b^2 + \hat{v}_c^2) - \frac{k^2L}{v_o}\left(\hat{v}_a\frac{d\hat{v}_a}{dt} + \hat{v}_b\frac{d\hat{v}_b}{dt} + \hat{v}_c\frac{d\hat{v}_c}{dt}\right) - i_o. \quad (13.46)$$

By assuming the grid voltages have a sinusoidal waveform defined by:

$$\hat{v}_a = V_p \sin(\omega_o t) \quad (13.47)$$

$$\hat{v}_b = V_p \sin\left(\omega_o t - \frac{2\pi}{3}\right) \quad (13.48)$$

$$\hat{v}_c = V_p \sin\left(\omega_o t - \frac{4\pi}{3}\right) \quad (13.49)$$

where V_p is the grid-voltage peak value, and by using (13.47)–(13.49) in (13.46), the following equation is deduced:

$$C\frac{dv_o}{dt} = \frac{3V_p^2k}{2v_o} - i_o. \quad (13.50)$$

The output voltage differential equation (13.50) is clearly nonlinear, and the output voltage dynamics depends on a proper design of k. In order to linearize the output voltage dynamics, the following nonlinear control of k is proposed:

$$k = \frac{2v_o}{3V_p^2}\left[k_i\int_{-\infty}^{t}(v_o^* - v_o)d\tau + k_p(v_o^* - v_o) + k_0 i_o\right] \quad (13.51)$$

where k_p and k_i are the proportional and integral gains, respectively, and k_0 can take the values 0 or 1. Now, by replacing (13.51) in (13.50) yields

$$C\frac{dv_o}{dt} = \left[k_i\int_{-\infty}^{t}(v_o^* - v_o)d\tau + k_p(v_o^* - v_o) - (1 - k_0)i_o\right]. \quad (13.52)$$

The dynamics of v_o can be obtained by taking the time derivative of (13.52), which in the case of resistive load takes the following form:

$$C\frac{d^2v_o}{dt^2} + \left(k_p + \frac{1-k_0}{R_L}\right)\frac{dv_o}{dt} + k_i v_o = k_i v_o^*. \quad (13.53)$$

The aforementioned equation is a linear differential equation which depends on the load value. This dependence can be removed by taking $k_0 = 1$. With this value, the current feedforward term is introduced and the system is now robust against sudden load changes. For $k_0 = 0$, only a PI controller is used and the output voltage dynamics is affected by the output current variations, which produces an excessive output voltage drop as it will be seen in Section 13.8. The stability can be assured if k_p and k_i are chosen as positive values. It must be noted that in order to obtain this controller, unlike other controllers presented in the literature [20, 21], the use of small-signal models are not required.

13.7 Sliding mode control operating at fixed switching frequency

13.7.1 Variable hysteresis band calculation

The conventional SMC is characterized by a variable switching frequency, which is not desired in most industrial applications. A fully digital hysteresis modulator is designed in this section whose main work is to make the switching frequency fixed by modifying the hysteresis band.

For analog hysteretic controllers, the expression of the hysteresis band to achieve constant switching frequency in a phase-leg is derived by using the inductor current differential equation given by:

$$L\frac{d\hat{i}_i}{dt} = \hat{v}_i - \frac{v_o}{2}u_i \tag{13.54}$$

Figure 13.5 represents the inductor current and the control signal according to 13.54. Using the two possible values of $u_i = \pm 1$, two different cases can be analyzed:

- When the slope of the inductor current is positive ($u_i = -1$):

$$\frac{d\hat{i}_i}{dt} = \frac{1}{L}\left(\hat{v}_i + \frac{v_o}{2}\right) = \frac{2h_i}{t_1} \tag{13.55}$$

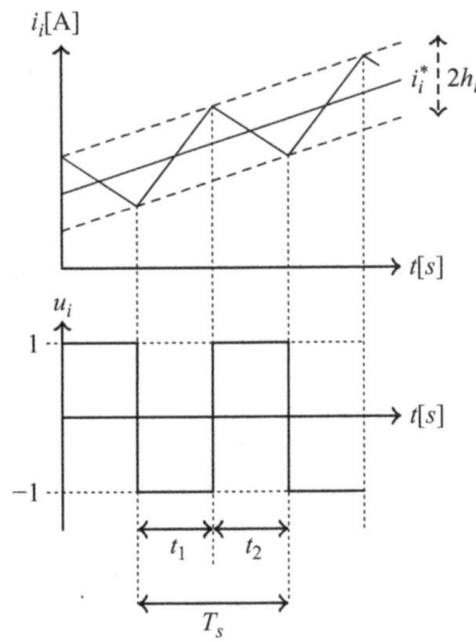

Figure 13.5 Inductor current and control signal waveforms

- When the slope of the inductor current is negative ($u_i = 1$):

$$\frac{d\hat{i}_i}{dt} = \frac{1}{L}\left(\hat{v}_i - \frac{v_o}{2}\right) = \frac{-2h_i}{t_2} \tag{13.56}$$

Defining the switching period as:

$$T_s = t_1 + t_2 \tag{13.57}$$

and using (13.55)–(13.57), the expression for the hysteresis band, h_i, is obtained, resulting in:

$$h_i = \frac{v_o}{8Lf_{sw}}\left[1 - \left(\frac{2\hat{v}_i}{v_o}\right)^2\right] \tag{13.58}$$

where f_{sw} is the switching frequency. Therefore, the hysteresis band should be varied according to (13.58) in order to obtain the constant switching frequency. Figure 13.6 shows the time evolution of the hysteresis band and switching surface in a digital implementation of the control scheme.

Due to the finite sampling frequency, the number of samples in one switching period is a finite number obtained from the relation

$$N = \frac{f_s}{f_{sw}} \tag{13.59}$$

While the sliding surface is moving inside the hysteresis bands $\pm h_i$, the switching frequency f_{sw} will remain constant. If a digital control is used, an error in the switching frequency is produced due to the sampling process. Figure 13.8 shows this phenomena, where some samples can be out of the hysteresis bands and the switching frequency obtained will be lower than the desired one.

The switching surface slope in Figure 13.8 can be calculated using the Euler approximation, obtaining:

$$m_k = \frac{S_k - S_{k-1}}{T_s} \tag{13.60}$$

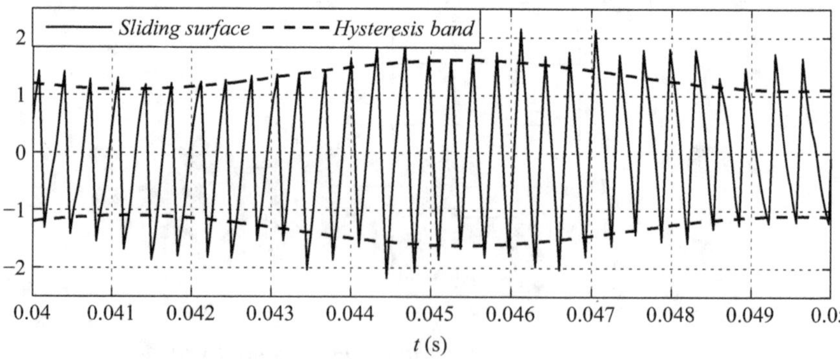

Figure 13.6 Switching surface and its hysteresis band

where m_k depends on the inductor value. From the previous expression, we can derive

$$S_k = S_{k-1} + m_k T_s \tag{13.61}$$

As it can be deduced from (13.60), if m is large the distance between two consecutive samples is large provoking that some samples appear out of the hysteresis band. As a consequence an error and a reduction of the desired switching frequency f_{sw} is provoked. This problem can be avoided if an SDA is implemented. This algorithm is based on computing the time between the actual sample and hysteresis band limit to ensure that most part of the samples will be inside the hysteresis limits. If this time is less than half switching period, the algorithm forces the system to switch. With this solution, the number of samples out of the hysteresis limits is drastically reduced and the quantization error due to the limited sampling frequency is reduced to a half.

13.7.2 Switching decision algorithm

The SDA is explained in this section. Assuming the hysteresis band has a slow variation in a sampling period, $h_i(t_k) \simeq h_i(t_{k+1})$, the algorithm can be implemented using the following steps:

- When a sample is acquired, the time between the sample and the hysteresis band is calculated as follows:
  - If the sample corresponds to the positive slope of the switching surface, m_1, where:

$$m_1 = \frac{1}{L}\left(\hat{v}_i + \frac{v_o}{2}\right) \tag{13.62}$$

  the expected switching time t_α can be computed as:

$$t_\alpha = \frac{h_i - S_i(t_k)}{m_1}. \tag{13.63}$$

  - Otherwise, if the sample corresponds to the negative slope of the switching surface, m_2, where:

$$m_2 = -\frac{1}{L}\left(\frac{v_o}{2} - \hat{v}_i\right) \tag{13.64}$$

  the expected switching time t_β can be computed as:

$$t_\beta = -\frac{S_i(t_{k+j}) + h_i}{m_2} \tag{13.65}$$

  being S_i is the switching surface for the phase-leg i.

- Once the estimated times t_α and t_β are known, the SDA is as follows:
  In the case of positive slope:
  - If $t_\alpha < \frac{T_s}{2}$, the control signal u_i is changed from -1 to 1, otherwise the control signal is not changed.

In the case of negative slope:

- If $t_\beta < \frac{T_s}{2}$, the control signal u_i is changed from 1 to -1, otherwise the control signal is not changed.

This procedure is carried out recursively for each sampling period where the times t_α and t_β are computed. Figure 13.7 shows the time representation of the hysteresis bands with the sliding surface for phase leg a. As it can be seen, time t_α is computed in each sampling period during the positive slope and t_β during the negative slope. As mentioned above, these times are compared with $T_s/2$ in order to decide the switching time.

Figure 13.8 shows the outcome of this algorithm. As it can be seen, the sliding surface is the most part of the time inside the hysteresis bands, and the error in the desired switching frequency is highly reduced. Finally, the control signals and the sliding surfaces when the system works at fixed switching frequency are shown in Figure 13.9.

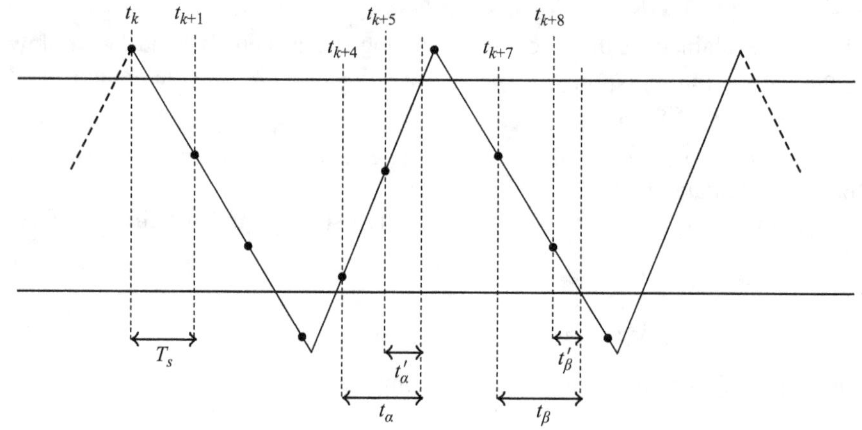

Figure 13.7 Hysteresis band with the switching surface

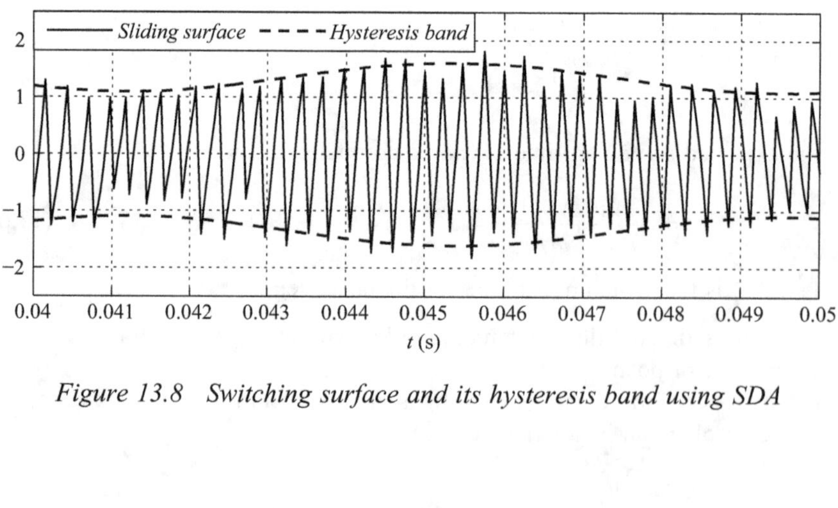

Figure 13.8 Switching surface and its hysteresis band using SDA

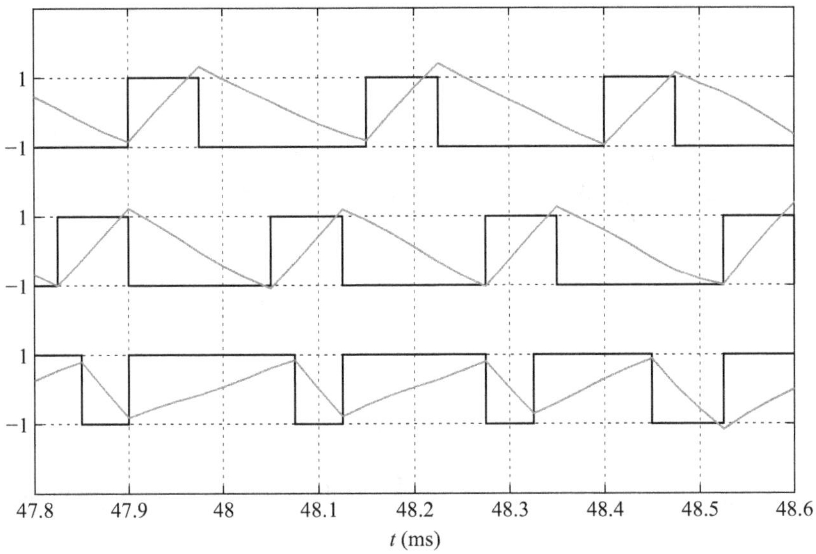

Figure 13.9 *Three-phase control signals with corresponding sliding surfaces when the SDA is in operation: top u_a and S_a, middle u_b and S_b, and bottom u_c and S_c*

Table 13.1 *System Parameters*

Symbol	**Description**	**Value**
L	Filter input inductance	5 mH
C	Output capacitor	340 μF
V_{dc}	DC-link voltage	250 V
f_s	Sampling frequency	40 kHz
f_{sw}	Switching frequency	4 kHz
f_{grid}	Grid frequency	60 Hz
V_{grid}	Grid voltage	50 V_{rms}
k_p	Proportional gain	0.03
k_i	Integral gain	2
R_L	Load	135 Ω

13.7.3 Switching frequency spectrums

This subsection analyses the properties of the studied SMC schemes by comparing the switching frequency spectrums. Table 13.1 lists the system parameters used in the comparison. Figure 13.10(a) shows the switching frequency spectrum in the case of conventional SMC without using the KF algorithm. Note that a wideband frequency spectrum is obtained in this case.

Figure 13.10(b) shows the spectrum using estimated currents but without the hysteresis modulator. As it can be seen the spectrum is also a wideband frequency

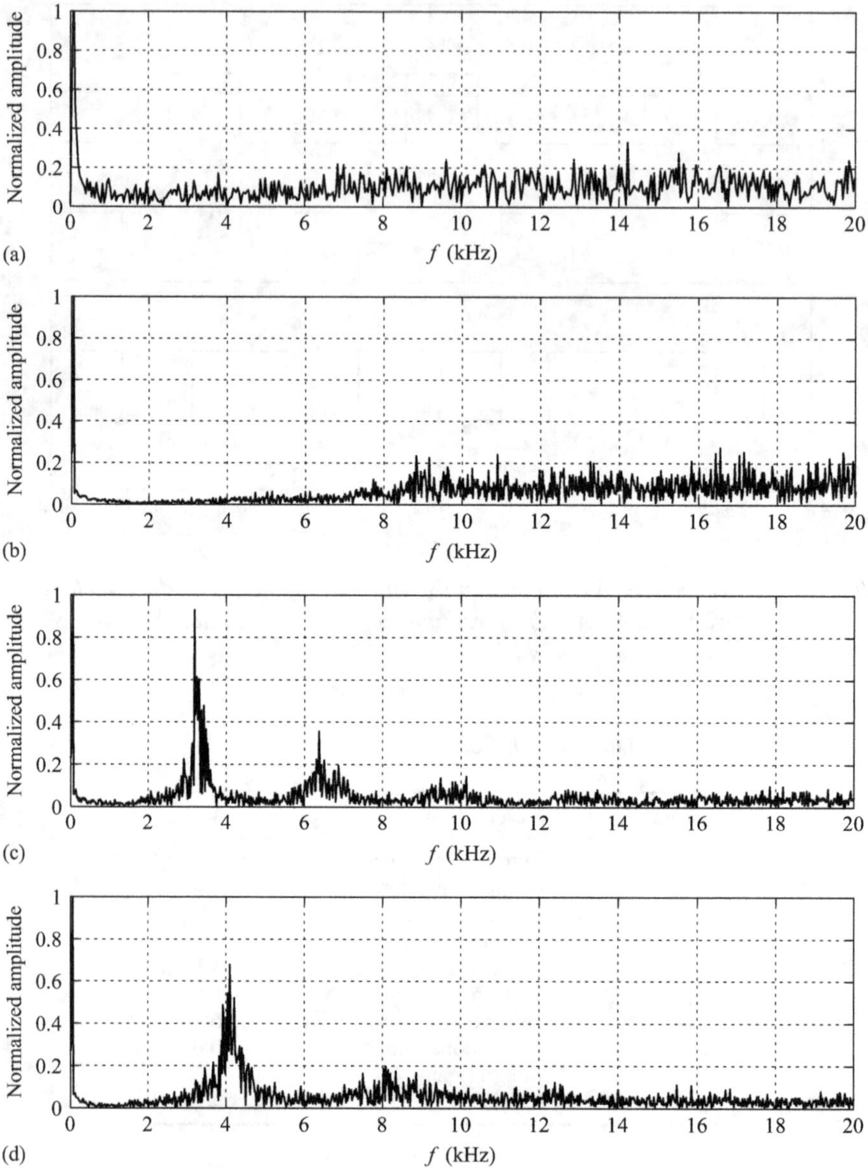

Figure 13.10 Experimental switching frequency spectrum with amplitude relative to fundamental (a) conventional SMC (b) SMC with estimated signals but without hysteresis modulator, (c) SMC with estimated signals and hysteresis modulator, and (d) with estimated signals, hysteresis modulator, and SDA algorithm

spectrum but concentrated between 11 and 20 kHz since the sliding motion is improved.

Figure 13.10(c) shows the switching spectrum with hysteresis bands but without the SDA. The spectrum is concentrated around 3 kHz due to the error produced by the sampling process. This problem is overcome by using the SDA algorithm, and the result is depicted on Figure 13.10(d), where the switching frequency is now around 4 kHz.

13.8 Experimental results

The experimental setup consists of a 4.5-kVA SEMIKRON full-bridge converter as the three-phase rectifier and a TMS320F28M35 floating-point DSP as the control platform with a sampling frequency of 40 kHz. The grid voltages have been generated using a PACIFIC 360-AMX AC source. Figure 13.4 shows the proposed control scheme, and the system parameters are listed in Table 13.1.

The experimental results are obtained for an SMC operating at a fixed switching frequency using SDA and a KF algorithm. The outer control loop is designed according to (13.53). In order to analyze the performances of the SMC when a sudden change in the load appears, two cases are presented.

Case 1: Figure 13.11 shows the output voltage, the inductor currents, and the amplitude control gain k, without using the feedforward output-current control ($k_0 = 0$). An important voltage drop appears in the output-voltage when the output power changes suddenly from 460 to 920 W since only a PI controller is used. As can be seen in Figure 13.11(b, c), the inductor currents amplitude and the amplitude control gain k have a slow variation.

Case 2: The same waveforms are shown in Figure 13.12 but using the feedforward term ($k_0 = 1$). In this case, the output-voltage dynamics is robust against load changes since the dynamics does not depend on the load resistor, as it was theoretically predicted by (13.53). For this reason the settling time of the amplitude control gain k is drastically reduced, and the voltage drop is inappreciable.

13.9 Summary

A digital SMC for a UPFR in natural frame has been proposed using a linear discrete model implemented using a KF algorithm. The proposed method uses a decoupled model of the converter obtained by removing the neutral point voltage from the physical model. This fact has a negligible effect in the low-frequency dynamics of the closed-loop system since the neutral point voltage is a high-frequency signal with nearly zero-average value. With this technique the interference between controllers is completely removed allowing to conceive three sliding mode controllers operating at fixed switching frequency. This proposal can be extended to other three-phase converters allowing the use of the best properties of SMC in three-phase converter systems.

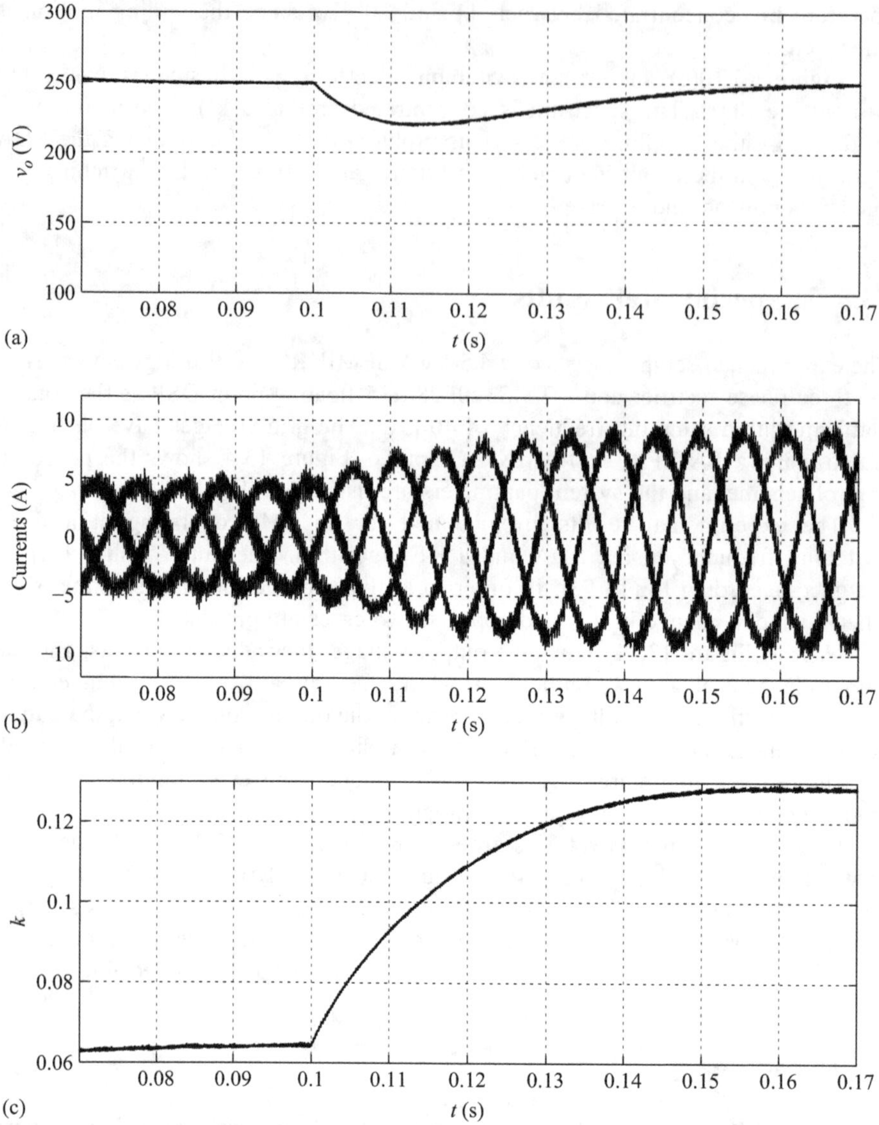

Figure 13.11 Experimental results without feedforward term for a load step change from 460 to 920 W: (a) output voltage, (b) three inductor currents, and (c) value of the amplitude control gain k

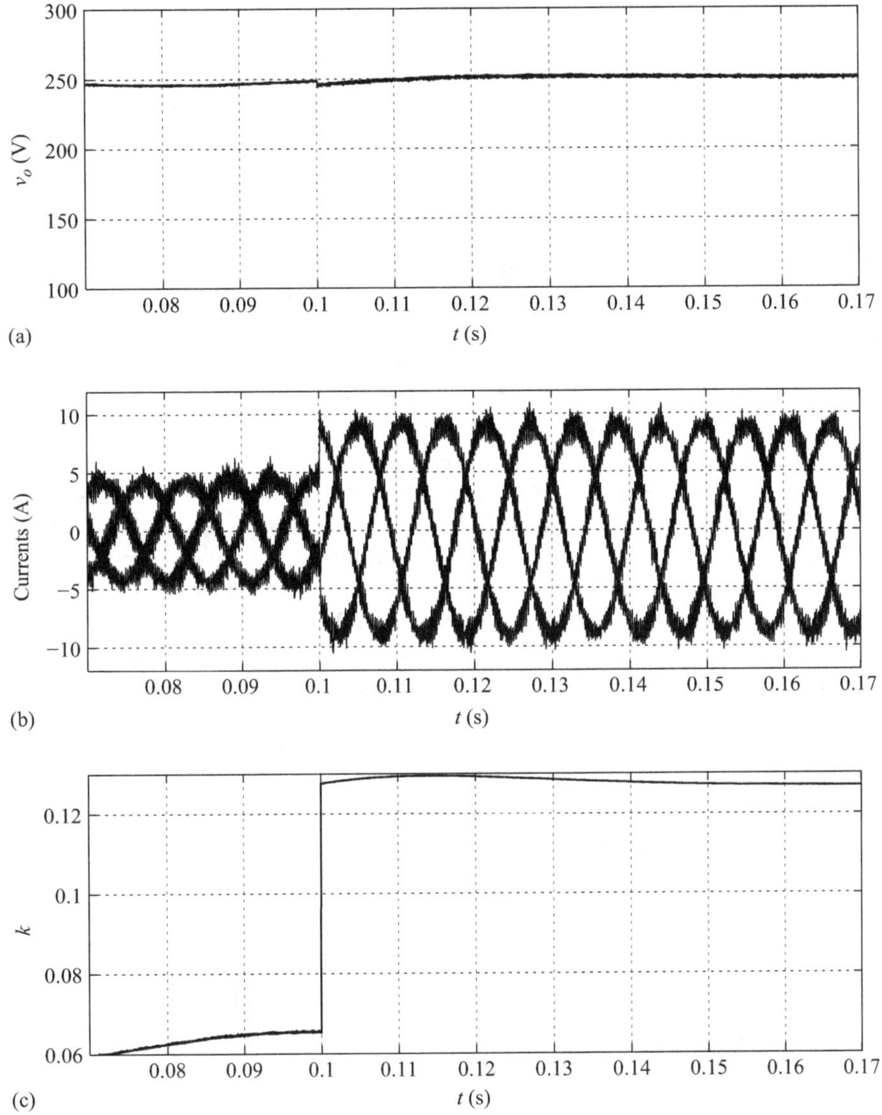

Figure 13.12 Experimental results with feedforward term for a load step change from 460 to 920 W: (a) output voltage, (b) three inductor currents, and (c) value of the amplitude control gain k

References

[1] Z. Chen, "PI and sliding mode control of a Cuk converter," *IEEE Trans. Power Electron.,* vol. 27, no. 8, pp. 3695–3703, 2012.

[2] X. Hao, X. Yang, T. Liu, L. Huang, and W. Chen, "A sliding-mode controller with multiresonant sliding surface for single-phase grid-connected VSI with an LCL filter," *IEEE Trans. Power Electron.,* vol. 28, no. 5, pp. 2259–2268, 2013.

[3] M. S. Grewal and A. P. Andrews, *Kalman Filtering: Theory and practice using MATLAB.* John Wiley & Sons Inc., New York, 2001.

[4] Y.-W. Cho, J.-M. Kwon, and B.-H. Kwon, "Single power-conversion ac–dc converter with high power factor and high efficiency," *IEEE Trans. Power Electron.,* vol. 29, no. 9, pp. 4797–4806, 2014.

[5] B. Mather and D. Maksimovic, "A simple digital power-factor correction rectifier controller," *IEEE Trans. Power Electron.,* vol. 26, no. 1, pp. 9–19, 2011.

[6] E. Mehl and I. Barbi, "An improved high-power factor and low-cost three-phase rectifier," *IEEE Trans. Ind. Appl.,* vol. 33, no. 2, pp. 485–492, 1997.

[7] A. Maswood, A. Yusop, and M. Azizur Rahman, "A novel suppressed-link rectifier-inverter topology with near unity power factor," *IEEE Trans. Power Electron.,* vol. 17, no. 5, pp. 692–700, 2002.

[8] J. Kolar and T. Friedli, "The essence of three-phase PFC rectifier systems-part I," *IEEE Trans. Power Electron.,* vol. 28, no. 1, pp. 176–198, 2013.

[9] T. Friedli, M. Hartmann, and J. Kolar, "The essence of three-phase PFC rectifier systems-part II," *IEEE Trans. Power Electron.,* vol. 29, no. 2, pp. 543–560, 2014.

[10] H. Kanaan and K. Al-Haddad, "Three-phase current-injection rectifiers: competitive topologies for power factor correction," *IEEE Magazine Ind. Electron.,* vol. 6, no. 3, pp. 24–40, 2012.

[11] A. Maswood and F. Liu, "A unity power factor front-end rectifier with hysteresis current control," *IEEE Trans. Energy Conv.,* vol. 21, no. 1, pp. 69–76, 2006.

[12] A. Maswood, E. Al-Ammar, and F. Liu, "Average and hysteresis current-controlled three-phase three-level unity power factor rectifier operation and performance," *IET Power Electron.,* vol. 4, no. 7, pp. 752–758, 2011.

[13] R. Guzman, L. Garcia de Vicuna, J. Morales, M. Castilla, and J. Matas, "Sliding-mode control for a three-phase unity power factor rectifier operating at fixed switching frequency," *IEEE Trans. Power Electron.,* vol. 31, no. 1, pp. 758–769, Jan. 2016.

[14] D. Holmes, R. Davoodnezhad, and B. McGrath, "An improved three-phase variable-band hysteresis current regulator," *IEEE Trans. Power Electron.,* vol. 28, no. 1, pp. 441–450, 2013.

[15] W. Stefanutti and P. Mattavelli, "Fully digital hysteresis modulation with switching-time prediction," *IEEE Trans. Ind. Appl.,* vol. 42, no. 3, pp. 763–769, 2006.

[16] L. Malesani, P. Mattavelli, and P. Tomasin, "Improved constant-frequency hysteresis current control of VSI inverters with simple feedforward bandwidth prediction," *IEEE Trans. Ind. Applications.,* vol. 33, no. 5, pp. 1194–1202, 1997.

[17] S. Mariethoz and M. Morari, "Explicit model-predictive control of a PWM inverter with an LCL filter," *IEEE Trans. Ind. Electron.,* vol. 56, no. 2, pp. 389–399, 2009.

[18] M. L. W. Gene F. Franklin, J. David Powell, *Digital Control of Dynamic Systems*. Addison-Wesley, Menlo Park, California, 1997.

[19] A. Isidori, *Nonlinear Control Systems.* New York: Springer Verlag, 1995.

[20] E. Figueres, J.-M. Benavent, G. Garcera, and M. Pascual, "Robust control of power-factor-correction rectifiers with fast dynamic response," *IEEE Trans. Ind. Electron.,* vol. 52, no. 1, pp. 66–76, 2005.

[21] E. Figueres, J. Benavent, G. Garcera, and M. Pascual, "A control circuit with load-current injection for single-phase power-factor-correction rectifiers," *IEEE Trans. Ind. Electron.,* vol. 54, no. 3, pp. 1272–1281, 2007.

Chapter 14
Dual-core DSP for control and communication in AC microgrids

Miguel Castilla[1] and Jaume Miret[1]

14.1 Introduction

The deregulation of the electric market and the increasing number of renewable energy sources connected to the public grid are gradually attracting more attention to the concept of AC microgrids. These power networks are active electricity distribution systems with the capacity to operate as controlled electrical islands [1]. These systems integrate distributed energy sources (including photovoltaic, wind power, mini hydro, gas turbine, etc.) and energy storage devices (such as batteries, ultra-capacitors, flywheels) into the main power system. AC microgrids can meet electrical requirements locally, supply uninterruptible power, improve power quality, reduce power losses and provide voltage support [2]. Furthermore, these systems can reduce environmental pollution and global warming by using low-carbon technology [3].

Control and communication are two key elements in the operation and performance of AC microgrids. For instance, the power production of distributed generators is optimised by the control schemes of their power converters [4]. The state-of-charge of energy storage devices is also governed by the control schemes of their power converters. In the case of loads, both the activation of controllable loads and the uninterruptible supply of critical loads are ensured by load control schemes. In addition, the energy management in the microgrid as a whole system is performed by the microgrid central controller (uGCC) which is located in the point of common coupling (PCC) [5]. All these functionalities require the use of a proper communication infrastructure to interchange data (state of devices, power generation and consumption and storage capacity) and control commands. In microgrids, the real-time communication between all the controllers associated to generators, storage devices, loads and the energy management central unit is normally based on Ethernet connectivity. Thus, in this application, the use of a certified industrial communication protocol is a requisite [6].

[1]Department of Electronic Engineering, Technical University of Catalonia, Vilanova i la Geltrú, Spain

Powerful digital control solutions are available nowadays for applications with real-time communication requirements as needed in AC microgrids. Specifically dual-core microcontroller architectures with a floating-point digital signal processor (DSP) and a companion communication processor are a very attractive solution [7]. In these architectures, the dual-core DSPs are capable of handling multiple complex control tasks simultaneously [8]. With fast ADC and PWM-specific peripherals, complete closed-loop control systems can be implemented including real-time feedback, control algorithm processing and driving signal generation. In addition, the connectivity capabilities are guaranteed by the communication processors allowing the sharing of data and control commands to manage the generators, storage devices and loads remotely.

This chapter presents the deployment of the control and communication infrastructure of an AC microgrid based on dual-core DSP devices. The essential concepts of an AC microgrid operating as a controlled electrical island are presented in Section 14.2. The control of power converters interfacing distributed generators is introduced in Section 14.3. The communication architecture is presented in Section 14.4 including service requirements and industrial protocols. A representative dual-core floating-point DSP device is described in Section 14.5. Selected experimental results on a low-scale laboratory microgrid are reported in Section 14.6 validating the control and communication concepts presented in this chapter. Section 14.7 is the conclusion.

14.2 Control in AC microgrids

This section focuses on the description of architecture, operation and control of AC microgrids.

14.2.1 Microgrid architecture

Figure 14.1 illustrates the typical architecture of an AC microgrid including distributed generators, storage devices and loads. The power converters are used to interface with these elements to the power bus and to provide flexibility of operation. Local controllers govern the operation of the power converters and are a fundamental element in the operation and performance of the AC microgrid. A communication link is also depicted in the figure as a data channel to interconnect all these controllers with the uGCC.

The central controller is responsible for managing the operation of the microgrid as a single power entity. It receives data from both the microgrid elements (generation, consumption, state-of-charge) and the main power system, and generates signals to remotely command the energy sources, storage devices and loads. The control commands are usually decided from a combination of factors including power costs, system efficiency, pricing schedules and emission penalties [2, 4]. For this task, the minimisation of a cost function defined as a weighted combination of these factors is normally used [5]. The uGCC is also in charge to connect and disconnect the microgrid to the main power system as well as to

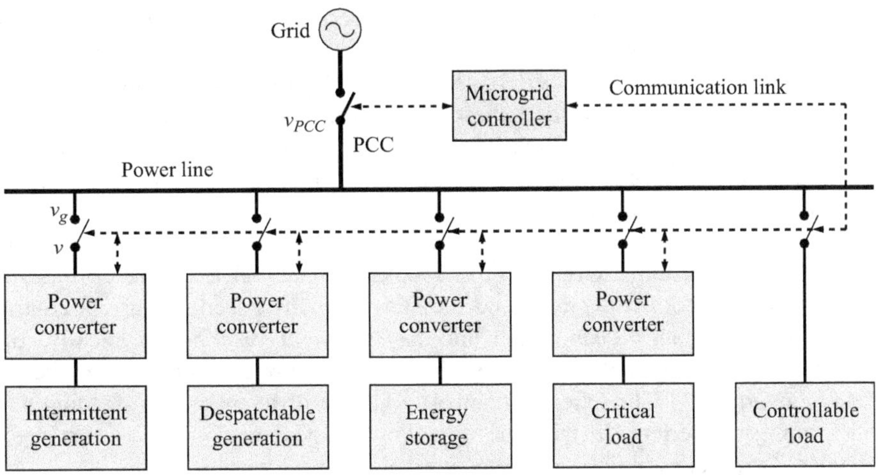

Figure 14.1 Typical AC microgrid architecture

connect and disconnect other microgrid elements such as distributed generators, storage devices and loads. Figure 14.1 emphasizes this flexible operation by showing the switches associated to each element of the microgrid.

The energy management in the microgrid is subjected to variable conditions. The intermittent generators (photovoltaic and wind turbines among others) produce power irregularly according to the meteorological conditions. The participation of the despatchable generators (such as gas turbines) strongly relies on the dynamic price of fuel. The load is by nature unpredictable, but peak shaving and load shifting techniques can be implemented in the controllable loads to improve the system performance. Critical loads require uninterruptible power supply even during these variable conditions.

AC microgrids can operate connected to the main power system or in islanded mode. In this last mode, microgrids are able to run independently from the main power system providing power to individual customers. This is an interesting property of microgrids, especially during extreme weather events or emergencies, due to the reliable operation for powering crucial public services such as hospitals, first responders and water treatment facilities.

14.2.2 Power converters in AC microgrids

Power converters operating in an AC microgrid can be broadly classified into grid-feeding and grid-forming converters. On the one hand, grid-feeding converters are responsible for exporting the active and reactive power specified by the control commands. To this end, these converters operate equivalently as AC current sources with a high output impedance connected in parallel. The reference signal for the output current is generated by measuring the voltage at the output terminals, and, for this reason, these converters require the presence of a stable and well-regulated voltage in the microgrid. Grid-feeding power converters cannot operate properly when the quality of the voltage is not good enough, for instance, during a black start when the microgrid is partially or totally de-energised [9]. On the other hand, grid-forming power converters are responsible for regulating the voltage at their output terminals. Therefore, these converters remain connected during the black start exporting active and reactive power while regulating the output voltage. To this end, grid-forming converters operate equivalently as AC voltage sources with a low output impedance in series [9].

The control schemes of both grid-feeding and grid-forming converters are divided in internal and external control loops. The internal loops have the same configuration for both converters, and they are normally based on current control loops with high control bandwidth and fast transient response. Typically, the bandwidth of these internal current loops is a decade below the switching frequency of the converter [9]. The external control loops are different in grid-feeding and grid-forming converters. In fact, the operation of these converters as current or voltage sources relies mainly on these external control loops.

In grid-feeding converters, the external control loops regulate active and reactive power with low control bandwidth. In order to avoid interactions with the fast current loop, the bandwidth of these external loops is usually a decade below to the grid frequency [9]. These external loops have been described in Chapter 8, and thus no further details are included here. In grid-forming converters, the reference signal for the internal control loop is generated by a voltage control loop. This voltage regulation loop is responsible for imposing the behaviour of the converters as AC voltage sources, as mentioned above. The bandwidth of the voltage control loop is slightly lower than the bandwidth of the internal current control loop [9]. The essential element in the control of grid-forming converters is the voltage reference generator. The aim of the voltage control loop is to track the output of the voltage reference generator; however, it is the generation of this signal that provides the desired properties to these power converters. Section 14.3 will present in detail the external control loops for grid-forming converters.

A power converter can operate as a grid-feeding or grid-forming converter by merely interchanging the external control loops. This fact can be clearly observed in the control diagram shown in Figure 14.2. Note that the reference signal for the

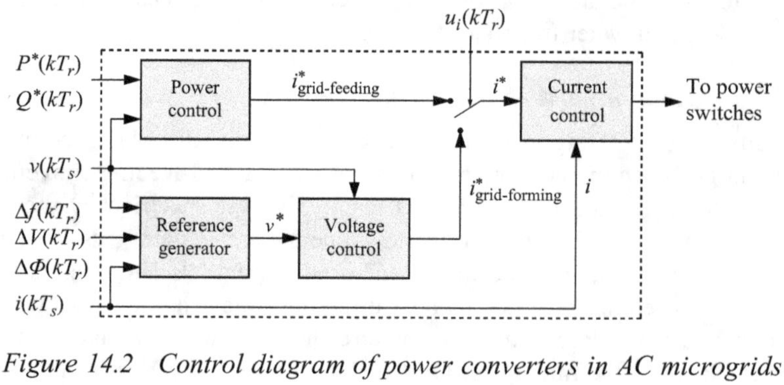

Figure 14.2 Control diagram of power converters in AC micrograms

internal current loop (denoted as i^*) is provided by the power control loop (grid-feeding reference) or the voltage control loop (grid-forming reference) according to the command received from the central controller. Other inputs of the control diagram are the output voltage and current measured locally in the power converter and the power set points received from the uGCC. The outputs are the driving signals for the power switches.

The control scheme of grid-feeding and grid-forming power converters is programmed in digital control devices, typically in DSPs. In these platforms, the sampling period T_s is defined as the time at which the local measures are updated. In addition, the transmission period T_r is the time instant at which signals are received from the central controller. Figure 14.2 clearly indicates the periods of the input signals to the control scheme (beings measured locally at T_s or received from the uGCC at T_r). As discussed in the next sections, the period T_r is several orders of magnitude greater than T_s in AC microgrid applications.

14.2.3 Microgrid scenarios

As stated above, AC microgrids can operate connected to the main power system or in islanded mode. In grid-connected mode, it is convenient to operate all the power converters as grid-feeding converters (i.e. as equivalent AC current sources controlled by the active and reactive power set points). The reason is that, in this mode, the main power system is in operation and the voltage at the PCC is fixed by this bulk power system. Therefore, it is normally not necessary to regulate the amplitude and frequency of the microgrid voltage.

In islanded mode, the microgrid is not connected to the main power system and the voltage in the microgrid must be regulated by grid-forming converters (i.e. converters operating as equivalent AC voltage sources). Thus, despatchable generators and energy storage devices are interfaced in islanded mode by grid-forming converters [5]. However, intermittent generators have irregular power production and probably insufficient power capacity in some circumstances to contribute to the voltage regulation. For this reason, it is preferable to interface intermittent generators with grid-feeding converters in both grid-connected and islanded modes [5]. Table 14.1 sums up this discussion.

Traditionally distributed generators are interfaced by grid-feeding converters. The challenge nowadays is the deployment of AC microgrids with the capacity to operate separately from the main power system. In this scenario, grid-forming power converters play an especial role. Section 14.3 is dedicated to this topic.

Table 14.1 Application scenarios of grid-feeding and grid-forming power converters

Operation and sources	Intermittent energy source	Controllable energy source
Grid-connected microgrid	Grid-feeding converters	Grid-feeding converters
Islanded microgrid	Grid-feeding converters	Grid-forming converters

14.3 Control of grid-forming power converters

In AC microgrids, grid-forming power converters are employed to regulate the output voltage in islanded mode. The control tasks of these converters are grouped in a hierarchical control architecture including primary, secondary and tertiary control layers [10, 11]. As a general rule, higher control layers require lower control bandwidth (and, thus, slower transient response). A description of these control layers is presented below.

14.3.1 Primary control

The primary control is implemented in the voltage reference generator as shown in Figure 14.2. The main control loop in the primary control is based on the droop method [10]. In this method, the frequency f and the amplitude V of the output voltage are fixed according to the averaged value of the active P and reactive Q powers supplied by the power converter to the microgrid [11]

$$f = f_o - mP \tag{14.1}$$

$$V = V_o - nQ \tag{14.2}$$

where the variables f_o and V_o correspond to the frequency and amplitude of the output voltage when the converter is inactive ($P=0$, $Q=0$). The control parameters m and n are the slopes of the frequency and amplitude functions, respectively. The design of these parameters affects the load sharing in the islanded microgrid.

Figure 14.3 shows a graphical representation of (14.1) and (14.2) for different values of m and n. A simple microgrid with only two grid-forming power converters and a common load is considered here to simplify the discussion of the load sharing concept. In Figure 14.3(a), the slope parameters coincide, $m_1 = m_2$ and $n_1 = n_2$. In this case, a perfect sharing of the active power supplied by each power converter is reached since the frequency in steady state matches in both converters $f_1 = f_2$. It is worth mentioning that (14.1) acts as a negative feedback mechanism that imposes the injection of the same active power by two separated converters using only local data (the communication infrastructure is not used to implement this power sharing mechanism). Even with $n_1 = n_2$, the reactive power supplied by the converters does not match as the amplitude of the output voltage of each converter is slightly different. The power flow through the impedances of the distribution lines and other elements (such as transformers) causes the mismatch between the amplitudes. Therefore, the amplitude is a local variable in the microgrid (the value is different according to the particular point of measure), while the frequency is a global variable (in steady state, all the microgrid points have the same frequency).

Figure 14.3(b) shows the operation of the droop method for different values of the slopes. In this case, the active and reactive powers supplied by the converters do not coincide and thus other power sharing strategies (different to reach active and

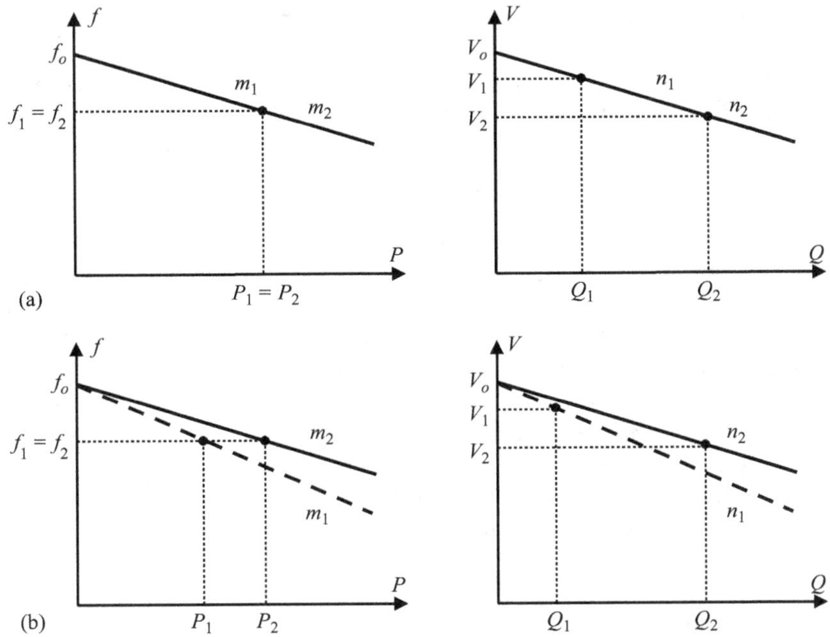

Figure 14.3 Frequency and amplitude control by means of the droop method. (a) Slopes $m_1 = m_2$ and $n_1 = n_2$; (b) Slopes $m_1 > m_2$ and $n_1 > n_2$

reactive power equalisation as desired with equal slopes) can be devised. From (14.1) and (14.2), the general active and reactive power expressions can be written as

$$P_1 = \frac{m_2}{m_1 + m_2} P_L \tag{14.3}$$

$$P_2 = \frac{m_1}{m_1 + m_2} P_L \tag{14.4}$$

$$Q_1 = \frac{n_2}{n_1 + n_2} Q_L - \frac{V_1 - V_2}{n_1 + n_2} \tag{14.5}$$

$$Q_2 = \frac{n_1}{n_1 + n_2} Q_L + \frac{V_1 - V_2}{n_1 + n_2} \tag{14.6}$$

where P_L and Q_L are the active and reactive power consumed by the common load. An alternative power sharing strategy to power equalisation can be the minimisation of power losses in the microgrid. This high-level control objective in a realistic microgrid with dispersed generators and loads is nearly reached when the local loads are supplied by their closest generators. The aim is to minimise the power flow in the microgrid by trying to supply loads with locally produced power. This strategy can be implemented in the uGCC using (14.3) to (14.6). In this strategy, the droop slopes are calculated in real time with the data measured locally (consumption and voltage amplitudes). Therefore, the particular slopes are provided to the local controllers of the grid-forming power converters using the communication service.

The droop method presented above is essential to the operation of the microgrid in islanded mode. However, for proper operation during transient response, two additional control loops are included in the primary control. First, a phase locked loop (PLL) is used to synchronise the phase of the output voltage v of a disconnected converter with the phase of the microgrid voltage v_g [11]. Note that the nodes with voltages v and v_g are shown in Figure 14.1 for the power converter interfacing with the intermittent distributed generator. Once both signals are in phase, the converter can be activated to participate in the power sharing. And, second, the converter closed-loop output impedance is programmed by a virtual impedance control loop [12]. The purpose of this loop is that the total equivalent impedance formed by the series connection of the grid impedance seen by the converter and the virtual impedance be dominantly inductive. This behaviour is strictly necessary to avoid controllability problems using (14.1) and (14.2) as equations for the droop method. The virtual impedance is programmed by including an extra term in the voltage reference as follows [12]

$$v^* = e^* - Z_v i \tag{14.7}$$

where e^* is the reference voltage generated by the droop method, Z_v is the virtual impedance and i is the converter output current.

14.3.2 Secondary control

The secondary control is responsible for eliminating the deviations in frequency and amplitude introduced inherently by the droop method [10, 11]. Note in (14.1) and (14.2) or in Figure 14.3 that these deviations raise as the power supplied by the converters increases, being thus a more significant problem in this case. The secondary control loop is implemented in the uGCC and tries to restore the frequency and amplitude to their nominal values f_n and V_n. To this end, it uses proportional and integral regulators with the frequency and amplitude deviations as inputs

$$\Delta f = k_{pf}(f_n - f_{PCC}) + k_{if}\int (f_n - f_{PCC})dt \tag{14.8}$$

$$\Delta V = k_{pv}(V_n - V_{PCC}) + k_{iv}\int (V_n - V_{PCC})dt \tag{14.9}$$

Note that the frequency f_{PCC} and amplitude V_{PCC} are measured precisely at the PCC (Figure 14.1), so that the errors in these variables will be eliminated in steady state at this particular point of the microgrid.

The outputs of these regulators are sent periodically with a transmission period T_r to the local controllers of the grid-forming power converters. This fact modifies the droop method equations, which are calculated every sampling period T_s, as follows

$$f(kT_s) = f_o - mP(kT_s) + \Delta f(kT_r) \tag{14.10}$$

$$V(kT_s) = V_o - nQ(kT_s) + \Delta V(kT_r) \tag{14.11}$$

The local controllers of grid-forming converters use expressions (14.10) and (14.11) to both implement the droop method and to correct the deviations. It is worth mentioning that the secondary control makes use of the communication service to provide the power converters with the necessary corrections to eliminate the frequency and amplitude deviations in steady state. An infrastructure with low transmission rate is enough to perform this control task, which typically updates the corrections every few seconds [10, 11].

A further objective in the secondary control is to prepare the microgrid to be connected to the main power system when the microgrid operates in islanded mode. To this end, the microgrid voltage at the PCC must accurately track the voltage of the main power system (Figure 14.1). This is ensured basically by matching amplitude, frequency and phase of both voltages. Note that the controllers expressed in (14.8) and (14.9) are specifically in charge of regulating the frequency and amplitude of the PCC voltage. Therefore, an easy way to reach this objective is by replacing the nominal frequency and amplitude references used in (14.8) and (14.9) by the frequency and amplitude of the measured voltage in the main power system. In addition, a central PLL is implemented in the uGCC in order to match the phases of both voltages. The PLL output $\Delta\Phi(kT_r)$ is a phase increment that must be sent periodically to the local controllers of the power converters in the same message used to communicate the other control commands, namely the deviations $\Delta f(kT_r)$ and $\Delta V(kT_r)$. Once both voltages are synchronised, the main switch can be closed and the central PLL disconnected. Obviously the microgrid must be adapted to the main power system before the connection (and not vice versa).

Figure 14.4 shows the complete diagram of the voltage reference generator. The inputs of this diagram are the synchronization variables received from the uGCC, the microgrid voltage and the converter output voltage and current. Instantaneous active and reactive powers are calculated using the local measures. A first-order low-pass filter is used to extract the averaged value of active and reactive powers, as needed by (14.10) and (14.11). The design of the cut-off frequency of this filter is essential to both reject harmonics coming from the measured voltage and current and provide a slow transient response. Thus, as a practical design guideline, the cut-off frequency f_c is chosen between one and two decades below to the microgrid frequency. The frequency and amplitude are generated using the droop equations with the averaged active and reactive powers and the correction increments received periodically from

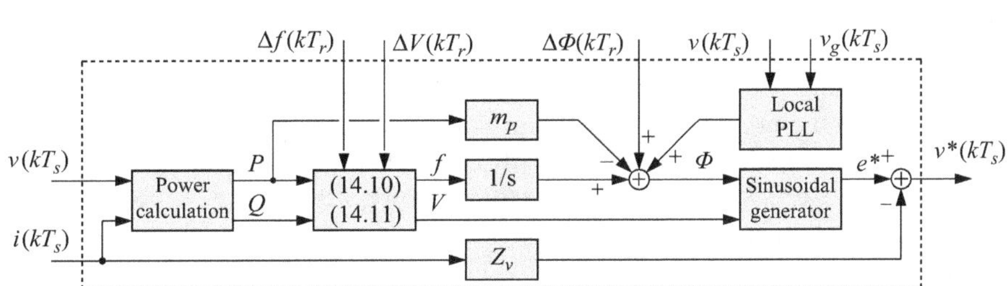

Figure 14.4 Detailed diagram of the voltage reference generator

the central controller (secondary control). The phase signal is calculated by integrating the frequency from (14.10) and adding the output of the local PLL, the phase correction $\Delta\Phi$ generated by the central PLL and a feed-forward term proportional to the active power. This last term improves the transient response of the power converter by introducing a degree of freedom in the location of the closed-loop poles of the system. The voltage reference of the droop method e^* is generated using the amplitude and phase signals. The output of the diagram is the voltage reference v^* obtained from (14.7) by using the virtual impedance concept.

An interesting issue in the operation of the microgrid is the uncoupled behaviour of the different control loops of local and central controllers. As a general design guideline, the unwanted interaction between these loops is avoided by designing the bandwidth of the secondary control one decade below to the bandwidth of the primary control (droop method, local PLL and virtual impedance). Therefore, the frequency and amplitude deviations are slowly corrected with a transient response 10 times slower than the transient response of the power sharing. Note that the dynamics of the secondary control is fixed by properly designing the gains of the proportional and integral terms included in (14.8) and (14.9). In addition, the control bandwidth for the central PLL is designed one decade below to the control bandwidth of the frequency and amplitude regulators of the secondary control. Following this design consideration, the phase deviation is slowly corrected once the microgrid frequency has reached the grid frequency.

14.3.3 Tertiary control

The tertiary control is responsible for the energy management during grid-connected mode. It coordinates the energy flow between the elements inside the microgrid and the main power system including other neighbouring microgrids. To this end, both centralised and decentralised approaches for tertiary control implementation have been presented in literature [11, 13, 14].

In centralised control, market and economic profits are the main goals that govern the microgrid operation and management. In this case, a single controller is used to program the tertiary control, usually located in the uGCC. This results in a complex control system but with an easy communication infrastructure. In addition the local controllers are simple control units, due to their limited intelligence. They are driven exclusively by the control commands provided by the uGCC. It is well recognised that at the present time the centralised approach is the preferred option in industry and therefore next sections are exclusively focused on this approach.

In decentralised control, the basic objective is to improve the overall performance of the microgrid. This is a hot topic of research nowadays, in particular the decentralised tertiary control based on multi-agent intelligent systems (MAS) [6, 14]. The main feature of the MAS approach is that the software running inside each agent presents local intelligence and can take some decisions independently [13]. Thus, a complex centralised controller can be divided into simple software agents located dispersedly in the local controllers of the microgrid. Following this approach, the microgrid and the whole main power system can operate like an artificial nervous system, where each element-agent can take decisions or receive control commands to accomplish horizontal collaboration between neighbours [14]. An ancillary benefit of using MAS in the tertiary control layer is the flexibility when adding new functionalities, which require minimum changes in the running agent-based software [13]. This distributed control approach is part of the Smart Grid concept that, in the near future, pretends to integrate all distributed generation in the main grid following IEC 61850, the standard of the International Electrotechnical Commission that regulates communication within electric power systems [15, 16].

14.4 Communication in AC microgrids

AC microgrids require a communication system to provide real-time data and control commands to the different control and management elements of the power system, including local and central controllers, the supervisory control and data acquisition (SCADA) system, and other grid and microgrid information entities such as the distribution network operator and market operator. This section focuses on the description of the communication infrastructure in AC microgrids and presents the deployment of the communication service in a low-scale laboratory microgrid.

14.4.1 Communication protocols

Until today DSP controllers used in industry had different (but limited) possibilities to implement communication services. Conventional DSPs carry built-in communication interfaces commonly called controller area network (CAN) standards such as CANopen, ControlNet, DeviceNet, Fieldbus, Modbus, Optomux, Profibus and Profinet. These applications were mainly based on custom communication protocols, which must be extended with great effort for each new improvement. In addition, these technologies are considered very limited in terms of throughput compared with recent non-real-time technologies [17].

Nowadays, it is preferable to use standards that permit to freely use (or buy) software instead of creating it. Also the possibility of connecting the system to the World Wide Web or to local area networks (LAN) using common protocols and standards must be taken into account in an increasingly connected world. The networks used in business and personal computing for web-connection are non-deterministic and asynchronous. The main need of these networks is a reliable delivery being secondary a precisely timed delivery. Network communications between these non-real-time systems are standardised as LAN-based topologies such as Ethernet, resulting in the current transmission control protocol and internet protocol (TCP/IP). Ethernet is the dominant and almost standard LAN being widely used, fast, easy to install, cheap and powerful. In this way, some DSP also carry built-in Ethernet interfaces additionally to conventional CAN capabilities. As a drawback, the most limiting feature of these protocols for using in industrial automation and control is its non-real-time behaviour. To overcome the challenge of real-time control with non-deterministic asynchronous networks, IEEE 1588 standard was developed. In particular, IEEE 1588-2002 (standard for a precision clock synchronization protocol for networked measurement and control systems), also known as precision time protocol (PTP), is a protocol used to synchronise clocks throughout a computer network. PTP provides clock accuracy in the sub-microsecond range being then suitable for measurement and control systems [18]. Consequently the possibility of using Ethernet for real-time communications in an AC microgrid seems an interesting option at a first glance.

The most widely used protocol over Ethernet backbone is TCP. This protocol can deal with big amounts of data in long messages although with the drawback of a latency that can reach hundreds of milliseconds. Furthermore, this protocol guarantees that all the information packets will be delivered, thanks to the re-transmission mechanism. However, this protocol is not practical for real-time networked control systems since the re-transmission of old data is generally not very useful [19]. For real-time control, a non-reliable protocol as UDP (user datagram protocol) is preferable, thanks to its low latency (some milliseconds) and its optimised transmission of short messages. UDP allows the datagram interchange without the need of establishing a previous connection because the encapsulated data has enough addressing information in its light header. As a major drawback, if a packet is lost in the network due to congestion or channel noise, it will never reach its destination because there is no re-transmission mechanism.

14.4.2 Example of a low-scale laboratory microgrid

This section presents the deployment of the communication service in a low-scale laboratory microgrid. This microgrid is used to experimentally validate the control and communication concepts presented in this chapter (Section 14.6).

Figure 14.5 shows the diagram of the laboratory microgrid. Each node of the microgrid consists of a grid-forming power converter interfacing a despatchable energy source with the power buses. The control system is based on a hierarchical architecture with centralised control. The central and local controllers are programmed in dual-core DSPs. Also each node of the microgrid has a local load in parallel with the energy source. A microgrid load, located near to the PCC, is supplied by the complete power system. Line impedances modelling the parasitic elements of the power cables and transformers are considered in this example. The nominal values of the microgrid components are listed in Table 14.2.

The microgrid uses both TCP and UDP protocols over an Ethernet link. Ethernet network has been chosen for two reasons: first, it is by far the most widely installed network technology, even though it was not designed for critical control applications; and second, it is the underlying LAN of the reduced OSI (open systems interconnection) stack defined by the IEC 61850 [16].

In the laboratory microgrid, the traffic of data and control commands in the communication channel is low. First, the transmission rate is low. As shown in Table 14.2, a transmission period of 1 s is selected in order that the secondary control cooperates with the primary control slowly. Undesired interactions between both control layers are also avoided with this low transmission rate. Second, the amount of data and control commands sent by the uGCC (amplitude, frequency and phase

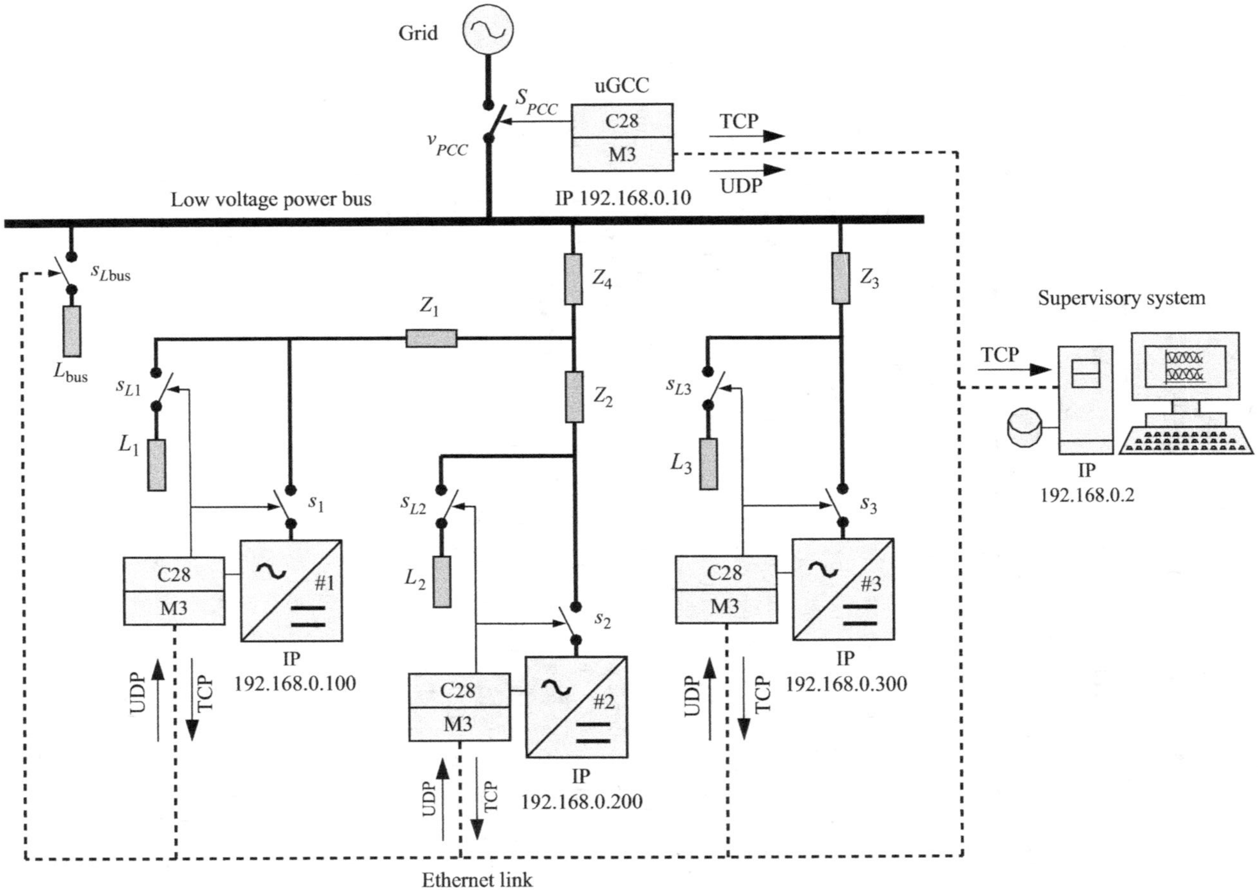

Figure 14.5 Diagram of the laboratory microgrid

Table 14.2 Nominal values of the microgrid under test

Grid voltage	**Load L_1**	**Load L_2**	**Load L_3**	**Load L_{bus}**
110 V_{rms}, 60 Hz	500 W, 0 VAr	500 W, 0 VAr	500 W, 0 VAr	1500 W, 0 VAr
T_s and T_r periods	**Line impedance Z_1**	**Line impedance Z_2**	**Line impedance Z_3**	**Line impedance Z_4**
100 μs, 1 s	0.5 Ω, 3 mH	0.5 Ω, 1 mH	1.13 Ω, 0.6 mH	0 Ω, 1 mH

increments for the local controllers, and state of switches for distributed generators and loads) is also low. This statement is also true for larger microgrids with a high number of nodes. The main reason is the centralised approach adopted for the control system which makes use of the one-to-all traffic scheme. Therefore, a low payload protocol as UDP is a good option for centralised controlled microgrids. In this case the low latency of UDP is even not fully exploited due to the low transmission rate. Furthermore, the high clock accuracy PTP protocol is actually not necessary in this microgrid due to the same reason (i.e. few data must be sent at a low transmission rate).

The state of the whole microgrid is monitored by a supervisory system. For this purpose, a TCP protocol is preferable to communicate the active control units (local and central controllers) with a SCADA system programmed in a computer, as illustrated in Figure 14.5. This computer node is normally equipped with a database to store data and events from the microgrid. Basic stored data are the power generation, energy storage and consumption in the different nodes of the microgrid. Other interesting measures to be stored and related to the power quality of the microgrid are the amplitude and frequency of the fundamental component of the node voltages, the amplitude of positive and negative sequence components at harmonic frequencies and the total harmonic distortion of voltages and currents. In addition, in grid-connected mode, transient events such as voltage sags and swells can be stored. The transmission of all these supervisory data does not need real-time communication and, therefore, TCP protocol is a good choice for the SCADA system.

14.5 Dual-core DSP for control and communication

Dual-core DSP technology is introduced in this section as an interesting approach to deploy the control and communication services in AC microgrids. A representative dual-core device is chosen to present its architecture and some examples of algorithms developed for centralised control and one-to-all traffic communications. The extension to other control and communication approaches is also discussed.

14.5.1 Control and communication in DSP technology

The implementation of real-time control and communication on a single-core DSP introduces significant software complexity. Apart from this difficulty, developers also have to manage other shared resources such as memory and access to peripherals. In addition, the communications link is normally served with secondary priority compared to the main control loops, thus limiting its bandwidth and penalising the communication performance.

One alternative option to the single-core DSP is to use two separate DSP devices. The first processor is in charge of the control tasks, while the second one is used for managing the communication and supervisory tasks. This approach improves the performance of the single-core DSP solutions but at the expense of increasing the system cost and size [20].

Nowadays the state-of-the-art solution for control and communication in a single DSP device is the dual-core embedded system. With dual-core architecture, control and communication are executed separately while sharing some memory area. This architecture simplifies the software development and reduces the system cost. Section 14.5.2 presents a representative dual-core DSP, the Texas Instruments Concerto dual-core with C28 and ARM Cortex-M3 processors.

14.5.2 Description of the dual-core system architecture

Dual core systems have recently appeared on the market in order to solve the tradeoffs for embedded control, removing the compromise of performing real-time control and host communications at the same time [21]. Figure 14.6 shows a simplified schematic of a Concerto microcontroller with dual hybrid architecture comprised a Texas Instruments C28 core and an ARM® Cortex™-M3 core. The C28 and M3 processors are implemented as independent subsystems, each with its

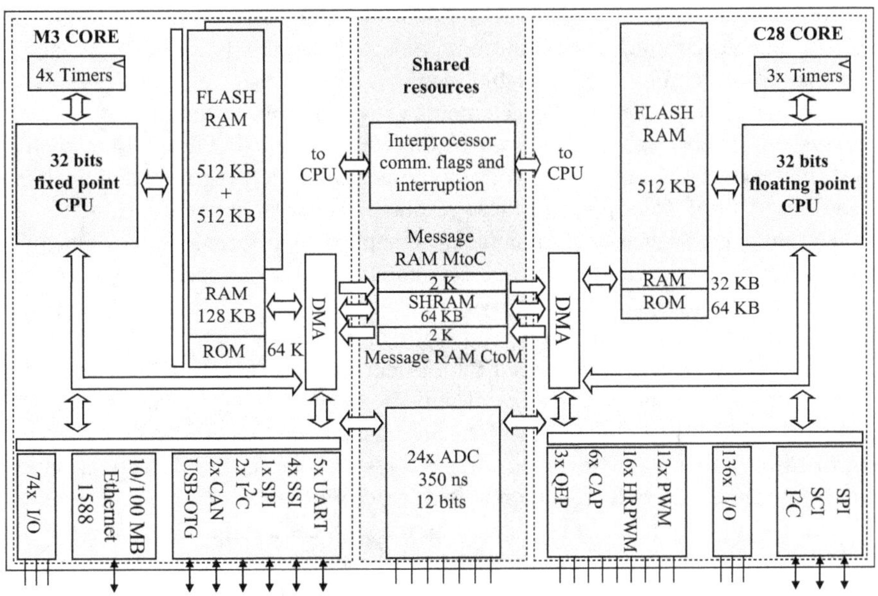

Figure 14.6 Diagram of the Concerto dual-core device, F28M36P63C

own memory, peripherals and interruptions controller. Each core disposes of its dedicated ROM and flash RAM and can execute code separately in each device in order to maximise their performance. In this way, the communication side of the system does not affect the real-time control processes and vice versa. This partitioning greatly simplifies application development because the software libraries can be executed independently on the separated cores. Moreover, the design has also independency in terms of clock and power management.

The C28 core is based on an integrated floating-point central processing unit (FPCPU) with a 32 × 32-bit multiply-accumulate processing capability. As a single core, it has been widely used in industrial control applications due to its powerful power processing and peripherals availability. The FPCPU is enhanced with a Viterbi mathematical-unit instruction accelerator that implements efficiently complex arithmetic calculations, fast Fourier transforms, cyclic redundancy check, etc. In addition, this core has a wide set of control-oriented enhanced peripherals such as PWMs, capture modules, and quadrature encoder pulse modules.

The ARM Cortex-M3 architecture is conceived for applications requiring a deterministic interruption management and communications with high flexibility. The M3 core is used widely throughout the industry for host communications, advanced human–machine interfaces and graphical user interfaces. This processor has a wide variety of communication peripherals, including Ethernet 1588, USB, CAN, some universal asynchronous receiver–transmitters, several serial ports, etc. The processor includes one Ethernet media access controller that conforms to IEEE 802.3 specifications and fully supports 10BASE-T and 100BASE-TX standards. The Ethernet controller enables network connectivity and embedded web servers to run on a device that can also execute embedded control loop code. This module also supports full and half-duplex operational modes both at 10 or 100 Mbps. M3 core can handle complex software communication protocols (as TCP/IP) at full scale thanks to its amount of available memory.

The shared resources are key elements providing high reliability to the dual-core system. These resources, shown in the central area of Figure 14.6, include RAM, message RAM, interprocessor communication (IPC) registers and analogue subsystem. A brief description of these resources is presented below.

The shared RAM (SHRAM) memory is mapped in different blocks that can be owned by either of the Concerto processors and be used for a specific application or for transferring messages and data. By default, on reset, the M3 owns all the shared memory blocks. The M3 can freely read, write or execute from the SHRAM blocks. The SHRAM speed-use is enhanced by a direct memory access (DMA) bus. The C28 is limited to read these blocks. Then, the SHRAM can be used by the M3 processors to send messages to the C28 processor, which is referred as master to control (MtoC) message. However, after the reset process, the M3 can assign the ownership to the C28, which is then able to read, write or execute on these blocks and the M3 can only read. In this situation, the C28 can send messages to the M3, which is referred as a control to master (CtoM) message. For safety questions, the SHRAM can be further configured to disable master DMA write, CPU write and/or CPU execution.

The messaging RAM is intended as a fast mechanism for passing messages between both processors. Each processor can write to its own messaging RAM and afterwards indicate to the other processor that data is just ready for reading. Then the processors can communicate to each other easily and with a minimal latency, without needing tokens, clocks or signal semaphores. The dual-core system has dedicated message RAM blocks for unidirectional communication-only purposes and that cannot be reconfigured. One is for the MtoC and the other for the CtoM. In the MtoC block, the M3 can read and write the block, while the C28 can only read and vice versa in the CtoM. After the message sending, the corresponding CPU can inform to the other via an interruption or a flag through the IPC.

The IPC is the mechanism that the two subsystems employ to share information. The IPC has a set of message registers intended to provide very simple and flexible scheme that can be used like in a mailbox to send messages back and forth. These registers are mapped to both CPUs. When a CPU sends a message, it can inform the other processor by using an interruption or a flag. There are identical IPC interruption and flag resources on both subsystems.

The analogue subsystem is formed by 12-bit ADC channels, running at 2.9 mega samples per second. Both processors have access to the ADC result registers in order to use this peripheral for control, communication and system supervision. Also, for high-speed protection purposes, asynchronous analogue comparators are shared in order to trigger on in an over-current condition. Both processors can trip immediately these high-speed comparators.

14.5.3 Control functions implemented in the C28 core

This section presents some examples of control algorithms implemented in local and central controllers of the laboratory microgrid as shown in Figure 14.5. As stated above, the primary and secondary layers of the hierarchical control system are programmed in the local and central controllers, respectively. Moreover, islanded mode is considered for the microgrid due to its challenging operation.

Local controllers for grid-forming power converters are programmed in the C28 core, as shown in Figure 14.5. The basic functions executed in the local controllers are:

- An empty Main() executes the C28 setup of input/output, peripherals and interruptions. Once the configuration is complete, the C28 core enters in an infinite loop whereas is waiting for periodical interruptions.
- A periodic interruption is configured at 10 kHz sampling frequency f_s in this example. The interruption service routine calls a state machine function, which is the core of the smart behaviour of the DSP. When the state machine reaches the status *running* the local controller routine is being served periodically (at 100 μs).
- The control routine begins sensing the output voltage $v(t)$ and current $i(t)$ of the power converter through its 12 bits ADC channels. These variables are used to implement the control loops shown in Figure 14.2. The control routine also receives periodically the secondary control signals (frequency Δf, amplitude ΔV and phase $\Delta \Phi$) sent by the uGCC. The transmission period is 1 s in this example.

- Once a control routine is executed, the duty cycle is provided to the space vector modulator which generates the driving signals for the power switches.

As shown in Figure 14.5, the uGCC employs the same dual-core DSP that the local controllers use. In this case, the control functions executed in the C28 core corresponding to the secondary control layer are:

- The C28 has an empty Main() infinite loop which is waiting for periodical interruptions. The interruption routine is being served at 100 μs.
- The control routine begins calculating the frequency Δf and amplitude ΔV restoration signals making use of (14.8) and (14.9). Next, the phase correction $\Delta\Phi$ signal is generated by the central PLL. All these signals are computed at 100 μs for reaching high accuracy; however, they are sent to local controllers at a transmission period of 1 s.
- Thus, once the control routine is executed, the resulting signals are provided to the M3 core at a sampling period of 1 s.

14.5.4 Communication procedures implemented in the M3 core

An UDP protocol with a one-to-all traffic scheme is used to communicate data and control commands from the uGCC to the local controllers as shown in Figure 14.5. In this configuration, the M3 core of the uGCC is configured as the UDP server, while the M3 cores of the local controllers are configured as UDP clients.

Figure 14.7 shows a diagram including control and communication functions performed by the uGCC dual-core DSP. The diagram presents specific functions for operating the uGCC as the UDP server. The C28 core generates the secondary control signals (frequency Δf, amplitude ΔV and phase $\Delta\Phi$) and writes the updated values into the shared SHRAM. A timer interruption is configured in the M3 core to periodically read these signals from the SHRAM at a transmission period of 1 s, pack them into a UDP datagram and send a broadcast communication to all the local controllers.

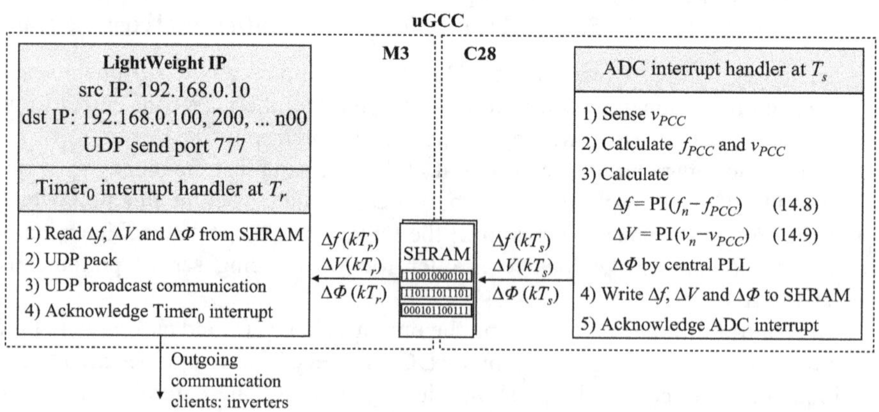

Figure 14.7 Dual-core DSP functions for operating the uGCC as the UDP server

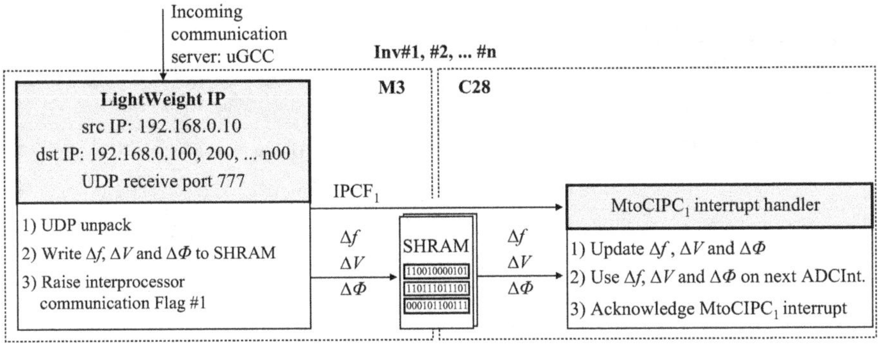

Figure 14.8 Dual-core DSP functions for operating the local controllers as UDP clients

Figure 14.8 shows a diagram with dual-core control and communication functions for operating the local controllers as UDP clients. When a UDP datagram is received in any client, first the M3 core unpacks the Δf, ΔV and $\Delta\Phi$ data, writes them into the SHRAM and raises an IPC interruption from M3 to C28. This interruption will be attained by an interrupt handler in the C28 core. The C28 core reads Δf, ΔV and $\Delta\Phi$ from the SHRAM and updates its values in the local controller. In the next execution of (14.10) and (14.11), these updated values will be used to generate the outputs of the primary control layer.

A TCP protocol is used to communicate the measured variables and data from the local and central controllers to the supervisory computer. For the power converters, the supervisory variables can be the active and reactive powers, amplitude and frequency of the output voltage, amplitude of the output current, among others. For the uGCC, the variables of interest are the amplitude and frequency of both the PCC voltage and the main power system voltage (measured at both sides of the main switch), among others. In this case, the M3 cores of the local and central controllers are configured as TCP servers, while the supervisory computer is configured as the UDP client.

Figure 14.9 shows a diagram of the dual-core DSP communication procedures for operating the local and central controllers as TCP servers. The C28 core executes the control algorithm and writes the chosen supervision variables to a buffer mapped in the SHRAM just before acknowledging the timer interruption. When a counter reaches end-of-buffer, an IPC interruption from C28 to M3 is raised. This interruption is attained by an interrupt handler in the M3 core that reads the SHRAM buffer, packs its data in a TCP packet and sends it to the supervisory computer. The computer receives this data and stores it in a database. Usually this information is also shown in a graphical interface.

All the code used to implement the UDP and TCP protocols are part of Lightweight (lwIP) TCP/IP stack, as shown in Figures 14.7–14.9. This stack was originally developed by Adam Dunkels and nowadays is available as an open-source resource [22]. With lwIP stack, TCP/IP protocols can be used at its full scale

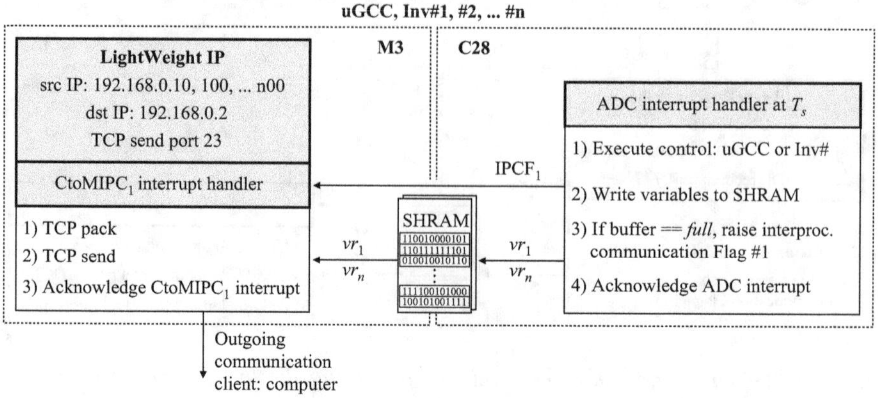

Figure 14.9 Dual-core DSP functions for operating the local and central controllers as TCP servers

in embedded systems with a typical code size of 25–40 KB while RAM requirements are approximately 15 to a few tens of kilobytes [23].

14.5.5 Extension to other control and communication schemes in AC microgrids

This section presented the functions and procedures to communicate local and centralised controllers executed in dual-core Concerto DSPs. The networked control system follows a hierarchical control approach, being only primary and secondary layers considered in the deployed microgrid. The basic ideas presented in this section can be extended to deal with tertiary control layers. In fact, the protocols and communication procedures can be directly employed for implementing the centralised tertiary control. This centralised control loop is normally located in the uGCC. Typically it is based on the minimisation of a cost function defined as a weighted combination of several factors including power costs, system efficiency, pricing schedules, and emission penalties [5]. The control commands generated by this cost function can be sent to the local controllers following the one-to-all traffic scheme with a UDP protocol, as described above for the secondary control layer. The only significant difference in this case is the lower transmission period required for tertiary control. With this low period, the interactions between secondary and tertiary control layers are totally avoided.

Unlike centralised control, more sophisticated communication approaches must be developed for decentralised networked control systems. This statement is especially true for large microgrids using multi-agent intelligent systems for implementing the tertiary control layer [6]. In this case, although dual-core Concerto DSPs can be maintained for the network physical layer, the deployment of the application layer is normally based on peer-to-peer networks. In contrast to the traditional client-server model employed in Figures 14.7–14.9, peer-to-peer networks are based on the functional equality of the communication nodes, which can behave simultaneously as both clients and servers. This property provides dynamical communication infrastructures with scalability and self-organization as main features. Additionally peer-to-peer networks degrade gracefully in the presence of failures, restore automatically after a fault and can cope well with dynamic environment [24].

Different protocols for peer-to-peer networks have been developed during the last decade. The most well known is probably gossiping. This protocol performs by spreading data into the network in a manner similar to the spread of a virus in a biological community. In this way, one node sends data randomly around its neighbours and these nodes repeat the operation. This protocol is very useful to periodically exchange data among all devices in a dynamical infrastructure, in which nodes may appear, disappear or re-appear due to functional behaviour (for instance, the intermittent generation of a distributed energy source) or electrical faults. Further discussion on communication protocols based on peer-to-peer networks can be found in References 24 and 25.

14.6 Experimental tests in the low-scale laboratory microgrid

This section presents some selected experimental results obtained from the laboratory microgrid in order to validate the main proposal of this chapter: the deployment of control and communication services in AC microgrids using dual-core DSP devices.

The experimental results were captured from the laboratory microgrid as shown in Figure 14.5. The nominal values for the microgrid components including grid voltage, line impedances and loads were listed in Table 14.2. The control and communication systems were implemented using the Concerto dual-core device F28M36P63C. For rapid prototyping, the control hardware development was carried out using the Concerto control card H63C2 Experimenter Kit from Texas Instruments. This digital control platform is implemented in the industry-standard 188-pin DIMM form factor board, which is supplied by a single 5 V power rail. The docking station gives access to all the control card signals and boundary-scan connectors for debugging. It also includes micro-USB, Ethernet, and MicroSD communication interfaces. The code development for the M3 and C28 cores is completely segregated, although code programming is done within the shared integrated development environment, the Code Composer Studio™. The primary and secondary control layers presented in Section 14.3 were programmed with this software package. The nominal values of the control parameters are listed in Tables 14.3 and 14.4.

Table 14.3 Nominal values of the primary control parameters

Frequency at no load	**Frequency function slope**	**Feed-forward gain**	**Virtual resistance**
$f_o = 60$ Hz	$m = 1000$ μrad/(Ws)	$m_p = 200$ μrad/W	$R_v = 0\,\Omega$
Amplitude at no load	**Amplitude function slope**	**LPF cut-off frequency**	**Virtual inductance**
$V_o = 155$ V	$n = 50$ μV/(VAr)	$f_c = 1$ Hz	$L_v = 10$ mH

Table 14.4 Nominal values of the secondary control parameters

Nominal frequency	Frequency proportional gain	Frequency integral gain
$f_n = 60$ Hz	$k_{pf} = 0.01$	$k_{if} = 0.12$ Hz
Nominal amplitude	**Amplitude proportional gain**	**Amplitude integral gain**
$V_n = 155$ V	$k_{pv} = 0.01$	$k_{iv} = 0.08$ Hz

The microgrid test was performed in islanded mode emulating a black start event. The power converters were configured as grid-forming converters to facilitate the support of the microgrid voltage. The power converters and loads were connected sequentially during the different stages of the black start. Three different tests were carried out in order to evaluate: (1) the performance of the primary control, (2) the performance of the secondary control, and (3) the effects of packet loss in the communication network. The results are presented and discussed in detail below.

14.6.1 Performance evaluation of the primary control

The objective of the primary control is to regulate the output voltage of the active converters as well as to provide a good load power sharing. These objectives are accomplished without using the communication service of the microgrid.

Figure 14.10 shows the experimental results for validating the performance of the primary control. In this test, the secondary control is inactive. The black start begins at $t = 0$ s with the converter #1 supplying the microgrid load L_{bus}. The rest of converters and loads are disconnected from $t = 0$ s to $t = 10$ s. By looking at the bottom of Figure 14.10(a), it can be noticed that the frequency of the converter #1 reduces from its nominal value 60 Hz to 59.75 Hz, as a consequence of the active power delivered to the microgrid load; see the top of Figure 4.10(a). The output voltage amplitude of the converter #1 is also lower than the nominal value (1 p.u.), as shown in the bottom of Figure 14.10(b). Both the supplied reactive power and the voltage loss in the virtual impedance are responsible of this amplitude reduction. The frequency and amplitude deviations will be eliminated in steady state by the secondary control, as shown in the next subsection.

At $t = 10$ s, the load L_1 is connected. From $t = 10$ s to $t = 20$ s, the total load supplied by converter #1 increases which further reduces the frequency and amplitude of the output voltage. A more interesting behaviour is observed at $t = 20$ s when the converter #2 is connected. From $t = 20$ s to $t = 30$ s, the converters #1 and #2 are both active operating as equivalent voltage sources (grid-forming converters). In this case, the load active power is perfectly shared by these converters, as shown in the top of Figure 14.10(a). Note also that the frequencies of these converters match in steady state, being their values higher than in the previous interval due to the reduction of the total load seen by each converter. Another interesting property of this system is the excellent transient response of active powers and frequencies observed when a new power converter is connected.

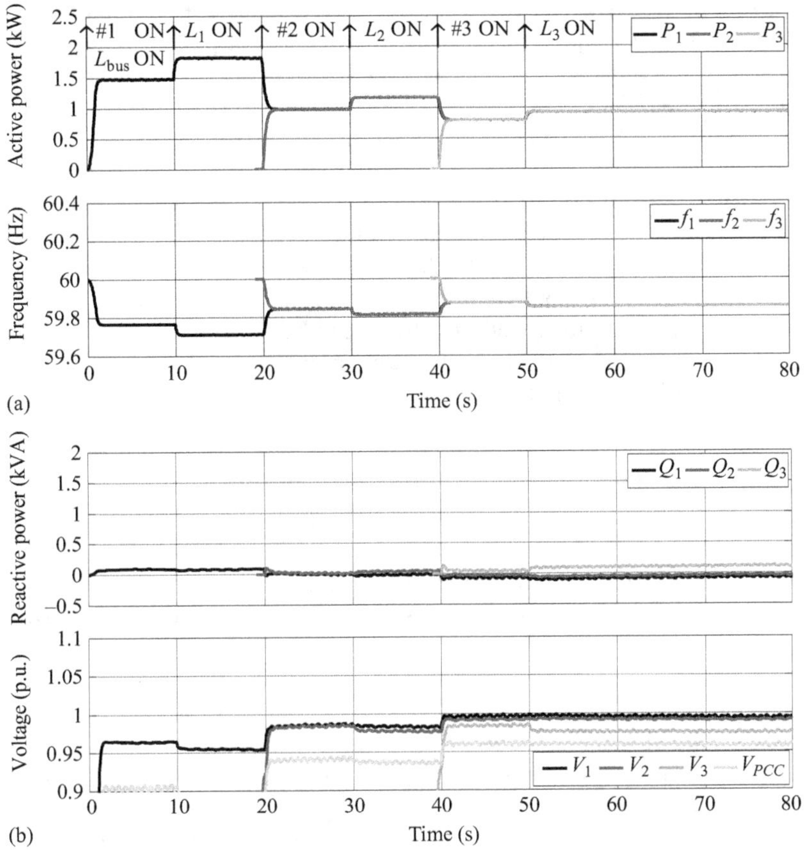

Figure 14.10 Experimental results for the primary control (the secondary control is inactive). (a) Top: Active powers supplied by the power converters; Bottom: Frequencies of the converter output voltages. (b) Top: Reactive powers supplied by the power converters; Bottom: Amplitudes of both the converter output voltages and the voltage at the PCC

In Figure 14.10(a), it can be seen that the transient response follows a first-order linear dynamics. In addition, the steady state is reached at roughly $t = 21.1$ s, thus the settling time is $t_s = 1.1$ s which is basically determined by the bandwidth of the primary control.

At $t = 30$ s, the load L_2 is connected. From $t = 30$ s to $t = 40$ s, the converters #1 and #2 are delivering power to the loads L_{bus}, L_1 and L_2. As mentioned above, an increase in load active power produces a reduction in the converter frequencies. In the two last intervals, the converter #3 is connected at $t = 40$ s and the load L_3 is connected at $t = 50$ s. A perfect sharing of active power is observed in the top of Figure 14.10(a) during these intervals when the three power converters operate simultaneously. The active power delivered by these converters match perfectly due to the frequency is a global variable in a power system and, as a consequence, in steady state all the frequencies coincide as shown in the bottom of Figure 14.10(a). Quite the opposite, the reactive powers delivered by the converters present a clear mismatch, as shown in the top of Figure 14.10(b). These divergences are caused by the different amplitudes of the converter output voltages observed in the bottom of Figure 14.10(b). In fact, this is the usual situation in a power system where the amplitude is a local variable with different value in separate nodes of the system. It must be noted that the loads are purely resistive, thus the reactive powers supplied to the grid is a result of the parasitic inductances of the interconnecting wires. Additionally, it can be noticed that the amplitude of the voltage at the PCC is lower than the nominal value (1 p.u.) in steady state when only the primary control is active.

14.6.2 Performance evaluation of the secondary control

The objective of the secondary control is to eliminate the deviations in frequency and amplitude in steady state introduced by the primary control. The communication service plays an important role in the accomplishment of this objective. This section presents the experimental results using the communication infrastructure presented in Sections 14.4 and 14.5.

The performance of the secondary control is evaluated here employing the same black start event used in the previous section. The converters and loads are connected sequentially following the time instants and intervals described above. Figure 14.11 shows the experimental results. By comparing this figure to Figure 14.10, it can be noticed that frequencies and amplitudes are now moving slowly towards the nominal values, 60 Hz and 1 p.u., respectively. These final values are reached when the secondary control arrives to steady state, roughly at $t = 70$ s, thus completely eliminating the frequency and amplitude deviations. As expected, the connection of a new converter or load produces a change in active and reactive powers (in order to balance the power sharing between the converters in operation) and, as a consequence, produces also a change in frequencies and amplitudes.

The bandwidth of the secondary control was designed more than one decade below to the bandwidth of the primary control. This design was carried out by selecting the proportional and integral gains listed in Table 14.4. With this design, both control loops operate correctly without undesired interactions. In other words, the steady-state deviations in frequency and amplitude are removed slowly without affecting the load power sharing. The settling time of the secondary control can be measured in the bottom subplot of Figure 14.11(b) by observing the amplitude of the PCC voltage. The last change in the microgrid is the connection of the load L_3 at $t = 50$ s. The amplitude of the PCC voltage reaches its final value, 1 p.u., at roughly $t = 70$ s. Therefore, the settling time of the secondary control is nearly 20 s, more than one decade above the settling time of the primary control $t_s = 1.1$ s.

The transient response of the isolated microgrid during the black start is excellent due to several reasons: (1) the design of the primary and secondary

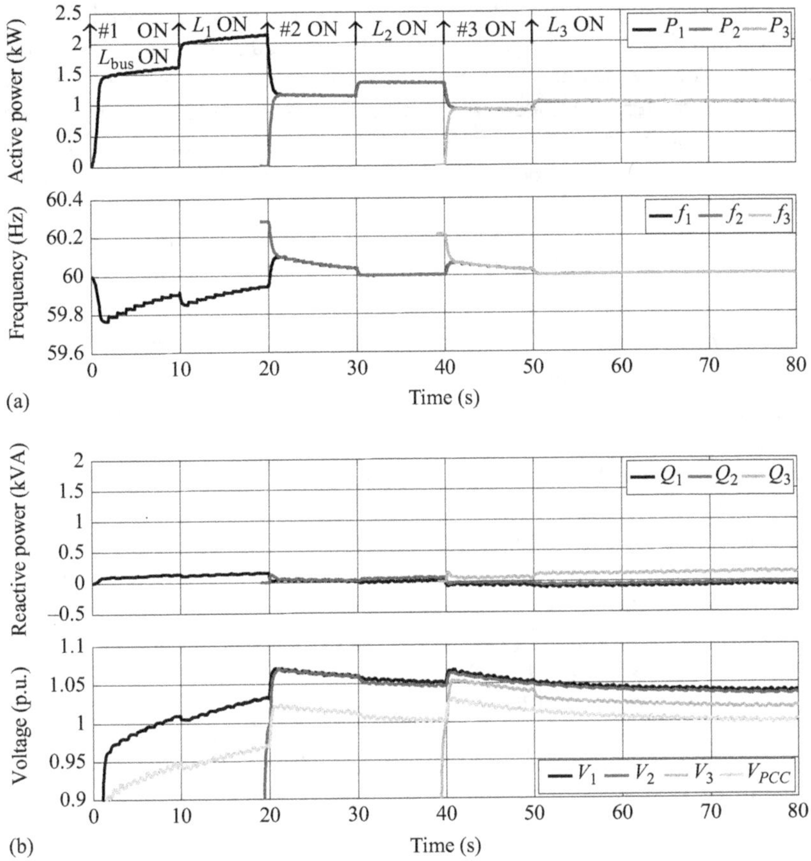

Figure 14.11 Experimental results for the secondary control in normal conditions. (a) Top: Active powers supplied by the power converters; Bottom: Frequencies of the converter output voltages. (b) Top: Reactive powers supplied by the power converters; Bottom: Amplitudes of both the converter output voltages and the voltage at the PCC

control bandwidths, (2) the sequential connection of converters and loads with large enough time intervals (it can be noticed that the primary control has enough time to reach the steady-state before a new connection is done) and, finally, (3) the simultaneous operation of the primary and secondary control layers from the time instant $t = 0$ s. The effects of this last point can be clearly observed in Figure 14.11. For instance, the converter #2 is activated at $t = 20$ s. This means that the switch S_2 (shown in Figure 14.5) is closed at $t = 20$ s. However, the local PLL of the primary control in this converter is working from $t = 0$ s, then adapting the phase of the converter output voltage to the phase of the microgrid voltage. By matching both phases, the connection of converter #2 has a smooth transition with no oscillations in active and reactive power. The effect of the secondary control is noticed in the initial condition of the frequency just before $t = 20$ s. At this time instant, the converter #2 does not supply power and therefore its frequency relies only on the secondary control term included in (14.10). This is why the frequency is abnormally high just before $t = 20$ s, as shown in the bottom of Figure 14.11(a). Once the switch S_2 is closed, the frequencies of the active converters equalise in about 1.1 s (i.e., the settling time of the primary control) and then they continue decreasing towards the nominal value by the action of the secondary control. A similar behaviour can be noticed around $t = 40$ s when the converter #3 is activated.

14.6.3 Effects of packet loss in the communication network

For control purposes, the communication service is mainly used to send data from the uGCC (secondary control) to the local controllers, as shown in Figure 14.5. The performance of this service in normal conditions was evaluated in previous section showing excellent results. In this section, the effects of packet loss in the communication network are analysed.

The data may be lost while in transit through the communication network due to transmission errors in the physical network links or to buffer overflows during congestion. In the current test, an oversized 20% of message loss probability is considered in order to evaluate the system in a more realistic scenario. This loss indicates that 2 out of every 10 messages (in average) are lost. It is important to point out that the loss probability is individual for each communication connection meaning that a message sent by the uGCC can be completely lost, or received only by one converter, or by two converters or by all of them.

Figure 14.12 shows the experimental results when the secondary control message loss probability was set to 20%. Comparing these results to those depicted in Figure 14.11 for ideal conditions, it can be noticed that the performance of the laboratory microgrid is robust against packet loss. In fact, only small differences between the experimental waveforms are observed from $t = 0$ s to $t = 20$ s, when only the converter #1 is active. For instance, for ideal conditions, the frequency is clearly updated every 1 s, as expected. In the new test, the packet loss is noticed by a constant frequency for more than one transmission period in the bottom of Figure 14.12(a). This fact can be clearly identified in three intervals: (1) $t = 1$ s to $t = 3$ s, (2) $t = 5$ s to $t = 7$ s and (3) $t = 10$ s to $t = 12$ s. Notice that the differences in the experimental waveforms are insignificant when more than one converter is active, from $t = 20$ s to $t = 80$ s.

The steady-state operation of the primary and secondary control layers are unaffected by the packet loss. As seen in the top of Figure 14.12(a), the load power sharing provided by the primary control is perfectly maintained even 20% of packets may be lost. The deviations in frequency and amplitude are correctly removed in steady state, confirming that the packets lost do not degrade the operation of the secondary control. It is worth mentioning that the performance of the transient response is not changed by the abnormal condition. In fact, the settling times of the primary and secondary control layers are preserved at 1.1 and 20 s, respectively, as also measured in the previous section for ideal condition.

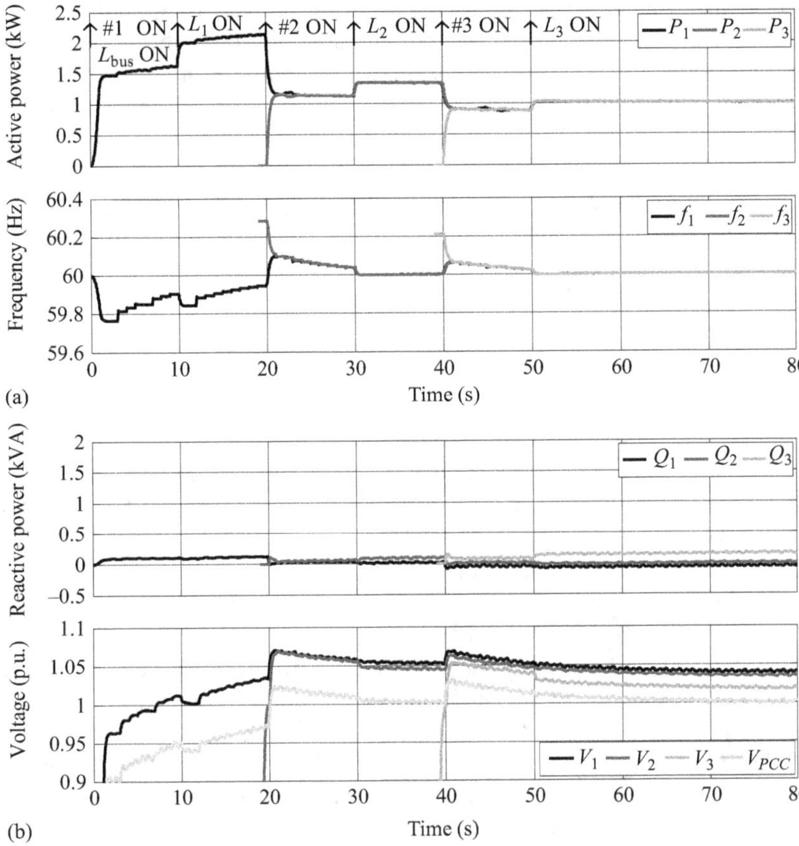

Figure 14.12 Experimental results for the secondary control when the message loss probability is 20%. (a) Top: Active powers supplied by the power converters; Bottom: Frequencies of the converter output voltages. (b) Top: Reactive powers supplied by the power converters; Bottom: Amplitudes of both the converter output voltages and the voltage at the PCC

14.7 Conclusions

AC microgrids are a promising type of active electricity distribution systems for the future smart grid scenario. At present, demonstration facilities are deployed around the world to show the great advantages that microgrids offer in relation to existing distribution networks. AC microgrids are plenty of power converters interfacing with distributed generators (energy sources), energy storage devices and loads. In this situation, control and communication are essential elements in the operation of these power systems.

This chapter introduced the basic concepts to control and manage the energy production, distribution and consumption in AC microgrids. The control platforms are located dispersedly around the microgrid. Some of them associated to the power converters of the energy sources and storage devices; others in the PCC between the microgrid and the main power system. A communication service with proper infrastructure is necessary in this scenario to interchange data and control commands in real-time between the controllers. The practical emphasis is placed on the description of a new generation of digital microcontrollers well suited for this application. These digital platforms are equipped with dual-core microcontrollers based on both a floating-point DSP device to execute complex control algorithms and a communication processor to guarantee real-time connectivity. Selected experimental results from a low-scale laboratory microgrid have been reported to validate the main ideas presented in this chapter.

References

[1] Pogaku N., Prodanovic M., Green T.C. 'Modeling, analysis and testing of autonomous operation of an inverter based microgrid'. *IEEE Trans. Power Electron.* 2007, 22(2), pp. 613–625

[2] Guo Z., Sha D., Liao X. 'Energy management by using point of common coupling frequency as an agent for islanded microgrids'. *IET Power Electron.* 2014, 7(8), pp. 2111–2122

[3] Chen C., Duan S. 'Optimal allocation of distributed generation and energy storage system in microgrids'. *IET Renewable Power Generation.* 2014, 8(6), pp. 581–589

[4] Chen C.L., Wang Y., Lai J., Lee Y.S., Martin D. 'Design of parallel inverters for smooth mode transfer microgrid applications'. *IEEE Trans. Power Electron.* 2010, 25(1), pp. 6–15

[5] Peças Lopes J., Moreira C., Madureira A.G. 'Defining control strategies for microgrids islanded operation'. *IEEE Trans. Power Syst.* 2006, 21(2), pp. 916–924

[6] Dimeas A.L., Hatziargyriou N.D. 'Operation of a multi-agent system for microgrid control'. *IEEE Trans. Power Syst.* 2005, 20(3), pp. 1447–1455.

[7] Texas Instruments, *Delfino F2837xD microcontrollers: Setting the MCU performance bar with the new dual-core F2837-D.* Texas Instruments white paper SPRT687A, 2014

[8] Sabapathy S. '*Designing the Next Generation of Industrial Drive and Control Systems*'. Texas Instruments White Paper SPRY258, 2014

[9] Rocabert J., Luna A., Blaabjerg F., Rodriguez P. 'Control of power converters in AC microgrids'. *IEEE Trans. Power Electron.* 2012, 27(11), pp. 4734–4749

[10] Vasquez J.C., Guerrero J.M., Miret J., Castilla M., Garcia de Vicuña L. 'Hierarchical control of intelligent microgrids'. *IEEE Ind. Electron. Mag.* 2010, 4(4), pp. 23–29

[11] Guerrero J.M., Vasquez J.C., Matas J., Garcia de Vicuña L., Castilla M. 'Hierarchical control of droop-controlled AC and DC microgrids – a general

approach towards standardization'. *IEEE Trans. Ind. Electron.* 2011, 58(1), pp. 158–172

[12] Jinwei He, Yun Wei Li. 'Analysis, design, and implementation of virtual impedance for power electronics interfaced distributed generation'. *IEEE Trans. Ind. Applic.* 2011, 47(6), pp. 2525–2538

[13] Katiraei F., Iravani R., Hatziargyriou N., Dimeas A. 'Microgrids management'. *IEEE Power and Energy Mag.* 2008, 6(3), pp. 54–65

[14] Zhabelova G., Vyatkin V. 'Multiagent smart grid automation architecture based on IEC 61850/61499 intelligent logical modes'. *IEEE Trans. Ind. Electron.* 2012, 59(5), pp. 2351–2362

[15] Timbus A., Larsson M., Yuen C. 'Active management of distributed energy resources using standardized communications and modern information technologies industrial electronics'. *IEEE Trans. Ind. Electron.* 2009, 56(10), pp. 4029–4037

[16] International Electrotechnical Commission's (IEC) Technical Committee 57, *Communication Networks and Systems in Substations.* IEC Std. 61850, 2003

[17] Decotignie J.D. 'Ethernet-based real-time and industrial communications'. *Proceedings of the IEEE.* 2005, 93(6), pp. 1102–1117

[18] I-Chun Chao, Kun-Yuan Tu, Shinn-Yan Lin, Fan-Ren Chang. 'Design and implementation of a switching controller for transient improvement in a time synchronization system'. *IEEE Trans. Instrum. and Meas.* 2011, 60(7), pp. 2184–2190

[19] Walsh G.C., Hong Ye. 'Scheduling of networked control systems', *IEEE Control Systems Mag.* 2001, 21(1), pp. 57–65

[20] Chon S., Novak B. '*Performance Without Compromise: Implementing Real-Time Control and Communications with a Dual-Subsystem Microcontroller*'. Texas Instruments Report SPRY174A, 2012

[21] Texas Instruments, *F28M36x Concerto Microcontrollers*. Texas Instruments report SPRS825C, 2012

[22] '*Lightweight IP* [online]'. 2015. Available from: http://savannah.nongnu.org/projects/lwip [Accessed 17 March, 2015]

[23] Texas Instruments, '*Using the Stellaris® Ethernet Controller with Lightweight IP (lwIP)*'. Texas Instruments Application Report SPMA025C, 2012

[24] Gu W., Liu W., Zhu J., *et al.* 'Adaptive decentralized under-frequency load shedding for islanded smart distribution networks'. *IEEE Trans. Sustainable Energy.* 2014, 5(3), pp. 886–895

[25] Banarjee S., Gopalan A., Das A.K., Shakkottai S. 'Epidemic spreading with external agents'. *IEEE Trans. Information Theory.* 2014, 60(7), pp. 4125–4138

Chapter 15

Use of computational intelligence for designing power electronics converters

Wenguan Wang[1], *Henry Shu-hung Chung*[1], *Jun Zhang*[2] *and Alan Wai-Lun Lo*[1]

15.1 Introduction

In the last decades, small-signal models have been widely adopted in power electronics circuits (PECs). Among various approaches, the state-space averaging and its variants [1–4] are the most common ones. Since the cutoff frequency of the output filter in a converter is much lower than the switching frequency, a linear time-invariant model approximates the time-variant PEC at its operating point. As a result, the feedback network (FN) can be designed by applying classical control theories. Although the procedures are simple and elegant, they are only applicable for specific circuits and control scheme [3, 4] that require comprehensive knowledge on the circuit operation. In addition, no exact waveform and the response profile can be obtained with state-space averaging technique, making it difficult to precisely predict the circuit responses under large-signal conditions [2].

As power electronics technology continues to develop, there is a growing need for automated synthesis that relies on a high-level statement of the desired behavior and optimizes the circuit component values accordingly. Considering the converter, shown in Figure 15.1, as an example, given a circuit topology, how can an engineer assign proper component values in order to obtain desired static and dynamic behaviors? To solve such problems, various analog circuit design automation techniques emerged. These methods incorporates heuristics [5], knowledge bases [6], and other algorithms to optimize the component values. Classical optimization techniques such as the gradient methods and hill-climbing techniques have been applied [7, 8]. However, some of those methods might be trapped into local minimum that leads to suboptimal parameter values, thus might not perform well in multimodal or noisy solution spaces.

Recently, some modern stochastic optimization techniques such as genetic algorithms (GAs) [9] have been shown to be effective when it comes to solving

[1]Department of Electronic Engineering, City University of Hong Kong, Kowloon, Hong Kong
[2]Department of Computer Science, Sun Yat-sun University, Guangzhou, China

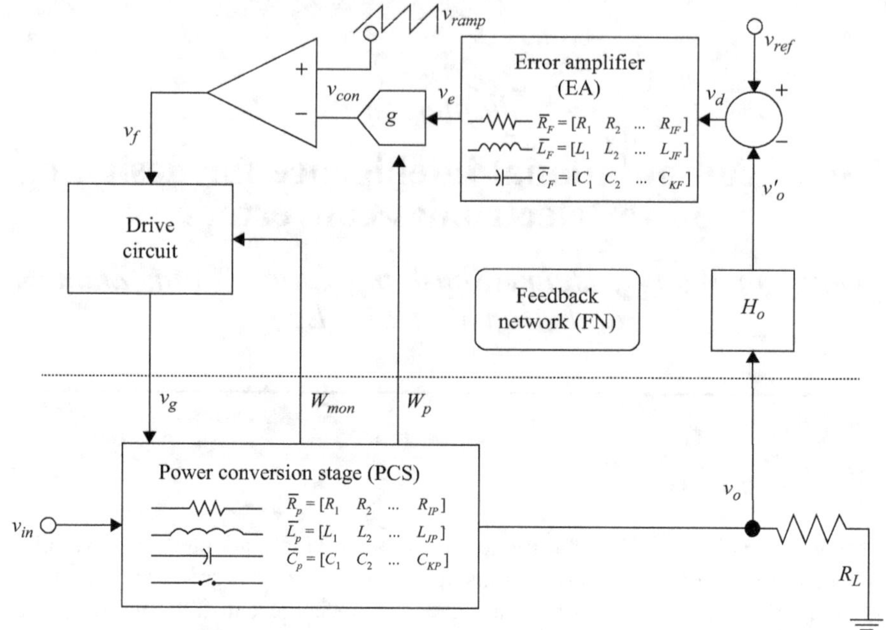

Figure 15.1 Converter diagram

optimization problems [10–14]. As one of the probabilistic algorithms, GAs do not search the solution space in a totally random manner as they combine elements of directed and stochastic search. In addition, they are more robust than existing directed search methods like hill-climbing. Another important property of GAs is that, unlike traditional techniques processing a single point in the search space, they maintain a population of potential solutions [14].

Many GA-based design schemes for analog circuits, like voltage reference circuit [11], transconductance amplifier [12], and analog circuit synthesis [13], have been proposed. Circuit behaviors are described by well-defined mathematical functions with unknown optimal coefficients. A set of guided stochastic searching procedures that are based loosely on the principles of genetics is formulated. The procedures are flexible, allowing mixed type, bounded decision variables, and complex multifaceted goals.

However, the values optimized by GA are sometimes not readily available for circuits and require composition. For example, it is sometimes necessary to connect several resistors to get the values given by GAs. So the designers have to choose the nearest discrete values available without considering component tolerance, which is an important issue for practical design. If the components used in the actual circuit are not optimal, the tolerance range will change. Thus, it is important to search the available values of the discrete-valued components.

To solve the optimization problems with discrete solutions, a new metaheuristic optimization method, ant colony optimization (ACO), has been proposed [15, 16].

A distinct feature of ACO is that it operates in discrete solution spaces rather than continuous ones. The algorithm simulates the strategies of ants finding paths from the colony to food. It is a multiagent approach for solving combinatorial optimization problems, such as traveling salesman problems [16], data mining [17], network routing [18], and controller design [19]. Recently, there is a growing trend of applying ACO to electrical engineering, such as power distribution system planning and load flow studies [20–22].

ACO optimizes circuits with discrete values. It uses many ants to search the best combination of the resistors and capacitors, such as $R_1, R_2, \ldots,$ in $\overline{R}_F$, and $C_1, C_2, \ldots,$ in $\overline{C}_F$ in Figure 15.1. The design parameters are interpreted as the nodes in the paths the ants explore. However, PEC generally consists of components such as inductors $L_1, L_2, \ldots,$ in $\overline{L}_F$, and some parameters like the voltage gains g and H_o in Figure 15.1, which have continuous values. Thus, the traditional ACO cannot be applied directly. Instead, the parameter with continuous value should be discretized first.

This chapter presents a decoupled optimization technique of designing switching regulators by GA and ACO. The mechanisms of the two popular computational intelligence (CI) techniques are briefly described. This chapter also emphasizes implementation issues, as GA is popularly chosen for optimization in continuous domain and ACO is more preferable to discrete domain. The optimization process selects component values in a switching regulator in order to meet both static and dynamic requirements. Although the methods inherit some characteristics of evolutionary computations, they do not optimize the circuit as a whole. Instead, the regulator is decoupled into two parts: the power conversion stage (PCS) and the FN. The PCS is optimized for static criteria and then the FN is optimized with the required static and dynamic behaviors of the whole system. Thus, intensive computations usually found in stochastic optimization techniques can be alleviated. The optimization procedures are described systematically and the techniques are illustrated with the design of a buck converter with overcurrent protection.

15.2 Formulation of fitness function

A fitness function assigns a value to each solution. The value relates to optimization objectives. This section demonstrates a multiobjective function for optimization of PCS and FN. Two types of fitness functions will be discussed here.

15.2.1 Type-one fitness function

This one is suitable for objectives that should be minimized, e.g., the steady-state error. The fitness function has a maximum value of K. For instance, a candidate solution gives a steady-state error of E_S during the searching process and a linear fitness function f is defined as follows:

$$f = K(1 - mE_S) \tag{15.1}$$

where m is the slope of the linear fitness function. The function f decreases as E_S increases and $m = m_1$. At the beginning of the searching process, most candidates do not perform satisfactorily and their steady-state errors are large. To cope with a wide distribution of E_S, m in (15.1) has to be small. However, after several generations, many candidates in the population have attained some acceptable level of the fitness value (i.e., their E_S are close to zero). In order to differentiate each candidate effectively in this stage, m should be large and $m = m_2$. Formulating an adaptive fitness function is possible, but requires adaptive tuning of m.

A more simple approach is to use a piecewise fitness function. $m(=m_2)$ is large when E_S is close to zero. Conversely, $m(=m_1)$ is small when E_S is far away from zero. Instead of using this piecewise linear fitness function for f, an exponential function is used to perform. Mathematically

$$f = Ke^{(-E_S)/\tau} \tag{15.2}$$

where τ is rate of decay of the function. τ depends on the expected fitness value at $E_S = E_M$. For example, it is required to let f decays to ε when $E_S = E_M$. Hence, τ is obtained by (15.2):

$$\varepsilon = Ke^{(-E_M)/\tau} \Rightarrow \tau = \frac{E_M}{-\ln(\varepsilon/K)} \tag{15.3}$$

The major advantages of the exponential function lie in its simplicity and well-defined characteristics in practice.

15.2.2 Type-two fitness function

Another form of the fitness function f_2 is based on the sigmoid function

$$f_2 = \frac{K}{1 + e^{(T_S - T_S^*)/\tau}} \tag{15.4}$$

Apart from the two-slope characteristics as in (15.2), f_2 becomes K when $T_S \ll T_S^*$. Equation (15.4) is suitable for specifications like the settling time, maximum overshoot, and undershoot.

15.2.3 Fitness function for the PCS

The fitness function ϕ_P of the solutions in PCS population is based on the following consideration, including

- the steady-state error of v_o within the input voltage range $v_{in} \in [V_{in,min}, V_{in,max}]$ and output load range $R_L \in [R_{L,min}, R_{L,max}]$,
- the operational constraints on circuit components, such as maximum voltage stresses and ripple current,
- the steady-state ripple voltage on v_o, and
- the intrinsic factors concerning the components, such as the total cost, physical size.

Each objective is measured by a function OF. For the nth solution in the population, the objective function is

$$\Phi_P(CP_n) = \sum_{\substack{R_L \in \Re_{\mathfrak{L}} \\ v_{in} \in \mathfrak{v}_{in} \\ k \in \mathfrak{k}}} OF_k(R_L, v_{in}, CP_n) \tag{15.5}$$

where $\Re_{\mathfrak{L}}$, $\mathfrak{v}_{in}$ and $\mathfrak{k}$ contains some representative values of R_L, v_{in} and k, respectively. The definition of all OFs in (15.5) are defined as follows.

- OF_1: The steady-state v_o affects the suitability of a solution Θ_{PCS} in the population. The implied goal is to find whether there exists a v_{con} in Figure 15.1 such that v_o', the value of v_o after the signal conditioning of H_o, equals v_{ref}. An iterative secant method [23] is applied to determine the steady-state waveform whose error is defined as

$$E = \sum_{m=1}^{N_s} [v_o'(m) - v_{ref}]^2 \tag{15.6}$$

in which v_o' is obtained by performing time-domain simulation. The initial conditions for steady state are obtained by the method discussed in Reference 24. The objective OF_1 is based on (15.2) and defined as

$$OF_1 = K_1 e^{-E/K_2} \tag{15.7}$$

where K_1 is the maximum value of OF_1 and K_2 adjusts the sensitivity of OF_1 with respect to E.

- OF_2: Under the steady-state condition, there are constraints controlling the operating limits of some waveforms. For example, if $\lambda_{C,m}$ is the limit of a considered quantity q_m in the mth constraint, the fitness function OF_2 will be based on (15.4) and is defined as

$$OF_2 = \sum_{m=1}^{N_C} \frac{K_{3,m}}{1 + e^{(q_m - \lambda_{C,m})/K_{4,m}}} \tag{15.8}$$

where N_C is the number of constraints, $K_{3,m}$ is the maximum value of the mth constraint, and $K_{4,m}$ determines the sensitivity of the considered quantity. For example, if λ_C represents the maximum voltage rating of a switch and q is the actual voltage stress, OF_2 is large when q is much smaller than λ_C.

- OF_3: The ripple voltage on v_o has to stay within a limit of $\pm\Delta v_o$ around the expected output $v_{o,exp}$. A measurement of CP_n in this objective is to count the area of v_o outside $v_{o,exp} \pm \Delta v_o$ in N_S-simulated samples. OF_3 is based on (15.2) and is defined as

$$OF_3 = K_5 e^{-A_1/K_6} \tag{15.9}$$

where K_5 is the maximum attainable value for this objective, K_6 is a decay constant, and A_1 is the ripple area outside the tolerance band. Similar to OF_1, OF_3 shall decrease as A_1 increases.

- OF_4: Apart from the electrical performance of the PCS, the objective function could also include some intrinsic factors like the cost, physical size, lifetime, etc. Thus OF_4 is based on (15.4) and can be expressed as

$$OF_4 = \sum_{i=1}^{I_P} \phi_R(R_i) + \sum_{j=1}^{J_P} \phi_L(L_j) + \sum_{k=1}^{K_P} \phi_C(C_k) \tag{15.10}$$

where ϕ_R, ϕ_L, and ϕ_C, are the objective functions measuring individual component type, defined as

$$\begin{aligned} \phi_R(R_i) &= \frac{K_{7,i}}{1 + e^{(R_i - R_{i,max})/\tau_R}} \\ \phi_L(L_j) &= \frac{K_{8,j}}{1 + e^{(L_j - L_{j,max})/\tau_L}} \\ \phi_C(C_k) &= \frac{K_{9,k}}{1 + e^{(C_k - C_{k,max})/\tau_C}} \end{aligned} \tag{15.11}$$

where $K_{7,i}$, $K_{8,j}$, $K_{9,k}$ are the maximum attainable values of ϕ_R, ϕ_L, ϕ_C. $R_{i,max}$, $L_{j,max}$, $C_{k,max}$ are the maximum values for R_i, L_j, C_k, respectively.

15.2.4 Fitness function for FN

Similar to PCS, the fitness function Φ_F that evaluates FN is based on the following considerations:

- the steady-state error of v_o within the required input voltage range and output load range,
- the maximum overshoot and undershoot, and the settling time of v_o (or v_d) during start-up,
- the steady-state ripple voltage on v_o, and
- the dynamic behaviors as in (2) during the input voltage and output load disturbances.

Φ_F measures CF for these objectives. Specifically, for the hth solution in the population, Φ_F is

$$\Phi_F(CF_h) = \sum_{\substack{R_L \in \Re_{\mathfrak{L}} \\ v_{in} \in \mathfrak{v}_{in} \\ k \in \{5,6,7\}}} OF_k(R_L, v_{in}, CF_h) + OF_8(CF_h) \tag{15.12}$$

- OF_5: With a set of component values in the PCS, the steady-state condition of the whole system is determined by the dual loop iteration method in Reference 23. As this objective is similar to OF_1, formulation of OF_5 is also based on (15.2) and is defined as $OF_5 = OF_1 = K_1 e^{-E_2/K_2}$

- OF_6 and OF_8: During start-up or external disturbances, a transient response appears at $v_d = v_{ref} - v_o^{'}$. OF_6 and OF_8 measure the transient response of v_d, including the maximum overshoot, the maximum undershoot, and the settling time. The general form of OF_6 and OF_8 is

$$OF_6 = OV(R_L, v_{in}, CF_h) + UV(R_L, v_{in}, CF_h) + ST(R_L, v_{in}, CF_h) \tag{15.13}$$

and

$$OF_8 = \sum_{i=1}^{N_T} [OV(R_{L,i}, v_{in,i}, CF_h) + UV(R_{L,i}, v_{in,i}, CF_h) + ST(R_{L,i}, v_{in,i}, CF_h)] \tag{15.14}$$

where N_T is the number of the input and load disturbances in the performance test. In the above expressions, OV, UV, and ST are the objective functions for minimizing the maximum overshoot, maximum undershoot, and settling time of v_d. Thus, (15.4) is applied and the functions in (15.13) and (15.14) are defined as follows:

$$\text{OV} = \frac{K_{10}}{1 + e^{(M_p - M_{p0})/K_{11}}} \tag{15.15}$$

where K_{10} is the maximum attainable value of this objective function, M_{p0} is the maximum overshoot, M_p is the actual overshoot, and K_{11} is a passband constant.

$$\text{UV} = \frac{K_{12}}{1 + e^{(M_v - M_{v0})/K_{13}}} \tag{15.16}$$

where K_{12} is the maximum attainable value of this objective function, M_{v0} is the maximum undershoot, M_v is the actual undershoot, and K_{13} is a passband constant.

$$\text{ST} = \frac{K_{14}}{1 + e^{(T_s - T_{s0})/K_{15}}} \tag{15.17}$$

where K_{14} is the maximum value of this objective function, T_{s0} is a constant, T_s is the actual settling time, and K_{15} adjusts the sensitivity. T_S is defined as the settling time of v_d that falls within a $\pm\sigma\%$ band. That is $|v_d(t)| \leq 0.01\sigma, t \geq T_S$.

- OF_7 is the same as the criteria in the PCS optimization. The number of samples that are outside the tolerance band of v_o are calculated. OF_7 is the same as (15.9). That is, $OF_7 = OF_3 = K_5 e^{-A_1/K_6}$.

15.3 Description of GA

GA, differing from conventional search techniques, begins with a set of random solutions called population. In other words, population is a group of potential solutions for the design. Each individual in the population is called a chromosome, representing a solution to the optimization problem. Each chromosome has a set of genes, which encode the value of components. An index of merit (fitness value) is assigned to each chromosome, according to a predefined fitness function. A new generation is evolved by a selection technique, in which a better individual would has higher chance of being selected. Pairs of chosen chromosomes act as parents in the construction of the next generation. A new generation is produced as a result of reproduction operators applied on parents, namely mutation and crossover. New generations are repeatedly produced until a predefined convergence level is reached.

The optimization procedures for the PCS and FN are similar. Their major differences are the definitions of the fitness functions and population. Thus, with the aid of the flowchart in Figure 15.2, only the steps of optimizing the PCS in one generation are illustrated.

- Step 1: Initialization: The population size N_p, the maximum number of generations G_{max}, the probability of crossover p_x, the probability of mutation p_m, and the generation counter *gen* are initialized. All chromosomes are initialized with random numbers within design limits. By using (15.5) for PCS or (15.12) for FN, the fitness values of all chromosomes are calculated. The best chromosome in the initial generation $CP_b(0)$ having the highest fitness value is then selected as a reference for the next generation.
- Step 2: Selection of chromosomes: A selection process based on roulette wheel rule is performed in this step. After updating *gen*, the algorithm proceeds with the calculation of the fitness value $\Phi_p[CP_n(gen)]$, the relative fitness value $\Phi_{p,r}[CP_n(gen)]$, and the cumulative fitness value $\Phi_{p,c}[CP_n(gen)]$ for the $CP_n(gen)$:

$$\Phi_{p,r}[CP_n(gen)] = \frac{\Phi_p[CP_n(gen)]}{\sum_{z=1}^{N_p} \Phi_p[CP_z(gen)]}, \tag{15.18}$$

and

$$\Phi_{p,c}[CP_n(gen)] = \sum_{z=1}^{n} \Phi_{p,r}[CP_z(gen)]. \tag{15.19}$$

A random number $p \in [0, 1]$ is generated and is compared with $\Phi_{p,c}[CP_n(gen)]$ for $n = 1, \ldots, N_p$. If $\Phi_{p,c}[CP_{z-1}(gen)] < p < \Phi_{p,c}[CP_z(gen)]$, CP_z is selected to be a member of the new population. This selection process is repeated until N_p members have been selected for the new population. Chromosomes with higher fitness values will have higher probability to survive and might appear repeatedly in the new population.
- Step 3: Reproduction operations: New chromosomes emerge from crossover and mutation. The crossover operation is illustrated in Figure 15.3(a). Two chromosomes are selected from the population. In order to determine whether a chromosome will undergo a crossover operation, a random selection test (RST) is

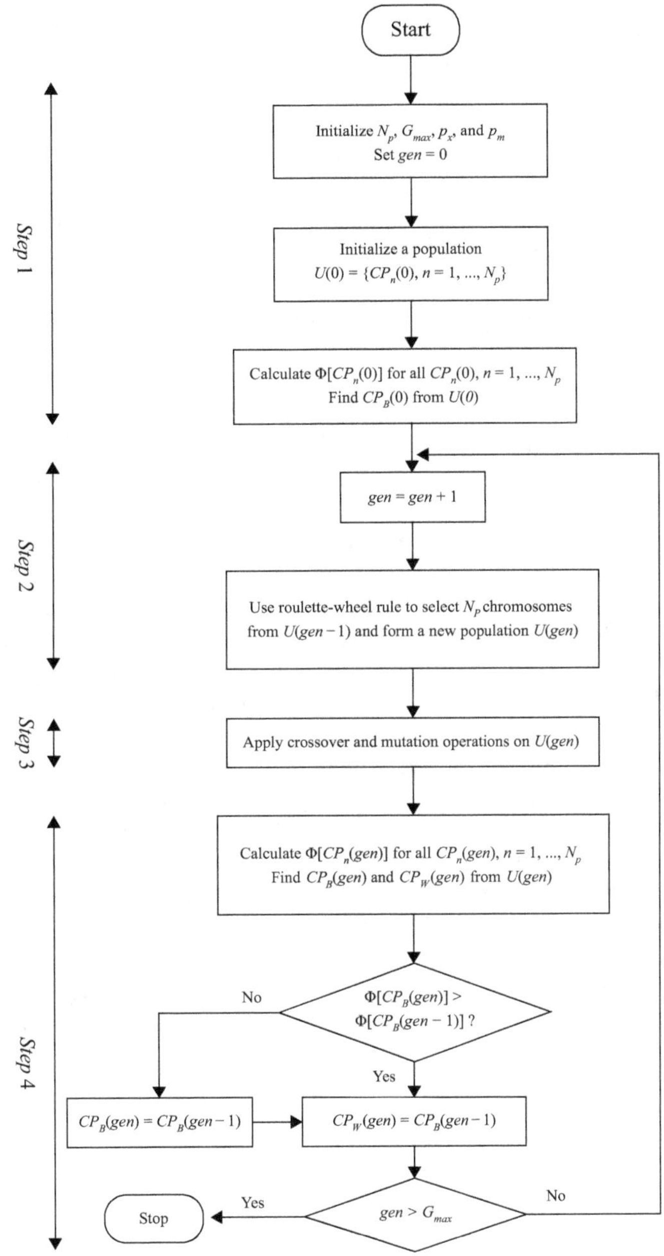

Figure 15.2 Flowchart for GA

performed. The RST is based on generating a random number $p \in [0, 1]$. If p is smaller than an assigned crossover probability p_x, the chromosome will be selected. Another chromosome will be chosen with the similar procedure. A crossover point is selected randomly with equal probability from 1 to the total

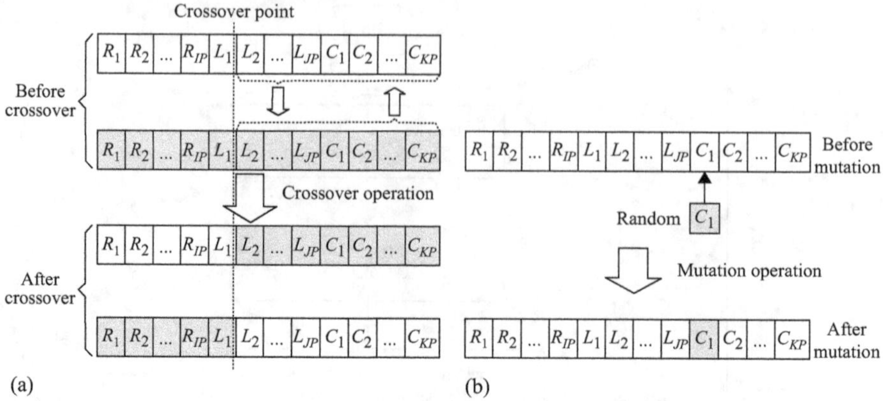

Figure 15.3 Genetic operations. (a) Crossover (b) Mutation © 2001 IEEE. Reprinted with permission from Reference 25

number of components in the chromosomes. The genes after the crossover point will be exchanged to create two new chromosomes. The operations repeat until all members in the population have been considered. The mutation operation in Figure 15.3(b) also starts with an RST for each chromosome. If a generated random number $p \in [0, 1]$ for a chromosome is smaller than an assigned mutation probability p_m, the chromosome will undergo mutation. A random number will be generated for the chosen component with a value lie within the component limits. The procedures will be repeated until all members have been considered.

- Step 4: Elitist function: After finishing the reproduction operation and the calculation of the fitness value of each chromosome, the best member $CP_B(gen)$ that has the largest fitness value and the worst member $CP_w(gen)$ that has the smallest fitness value will be identified. $CP_B(gen)$ will be compared with the best one in the last generation [i.e., $CP_B(gen-1)$]. If the fitness value of $CP_B(gen)$ is smaller than the one of $CP_B(gen-1)$, the chromosome content of $CP_B(gen-1)$ will replace the content of $CP_B(gen)$. Afterwards, the chromosome content of $CP_B(gen-1)$ will be substituted into $CP_w(gen)$ and the next GA cycle repeats from Step 2.

15.4 Description of ACO

ACO is typically used to search paths through graphs, it simulates the behavior of ants in finding paths from the colony to food. The searching mechanism is applicable for optimizing electric circuits with components having discrete values like resistors and capacitors. However, power electronic circuits generally have some components manufactured in continuous values, such as inductors. Therefore, the traditional ACO algorithm cannot be applied directly. In this chapter, an extended ACO (eACO) that searches the optimal values of components manufactured in discrete and continuous values is present. In this approach, the orthogonal design method (ODM) [26, 27] dynamically updates the database of the

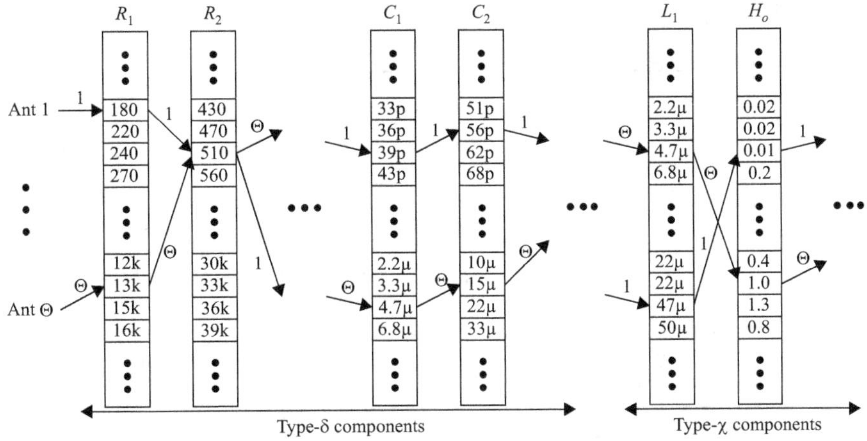

Figure 15.4 Illustration of the ACO in finding various combination of component values © 2009 IEEE. Reprinted with permission from Reference 28

components available with continuous values, so that those components will have pseudodiscrete values in the search space. The algorithm has been used to optimize the design of a buck regulator.

15.4.1 Data structure

As illustrated in Figure 15.4, there are Θ ants to search different combinations of the components in the optimization algorithm. The data structure of each ant is described as follows.

The circuit components are categorized into type-δ and type-χ. The type-δ components have discrete values in practice. Examples of this type of components are resistors and capacitors that are manufactured accurately within a specified tolerance of their nominal values. The type-χ components have continuous values. Each component can be of any value within the specified lower and upper limits. Examples of this type of components are inductors and voltage gains, which are manufactured rather accurately.

Each ant contains entries pointing to available values stored in the database for the chosen type-δ and type-χ components and its fitness value. Each available component has two parameters: the component value and the associated pheromone.

15.4.2 Procedures

The optimization procedures consist of eight steps as shown in Figure 15.5.

- Step 1: Initialization: The inventory of components is initialized at the beginning of the procedure. Also, pheromone is uniformly distributed over the search space.
- Step 2: Formation of the probability density table for choosing the components: probability density tables containing the probabilities of all entries in each component set are constructed in the same way as the roulette wheel table in the GA (Section 15.3).

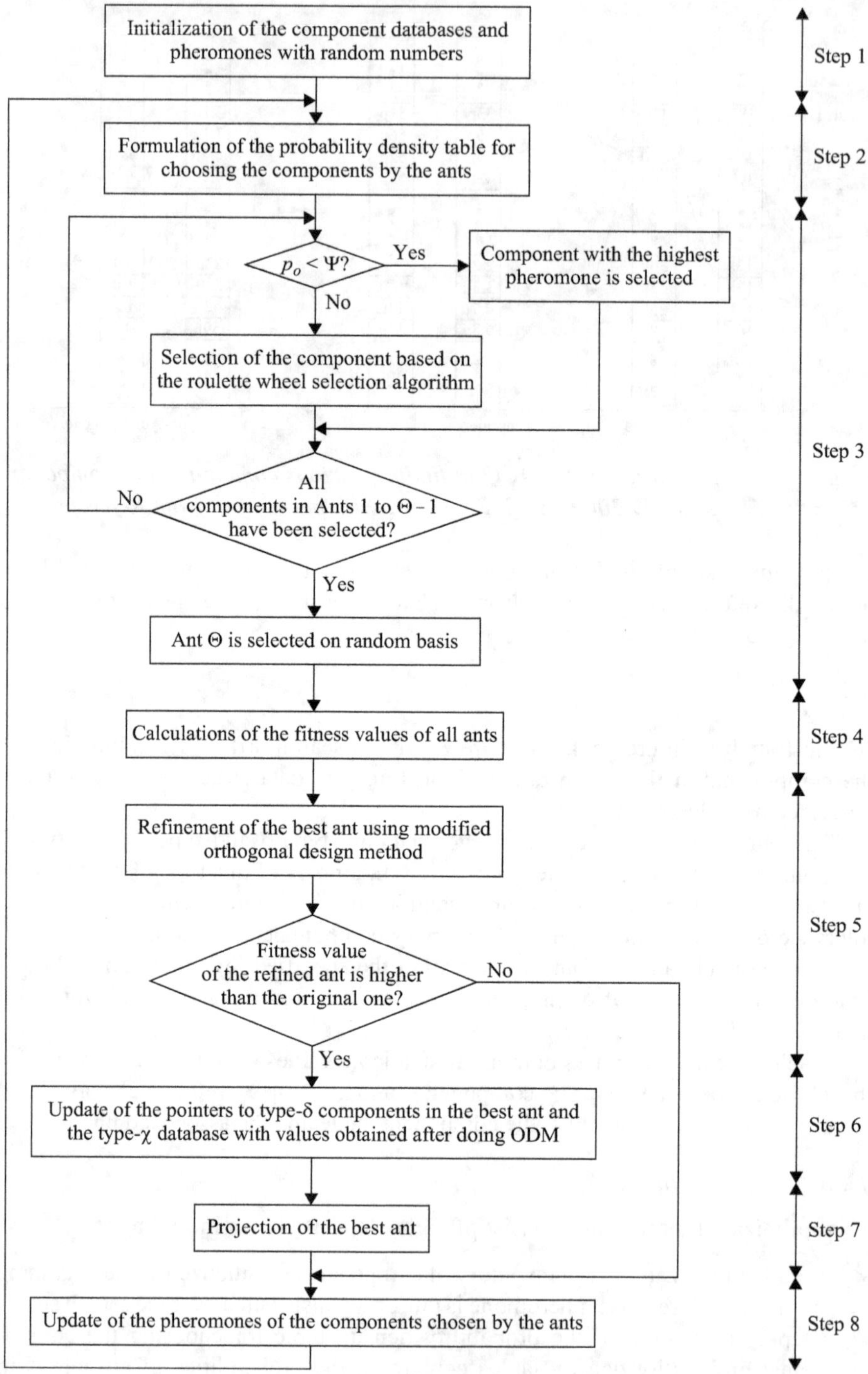

Figure 15.5 Flowchart of the ACO procedures © 2009 IEEE. Reprinted with permission from Reference 28

- Step 3: Selection of the components for all ants: In a set of components, every ant except the last one selects the component having the highest pheromone value with probability Ψ, or use roulette wheel table to select one component with probability $1 - \Psi$; for the last ant, the components are selected randomly.
- Step 4: Calculations of the fitness values of all ants once their values have been assigned.
- Step 5: Refinement of the best ant: The modified ODM [26–28] is performed on the best ant A_b, resulting a group of ants, among which $A_b^{'}$ is the best. If A_b is better than $A_b^{'}$, go to step 8.
- Step 6: A new ant $A_b^{''}$ is extrapolated from A_b and $A_b^{'}$.
- Step 7: The best among A_b, $A_b^{''}$, and $A_b^{''}$ is chosen as the new A_b.
- Step 8: Update of the pheromones of the components chosen by the ants: The pheromones of the components chosen are updated in accordance with the formula $p = (1 - \beta)p + \beta T$, where p is the pheromone of the component, β is the rate of updating, and T is the amount of newly added pheromone. T is set higher for the best ant than that for other ants.

15.5 Design examples and implementation issues

This section applies the GA and the eACO to design a converter according to its schematics and required specifications. The simulated results and measured results of the optimized circuit are shown.

15.5.1 Design objectives

Figure 15.6 shows the schematic of a buck regulator with overcurrent protection [29]. It consists of a buck converter and a proportional-plus-integral (PI) controller. The overcurrent protection is implemented in the driver circuit by monitoring inductor current and shorting the op-amp output to ground if necessary [29]. The required specifications are as follows.

- Input voltage range: 40 ± 20 V
- Output load range: 5 Ω ~ 10 Ω
- Nominal output voltage: 5 V $\pm$ 1%
- Switching frequency: 20 kHz
- Maximum settling time: 20 ms.

In Figure 15.6, v_{ramp} and v_{ref} are 0.2 V/μs and 5 V, respectively. For the PCS, L and C are design parameters and R_L, r_C, r_L, and r_{in} are assumed to be known *a priori*. For the FN, all components are the design parameters. All fitness functions except OF_2 are used in the optimization. OF_2 is not considered because no special constraints are imposed on the waveform of the converter. The maximum attainable value of each fitness function is chosen to be two arbitrarily. Thus, $K_1, K_5, K_8, K_9, K_{10}, K_{12}$, and K_{14} equal two. Other coefficients are determined as follows.

- OF_1 and OF_5: As these two objective functions govern the steady-state output, this requirement should be tight. OF_1 and OF_1 are made equal 0.2 (i.e., 10% of

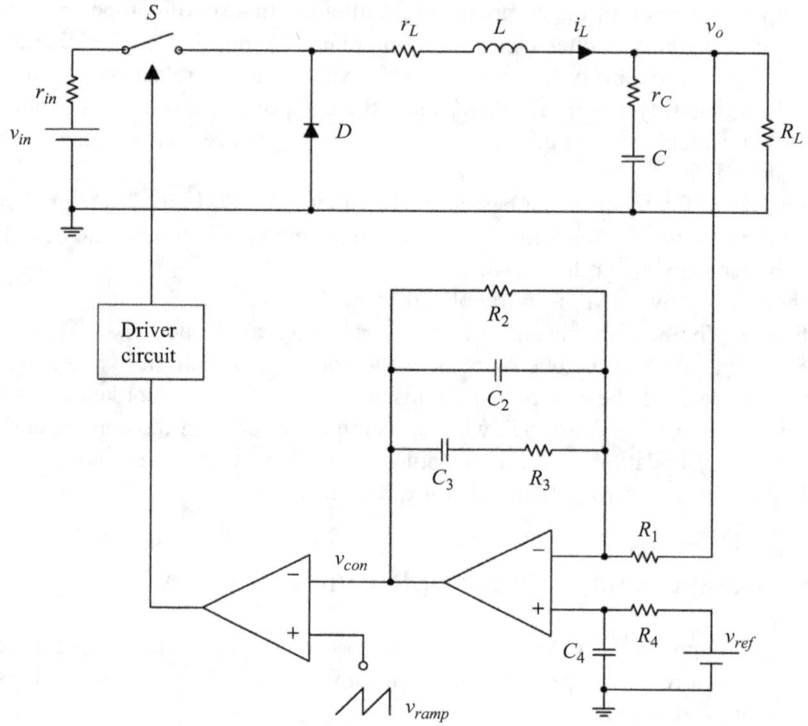

Figure 15.6 A buck converter

the maximum value) if the steady-state value of the N_S samples in (15.6) has 5% deviation from the expected 5 V output. N_S is equal to 15,000. Thus, based on (15.3), $K_2 = 400$.

- OF_3 and OF_7: This objective function ensures that the output voltage is within the $\pm 1\%$ tolerance band. A very tight arrangement is that OF_3 becomes 0.2 if the total output voltage samples has $\pm 0.1\%$ outside the tolerance band. Thus, based on (3), $K_6 = 32$.
- OF_4: Only ϕ_L and ϕ_C are significant in this objective function. $L_{j,max}$ and $C_{k,max}$ are chosen to be 5 mH and 4700 μF. ϕ_L equals 0.2 if L_j is five times larger than $L_{j,max}$. ϕ_C equals 0.2 if C_k is twice $C_{k,max}$. Thus, $\tau_L = 4.55 \times 10^{-3}$ and $\tau_C = 2.14 \times 10^{-3}$ in (15.11).
- OF_6 and OF_8: OV and UV are determined in the same manner. M_{po} and M_{vo} are chosen to be 4 V during disturbances. OV and UV will be less than 0.2 if M_p and M_v are larger than 5 V. Thus, $K_{11} = K_{13} = 0.455$. ST becomes 0.2 if T_s is 30 ms. T_{s0} is taken to be $(20 + 30)/2$ ms = 25 ms. Thus, $K_{15} = 2.28$ ms.

15.5.2 Design using GA

All coefficients are tabulated in Table 15.1. The computer program continuously monitors the fitness value and stops when the fitness value has close to a relatively

Table 15.1 Parameters used in the optimization © 2001 IEEE. Reprinted with permission from Reference 25

Power conversion stage (PCS)		Feedback network (FN)	
Parameter	**Value**	**Parameter**	**Value**
P_x	0.85	P_x	0.85
P_m	0.25	P_x	0.25
G_{max}	500	G_{max}	500
N_p	30	N_p	30
N_S	15,000	N_S	15,000
N_T	6	N_T	6
K_1	2	K_{10}	2
K_2	400	K_{11}	0.455
K_5	32	K_{12}	2
K_6	32	K_{13}	0.455
K_8	2	K_{14}	2
K_9	2	K_{15}	2.28×10^{-3}
τ_L	4.55×10^{-3}		
τ_C	2.14×10^{-3}		
δv_{in}	20 V		
δR_L	3 Ω		

Table 15.2 Results obtained by GA © 2001 IEEE. Reprinted with permission from Reference 25

Component	Initial values	Optimized values
L	200 μH	194 μH
C	1000 μF	1054 μF
R_1	0.6 kΩ	1.1 kΩ
R_2	300 kΩ	767 kΩ
R_3	4.7 kΩ	3.5 kΩ
R_4	1 kΩ	6.5 kΩ
C_2	2 μF	5.9 μF
C_3	3.3 μF	0.46 μF
C_4	1.8 μF	1.1 μF

constant value. In this example, it was found that the fitness value has been steady after 500 generations. Table 15.2 shows the initial and resulting values of L and C. The optimized values of the inductor and capacitor in the buck converter were found to be 194 μH and 1054 μF, respectively. These two values are close to the ones in Reference 29. This means that the original L and C have shown satisfactory performance within the requirements. In the actual implementation, an inductor of 200 μH and a capacitor of 1000 μF are used. The PI controller is then optimized after the PCS optimization. Table 15.2 also shows the initial and optimized component values for the controller. These values differ from the ones in Reference 29 even if the components of the PCS are similar. On a Pentium III 500 MHz machine, the decoupled GA technique took 5 h to complete while the original method required 8 h. The startup transients when the input voltage is 20 V and the output load is 5 Ω are shown in Figure 15.7. For more details, the readers are referred to Reference 25.

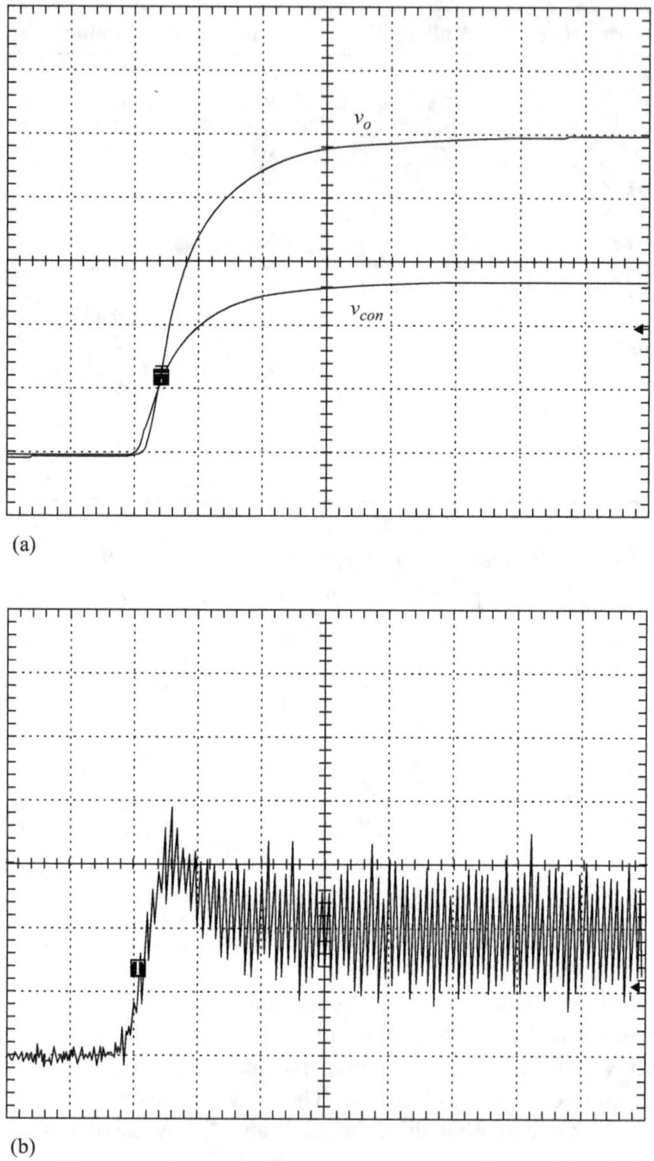

Figure 15.7 Start-up transient of the converter designed by GA © 2001 IEEE. Reprinted with permission from Reference 25 (a) v_o (1 V/div) and v_{con} (1 V/div) (timebase: 5 ms/div); (b) i_L (0.5 A/div) (timebase: 2 ms/div)

Table 15.3 Results obtained by ACO © 2009 IEEE. Reprinted with permission from Reference 28

Component	Results from ACO	Results from GA
L	265 μH	194 μH
C	1000 μH	1054 μF
R_1	82 Ω	1.1 kΩ
R_2	82 kΩ	767 kΩ
R_3	470 Ω	3.5 kΩ
R_4	560 Ω	6.5 kΩ
C_2	0.22 μF	5.9 μF
C_3	3.3 μF	0.46 μF
C_4	2.2 μF	1.1 μF

15.5.3 Design using ACO

The optimized values of the components are given in Table 15.3. Compared with the results obtained by GA, the results from ACO are available practically, so it is not necessary to use several resistors or capacitors to fabricate the optimized values. Additionally, the circuit designed by ACO has slightly better transient performance, as shown in Figure 15.8. More details about the experiment can be found in Reference 28.

15.6 Summary

This chapter presents an implementation of a decoupled optimization technique for design of switching regulators using two CI techniques, GA and ACO. A buck converter has been decoupled into PCS and FN. The PCS is optimized with the required static characteristics, while the FN is optimized with the required static and dynamic behaviors of the whole system.

In the presented example, both GA and eACO produce satisfactory results. Though the circuit designed by ACO is slightly better than GA here in this chapter, it should be noted that GA has its merit of being simpler and more suitable for finding solutions in a continuous domain. On the other hand, ACO is more suitable for discrete domains. Meanwhile, CI techniques require much computational resources, as implied in their name. Thus, there should be careful consideration of the objective function when employing CI algorithms.

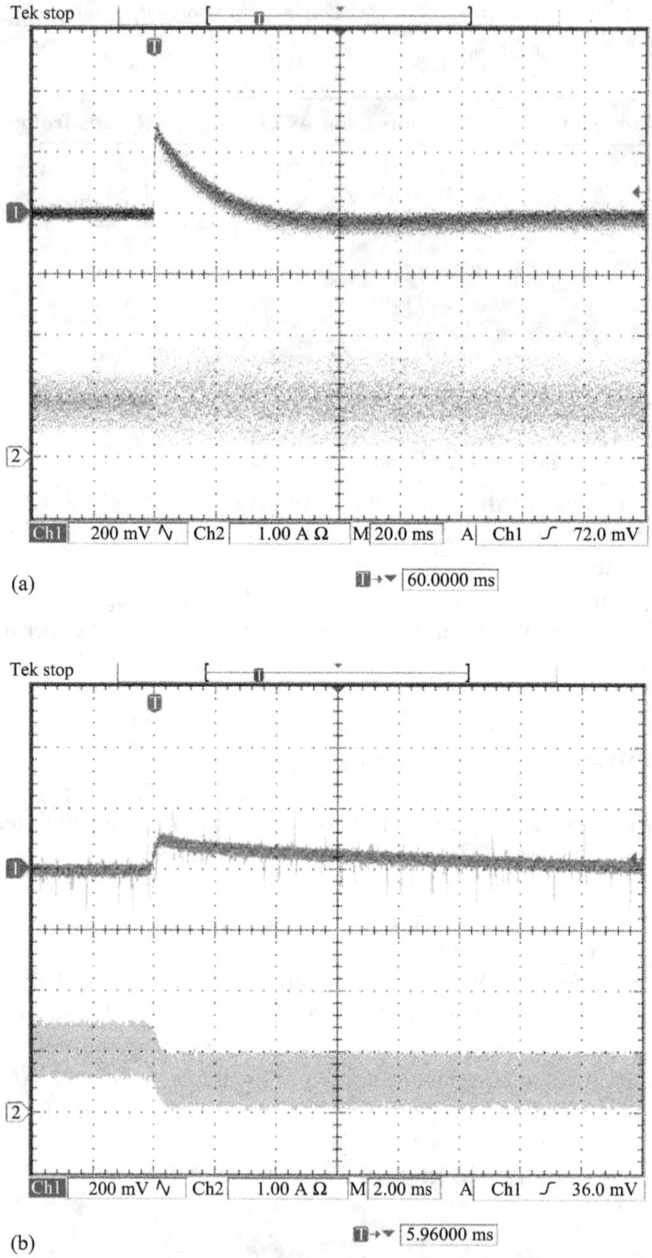

(a)

(b)

Figure 15.8 Experimental transient responses of v_o (200 mV/div) and i_L (1 A/div) when v_{in} changed from 20 to 40 V. © 2009 IEEE. Reprinted with permission from Reference 28 (a) circuit designed by GA; (b) circuit designed by ACO

References

[1] R.D. Middlebrook and S. Ćuk. Advances in switched-mode power conversion. Number 1–2 In: Power electronics series. TESLAco, Pasadena, 1983.

[2] Y.-S.S. Lee. *Computer-aided analysis and design of switch-mode power supplies.* Marcel Dekker, Inc., New York, 1993.

[3] R.D. Middlebrook. Modeling current-programmed buck and boost regulators. *IEEE Transactions on Power Electronics,* 4(1):36–52, 1989.

[4] G.C. Verghese, C.A. Bruzos, and K.N. Mahabir. Averaged and sampled-data models for current mode control: a re-examination. In: *Power Electronics Specialists Conference, 1989. PESC '89 Record, 20th Annual IEEE,* volume 1, pages 484–491, 1989.

[5] G.J. Sussman and R. Stallman. Heuristic techniques in computer-aided circuit analysis. *IEEE Transactions on Circuits and Systems,* 22(11):857–865, 1975.

[6] R. Harjani, R.A. Rutenbar, and L.R. Carley. OASYS: a framework for analog circuit synthesis. *IEEE Transactions on Computer-Aided Design of Integrated Circuits and Systems,* 8(12):1247–1266, 1989.

[7] L.P. Huelsman. Optimization—a powerful tool for analysis and design. *IEEE Transactions on Circuits and Systems I: Fundamental Theory and Applications,* 40(7):431–439, 1993.

[8] R.E. Massara. *Optimization methods in electronic circuit design.* Longman Publishing Group, 1991.

[9] D.E. Goldberg and J.H. Holland. Genetic algorithms and machine learning. *Machine Learning,* 3(2):95–99, 1988.

[10] V. Petridis, S. Kazarlis, and A. Bakirtzis. Varying fitness functions in genetic algorithm constrained optimization: the cutting stock and unit commitment problems. *IEEE Transactions on Systems, Man, and Cybernetics, Part B: Cybernetics,* 28(5):629–640, 1998.

[11] D. Nam, Y.D. Seo, L.-J. Park, C.H. Park, and B. Kim. Parameter optimization of a voltage reference circuit using evolutionary programming. In: *The 1998 IEEE International Conference on Evolutionary Computation Proceedings, World Congress on Computational Intelligence,* pages 301–305. IEEE, 1998.

[12] M. Wojcikowski, J. Glinianowicz, and M. Bialko. System for optimisation of electronic circuits using genetic algorithm. In: *Proceedings of the Third IEEE International Conference on Electronics, Circuits, and Systems, 1996. ICECS'96*, volume 1, pages 247–250. IEEE, 1996.

[13] N.R. Dhanwada, A. Nunez-Aldana, and R. Vemuri. A genetic approach to simultaneous parameter space exploration and constraint transformation in analog synthesis. In: *Proceedings of the 1999 IEEE International Symposium on Circuits and Systems, 1999. ISCAS'99*, volume 6, pages 362–365. IEEE, 1999.

[14] Z. Michalewicz. *Genetic algorithms + data structures = evolution programs.* Springer, New York, 1996.

[15] M. Dorigo, V. Maniezzo, and A. Colorni. Ant system: optimization by a colony of cooperating agents. *IEEE Transactions on Systems, Man, and Cybernetics, Part B: Cybernetics,* 26(1):29–41, 1996.

[16] M. Dorigo and L.M. Gambardella. Ant colony system: a cooperative learning approach to the traveling salesman problem. *IEEE Transactions on Evolutionary Computation,* 1(1):53–66, 1997.

[17] R.S. Parpinelli, H.S. Lopes, and A.A. Freitas. Data mining with an ant colony optimization algorithm. *IEEE Transactions on Evolutionary Computation,* 6(4):321–332, 2002.

[18] K.M. Sim and W.H. Sun. Ant colony optimization for routing and load-balancing: survey and new directions. *IEEE Transactions on Systems, Man and Cybernetics, Part A: Systems and Humans,* 33(5):560–572, 2003.

[19] C.-F. Juang, C.-M. Lu, C. Lo, and C.-Y. Wang. Ant colony optimization algorithm for fuzzy controller design and its FPGA implementation. *IEEE Transactions on Industrial Electronics,* 55(3):1453–1462, 2008.

[20] J.F. Gomez, H.M. Khodr, P.M. De Oliveira, L. Ocque, J.M. Yusta, R. Villasana, and A.J. Urdaneta. Ant colony system algorithm for the planning of primary distribution circuits. *IEEE Transactions on Power Systems,* 19(2):996–1004, 2004.

[21] S. Favuzza, G. Graditi, M.G. Ippolito, and E.R. Sanseverino. Optimal electrical distribution systems reinforcement planning using gas micro turbines by dynamic ant colony search algorithm. *IEEE Transactions on Power Systems,* 22(2):580–587, 2007.

[22] J.G. Vlachogiannis, N.D. Hatziargyriou, and K.Y. Lee. Ant colony system-based algorithm for constrained load flow problem. *IEEE Transactions on Power System,* 20(3):1241–1249, 2005.

[23] B.K.H. Wong and H.S.-H. Chung. Steady-state analysis of PWM DC/DC switching regulators using iterative cycle time-domain simulation. *IEEE Transactions on Industrial Electronics,* 45(3):421–432, 1998.

[24] B.K.H. Wong and H.S.-H. Chung. An efficient technique for the time-domain simulation of power electronic circuits. *IEEE Transactions on Circuits and Systems I: Fundamental Theory and Applications,* 45(4): 364–376, 1998.

[25] J. Zhang, H.S.-H. Chung, W.-L. Lo, S.Y. Hui, and A.K.-M. Wu. Implementation of a decoupled optimization technique for design of switching regulators using genetic algorithms. *IEEE Transactions on Power Electronics,* 16(6):752–763, 2001.

[26] A.S. Hedayat, N.J.A. Sloane, and J. Stufken. *Orthogonal arrays: theory and applications.* Springer, New York, 1999.

[27] C.R. Hicks. *Fundamental concepts in the design of experiments.* New York: Holt, Rinehart and Winston, 1964.

[28] J. Zhang, H.S.-H. Chung, A.W.-L. Lo, and T. Huang. Extended ant colony optimization algorithm for power electronic circuit design. *IEEE Transactions on Power Electronics,* 24(1):147–162, 2009.

[29] D. Bedrosian and J. Vlach. Time-domain analysis of networks with internally controlled switches. *IEEE Transactions on Circuits and Systems I: Fundamental Theory and Applications,* 39(3):199–212, 1992.

Index

www.ingramcontent.com/pod-product-compliance
Lightning Source LLC
LaVergne TN
LVHW011902200926
848484LV00002B/506
9781849198226